GEOLOGIC FRAMEWORK OF THE CAPITAN REEF

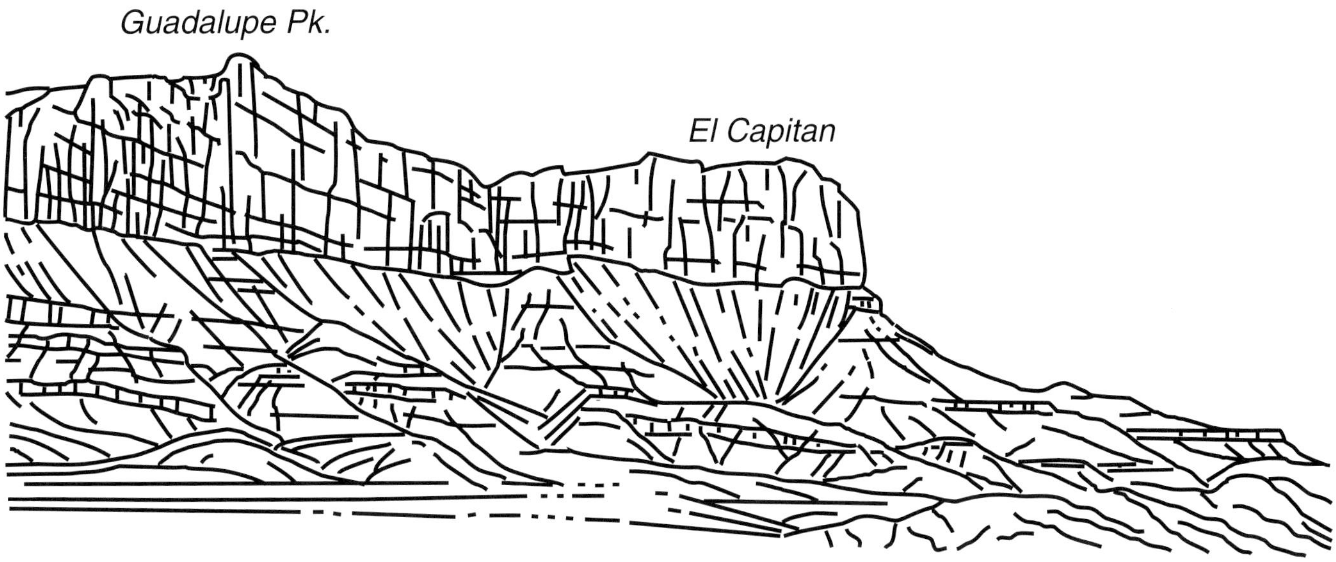

Arthur H. Saller, Unocal Corporation

Paul M. (Mitch) Harris, Chevron Petroleum Technology Company

Brenda L. Kirkland, University of Texas

and

S. J. Mazzullo, Wichita State University

SEPM (Society for Sedimentary Geology)

Robert W. Dalrymple, Editor of Special Publications

Special Publication No. 65

Tulsa, Oklahoma, U.S.A. *December, 1999*

SEPM thanks the following for their generous contributions to
Geologic Framework of the Capitan Reef:

Chevron Petroleum Technology Company

Unocal Corporation

ON THE COVER: *El Capitan, the geomorphic feature that lends its name to the Capitan Formation, as seen looking north from the road to Williams Ranch House, Guadalupe Mountains National Park, Texas.*

PHOTOGRAPHER: *Brenda L. Kirkland*

ISBN: 1-56576-063-8

SEPM (Society for Sedimentary Geology)

1731 E. 71st Street

Tulsa, OK 74136-5108

1-800-865-9765

Printed in the United States

CONTENTS

STRATIGRAPHIC/DEPOSITIONAL CONTRIBUTIONS

PALEOECOLOGIC CONTRIBUTIONS

DIAGENETIC CONTRIBUTIONS

GEOLOGIC FRAMEWORK OF THE CAPITAN DEPOSITIONAL SYSTEM—PREVIOUS STUDIES, CONTROVERSIES, AND CONTENTS OF THIS SPECIAL PUBLICATION

ARTHUR H. SALLER
Unocal Corporation, Sugar Land, Texas 77478

PAUL M. (MITCH) HARRIS
Chevron Petroleum Technology Company, La Habra, California 90633

BRENDA L. KIRKLAND
University of Texas, Austin, Texas 78712

AND

S. J. MAZZULLO
Wichita State University, Wichita, Kansas 67260

INTRODUCTION

The Guadalupe and Delaware mountains of west Texas and southeastern New Mexico contain superb outcrops of the Capitan depositional system (Fig. 1). The Capitan depositional system as used here includes the reef/forereef of the Capitan Formation, shelfal equivalents (Seven Rivers, Yates and Tansill formations), and the basinal equivalent (Bell Canyon Formation) (King, 1948; Newell et al., 1953; Hayes, 1964) (Fig. 2). Exposures in the Guadalupe Mountains are immense (1000 m vertical; 5–10 km in dip direction; 70 km along strike), and have suffered minimal structural deformation. The Capitan depositional system of the Guadalupe Mountains has been visited and studied by many of the world's premier geologists. In addition, it serves as one of the primary training grounds for earth scientists from the petroleum industry and academia. As a result, the Capitan depositional system has been the focus of numerous stratigraphic, facies, paleontologic, and diagenetic studies, many of which have become classic works in their respective fields. Many facets of the Capitan system have been and continue to be controversial, and much excellent geological research is still being done on the Capitan reef and related units.

An informal meeting of stratigraphers, sedimentologists, and paleontologists was held in Carlsbad, New Mexico, and the Guadalupe Mountains in October of 1996. This international conference was cosponsored by the University of Texas at Austin and Cambridge University in England. The purpose of the conference was to let Capitan researchers meet each other and to share recent findings regarding the stratigraphy and paleoecology of the Capitan reef. This publication is an outgrowth of that conference. The purpose of this SEPM Special Publication is to present different aspects of recent research on the Capitan depositional system. Hopefully, this volume can serve as a compilation of recent studies on the Capitan system for use by geologists visiting the Guadalupe Mountains for the first time, geologists who have been studying the Capitan system for years, and geologists working on similar depositional systems in other parts of the world. In this introductory paper, we will (1) review previous studies of the Capitan depositional system, (2) discuss some of the current controversies regarding the Capitan system, and (3) introduce the contributions of this SEPM Special Publication.

PREVIOUS STUDIES

Historically, outcrops of the Capitan depositional system have been studied in the Guadalupe, Glass, Apache, and Delaware Mountains. Outcrop studies have been supplemented by subsurface data from the Northwest Shelf, Central Basin Platform, and Delaware Basin (Fig. 1). As far as we are aware, the earliest geological study in the Guadalupe Mountains area was that of Shumard (1858). In the early 1900s, the Capitan system in the Glass and Guadalupe Mountains was part of several reconnaissance geologic and paleontological studies (see Hill, 1996, for a comprehensive review). The "Capitan Limestone" was named by Richardson (1904) for the massive limestone outcropping in the Guadalupe Mountains, especially at "El Capitan." Early studies of the Capitan include Blanchard and Davis (1929), Crandall (1929), Lloyd (1929), and Lang (1937). From the 1920s through 1940s, P. B. King (1930, 1934, 1942 and 1948) studied the Capitan as well as other formations in the Glass, Guadalupe and nearby mountain ranges. He did a superb job of describing the Capitan and related formations. King made fundamental observations about the response of the system to relative changes in sea level, recognized patterns of progradation and aggradation, and produced exquisite line drawings that are still the best documentation of many facets of Capitan stratigraphy.

After World War II, an excellent team of paleontologists, stratigraphers, and sedimentologists conducted a thorough examination of the Capitan reef complex that culminated in a book entitled *The Permian Reef Complex of the Guadalupe Mountains Region, Texas and New Mexico* (Newell et al., 1953). This book includes comprehensive descriptions of the flora and fauna of the reef and adjacent strata, physical stratigraphy, biostratigraphy, depositional features, and diagenetic features. Newell et al. (1953) interpreted the Capitan as a barrier reef complex whose current basinward dip was the result of post-depositional structural deformation. Although some of their interpretations were later challenged, the publications of Newell

Geologic Framework of the Capitan Reef, SEPM Special Publication No. 65
 ISBN 1-56576-063-8, p. 1-13

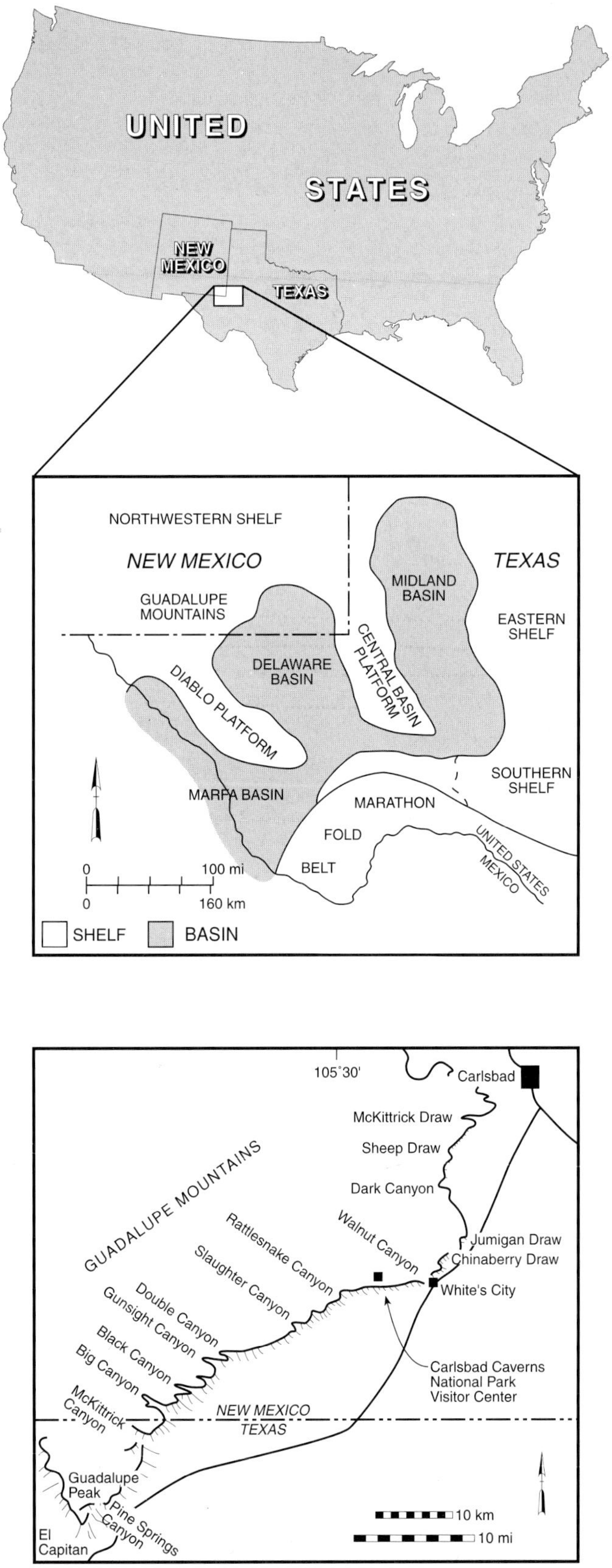

FIG. 1.—Maps showing locations of the Guadalupe Mountains.

et al. (1953) and Newell (1955) gave subsequent workers a solid descriptive and interpretative base from which to work.

Many other individuals also worked on various aspects of the Capitan depositional system during the 1950s and 1960s. Geologic mapping and stratigraphic interpretations of the area were published by Hayes and Gale (1957), Hayes and Koogle (1958), Motts (1962, 1972), and Hayes (1964). Stratigraphic and depositional features of the Capitan depositional system were described and interpreted by Adams and Frenzel (1950), Tyrrell (1964, 1969), Silver and Todd (1969), Kendall (1969), and Achauer (1969). Basinal clastics of the Delaware Mountain Group were described and interpreted by Newell et al. (1953), Hull (1957), Rigby (1958), Nottingham (1960), Hayes (1964), Grauten (1965), Jones and Smith (1965), Beck (1967), and others.

R. J. Dunham profoundly influenced interpretations of the Capitan depositional system in the late 1960s and 1970s. Although many of Dunham's interpretations have been modified or refuted, his observations are arguably still the most detailed and insightful published to date (Dunham, 1969, 1970, 1972). His hypotheses reshaped the way carbonate sedimentologists viewed the Capitan depositional system and continue to influence thought about the development of Permian carbonate systems. Dunham's publications and perhaps more importantly, the field trips he led, brought new ideas and insights to those who accompanied him (Pray, 1989a).

Although the term "sequence stratigraphy" had not yet been coined, the model for "reciprocal sedimentation" of Meissner (1972) illustrated "sequence stratigraphy" superbly. According to the reciprocal sedimentation model, carbonates and evaporites accumulated on the shelf during highstands of sea level, whereas clastics largely bypassed the shelf and were deposited in the basin during lowstands of sea level. Meissner's use of the concept provided great insight into the origin of interbedded and interfingering clastics and carbonates. It also helped explain how carbonates could be dominant on the shelf, whereas clastics were dominant in the basin (similar to the reciprocal sedimentation of Wilson [1967] in the Sacramento Mountains). The shelfal carbonates exposed in the Guadalupe Mountains and the siliciclastics of the Delaware Basin provide a classic example of this process.

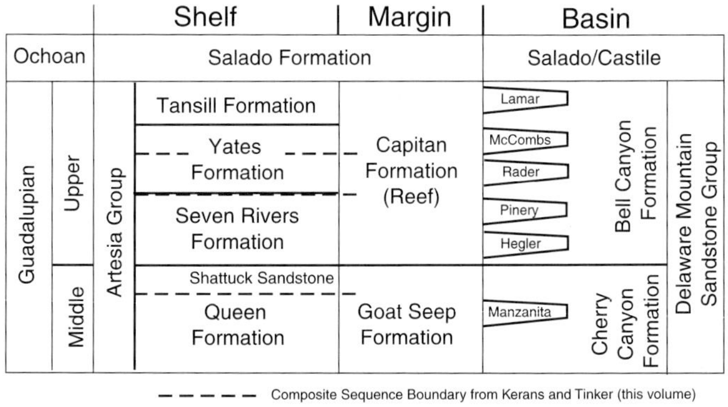

FIG. 2.—Correlation chart showing stratigraphic components of the Capitan depositional system (modified from Pray and Esteban, 1977; Garber et al., 1989; Tinker, 1998). Correlations of shelf strata to basinal strata are still debated.

A major surge of research on the Capitan complex occurred during the 1970s and 1980s. Published work focused on shelf-to-basin correlations (Kelley, 1972; Smith, 1973; Sneed, 1977), environments and cycles of shelf deposits (Motts, 1972; Smith, 1974), tepee structures (Asserto and Kendall, 1977), and comprehensive field guide overviews (Scholle and Halley, 1980; Toomey and Babcock, 1983). Graduate students under the supervision of L. C. Pray at the University of Wisconsin, Madison, studied Capitan sedimentology, stratigraphy, and paleoecology, first in fairly accessible areas, and later in increasingly remote parts of the Guadalupe Mountains. The first generation of Pray's students published their results along with the work of other authors in SEPM Permian Basin Section Publication 77-16, *Upper Guadalupian Facies, Permian Reef Complex, Guadalupe Mountains, New Mexico and West Texas* (Hileman and Mazzullo, 1977). That two-volume compilation and field trip guide discussed many aspects of the Capitan system including: Capitan reef textures and paleo-ecology (J. A. Babcock, 1977; Toomey, 1977; Toomey and Cys, 1977; Yurewicz, 1977; Schmidt and Klement, 1977); back-reef sedimentology and stratigraphy (Neese and Schwartz, 1977; Esteban and Pray, 1977; Sarg, 1977); Capitan and back-reef diagenesis (Mazzullo, 1977; Mazzullo and Cys, 1977; Schmidt, 1977); and basinal carbonates and clastics (L. C. Babcock, 1977; Williamson, 1977). The volume also contained an article by Cys et al. (1977), which summarized many aspects of the Capitan system.

Studies of the Capitan system continued in the late 1980s and 1990s. SEPM Core Workshop Number 13, "Subsurface and Outcrop Examination of the Capitan Shelf Margin, Northern Delaware Basin" (Harris and Grover, 1989), was built around descriptions of the Gulf PDB-04 well, which continuously cored the Bell Canyon, Capitan, Seven Rivers, Yates and Tansill formations in the northern Delaware basin. The descriptions (Garber et al., 1989) are a unique documentation of lithologies, facies, and diagenesis of the Capitan in the subsurface. That workshop volume contained many other excellent articles on depositional features of the Capitan Formation (Babcock and Yurewicz, 1989; Harwood, 1989), back-reef and shelf equivalents (Parsley and Warren, 1989; Mazzullo et al., 1989; Neese, 1989; Borer and Harris, 1989; Candelaria, 1989; Hurley, 1989; Sarg, 1989; Wheeler, 1989), and diagenesis of the Capitan Formation (Mruk, 1989; Melim and Scholle, 1989). *The Guide to the Permian Reef Geology Trail, McKittrick Canyon, Guadalupe Mountains National Park, West Texas* (Bebout and Kerans, 1993) is another important collection of descriptions and interpretations of the Capitan reef complex at a locality where excellent exposures allow a traverse from basinal strata, through slope, reef, and outer shelf strata in a single day's hike. During the last decade, Capitan reef paleoecology has been the subject of many insightful articles including Noé and Mazzullo (1992), Weidlich (1996), Wood et al. (1994, 1996), Senowbari-Daryan and Rigby (1996), Noé (1996), and Weidlich and Fagerstrom (1998). An excellent and comprehensive review of most aspects of the Capitan system including outcrop and subsurface data is available in Hill's (1996) volume on the geology of the Delaware basin.

Recently a number of detailed stratigraphic studies have been conducted on the Capitan depositional system. An understanding of high-resolution stratigraphic relationships for the Capitan system generally suffers from limited resolution of biostratigraphic control and the inability to trace beds or time lines from the shelf into the basin. However, innovative sequence stratigraphic approaches are adding significant details to the stratigraphic framework of the Capitan. Studies in McKittrick Canyon (Brown and Loucks, 1993a and b; Kerans and Harris, 1993; Brown, 1996; Tinker, 1996, 1998), Slaughter Canyon (Rankey and Lehrmann, 1996; Osleger, 1998), and in basinal strata (Kerans et al., 1992, 1993; Borer and Harris, 1995) are improving views of chronostratigraphy, interrelationships between depositional facies, and shelf-to-basin relations.

CONTROVERSIES OF THE CAPITAN SYSTEM

Although the Capitan depositional system occurs in spectacular outcrops that have been studied by literally hundreds, if not thousands, of geologists, many controversies have developed and many still remain. Some controversies appear to be converging on a single interpretation, whereas others continue to be debated, and new ones are being introduced. Below we discuss the "current" status of some of the more important controversies. For more detailed discussions, see Hill (1996) and the original works cited below.

Origin of the Capitan "Reef"

One of the longstanding, fundamental debates regarding the Capitan system is whether the Massive Member of the Capitan Formation is an "ecologic reef," a "diagenetic reef," or a "stratigraphic reef" (Dunham, 1969). It is ironic, but one of the problems in studying the Capitan reef has been a lack of exposure in which depositional fabrics can be clearly seen. Although large sections of canyons have outcrops of sparsely vegetated Capitan, surficial weathering has resulted in outcrop surfaces that reveal little or, more commonly, nothing of the underlying rock fabrics. For many years, Capitan fabrics could be clearly seen only on a few naturally etched outcrop "windows" and from samples collected and slabbed. Lithologic descriptions of those outcrops and samples varied from wackestone to boundstone. Workers who saw a dominance of wackestone concluded that although the upper Capitan was a massive carbonate buildup ("stratigraphic reef"), it was not an ecologic reef (Baars, 1964; Achauer, 1969; Dunham, 1970; Tyrrell, 1969). Others saw a dominance of syndepositional cement and concluded that inorganic cement was the critical binding agent in the reef (Schmidt, 1977; Mazzullo and Cys, 1977, 1978); and still others (Cronoble, 1974; J. A. Babcock, 1977; Yurewicz, 1977; Cys et al., 1977; Scholle and Halley, 1980) observed substantial amounts of organic boundstone and concluded that much of the Capitan was an organic reef.

Construction of the McKittrick Canyon Geology Trail in the early 1980s resulted in the exposure of many rock faces with naturally etched surfaces. These exposures along with artificially etched reefal surfaces at localities such as the mouth of Walnut Canyon and Dark Canyon have allowed viewing of larger surfaces of reefal fabric than was previously possible. The etched outcrops clearly show that much of the Capitan has a reefal framework with bryozoa, calcareous sponges, and other organisms in growth position and with internal cavities (Kirkland et al., 1993; Wood et al., 1994, 1996; Weidlich and Fagerstrom, 1998). In many areas, the original organic framework formed by fenestrate bryozoa constitutes substantially less than 5% of the rock volume (see photos and drawings in Wood et al., 1996). The framework was subsequently bound together by *Archaeolithoporella* and microbial micrite, and finally large volumes of botryoidal aragonite filled the internal cavities created originally by framebuilding organisms (J. A. Babcock, 1977; Mazzullo and Cys, 1977; Yurewicz, 1977; Wood et al., 1994, 1996; Kirkland et al., 1998).

Depth of Reef/Dip of the "Fall-in" Beds

The most intense debates about the Capitan reef have involved four related questions: (1) the origin of dip in the back-reef "fall-in beds," (2) water depth of the top of the reef, (3) whether the Capitan was a barrier reef, and (4) whether the reef was periodically subaerially exposed. Four main paleobathymetric profiles or models have been proposed for the Capitan shelf margin (Fig. 3): (1) the uninterrupted slope from shelf to basin (King, 1948), (2) a shelf-margin barrier reef (Newell et al., 1953), (3) a bathymetric high at pisolitic shelf crest and continuous basinward dip from the pisolite shoal through the reef and slope (the marginal mound profile of Lang, 1937; Dunham, 1969, 1972); and (4) shoaling at both a shelf-margin barrier reef and pisolite shelf crest (Kirkland-George, 1992). These different profiles highlight the uncertainty about the original dip of the fall-in beds and depth of the reef.

Stratal geometries of the Capitan shelf-to-reef transition are characterized by a change from gently dipping, bedded shelf crest strata, to more steeply dipping, crudely stratified outer shelf beds that pass gradationally into massive reef facies (Bebout et al., 1993; Osleger, 1998; Tinker, 1998). The basinward-dipping geometry of back-reef strata, referred to locally as the "fall-in bed" profile (Hurley, 1978, 1989), is best developed in older Seven Rivers strata, but is also apparent in the Yates Formation and, to a lesser degree, in the Tansill Formation. Arguments regarding the paleobathymetry of the reef are in part related to whether back-reef beds have undergone significant differential tilting after deposition. If the current dip of the back-reef beds is largely depositional, then the top of the Capitan must have been deposited in tens of meters of water; hence, the Capitan could not have been a barrier reef, and subaerial exposure would have required a large fall in relative sea level.

Newell et al. (1953) interpreted the Capitan based on sedimentologic and stratigraphic models for modern reefs. They viewed the Capitan as a "barrier reef" complex, and thought the current stratigraphic configuration was the result of postdepositional deformation. Lack of obvious tectonic deformation in overlying back-reef strata caused many later workers to dismiss the barrier reef concept of Newell et al. (1953). Many subsequent workers proposed instead that current stratigraphic profiles were largely depositional (Dunham, 1972; Pray and Esteban, 1977; Yurewicz, 1977; Hurley, 1978, 1989; Kerans and Harris, 1993; Osleger, 1998; Tinker, 1998; Osleger and Tinker, in press., 1998), and therefore interpreted the Capitan as a deep reef or bank. Smith (1973), however, also argued that post-depositional basinward tilting had substantially modified the Capitan depositional profile.

Different researchers have considered the distribution of depositional facies in the outer shelf area as evidence that the Capitan was deposited as a barrier reef near sea level or at depths of tens of meters. Newell et al. (1953), for example, believed the characteristics of back-reef facies were similar to modern barrier reefs. Kirkland and Moore (1996) described an abrupt lateral facies change from the upper Capitan reef to angular, reef-

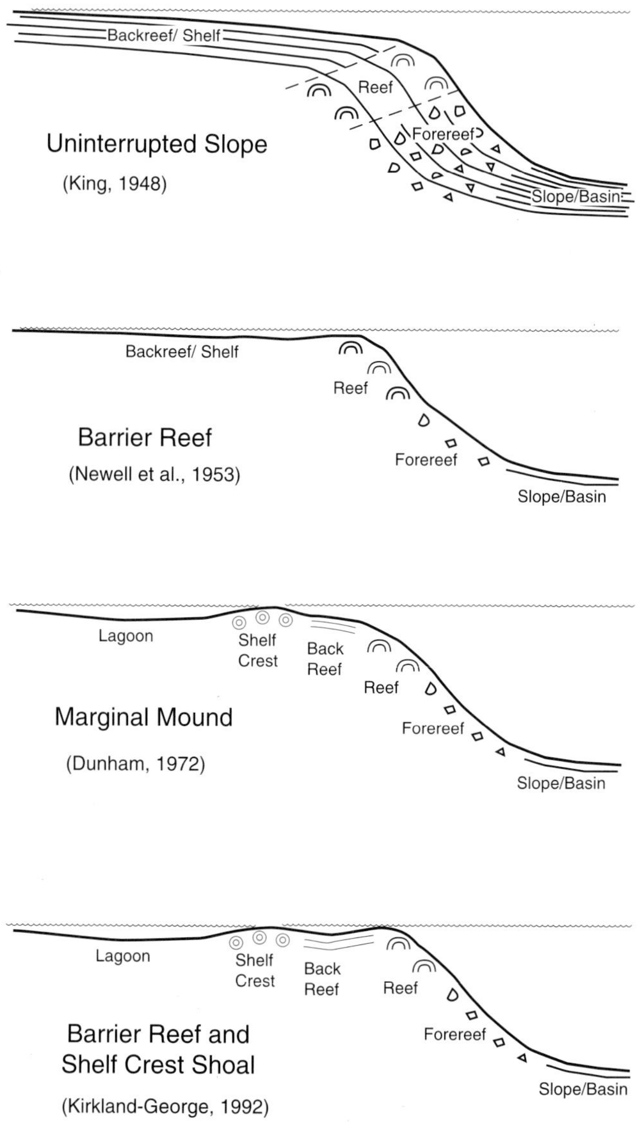

FIG. 3.—Contrasting paleobathymetric and depositional profiles proposed for the Capitan depositional system during highstands of sea level.

derived sediments of the outer shelf and noted that a traverse from the reef across the outer shelf reveals an increase in micrite and a marked decrease in biotic diversity shelfward. They argued these changes suggest that the reef was shallow enough (a "barrier") to reduce current energy and affect deposition on the outer shelf. In contrast, Dunham (1969) proposed that pisoids formed in the vadose zone at the crest of the Capitan depositional profile, and that paleobathymetry deepened toward the reef, and therefore the reef accumulated in an upper slope environment. The group of researchers from Wisconsin concluded that the crest of the Capitan reef (the marginal mound shelf profile of Dunham, 1969) was in generally deeper water, but shoaled from a depth of 45 m during deposition of the Seven Rivers Formation to 10 m during deposition of the Tansill (Neese and Schwartz, 1977; J. A. Babcock, 1977; Yurewicz, 1977; Hurley, 1978, 1989; Esteban and Pray, 1983; Neese, 1989; Babcock and Yurewicz, 1989). Those same researchers proposed that the distribution and characteristics of back-reef facies indicated a consistent deepening of depositional environments from the pisolitic grainstone shoals at the shelf crest to an upper slope reef. Back-reef facies change from cross-stratified grainstones with pisoids at the shelf crest to burrowed bioclastic packstones with outer shelf biota near the reef, which were interpreted to reflect decreasing depositional energy and increasing water depth toward the reef (Neese and Schwartz, 1977; J. A. Babcock, 1977; Yurewicz, 1977; Hurley, 1978, 1989; Esteban and Pray, 1983; Neese, 1989; Babcock and Yurewicz, 1989; Kerans and Harris, 1993; Tinker, 1996, 1998; Osleger, 1998; Osleger and Tinker, in press).

One method to determine whether differential compaction has substantially modified the depositional profile is analysis of geopetal structures within the reef and fall-in beds. In general, geopetal cavity fills are thought to be emplaced with their top surface approximately horizontal. If emplaced penecontemporaneously with deposition, the attitude of geopetal fills should indicate whether post-depositional tilting of strata has occurred. Working in north McKittrick Canyon, Hurley (1978, 1989) found eight roughly horizontal geopetal surfaces in Seven Rivers "fall-in beds," which dipped basinward (toward the reef) at approximately 8°. If those geopetals formed penecontemporaneously with deposition, their attitude suggests that the current dip of the fall-in beds is approximately depositional. In contrast, Saller (1996) measured the attitude of approximately 30 geopetals in reefal cavities of the Yates-age Capitan reef along the McKittrick Canyon Reef Trail. The tops of those geopetal fills showed an average basinward dip of approximately 10° (average dip azimuth of 124°), suggesting post-depositional tilting of the reef at approximately the same amount and direction as the dip of the overlying fall-in beds (Saller, 1996). Those data suggested that the dip of the fall-in beds was related to differential compaction. Very recently, Fitchen and Hunt (personal communication, 1998) found 14 measurable geopetal structures in fall-in beds of the Seven Rivers Formation at Slaughter Canyon. Those geopetals have an average basinward dip of 14° at an azimuth of 123°, which is very similar to geopetals that Saller (1996) observed 25 km to the west-southwest in McKittrick Canyon, lending further support for post-depositional tilting of the fall-in beds (Fitchen and Hunt, pers. comm., 1998).

Although there is ample evidence for periodic subaerial exposure of the platform (e.g., Dunham, 1969, 1972; Kendall, 1969; Smith, 1974; Cys et al., 1977; Pray, 1985a; Noé, 1996), another long-standing controversy has been whether or not the Capitan reef was ever exposed subaerially between periods of deposition. Dunham (1969, 1972) and Kendall (1969) believed that the reef was exposed subaerially at times, and they cited the presence of siliciclastic silt within fractures in the reef, and presumed caliche pisolites (most of which are in back-reef strata), as evidence of vadose exposure. However, many of the early (that is, pre-Tertiary) fractures in the reef are largely filled by crusts of interlaminated *Archaeolithoporella* and former aragonite marine cements that are encrusted locally by *Shamovella* (*Tubiphytes*) (e.g., J. A. Babcock, 1977); hence, they were present during marine deposition, and are possibly neptunian, although the significance of the silt therein remains equivocal (e.g., Pray, 1985b). Jacka et al. (1960) suggested that platform by-passing of sands into the Delaware basin indicates that the reef was subaerially exposed during sea-level lowstands, whereas others have invoked exposure of only the back-reef platform during lowstands (e.g., Meissner, 1972; Pray, 1985c).

Since the pioneering study of Newell et al. (1953), unequivocal sedimentologic or especially diagenetic evidence that the Capitan reef was ever subaerially exposed during deposition has not been presented. Many studies have confirmed that the bulk of the cements in the reef are marine in origin, and marine cements become volumetrically more abundant from the lower Capitan to the upper Capitan (e.g., Cys et al., 1977; Mazzullo and Cys, 1977; Schmidt, 1977; Yurewicz, 1977; Mazzullo and Cys, 1978; Given and Lohmann, 1985; Mruk, 1989; Kirkland et al., 1993; Noé, 1996). The coarsely crystalline calcites that are common in outcrops are not syndepositional but, rather, precipitated later, probably during deep burial (e.g., Mruk, 1989; Scholle et al., 1992; Hill, 1996). Vadose or unequivocal phreatic meteoric cements have not been identified in these rocks except, perhaps, for some presumed meteoric phreatic cements that may have precipitated locally in reef and fore-reef deposits in late Guadalupian time (Given and Lohmann, 1986; Mruk, 1989). The timing of neomorphism of the aragonitic allochems and cements in the reef has not been adequately explored.

In summary, the depth, nature and subaerial exposure of the Capitan reef remain subjects of debate. Although the Capitan has been studied carefully by numerous researchers, the actual amount of Capitan strata studied in detail by those workers is minute compared with the volume of Capitan present. Different areas studied by different workers represent reef margins of varying ages and locations. The Capitan system probably varied laterally (Pray, 1989b) and with time (Borer and Harris, 1991a, 1991b, 1995; Osleger, 1998; Tinker, 1998; Kerans and Tinker, this volume). Sometimes the Capitan was a deeper reef, and at other times it was a shallow reef system. One needs only to examine exposures along the north wall of McKittrick or Slaughter canyons (Bebout et al., 1993; Osleger, 1998; Tinker, 1998) to demonstrate that the shelf profile and depth of water over the reef clearly varied significantly during development of the Capitan shelf margin.

Origin of Pisolites and Tepees

Although numerous ideas have been proposed for the origin of pisolites in back-reef (shelf crest) strata of the Capitan system, those ideas seem to be converging. Johnson (1942) interpreted the pisolite's laminations as blue-green algal coatings. Pisolites were interpreted by Newell et al. (1953) as lagoonal and also algal in origin. In contrast, Dunham (1969) interpreted many of the pisolites as having formed in a vadose environment during periods of subaerial exposure. Esteban and Pray (1977, 1983) also interpreted pisolites as forming in current agitated subtidal to possibly peritidal depositional environments, often occurring in inter-tepee depressions. Later laminations and cements precipitated around many of the original pisolites, and stable isotope compositions indicate that most of those coatings formed in marine or evaporated marine waters (Mutti and Simo, 1994). The general consensus in the literature appears to be that pisolites formed in an agitated subtidal to peritidal environment and that some later coatings may have formed in subtidal, peritidal, and/or vadose environments after deposition of the initial pisolite.

Tepees are another distinctive structure found on the Capitan shelf that have been difficult to interpret. Tepees are commonly found in association with pisolite shoal and tidal flat facies. Debate has largely focused on the mechanism responsible for pushing up the peritidal carbonates into broad polygonal patterns that are described in cross-section as teepees. Adams and Frenzel (1950) who coined the term "tepee," thought the structures were formed by compression during deposition, whereas Newell et al. (1953) proposed that tepees formed in association with crystallization of evaporite minerals. Kendall (1969), Smith (1974), and Assereto and Kendall (1977) interpreted the tepees as large "desiccation cracks" that were followed by cement precipitation in the cracks which aided in forcing up the tepees. Burri et al. (1973) also thought thermal expansion and contraction were important in tepee formation. Using "analogues" from south Australia, Warren (1983) proposed that tepees formed due to pressure associated with upward-moving groundwater. Handford et al. (1984) also suggested that tepees and related precipitates might be forming in association with discharging groundwaters. Tinker (1996) proposed that desiccation occurred during highstands of sea level, and that the siliciclastic sediment found in the core of tepee structures was deposited during lowstands of sea level. Pisolite grainstones and tepees are characteristic of the top of meter-scale cycles at the shelf crest (Borer and Harris, 1991a, b; Kerans and Harris, 1993; Osleger, 1998; Tinker, 1998). Unfortunately, a consensus has not been reached regarding the mechanism responsible for tepee development.

Deposition of Basinal Siliciclastics

Depositional models for Capitan-equivalent basinal sandstones and siltstones are still being debated; Hill (1996) presents a thorough overview. King (1948) and Newell et al. (1953) initially proposed deposition of the Brushy Canyon Sandstone (lowest in the Delaware Mountain Group) in a shallow-marine environment based on the abundance of sand and abundant ripples. The recognition of graded beds in the basin-filling deposits of the Delaware Mountain Group (Hull, 1957; Jacka et al., 1968) suggested deposition as deep-water turbidites. The stratigraphic position of the basin-filling sandstones has led most subsequent workers to invoke deeper water depositional environments (Payne, 1976; Bozanich, 1979; Williamson, 1977, 1979). Harms (1974) and Harms and Williamson (1988) proposed deposition by density currents created when high-salinity shelf waters flowed down into the less saline basin. Indirectly, this model requires the shelf to be flooded (that is, highstand of sea level) in order to generate significant amounts of hypersaline waters, which is not compatible with deposition during a lowstand of sea level.

Mazzullo et al. (1985) and Fischer and Sarnthein (1988) proposed deposition of sands and silts in the basin largely by eolian processes during base-level falls when the adjacent shelves were exposed above sea level. In this model, sands were carried across the shelf in dunes before deposition in subtidal environments at the basin margin. Those sands were episodically carried downslope and into the basin by gravity flows. In contrast, silts were transported largely as airborne dust, raining out in a number of areas including the Delaware basin. Kocurek and Kirkland (1998) proposed that these sediments were derived from eolian systems in the Whitehorse Group of the Anadarko Basin. Borer and Harris (1995) argued for repeated input of siliciclastic sands into the basin during high-frequency subaerial exposure and bypassing of the shelf. Mruk and Bebout (1993), Brown and Loucks (1993a, b), and Brown (1996) documented high-frequency siliciclastic input within the slope and toe-of-slope at McKittrick Canyon outcrops. Although they concentrate on stratigraphically lower sands of the Brushy Canyon Formation, stratigraphic and depositional analyses by Gardner and Sonnenfeld (1996) have clarified depositional processes, which are also probably responsible for Bell Canyon deposition. One of the remaining questions is the mechanism responsible for gravity flows to transport sand into the deep basin. The sediments are clay-poor, making a true "turbidity current" difficult to envisage. The stratigraphic consensus appears to be that deposition of basinal sands occurred during lowstands of sea level when the shelf was exposed, making dense hypersaline brines difficult to concentrate on the shelf.

Sequence Stratigraphic Relationships

As "sequence stratigraphy" gained popularity in the 1980s and 1990s, the Capitan system was a logical interval to study in a sequence stratigraphic context. Many new sequence stratigraphic studies are built on the concept of "reciprocal sedimentation," which involves sands being carried across the shelf and into the basin during lowstands of sea level, and shelfal carbonates being deposited during highstands of sea level (Jacka et al., 1960; Meissner, 1972). Fischer and Sarnthein (1988) added to the reciprocal sedimentation model for middle Permian carbonates and sandstones with analogues to recent Cenozoic systems. Controversies associated with sequence stratigraphy include: shelf to basin correlations, stratal geometries in Slaughter Canyon, and the nature of transitions between cyclic shelf sandstones and carbonates.

Although Capitan outcrops are spectacular, no single outcrop allows an unambiguous correlation between shelfal and basinal strata. Specific questions involve which sands or surfaces on the shelf correlate with which sands in the basin, and which basinal carbonates correlate with which shelfal carbonates. Based on Tinker (1996, 1998) and Kerans and Tinker (this volume), three composite sequences are identified within the Capitan depositional system (Fig. 2). Within this stratigraphic framework, they correlate: (1) the Manzanita, Hegler, and Pinery carbonate members of the Cherry and Bell Canyon formations with Seven Rivers shelfal strata; (2) Rader and McCombs carbonate members of the Bell Canyon Formation with Yates shelfal strata; and (3) the Lamar carbonate member of the Bell Canyon Formation with Tansill shelfal strata. These shelf-to-basin correlations are somewhat different from those of previous workers (Newell et al., 1953; Wilde, 1975; Pray and Esteban, 1977; Garber et al., 1989, and others) (Fig. 2). The precise correlation between shelf and basin is still evolving, and will likely continue to receive much attention as sequence stratigraphic and biostratigraphic studies continue.

Although outcrops of the prograding Capitan system are generally good, differences in interpretation of stratal geometries remain. Slaughter Canyon has become a "hot bed" of stratigraphic research where the interpretation of stratal geometries of prograding Seven Rivers strata are controversial. Rankey and Lehrmann (1996) describe toplap in the progradation, whereas Osleger (1998) and Longley (this volume) recognize only prograding sigmoidal patterns. Hunt and Fitchen (pers. comm., 1998) have also very recently recognized syndepositional growth faults in back-reef Capitan strata at Slaughter Canyon.

According to reciprocal sedimentation concepts, siliciclastic sands are carried across the shelf during lowstands of sea level when the shelf was subaerially exposed. However, most shelf sands in the Capitan depositional system are interpreted to be deposited in a subtidal environment (Pray, 1977; Candelaria, 1989). As a result, questions emerged as to whether some sands were deposited during highstands of sea level when the shelf was flooded (Pray, 1977). Our perspective of the shelf sands and carbonates was broadened greatly when subsurface data from inner and middle shelf environments (Borer and Harris, 1991a, b; Andreason, 1992) were added to descriptions of outcropping outer shelf deposits. Recently, interpretations of cyclic shelf sands and carbonates appear to be converging. Most shelf sands occur above subaerial exposure surfaces (unconformities) that probably represent the time when most basinal sands were carried across the shelf (Borer and Harris, 1995; Osleger, 1998; Osleger and Tinker, in press). These shelf sands are interpreted as being the transgressive portions of shelf cycles, with final deposition by a reworking of eolian dune sands and sand blown into adjacent subtidal environments (Kerans and Tinker, 1997; Osleger, 1998). With continued flooding of the shelf, carbonates form the cap to the shelf cycles. Kerans and Harris (1993), Tinker (1996, 1998), Osleger (1998), and Osleger and Tinker (in press) examined the lateral variation and stacking patterns of cycles in McKittrick and Slaughter canyons and showed that there is substantial variation in the nature of a cycle depending on its position along the depositional profile and within a sequence.

Origin of Caves

Numerous caves are present in the rocks of the Capitan system in the Guadalupe Mountains, with Carlsbad Caverns and Lechuguilla Cave being the most spectacular. The theories of cave formation in the Guadalupe Mountains have changed over the last 50 years, but seem to be converging. Dissolution was initially attributed to "normal cave processes," for example, carbonic acid occurring in rainwater (Bretz, 1949). During the last 20 years, speleologists who have been studying Carlsbad and other caverns in the Guadalupe Mountains have developed a more complex model for development of caverns in the Capitan depositional system (Jagnow, 1979, 1989; Hill, 1989; DuChene and McLean, 1989). Based on cave geometries and the geochemistry of the cave fill, Hill (1987, 1990, 1995, 1996) postulated four stages of cave development. The last and volumetrically most important dissolution event was sulfuric acid karst associated with basinal hydrogen sulfide mixing with oxidizing freshwater during the last 15 million years. This model has been substantiated and is now being considered for other cave systems around the world.

THIS VOLUME

The purpose of this publication is to present papers that represent progress in our understanding of the Capitan depositional system. The papers in this volume fall into three general categories: (1) stratigraphic/depositional, (2) paleoecologic, and (3) diagenetic, although some contributions fall into more than one category. Many of the contributions address one or more of the Capitan controversies mentioned above. One of the intentions of this volume is to "cross pollinate" between disciplines, especially to convey paleontologic ideas to sedimentologists, and give sedimentologic and stratigraphic ideas to paleontologists. Brief summaries of the contributions in this volume are given below.

Stratigraphic/Depositional

In their article, "Extrinsic Stratigraphic Controls on Development of the Capitan Reef Complex," Kerans and Tinker place the Capitan system into a comprehensive sequence stratigraphic framework. The Capitan is described as the last part of a long-term (third-order?) sequence. The Capitan includes the transgressive and highstand parts of three composite sequences and 12 high-frequency sequences. Kerans and Tinker argue that the complex biologic and diagenetic reef fabrics observed in scattered outcrop localities can be understood best when the Capitan is put into a sequence stratigraphic framework. Their discussions also touch on several of the Capitan controversies discussed above.

Outcrops of the Capitan depositional system are commonly visited by oil company geoscientists and engineers who make analogues to subsurface exploration objectives and producing fields. In a contribution entitled, "Subsurface Expression of the Capitan Depositional System and Implications for Hydrocarbon

Reservoirs, Northeastern Delaware Basin," Harris and Saller illustrate the subsurface expression of the Capitan with high-quality seismic data and well log cross sections. Their large-scale shelf-to-basin correlations give insight not available from partially eroded outcrops. In addition, they describe the occurrence of oil reservoirs in back-reef and basinal sandstones that are stratigraphically equivalent to the Capitan.

Statistical evaluation of depositional cycles has been an important avenue of research in recent years. Lehrmann and Rankey discuss statistical patterns of cyclic back-reef deposition in their article, "Do Meter-Scale Cycles Exist? A Statistical Evaluation from Vertical (1-D) and Lateral (2-D) Patterns in Shallow-Marine Carbonates-Siliciclastics of the 'Fall In' Strata of the Capitan Reef, Seven Rivers Formation, Slaughter Canyon, New Mexico." Although some statistical analysis indicates randomness in vertical successions, careful evaluation of facies, surfaces, and 2-D distributions of facies indicate statistically valid cyclicity.

Important questions remain regarding the biostratigraphy of the Capitan system. Wilde, Rudine, and Lambert in their article, "Formal Designation: Reef Trail Member, Bell Canyon Formation, and its Significance for Recognition of the Guadalupian-Lopingian Boundary," give new information on fusulinid and conodont biostratigraphic markers within the Capitan and on an upper boundary for the Guadalupian Series. Their study puts elements of Capitan biostratigraphy into a global context.

Longley presents results of his detailed outcrop study of the Capitan system in Slaughter, Rattlesnake, Walnut and Dark canyons in an article entitled, "Differential Compaction and its Effects on the Outer Shelf of the Permian Capitan Reef Complex, Guadalupe Mountains, New Mexico." He presents many intriguing ideas including: (1) micritic lower slope deposits episodically lost pore fluids and pressure which also caused differential compaction; (2) compaction substantially modified the bathymetric profile of the Capitan system during deposition; (3) compaction influenced deposition of outer shelf sediments and induced small-scale, relative sea-level changes near the shelf margin; and (4) pore fluids and pressure released during compaction were in part responsible for injecting sand into fractures.

"Paleoenvironments, Cyclicity, and Diagenesis in the Outer Shelf Tansill Formation in the Carlsbad Embayment (Dark Canyon), Northern Guadalupe Mountains, New Mexico" are described in great detail by Mazzullo. Three shallowing-upward cycles are recognized in outer shelf deposits, and define a prolonged period of low relative sea level. Those deposits consist of peritidal-flat islands surrounded by shallow subtidal carbonate sands. Diagenetic features include marine cements, meteoric cements, and dolomites, which are characterized petrographically and geochemically, and then placed into the stratigraphic context.

Paleoecologic

Recently, Wood et al. (1994, 1996) made great progress in defining the paleoecology of the Capitan reef. In "Paleoecology of the Capitan Reef," Wood summarizes the results of that recent work. As she shows, any given part of the Capitan reef typically contains a succession of paleoecologic reefal communities. Reefal boundstones commonly started with thin or platey framebuilders (bryozoa or platey sponges), which created large shelter cavities. Those shelter cavities were subsequently inhabited by cryptic organisms including encrusting microbes, which precipitated laminar micrite, and "dangling calcareous sponges." Many internal framework cavities were finally filled with micritic internal sediment and large amounts of submarine aragonitic cement.

Quantitative paleoecology is a rapidly advancing science, and large, clean exposures of the Capitan lend themselves to this style of paleoecologic analysis. In their article entitled, "Influence of Sea-Level Changes on Development, Community Structure, and Quantitative Composition of the Upper Capitan-Massive (Permian), Guadalupe Mountains, Texas and New Mexico," Weidlich and Fagerstrom present results in which different components of the Capitan reef are described, analyzed, and interpreted quantitatively. That information is then related to variations in sea level during deposition.

Kirkland, Longacre, and Stoudt put new paleoecologic interpretations into a stratigraphic framework in their contribution, "The Dynamic Capitan Reef: An Image of an Ancient Reef and Suggestions for Future Research." This article combines published observations (e.g. Newell, 1955; Babcock, 1977; Noé, 1996; Wood et al., 1994, 1996) with new observations and interpretations to derive a comprehensive depositional model for the Capitan reef and adjacent strata.

Diagenetic

The Capitan depositional system has been extensively modified by diagenesis, and marine cements comprise a major part of the Capitan reef. Rahnis and Kirkland, in their article "Distribution, Petrography and Geochemical Characterization of Radiaxial Calcite and Associated Diagenetic Events in the Capitan Formation, West Texas and New Mexico," describe the petrography and geochemistry of radiaxial calcite and other originally calcite cements in the Capitan reef. They then relate that cementation to the depositional and diagenetic history of the Capitan.

Although many publications have discussed diagenesis in the Capitan reef, few have dealt with the forereef. Melim and Scholle discuss many aspects of forereef diagenesis in their article, "Diagenesis of the Capitan Forereef Facies (Permian, West Texas and New Mexico)." Important diagenetic processes in the forereef include cementation, calcium sulfate emplacement and removal, as well as dolomitization. These processes occurred in marine, early burial, and late burial environments. Depositional facies and fabric were major controls on the distribution of the diagenetic products, and the Capitan was most altered during early burial diagenesis.

Caves like Carlsbad Caverns are some of the most spectacular features in the Capitan depositional system. Carol Hill has made many breakthroughs in her studies of Carlsbad Caverns and other caves in the Guadalupe Mountains. During the last 20 years, interpretation of Carlsbad Caverns has changed from dissolution mainly by carbonic acid in freshwater to a sequence of dissolution events to create the caves. The main stage of cave formation in Carlsbad Caverns and many other caves in the Guadalupe Mountains occurred by dissolution in sulfuric acid, which formed when basinal hydrogen sulfide mixed with oxidized freshwater. In her article "Origin of Caves in the Capitan," Hill summaries many of her key observations and interpretations.

ACKNOWLEDGMENTS

We thank SEPM for support in publishing this volume. What originated as an informal gathering of workers interested in the Capitan has resulted in a publication summarizing new research results. We sincerely thank the authors, who contributed to the volume in a timely manner. We are grateful to the following individuals who greatly improved this volume by helping to review manuscripts: Tom Algeo, Denise Cox, Bill Fitchen, Ray Garber, Brian Glenister, Mike Grammer, Robert Handford, Ron Johns, Chris Kendall, Charlie Kerans, Zakaria Lasemi, Jerry Lewis, Rick Major, Leslie Melim, John Mylroie, Merlynd Nestell, Dave Osleger, Gene Rankey, Emily Stoudt, and Oliver Weidlich.

REFERENCES

ACHAUER, C. W., 1969, Origin of Capitan Formation, Guadalupe Mountains, New Mexico and Texas: American Association of Petroleum Geologists Bulletin, v. 53, p. 2314-2323.

ADAMS, J. E., and FRENZEL, H. W., 1950, Capitan barrier reef, Texas and New Mexico: Journal of Geology, v. 58, p. 289-312.

ANDREASON, M. W., 1992, Coastal siliciclastic sabkhas and related evaporative environments of the Permian Yates formation, North Ward-Estes field, Ward County, Texas: American Association of Petroleum Geologists Bulletin, v. 76, p. 1735-1759.

ASSERETO, R. L., and KENDALL, C. G. St. C., 1977, Nature, origin, and classification of peritidal tepee structures and related breccias: Sedimentology, v. 24, p. 153-210.

BAARS, D. L., 1964, Modern carbonate sediments as a guide to old limestones: World Oil, v. 158, p. 95-100.

BABCOCK, J. A., 1977, Calcareous algae, organic boundstones, and the genesis of the Upper Capitan Limestone (Permian, Guadalupian), Guadalupe Mountains, West Texas and New Mexico, *in* Hileman, M. E., and Mazzullo, S. J., eds., Upper Guadalupian Facies, Permian Reef Complex, Guadalupe Mountains, New Mexico and West Texas: Midland, Society of Economic Paleontologists and Mineralogists, Permian Basin Section Publication 77-16, p. 3-44.

BABCOCK, L. C., 1977, Life in the Delaware Basin: The paleoecology of the Lamar Limestone, *in* Hileman, M. E., and Mazzullo, S. J., eds., Upper Guadalupian Facies, Permian Reef Complex, Guadalupe Mountains, New Mexico and West Texas: Midland, Society of Economic Paleontologists and Mineralogists, Permian Basin Section, Publication 77-16, p. 357-389.

BABCOCK, J. A., and YUREWICZ, D. A., 1989, The massive facies of the Capitan limestone, Guadalupe Mountains, Texas and New Mexico, *in* Harris, P. M., and Grover, G. A., eds., Subsurface and outcrop examination of the Capitan Shelf Margin, Northern Delaware Basin: Tulsa, Society of Economic Paleontologists and Mineralogists Core Workshop 13, San Antonio, Texas, p. 365-371.

BEBOUT, D. G., and Kerans, C., 1993, eds., Guide to the Permian Reef Geology Trail, McKittrick Canyon, Guadalupe Mountains National Park, West Texas, Guidebook 26: Austin, Bureau of Economic Geology, University of Texas, 48 p.

BEBOUT, D. G., KERANS, C., and HARRIS, P. M., 1993, Introduction, *in* Bebout, D. G., and Kerans, C., 1993, eds., Guide to the Permian Reef Geology Trail, McKittrick Canyon, Guadalupe Mountains National Park, West Texas, Guidebook 26: Austin, Bureau of Economic Geology, University of Texas, p. 1-4.

BECK, R. H., 1967, Depositional mechanics of the Cherry Canyon Formation, Delaware Basin, Texas: Unpublished M.S. Thesis, Texas Technological College, 107 p.

BLANCHARD, W. G., and DAVIS, M. J., 1929, Permian stratigraphy and structure of parts of southeastern New Mexico and southwestern Texas: American Association of Petroleum Geologists Bulletin, v. 13, p. 957-995.

BORER, J. M., and HARRIS, P. M., 1989, Depositional facies and cycles in Yates Formation outcrops, Guadalupe Mountains, New Mexico, *in* Harris, P. M., and Grover, G. A., eds., Subsurface and outcrop examination of the Capitan shelf margin, northern Delaware Basin: Tulsa, Society of Economic Paleontologists and Mineralogists Core Workshop 13, San Antonio, Texas, p. 305-317.

BORER, J. M., and HARRIS, P. M., 1991a, Lithofacies and cyclicity of the Yates Formation, Permian Basin—implications for reservoir heterogeneity: American Association of Petroleum Geologists Bulletin, v. 75, p. 726-779.

BORER, J. M., and HARRIS, P. M., 1991b, Depositional facies and model for mixed siliciclastics and carbonates of the Yates Formation, Permian basin, *in* Lomando, A. J., and Harris, P. M., eds., Mixed carbonate-siliciclastic sequences: Tulsa, Society of Economic Paleontologists and Mineralogists Core Workshop 15, p. 1-133.

BORER, J. M., and HARRIS, P. M., 1995, Computer simulation of the Yates Formation (Permian, Delaware basin)—sequence stratigraphy and shelf-to-basin correlation implications, *in* Martin, R. L., ed., In search of new Permian basin oil and gas fields: Using today's technologies and tomorrow's ideas for exploration, development, and 3D seismic in a mature basin: Midland, West Texas Geological Society Symposium, Publication 95-98, p. 111-132.

BOZANICH, R. G., 1979, The Bell Canyon and Cherry Canyon formations, eastern Delaware basin, Texas—lithology, environments, and mechanisms of deposition, *in* Sullivan, N. M., ed., Guadalupian Delaware Mountain Group: Midland, Society of Economic Paleontologists and Mineralogists, Permian Basin Section, Publication 79-18, p. 121-141.

BRETZ, J. H., 1949, Carlsbad Caverns and other caves of the Guadalupe block, New Mexico: Journal of Geology, v. 57, p. 447-463.

BROWN, A. A., 1996, Position of the *Polydiexodina* last-occurrence datum in Guadalupian strata: revised correlations at McKittrick Canyon, *in* DeMis, W. D., and Cole, A. G., eds., The Brushy Canyon Play in Outcrop and Subsurface: Concepts and Examples: Midland, Society of Economic Paleontologists and Mineralogists, Permian Basin Section, Publication 96-38, p. 75-83.

BROWN, A. A. , and Loucks, R. G., 1993a, Influence of sediment type and depositional processes on stratal patterns in the Permian basin-margin Lamar limestone, McKittrick Canyon, Texas, *in* Loucks, R. G., and Sarg, J. F., eds., Carbonate Sequence Stratigraphy, Recent Developments and Applications: Tulsa, American Association of Petroleum Geologists Memoir 57, p. 133-156.

BROWN, A. A., and LOUCKS, R. G., 1993b, Toe of slope, *in* Bebout, D. G., and Kerans, C., eds., Guide to the Permian Reef Geology Trail, McKittrick Canyon, Guadalupe Mountains National Park, West Texas, Guidebook 26: Austin, Bureau of Economic Geology, University of Texas, p. 5-13.

BURRI, P., DRESNAY, R., and WAGNER, C. W., 1973, Tepee structures and associated diagenetic features in intertidal carbonate sands (Lower Jurassic, Morocco): Sedimentary Geology, v. 9, p. 221-228.

CANDELARIA, M. P., 1989, Shallow marine sheet sandstones, Upper Yates Formation, Northwest Shelf, Delaware Basin, New Mexico, *in* Harris, P. M., and Grover, G. A., eds., Subsurface and outcrop examination of the Capitan shelf margin, northern Delaware Basin: Tulsa, Society of Economic Paleontologists and Mineralogists, Core Workshop 13, p. 319-324.

CRANDALL, K. H., 1929, Permian stratigraphy of southeastern New Mexico and adjacent parts of western Texas: American Association of Petroleum Geologists Bulletin, v. 13, p. 927-945.

CRONOBLE, J. M., 1974, Biotic constituents and origin of facies in Capitan reef, New Mexico and Texas: The Mountain Geologist, v. 11, p. 95-108.

CYS, J. M., TOOMEY, D. F., BREZINA, J. L., GREENWOOD, E., GROVES, D. B., KLEMENT, K. W., KULLMAN, J. D., MCMILLAN, T. L., SCHMIDT, V., SNEED, E.

D., and Wagner, L. H., 1977, Capitan Reef—evolution of a concept, *in* Hileman, M. E., and Mazzullo, S. J., eds., Upper Guadalupian facies, Permian Reef Complex, Guadalupe Mountains, New Mexico and west Texas, Guidebook Publication 77-16: Midland, Society of Economic Paleontologists and Mineralogists, Permian Basin Section, p. 201-321.

Duchene, H. R., and McLean, J. S., 1989, The role of hydrogen sulfide in the evolution of caves in the Guadalupe Mountains of southeastern New Mexico, *in* Harris, P. M., and Grover, G. A., eds., Subsurface and outcrop examination of the Capitan shelf margin, northern Delaware Basin: Tulsa, Society of Economic Paleontologists and Mineralogists Core Workshop 13, p. 475-481.

Dunham, R. J., 1969, Vadose pisolite in the Capitan reef (Permian), New Mexico and Texas, *in* Friedman, G. M., ed., Depositional Environments in Carbonate Rock: Tulsa, Society of Economic Paleontologists and Mineralogists Special Publication 14, p. 182-191.

Dunham, R. J., 1970, Stratigraphic reefs versus ecologic reefs: American Association of Petroleum Geologists Bulletin, v. 54, p. 1931-1950.

Dunham, R. J., 1972, Capitan reef, New Mexico and Texas—Facts and Questions to Aid Interpretation and Group Discussion: Midland, Society of Economic Paleontologists and Mineralogists, Permian Basin Section, Publication 72-14, 297 p.

Esteban, M., and Pray, L. C., 1977, Origin of the pisolite facies of the shelf crest, *in* Hileman, M. E., and Mazzullo, S. J., eds., Upper Guadalupian Facies, Permian Reef Complex, Guadalupe Mountains, New Mexico and West Texas, Guidebook Publication 77-16: Midland, Society of Economic Paleontologists and Mineralogists, Permian Basin Section, p. 479-484.

Esteban, M., and Pray, L. C., 1983, Pisoids and pisolite facies (Permian), Guadalupe Mountains, New Mexico and West Texas, *in* Peryt, T. M., ed., Coated Grains: Berlin, Springer-Verlag, p. 503-537.

Fischer, A. G., and Sarnthein, M., 1988, Airborne silts and dune-derived sands from the Permian of the Delaware basin: Journal of Sedimentary Petrology, v. 58, p. 637-643.

Garber, R. A., Grover, G. A., and Harris, P. M., 1989, Geology of the Capitan shelf margin—subsurface data from the northern Delaware basin, *in* Harris, P. M., and Grover, G. A., eds., Subsurface and outcrop examination of the Capitan shelf margin—a core workshop: Tulsa, Society of Economic Paleontologists and Mineralogists Core Workshop 13, p. 3-269.

Gardner, M. H., and Sonnenfeld, M. D., 1996, Stratigraphic changes in facies architecture of the Permian Brushy Canyon Formation in Guadalupe Mountains National Park, west Texas, *in* DeMis, W. D., and Cole, A. G., eds., The Brushy Canyon Play in Outcrop and Subsurface: Concepts and Examples: Midland, Society of Economic Paleontologists and Mineralogists, Permian Basin Section, Publication 96-38, p. 17-40.

Given, R. K., and Lohmann, K. C., 1985, Derivation of the original isotopic composition of Permian marine cements: Journal of Sedimentary Petrology, v. 55, p. 430-439.

Given, R. K., and Lohmann, K. C., 1986, Isotopic evidence for the early meteoric diagenesis of the reef facies, Permian Reef Complex of West Texas and New Mexico: Journal of Sedimentary Petrology, v. 56, p. 183-193.

Grauten, W. F., 1965, Fluid relationships in Delaware Mountain sandstones, *in* Young, A., and Galley, J. E., eds., Fluids in Subsurface Environments: Tulsa, American Association of Petroleum Geologists Memoir 4, p. 294-307.

Handford, C. R., Kendall, A. C., Prezbindowski, D. R., Durham, J. B., and Logan, B. W., 1984, Salina-margin tepees, pisoliths, and aragonite cements, Lake McLeod, Western Australia: Their significance in interpreting ancient analogs: Geology, v. 12, p. 523-527.

Harms, J. C., 1974, Brushy Canyon Formation, Texas—a deep water density current deposit: Geological Society of America Bulletin, v. 85, p. 1763-1784.

Harms, J. C., and Williamson, C. R., 1988, Deep-water density current deposits of Delaware Mountain group (Permian), Delaware Basin, Texas and New Mexico: American Association of Petroleum Geologists Bulletin, v. 72, p. 299-317.

Harris, P. M., and Grover, G. A., eds., 1989, Subsurface and outcrop examination of the Capitan shelf margin, northern Delaware Basin: Tulsa, Society of Economic Paleontologists and Mineralogists Core Workshop 13, San Antonio, Texas, 481 p.

Harwood, G. M., 1989, Large-scale channel development within the Capitan Reef Complex—evidence from Carlsbad Cavern, *in* Harris, P. M., and Grover, G. A., eds., Subsurface and outcrop examination of the Capitan shelf margin, northern Delaware Basin: Tulsa, Society of Economic Paleontologists and Mineralogists Core Workshop 13, San Antonion, Texas, p. 379-386.

Hayes, P. T., 1964, Geology of the Guadalupe Mountains, New Mexico: Washington, D.C., U. S. Geological Survey Professional Paper 446, 69 p.

Hayes, P. T., and Gale, B. T., 1957, Geology of the Carlsbad Caverns east quadrangle, New Mexico: Washington, D.C., U.S. Geological Survey, Geological Quadrant Map GQ-98, scale 1:62,000.

Hayes, P. T., and Koogle, R. L., 1958, Geology of the Carlsbad Caverns west quadrangle, New Mexico-Texas: Washington, D.C., U. S. Geological Survey, Geological Quadrangle Map GQ-112.

Hileman, M. E., and Mazzullo, S. J., eds., 1977, Upper Guadalupian Facies, Permian Reef Complex, Guadalupe Mountains, New Mexico and West Texas, 1977 Field Conference Guidebook, v. 1: Midland, Society of Economic Paleontologists and Mineralogists, Permian Basin section, Publication 77-16, 508 p.

Hill, C. A., 1987, Geology of Carlsbad Cavern and other caves in the Guadalupe Mountains, New Mexico and Texas: New Mexico Bureau of Mines and Mineral Resources Bulletin, v. 117, 150 p.

Hill, C. A., 1989, The evolution of the Delaware Basin—preliminary results, *in* Harris, P. M., and Grover, G. A., eds., Subsurface and outcrop examination of the Capitan shelf margin, northern Delaware Basin: Tulsa, Society of Economic Paleontologists and Mineralogists Core Workshop 13, San Antonio, Texas, p. 467-473.

Hill, C. A., 1990, Sulfuric acid speleogenesis of Carlsbad Cavern and its relationship to hydrocarbons, Delaware Basin, New Mexico and Texas: American Association of Petroleum Geologists Bulletin, v. 74, p. 1685-1694.

Hill, C. A., 1995, H_2S-related porosity and sulfuric acid oil-field karst, *in* Budd, D. A., Saller, A. H., and Harris, P. M., eds., Unconformities in carbonate strata: Their recognition and the significance of associated porosity: Tulsa, American Association of Petroleum Geologists Memoir 61, p. 301-306.

Hill, C. A., 1996, Geology of the Delaware basin, Guadalupe, Apache, and Glass mountains, New Mexico and West Texas: Midland, Society of Economic Paleontologists and Mineralogists, Permian Basin Section, Publication 96-39, 480 p.

Hull, J. P. D., Jr., 1957, Petrogenesis of Permian Delaware Mountain sandstone, Texas and New Mexico: American Association of Petroleum Geologists Bulletin, v. 41, p. 278-307.

Hurley, N. F., 1978, Facies mosaic of the lower Seven Rivers Formation (Permian, Guadalupian), Guadalupe Mountains, New Mexico: Unpublished M.A. Thesis, University of Wisconsin, Madison, 198 p.

Hurley, N. F., 1989, Facies mosaic of the lower Seven Rivers Formation, McKittrick Canyon, New Mexico, *in* Harris, P. M., and Grover, G. A., eds., Subsurface and outcrop examination of the Capitan shelf margin, northern Delaware basin: Tulsa, Society of Economic Paleontologists and Mineralogists Core Workshop 13, p. 325-346.

Jacka, A. D., Beck, R. H., St. Germain, L. C., and Harrison, S. C., 1968, Permian deep-sea fans of the Delaware Mountain Group (Guadalupian), Delaware basin, *in* Silver, B., ed., Guadalupian facies, Apache Mountains Area, West Texas: Midland, Society of Economic Paleontologists and Mineralogists, Permian Basin Section, Publication 68-11, p. 49-90.

Jacka, A. D., Thomas, C. M., Beck, R. H., Williams, K. W., and Harrison, S. C., 1960, Guadalupian depositional cycles of the Delaware Basin and Northwest Shelf, *in* Elam, J. G., and Chuber, S. , eds., Cyclic Sedimentation in the Permian Basin: Midland, West Texas Geological Society Publication 72-60, p. 151-195.

Jagnow, D. H., 1979, Cavern development in the Guadalupe Mountains: Adobe Press, 55 p.

Jagnow, D. H., 1989, The geology of Lechuguilla Cave, New Mexico, *in* Harris, P. M., and Grover, G. A., eds., Subsurface and outcrop examination of the Capitan shelf margin, northern Delaware basin: Tulsa, Society of Economic Paleontologists and Mineralogists Core Workshop 13, p. 459-466.

Johnson, J. H., 1942, Permian lime-secreting algae from the Guadalupe Mountains, New Mexico: Geological Society of America Bulletin, v. 53, p. 195-226.

Jones, T. S., and Smith, H. M., 1965, Relationships of oil composition and stratigraphy in the Permian Basin of West Texas and New Mexico, *in* Young, A., and Galley, J. E., eds., Fluids in subsurface environments: Tulsa, American Association of Petroleum Geologists Memoir 4, p. 101-224.

KELLEY, V. C., 1972, Geometry and correlation along Permian Capitan escarpment, New Mexico and Texas: American Association of Petroleum Geologists Bulletin, v. 56, p. 2192-2211.

KENDALL, C. G. St. C., 1969, An environmental reinterpretation of the Permian evaporite/carbonate shelf sediments of the Guadalupe Mountains: Geological Society of America Bulletin, v. 80, p. 2503-2526.

KERANS, C., FITCHEN, W. M., GARDNER, M. H., SONNENFELD, M. D., TINKER, S. W., and WARDLAW, B. R., 1992, Styles of sequence development within uppermost Leonardian through Guadalupian strata of the Guadalupe Mountains, Texas and New Mexico, *in* Mruk, D. H., and Curran, B. C., eds., Permian Basin exploration and production strategies: Applications of sequence stratigraphic and reservoir characterization concepts: Midland, West Texas Geological Society Publication 92-91, p. 1-6.

KERANS, C., FITCHEN, W. M., GARDNER, M. H., and WARDLAW, B. R., 1993, A contribution to the evolving stratigraphic framework of Middle Permian strata of the Delaware Basin, Texas and New Mexico, *in* Love, D. W., Hawley, J. W., Kues, B. S., Adams, J. W., Austin, G. S., and Barker, J. M., eds., Carlsbad region, New Mexico and West Texas: New Mexico Geological Society, Forty-fourth Annual Field Conference, p. 175-184.

KERANS, C., and HARRIS, P. M., 1993, Outer shelf and shelf crest, *in* Bebout, D. G., and Kerans, C., eds., Guide to the Permian Reef Geology Trail, McKittrick Canyon, Guadalupe Mountains National Park, West Texas, Guidebook 26: Austin, University of Texas, Bureau of Economic Geology, Guidebook 26, p. 32-43.

KERANS, C., and TINKER, S. W., 1997, Sequence stratigraphy and characterization of carbonate Reservoirs: Tulsa, Society of Economic Paleontologists and Mineralogists Short Course Notes 40, 130 p.

KING, P. B., 1930, The Geology of the Glass Mountains, pt. 1: Austin, Bureau of Economic Geology, University of Texas 3038, 167 p.

KING, P. B., 1934, Permian stratigraphy of Trans-Pecos Texas: Geological Society of America Bulletin, v. 45, p. 697-798.

KING, P. B., 1942, Permian of West Texas and southeastern New Mexico: American Association of Petroleum Geologists Bulletin, v. 26, p. 535-763.

KING, P. B., 1948, Geology of the southern Guadalupe Mountains, Texas: Washington, D.C., U. S. Geological Survey Professional Paper 215, 183 p.

KIRKLAND-GEORGE, B. L., 1992, Distinctions Between Reefs and Bioherms Based on Studies of Fossil Algae *Mizzia*, Permian Capitan Reef Complex (Guadalupe Mountains, Texas and New Mexico) and *Eugonophyllum*, Pennsylvanian Holder Formation (Sacramento Mountais, New Mexico): Unpublished Ph.D. Dissertation, Louisiana State University, 156 p.

KIRKLAND, B.L., DICKSON, J.A.D., WOOD, R.A., and LAND, L.S., 1998, Microbialite and microstratigraphy: The origin of encrustations in the middle and upper Capitan Formation, Guadalupe Mountains, Texas and New Mexico, U.S.A.: Journal of Sedimentary Research, v. 68, p. 956-969.

KIRKLAND, B. L., LONGACRE, S. A. and STOUDT, E. L., 1993, Reef, *in* Bebout, D. G., and Kerans, C., eds., Guide to the Permian Reef Geology Trail, McKittrick Canyon, Guadalupe Mountains National Park, West Texas, Guidebook 26: Austin, Bureau of Economic Geology, University of Texas, p. 23-31.

KIRKLAND, B. L. and MOORE, C. H., 1996, Microfacies analysis of the Tansill outer shelf, Permian Capitan Reef Complex: Midland, West Texas Geological Society Publication 96-101, p. 99-106.

KOCUREK, G., and KIRKLAND, B. L., 1998, Getting to the source: Aeolian influx to the Permian Delaware Basin region: Sedimentary Geology, v. 117, p. 143-149.

LANG, W. B., 1937, The Permian Formations of the Pecos Valley of New Mexico and Texas: American Association of Petroleum Geologists Bulletin, v. 21, p. 833-898.

LLOYD, E. R., 1929, Capitan Limestone and associated formations of New Mexico and Texas: American Association of Petroleum Geologists Bulletin, v. 13, p. 645-658.

MAZZULLO, S. J., 1977, Synsedimentary diagenesis of reefs, *in* Hileman, M. E., and Mazzullo, S. J., eds., Upper Guadalupian Facies, Permian Reef Complex, Guadalupe Mountains, New Mexico and West Texas: Midland, Society of Economic Paleontologists and Mineralogists, Permian Basin Section, Publication 77-16, p. 323-354.

MAZZULLO, S. J., and CYS, J. M., 1977, Submarine cements in Permian boundstones and reef-associated rocks, Guadalupe Mountains, West Texas and southeastern New Mexico, *in* Hileman, M. E., and Mazzullo, S. J., eds., Upper Guadalupian Facies, Permian Reef Complex, Guadalupe Mountains, New Mexico and West Texas: Midland, Society of Economic Paleontologists and Mineralogists, Permian Basin Section, Publication 77-16, p. 151-200.

MAZZULLO, S. J., and CYS, J.M., 1978, *Archaeolithoporella*-boundstones and marine aragonite cements, Permian Capitan reef, New Mexico and Texas, USA: Neues Jahrbuch für Geologie und Paläontologie, Monatshefte, p. 600-611.

MAZZULLO, S. J., MAZZULLO, J., and HARRIS, P. M., 1985, Eolian origin of quartzose sheet sands in Permian shelf facies, Guadalupe Mountains, (abs.), *in* Cunningham, B. K., and Hedrick, C. L., eds., Permian carbonate/clastic sedimentology, Guadalupe Mountains: Analogs for shelf and basin reservoirs: Midland, Society of Economic Paleontologists and Mineralogists, Permian Basin Section, Publication 85-24, p. 71.

MAZZULLO, S. J., BISHOFF, W. D., and Hedrick, C. L., 1989, Stacked island facies in Tansill outer-shelf platform, Dark Canyon, Guadalupe Mountains, New Mexico, *in* Harris, P. M., and Grover, G. A., eds., Subsurface and outcrop examination of the Capitan shelf margin, northern Delaware Basin: Tulsa, Society of Economic Paleontologists and Mineralogists Core Workshop 13, p. 287-293.

MEISSNER, F. F., 1972, Cyclic sedimentation in middle Permian strata of the Permian basin, *in* Elam, J. G., and Chuber, S., eds., Cyclic sedimentation in the Permian basin (2nd edition): Midland, West Texas Geological Society Publication 72-60, p. 203-232.

MELIM, L. A., and SCHOLLE, P. A., 1989, Dolomitization model for the forereef facies of the Permian Capitan Formation, Guadalupe Mountains, Texas-New Mexico, *in* Harris, P. M., and Grover, G. A., eds., Subsurface and outcrop examination of the Capitan shelf margin, northern Delaware Basin: Midland, Society of Economic Paleontologists and Mineralogists Workshop 13, p. 407-413.

MOTTS, W. S., 1962, Geology of the West Carlsbad Quadrangle, New Mexico: Washington, D.C., U.S. Geological Survey, Map GQ-167, scale 1:62,500.

MOTTS, W. S., 1972, Geology and paleoenvironment of the northern segment, Capitan Shelf, New Mexico and West Texas: Geological Society of America Bulletin, v. 83, p. 701-722.

MRUK, D. H., 1989, Diagenesis of the Capitan Limestone, Upper Permian, McKittrick Canyon, West Texas, *in* Harris, P. M., and Grover, G. A., eds., Subsurface and outcrop examination of the Capitan shelf margin, northern Delaware Basin: Tulsa, Society of Economic Paleontologists and Mineralogists Core Workshop 13, p. 387-406.

MRUK, D. H., and BEBOUT, D. G., 1993, Slope, *in* Bebout, D. G., and Kerans, C., eds., Guide to the Permian Reef Geology Trail, McKittrick Canyon, Guadalupe Mountains National Park, West Texas, Guidebook 26: Austin, Bureau of Economic Geology, University of Texas, p. 14-22.

MUTTI, M., and SIMO, J. A., 1994, Distribution, petrography and geochemistry of early dolomite in cyclic shelf facies, Yates Formation (Guadalupian), Capitan Reef Complex, USA, *in* Loucks, R. L., and Sarg, J. F., eds., Carbonate Sequence Stratigraphy—Recent Developments and Applications: Tulsa, American Association of Petroleum Geologists Memoir 57, p. 515-534.

NEESE, D. G., 1989, Peritidal facies of the Guadalupian shelf crest, Walnut Canyon, New Mexico, *in* Harris, P. M., and Grover, G. A., Subsurface and outcrop examination of the Capitan shelf margin, northern Delaware Basin: Tulsa, Society of Economic Paleontologists and Mineralogists Core Workshop 13, p. 295-303.

NEESE, D. G., and SCHWARTZ, H. H., 1977, Facies mosaic of the upper Yates and lower Tansill formations, Walnut and Rattlesnake Canyons, Guadalupe Mountains, New Mexico, *in* Hileman, M. E., and Mazzullo, S. J., eds., Upper Guadalupian Facies, Permian Reef Complex, Guadalupe Mountains, New Mexico and West Texas: Midland, Society of Economic Paleontologists and Mineralogists, Permian Basin Section, Publication 77-16, p. 437-450.

NEWELL, N.D., 1955, Depositional fabric in Permian reef limestone: Journal of Geology, v. 63, p. 301-309.

NEWELL, N. D., RIGBY, J. K., FISCHER, A. G., WHITEMAN, A. J., HICKOX, J. E., and BRADLEY, J. S., 1953, The Permian reef complex of the Guadalupe Mountains region, Texas and New Mexico, a study in paleoecology: San Francisco, W. H. Freeman and Company, 236 p.

NOÉ, S.U., 1996, Late-stage evolution of the Permian reef complex: Shelf margin and outer-shelf development of the Tansill Formation (Late Permian), northern Guadalupe Mountains, New Mexico, USA, *in* Reitner, J., Neuwieler, R., and Gunkel, F. eds., Global and Regional Controls on Biogenic Sedimentation, I, Reef Evolution, Research Reports. Göttinger Arbeiten zur Geologie und Paläontologie, Sonderband 2, p. 317-324.

Noé, S. U., and Mazzullo, S. J., 1992, Upper Tansill patch reef facies, Sheep Draw Canyon, Guadalupe Mountains, Eddy County, New Mexico: West Texas Geological Society Bulletin, v. 32, p. 5-11.

Nottingham, M. W., 1960, Recent Bell Canyon exploration in the north Delaware Basin, *in* Abilene Geological Society, Natural gas in the Southwest: Southwestern Federation of Geological Sciences, 3rd Annual Meeting, Abilene, Texas, p. 139-148.

Osleger, D. A., 1998, Sequence architecture and sea-level dynamics of upper Permian shelfal facies, Guadalupe Mountains, southern New Mexico: Journal of Sedimentary Research, v. 68, p. 327-346.

Parsley, M. J., and Warren, J. K., 1989, Characterization of an upper Guadalupian barrier-island complex from the middle and upper Tansill Formation (Permian), east Dark Canyon, Guadalupe Mountains, New Mexico, *in* Harris, P. M., and Grover, G. A., Subsurface and outcrop examination of the Capitan shelf margin, northern Delaware Basin: Tulsa, Society of Economic Paleontologists and Mineralogists, Core Workshop 13, p. 279-285.

Payne, M. W., 1976, Basinal sandstone facies, Delaware basin, west Texas and southeast New Mexico: American Association of Petroleum Geologists Bulletin, v. 60, p. 517-527.

Pray, L. C., 1977, The all wet constant sea level hypothesis of Upper Guadalupian shelf and shelf edge strata, Guadalupe Mountains, New Mexico and Texas, *in* Hileman, M. E., and Mazzullo, S. J., eds., Upper Guadalupian Facies, Permian Reef Complex, Guadalupe Mountains, New Mexico and West Texas: Midland, Society of Economic Paleontologists and Mineralogists, Permian Basin section, Publication 77-16, p. 433.

Pray, L. C., 1985a, The Capitan-massive and its proximal carbonate, and minor siliciclastics of the Capitan back-reef, Walnut Canyon area, Carlsbad Caverns National Park, New Mexico, *in* Cunningham, B. K., and Hedrick, C.L., eds., Permian Carbonate/Clastic Sedimentology, Guadalupe Mountains: Analogs for Shelf and Basin Reservoirs: Midland, Society of Economic Paleontologists and Mineralogists, Permian Basin Section, Publication 85-24, p. 25-41.

Pray, L. C., 1985b, Sea level and the Capitan reef complex, southwest U.S.A., a famous model "in flux", *in* Cunningham, B. K., and Hedrick, C. L., eds., Permian Carbonate/Clastic Sedimentology, Guadalupe Mountains: Analogs for Shelf and Basin Reservoirs: Midland, Society of Economic Paleontologists and Mineralogists, Permian Basin Section, Publication 85-24, p. 68.

Pray, L. C., 1985c, The Capitan Limestone and its carbonate and siliciclastic back-reef facies, *in* Cunningham, B. K., and Hedrick, C. L., eds., Permian Carbonate/Clastic Sedimentology, Guadalupe Mountains: Analogs for Shelf and Basin Reservoirs: Midland, Society of Economic Paleontologists and Mineralogists, Permian Basin Section, Publication 85-24, p. 23-24.

Pray, L. C., 1989a, Philip B. King and Robert J. Dunham–an appeciation, *in* Harris, P. M., and Grover, G. A., eds., Subsurface and outcrop examination of the Capitan shelf margin, northern Delaware Basin: Tulsa, Society of Economic Paleontologists and Mineralogists Core Workshop 13, p. x-xii.

Pray, L. C., 1989b, Lateral variability of the Capitan Reef Complex, West Texas and New Mexico, *in* Harris, P. M., and Grover, G. A., eds., Subsurface and outcrop examination of the Capitan shelf margin, northern Delaware Basin: Tulsa, Society of Economic Paleontologists and Mineralogists Core Workshop 13, p. 273-278.

Pray, L. C., and Esteban, M., 1977, Upper Guadalupian facies, Permian reef complex, Guadalupe Mountains, New Mexico and west Texas, 1977 field conference guidebook: Midland, Society of Economic Paleontologists and Mineralogists, Permian Basin Section, Publication 77-16, v. 2, 194 p.

Rankey, E. C., and Lehrmann, D. J., 1996, Anatomy and origin of toplap in a mixed carbonate-clastic system, Seven Rivers Formation (Permian, Guadalupian), Guadalupe Mountains, New Mexico, USA: Sedimentology, v. 43, p. 807-826.

Richardson, G. B., 1904, Reconnaissance in Trans-Pecos Texas North of the Texas and Pacific Railway: Austin, University of Texas, Mineral Survey Bulletin 9, 117 p.

Rigby, J. K., 1958, Mass movements in Permian rocks of Trans-Pecos Texas: Journal of Sedimentary Petrology, v. 28, p. 298-315.

Saller, A. H., 1996, Differential compaction and basinward tilting of the prograding Capitan reef complex, Permian, west Texas and southeast New Mexico, USA: Sedimentary Geology, v. 101, p. 21-30.

Sarg, J. F., 1977, Sedimentology of the carbonate-evaporite facies transition of the Seven Rivers Formation (Guadalupian, Permian) in southeast New Mexico, *in* Hileman, M. E., and Mazzullo, S. J., eds., Upper Guadalupian facies, Permian Reef Complex, Guadalupe Mountains, New Mexico and west Texas: Midland, Society of Economic Paleontologists and Mineralogists, Permian Basin Section, Publication 77-16, p. 451-478.

Sarg, J. F., 1989, Stratigraphy and sedimentology of the back-reef upper Queen–lower Seven Rivers strata, Goat-Seep-Capitan reef complexes (Middle-Late Guadalupian, Permian), southeast New Mexico, *in* Harris, P. M., and Grover, G. A., eds., Subsurface and outcrop examination of the Capitan shelf margin, northern Delaware Basin: Tulsa, Society of Economic Paleontologists and Mineralogists Core Workshop 13, p. 347-352.

Schmidt, V., 1977, Inorganic and organic reef growth and subsequent diagenesis in the Permian Capitan Reef Complex, Guadalupe Mountains, Texas, New Mexico, *in* Hileman, M. E., and Mazzullo, S. J., eds., Upper Guadalupian Facies, Permian Reef Complex, Guadalupe Mountains, New Mexico and West Texas: Midland, Society of Economic Paleontologists and Mineralogists, Permian Basin Section, Publication 77-16, p. 93-131.

Schmidt, V., and Klement, K. W., 1977, Diagenesis of the Capitan Formation, *in* Hileman, M. E., and Mazzullo, S. J., eds., Upper Guadalupian facies, Permian Reef Complex, Guadalupe Mountains, New Mexico and West Texas, Guidebook Publication 77-16: Midland, Society of Economic Paleontologists and Mineralogists, Permian Basin Section, p. 265-286.

Scholle, P. A., and Halley, R. B., 1980, Upper Paleozoic depositional and diagenetic facies in a mature petroleum province (a field guide to the Guadalupe and Sacramento Mountains of West Texas and New Mexico): Washington, D.C., U. S. Geological Survey Open File Report 80-383, 191 p.

Scholle, P. A., Ulmer, D. S., and Melim, L. A., 1992, Late-stage calcites in the Permian Capitan Formation and its equivalents, Delaware Basin margin, west Texas and New Mexico—evidence for replacement of precursor evaporites: Sedimentology, v. 39, p. 207-234.

Senowbari-Daryan, B. and Rigby, K., 1996, Brachiopod mounds not sponge reefs, Permian Capitan-Tansill Formations, Guadalupe Mountains, New Mexico: Journal of Paleontology, 70, 697-701.

Shumard, G. G., 1858, Observations on the geological formation of the country between the Rio Pecos and the Rio Grande in New Mexico: St. Louis Academy of Science, Transactions, p. 273-289.

Silver, B. A., and Todd, R. G., 1969, Permian cyclic strata, northern Midland and Delaware basins, west Texas and southeastern New Mexico: American Association of Petroleum Geologists Bulletin, v. 53, p. 2223-2251.

Smith, D. B., 1973, Geometry and correlation along Permian Capitan escarpment, New Mexico and Texas: American Association of Petroleum Geologists Bulletin, v. 57, p. 940-945.

Smith, D. B., 1974, Sedimentation of upper Artesia (Guadalupian) cyclic shelf deposits of northern Guadalupe Mountains, New Mexico: American Association of Petroleum Geologists Bulletin, v. 58, p. 1699-1730.

Sneed, E. D., 1977, Capitan shelf to basin correlations, *in* Hileman, M. E., and Mazzullo, S. J., eds., Upper Guadalupian Facies, Permian Reef Complex, Guadalupe Mountains, New Mexico and Texas, Guidebook Publication 77-16: Midland, Society of Economic Paleontologists and Mineralogists, Permian Basin Section, p. 286-311.

Tinker, S. W., 1996, Reservoir-scale sequence stratigraphy: McKittrick canyon and 3-D subsurface examples, west Texas and New Mexico: Unpublished Ph. D. Dissertation, University of Colorado, Boulder, 245 p.

Toomey, D. F., 1977, Is the "Capitan reef" a reef—or are we confused by semantics?, *in* Hileman, M. E., and Mazzullo, S. J., eds., Upper Guadalupian facies, Permian Reef Complex, Guadalupe Mountains, New Mexico and Texas, Guidebook Publication 77-16: Midland, Society of Economic Paleontologists and Mineralogists, Permian Basin Section, p. 226-233.

Toomey, D. F., and Babcock, J. A., 1983, Precambrian and Paleozoic algal carbonates, west Texas-southern New Mexico: Colorado Schol of Mines, Professional Contributions, v. 11, 345 p.

Toomey, D. F., and Cys, J. M., 1977, Rock/biotic relationships of the Permian Tansill-Capitan facies exposed on the north side of the entrance to Dark Canyon, Guadalupe Mountains, southeastern New Mexico, *in* Hileman, M. E., and Mazzullo, S. J., eds., Upper Guadalupian Facies, Permian Reef Complex, Guadalupe Mountains, New Mexico and West Texas: Midland, Society of Economic Paleontologists and Mineralogists, Permian Basin

Section, Publication 77-16, p. 133-150.

Tyrrell, W. W., Jr., 1964, Petrology and stratigraphy of near-reef Tansill-Lamar strata, Guadalupe Moutnains, Texas and New Mexico, *in* Geology of the Capitan Reef Complex of the Guadalupe Mountains, Guidebook: Roswell Geological Society, p. 66-82.

Tyrrell, W. W., Jr., 1969, Criteria useful in interpreting environments of unlike but time-equivalent carbonate units (Tansill-Capitan-Lamar), Capitan reef complex, West Texas and New Mexico, *in* Friedman, G. M., ed., Depositional Environments in Carbonate Rocks: Tulsa, Society of Economic Paleontologists and Mineralogists Special Publication 14, p. 80-97.

Warren, J. K., 1983, Tepees, modern (South Australia) and ancient Permian—Texas and New Mexico—a comparison: Sedimentary Geology, v. 34, p. 1-19.

Weidlich, O., 1996, Comparative analysis of Late Permian reefal limestones from the Capitan Reef (New Mexico, USA) and the Oman Mountains, *in* Reitner, J., Neuweiler, F., and Gunkel, F., eds., Global and regional controls on biogenic sedimentation, I. Reef evolution, research reports: Göttinger Arbeiten zur Geologie und Paläontologie, Sonderband 2, p. 329-332.

Weidlich, O., and Fagerstrom, J. A., 1998, Evolution of the Upper Capitan Massive (Permian), Guadalupe Mountains, New Mexico: Salt Lake City, Brigham Young University Geology Studies, v. 42, p. 167-188.

Wheeler, C.W., 1989, Stratigraphy and sedimentology of the Shattuck member (Queen Formation) and lowermost Seven Rivers formation (Guadalupian), North McKittrick and Dog canyons, Guadalupe Mountains, New Mexico and West Texas, *in* Harris, P. M., and Grover, G. A., eds., Subsurface and outcrop examination of the Capitan shelf margin, northern Delaware basin: Tulsa, Society of Economic Paleontologists and Mineralogists Core Workshop 13, p. 353-364.

Wilde, G. L., 1975, Fusulinid-defined Permian stages, *in* Cys, J. M. ,and Toomey, D. F., eds., Permian exploration, boundaries, and stratigraphy: Midland, West Texas Geological Society Publication 75-65, p. 67-83.

Williamson, C. R., 1977, Deep-sea channels of the Bell Canyon Formation (Guadalupian), Delaware Basin, Texas - New Mexico, *in* Hileman, M. E. and Mazzullo, S. J., eds., Upper Guadalupian Facies, Permian Reef Complex, Guadalupe Mountains, New Mexico and West Texas: Midland, Society of Economic Paleontologists and Mineralogists, Permian Basin Section, Publication 77-16, p. 409-431.

Williamson, C. R., 1979, Deep-sea sedimentation and stratigraphic traps, Bell Canyon Formation (Permian), Delaware Basin, Symposium and Field Conference Guidebook: Midland, Society of Economic Paleontologists and Mineralogists, Permian Basin Section, Publication 79-18, p. 39-74.

Wilson, J. L., 1967, Cyclic and reciprocal sedimentation in Virgilian strata of southern New Mexico: Geological Society of America Bulletin, v. 78, p. 805-818.

Wood, R., Dickson, J. A. D., and Kirkland-George, B., 1994, Turning the Capitan reef upside down: A new appraisal of the Permian Capitan reef, Guadalupe Mountains, Texas and New Mexico: Palaios, p. 422-427.

Wood, R., Dickson, J. A. D., and Kirkland, B. L., 1996, New observations on the ecology of the Permian Capitan reef, Texas and New Mexico: Paleontology, v. 39, p. 733-762.

Yurewicz, D. A., 1977, The origin of the massive facies of the lower and middle Capitan limestone (Permian), Guadalupe Mountains, New Mexico and west Texas, *in* Hileman, M. E., and Mazzullo, S. J., eds., Upper Guadalupian Facies, Permian Reef Complex, Guadalupe Mountains, New Mexico and West Texas: Midland, Society of Economic Paleontologists and Mineralogists, Permian Basin Section, Publication 77-16, p. 45-129.

EXTRINSIC STRATIGRAPHIC CONTROLS ON DEVELOPMENT OF THE CAPITAN REEF COMPLEX

CHARLES KERANS

Bureau of Economic Geology, The University of Texas at Austin, Austin, Texas 78713[1]

AND

SCOTT W. TINKER

Marathon Oil Company, Petroleum Technology Center, Littleton, Colorado 80122

ABSTRACT: The spectacular mixed siliciclastic/carbonate exposures of the Guadalupe Mountains include 30 high-frequency sequences (HFS) that stack together to form six composite sequences (CS), the CS9 through CS14. These sequences include carbonate ramps and reef-rimmed platforms as well as basin-restricted lowstand sequences. The Capitan Formation represents the shelf-margin and slope facies tracts of the upper 12 HFS. The Capitan is examined in the context of this late Leonardian-Guadalupian ramp-to-rimmed-shelf system by focusing on extrinsic controls on platform development.

Eustatic changes initiate and punctuate larger scale changes in platform evolution. Rapid shifts of large magnitude, such as the latest Leonardian (L7-L8 HFS) eustatic rise, are a first-order control on platform architecture and reef formation. The model for the late Permian eustatic curve based on the present stratigraphic framework suggests that by the time the Capitan was established, eustatic amplitudes were in the range of 20 m or less. This amplitude variation does not cause major shifts in the shelf-margin location but is sufficient to affect critical accommodation factors that influence reef depth and faunal composition.

Antecedent topography, whether of tectonic, depositional, erosional, or compactional origin, is the critical parameter in controlling the timing and development of the Capitan and other buildups in the Leonardian-Guadalupian sequences, as well as the primary control driving the ramp-to-rimmed-shelf transition. The shelf-slope break, whether a ramp or rimmed shelf, is only one of numerous geometric parameters that can be used to describe the dynamic evolution of carbonate platforms. Changing styles of carbonate-platform progradation and aggradation, which are responses to changes in platform and basin accommodation and sediment-supply, can be captured using P/A (ratio of progradation to aggradation for a given chronostratigraphic unit) and SMP/A (shelf-margin progradation/aggradation ratio) ratios. P/A values > 25 are characteristic of ramps and lowstand wedges, whereas P/A values < 25 are indicative of either transgressive-dominated ramps or reef-rimmed margins. SMP/A values within the Capitan-equivalent sequences can be used to document the complex but systematic and predictable progradational-aggradational-progradational response of the shelf margin to changing base level.

Within the high-frequency sequence framework, other analytical tools, including facies tract substitution and facies proportions, can be used to better constrain interpretations of the dynamic water-depth setting of the Capitan margin and factors controlling its position on the profile. This holistic approach, which draws on relationships from outer-shelf and shelf-crest facies tracts in the interpretation of the Capitan margin, demonstrates the power of a stratigraphic framework for sedimentologic analysis.

INTRODUCTION

One of the most significant contributions of high-frequency sequence stratigraphy is its ability to add an important tool for depositional facies analysis and reservoir prediction by providing a framework for comparative sedimentology within a stratigraphic context. This paper examines the shelf-margin facies tract of the Capitan reef within a stratigraphic context using data from the latest Leonardian through Guadalupian sequences exposed in the Guadalupe Mountains and nearby subsurface (Figs. 1 and 2). We will discuss the development of the shelf-margin reef by examining its sensitivity to key extrinsic stratigraphic variables. The combination of extensive outcrop exposure and a detailed sequence framework for this Permian section provides an ideal setting for examining the relationship between reef development and stratigraphic evolution of shelfal facies tracts.

The relevance of this stratigraphic perspective is made clear by examining the long-running debate concerning the water depth of Capitan reef development. Most studies have addressed this question through local detailed analysis of rock fabrics. Once put in a larger stratigraphic context, the often-asked question, "Did the Capitan reef form in shallow or deep water?," can be answered simply: "Yes." Depending on whether it is the Capitan facies that is part of the lower Seven Rivers transgressive sequence set (TSS) or the upper Yates highstand sequence set (HSS) that is being considered, the water depth reconstructed for the shelf-margin reef can vary from < 10 m to > 100 m. Current understanding of the spectacular Capitan reef exposures can be improved by placing them within the newly evolving sequence stratigraphic framework for the area (Kerans et al., 1992; S. W. Tinker, 1996, 1998; Osleger, 1998). Future studies of the Capitan carried out within this high-resolution sequence stratigraphic context should have a significantly improved focus for developing a predictive understanding of this classic facies tract.

Stratigraphic Framework

This paper uses as its data base the documentation of late Leonardian through Guadalupian buildup/reef facies by Kerans et al. (1992) and the detailed analysis of the Capitan-equivalent Seven Rivers and Yates formations in North McKittrick Canyon by Tinker (1996) (Figs. 3 and 4). The stratigraphic framework is

[1]Published with permission of the Director, Bureau of Economic Geology, The University of Texas at Austin.

Geologic Framework of the Capitan Reef, SEPM Special Publication No. 65

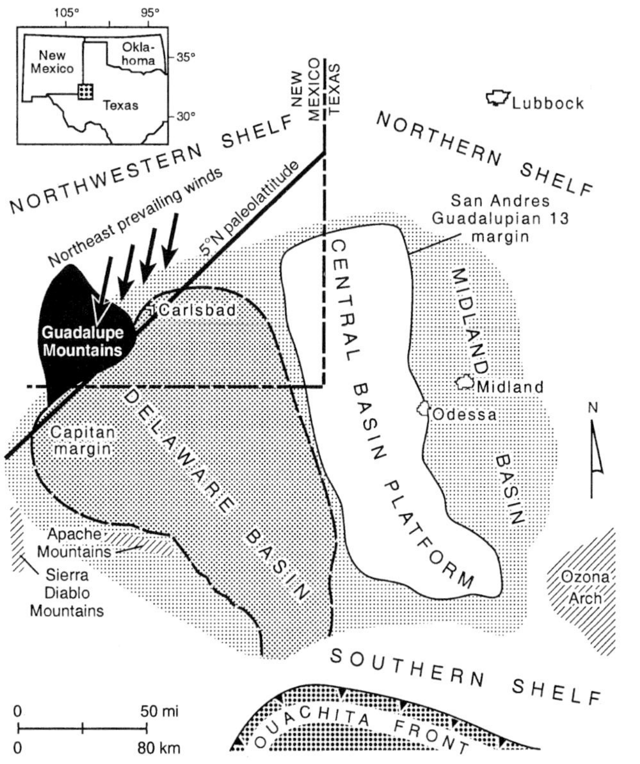

FIG. 1.—Regional paleogeography, paleolatitude, and prevailing wind direction as assumed for late Permian time (modified after Kerans and Fitchen, 1995, and Walker et al., 1995).

based on the hierarchical scheme of Mitchum and Van Wagoner (1991) for carbonate systems as modified by Kerans and Fitchen (1995). In order of descending time duration, composite sequences, high-frequency sequences, and high-frequency cycles are the main genetic units in this classification. Composite sequences are analogous to the depositional sequences of Vail (1987), with the exception that composite sequences are made up of unconformity-bound, high-frequency sequences (HFS) that are stacked into lowstand sequence sets (LSS), transgressive sequence sets (TSS), and highstand sequence sets (HSS). The magnitude of exposure atop each HFS varies depending on its setting within the composite sequence, as length of exposure and degree of karstification increases in HSS HFS. The building blocks of HFS are in turn high-frequency cycles or parasequences (Van Wagoner et al., 1988), which are below the level of resolution considered for this paper.

The total Wolfcampian-Leonardian-Guadalupian section includes 14 composite sequences, each containing between two and six HFS, that in turn contain between five and 20 high-frequency cycles (Kerans and Fitchen, 1995; Tinker, 1998; Osleger, 1998). This study focuses on those sequences exposed in the Guadalupe Mountains, specifically composite sequences 9 through 14 (CS9 through CS14) (Fig. 5). Relationships between the chronostratigraphically based sequence terminology and lithostratigraphically based formation/member names are given in Fig. 5. Both sets of terms are used throughout the text.

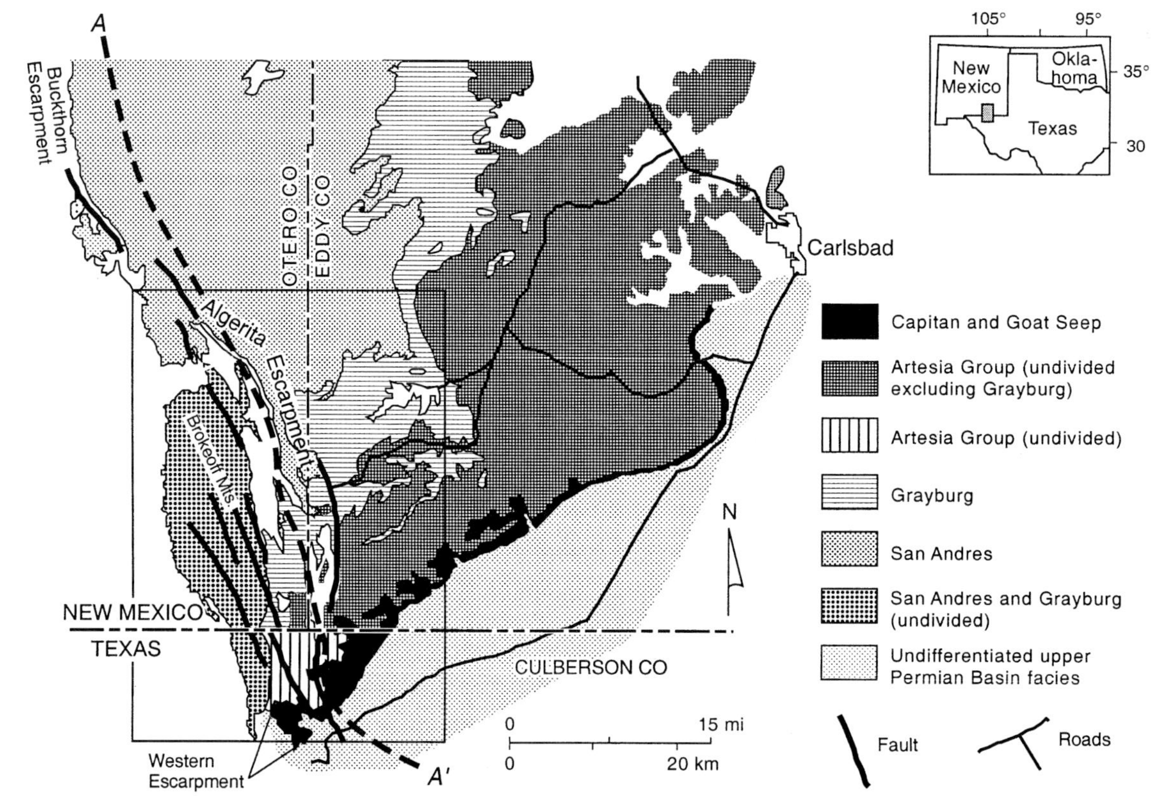

FIG. 2—Geologic map showing distribution of major lithostratigraphic units of the Guadalupe Mountains. Modified after King (1948) and Hayes (1964). Line A-A´ is the general line of section for Fig. 4, but data from outcrops including McKittrick and Dark Canyons have been projected onto the line. Inset box indicates the area of the Landsat image of Fig. 3.

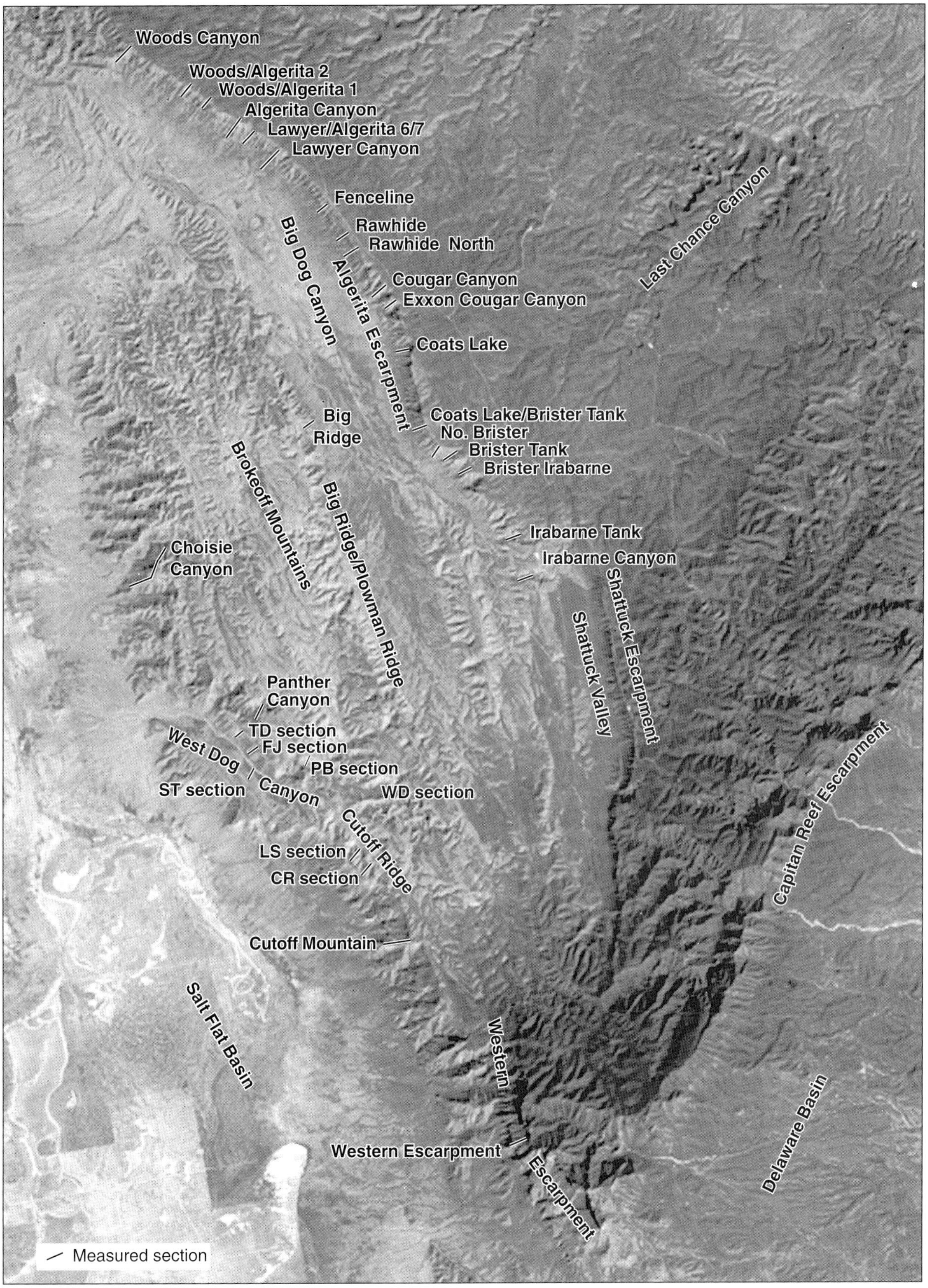

FIG. 3.—Landsat image of Guadalupe Mountains showing location of key sections used in stratigraphic model of Fig. 4. Modified from Kerans and Fitchen (1995).

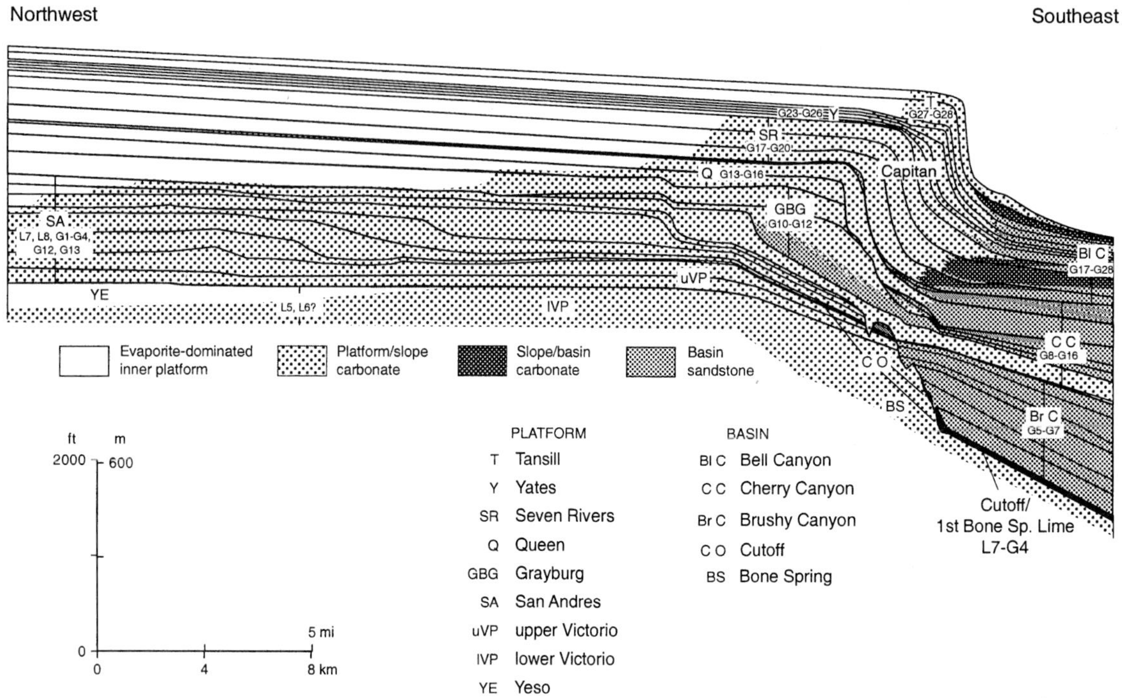

FIG. 4.—Generalized sequence framework of Guadalupe Mountains late Leonardian through Guadalupian section. Major lithofacies and formation terminology is given in relation to sequence framework. Modified from Kerans and Fitchen (1994).

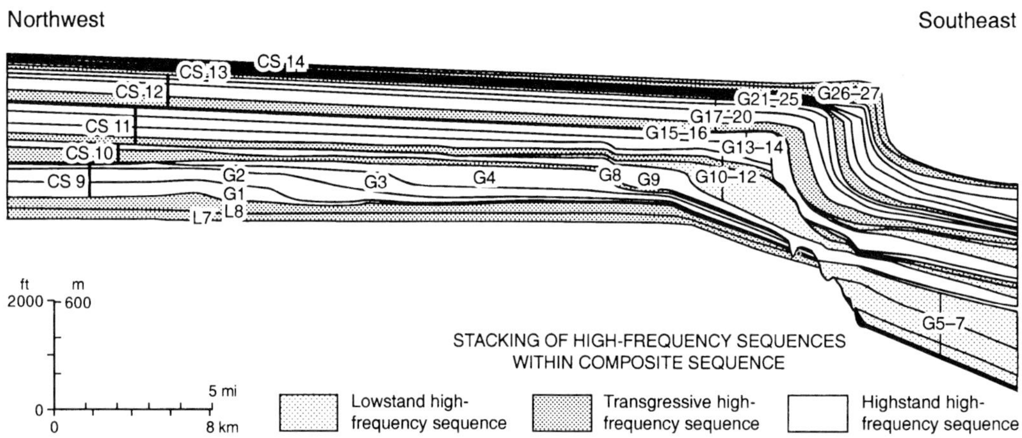

COMPOSITE SEQUENCE	HFS	NORTHWEST SHELF	SHELF MARGIN	FORE-SLOPE	DELAWARE BASIN		
CS 14	G27–28	Tansill	Upper	Capitan "massive"	Capitan "bedded"	Lamar	Bell Canyon	DELAWARE MOUNTAIN GROUP
	G25–26	Yates	Middle			McCombs		
CS 13	G21–24					Rader		
CS 12	G17–20	Seven Rivers	Lower			Pinery Hegler		
	G15–16	Shattuck				Manzanita	Cherry Canyon	
CS 11	G13–14 G10–12	Queen Grayburg	Goat Seep			South Wells		
CS 10	G8–9	upper San Andres				CCT		
	G5–7					Brushy Canyon		
CS 9	G1–4 L7–8	lower–middle San Andres				Cutoff		

FIG. 5.—Composite sequence framework of Guadalupe Mountain region and relationship to lithostratigraphic terminology.

Four main platform profiles characterize the latest Leonardian through Guadalupian (Fig. 6). These generalized illustrations portray the facies-tract terminology used within the context of the high-frequency sequences and their characteristic lithofacies. The importance of each facies tract can be analyzed in terms of its relative proportion within a sequence, which changes as a function of base level and sediment supply. Terminology used for reefs and buildups is generalized and informal. Facies characterized by massive appearance and dominated by baffling, binding, or frame-associated organisms are considered to be reefs or buildups. Buildups found on the shelf are called shelf buildups and those found at the shelf margin are called shelf-margin buildups or reefs.

Stratigraphic Parameters Controlling Reef Growth

Key parameters that can be used to characterize reef development include both intrinsic and extrinsic factors. Extrinsic factors affecting reef development include oceanographic effects, climate, subsidence, eustasy, siliciclastic sediment supply, turbidity, salinity, and antecedent topography. Antecedent topography controls the presence or absence and location of reef facies and is initially extrinsically controlled by preexisting depositional, erosional, or compactionally controlled features. Once reef growth and associated early lithification have been initiated, topography becomes, at least in part, an intrinsic factor. Several of the key extrinsic and intrinsic parameters associated with the CS9 through CS14 reef facies are summarized in Fig. 7.

Intrinsic factors are dominated by characteristics of the organisms involved in reef construction. Presence or absence of frame-builders and binders, rate of sediment production by organisms, sensitivity of reef-building organisms to light, salinity, sediment influx, and relative importance of marine cements in the binding process are all basic properties of the reef community. These factors determine the degree to which the reef facies will build its own topography and hence modify its environment to form buildup-controlled facies associations (back-reef bioclastic grainstones/rudstones, lagoonal mudstones and wackestones, and upper- and lower-slope sediment-gravity-flow aprons).

This paper deals primarily with the extrinsic parameters of eustasy and antecedent topography. A multivariate interpretation of these and other extrinsic parameters is required to incorporate and understand all of the observed stratigraphic variations.

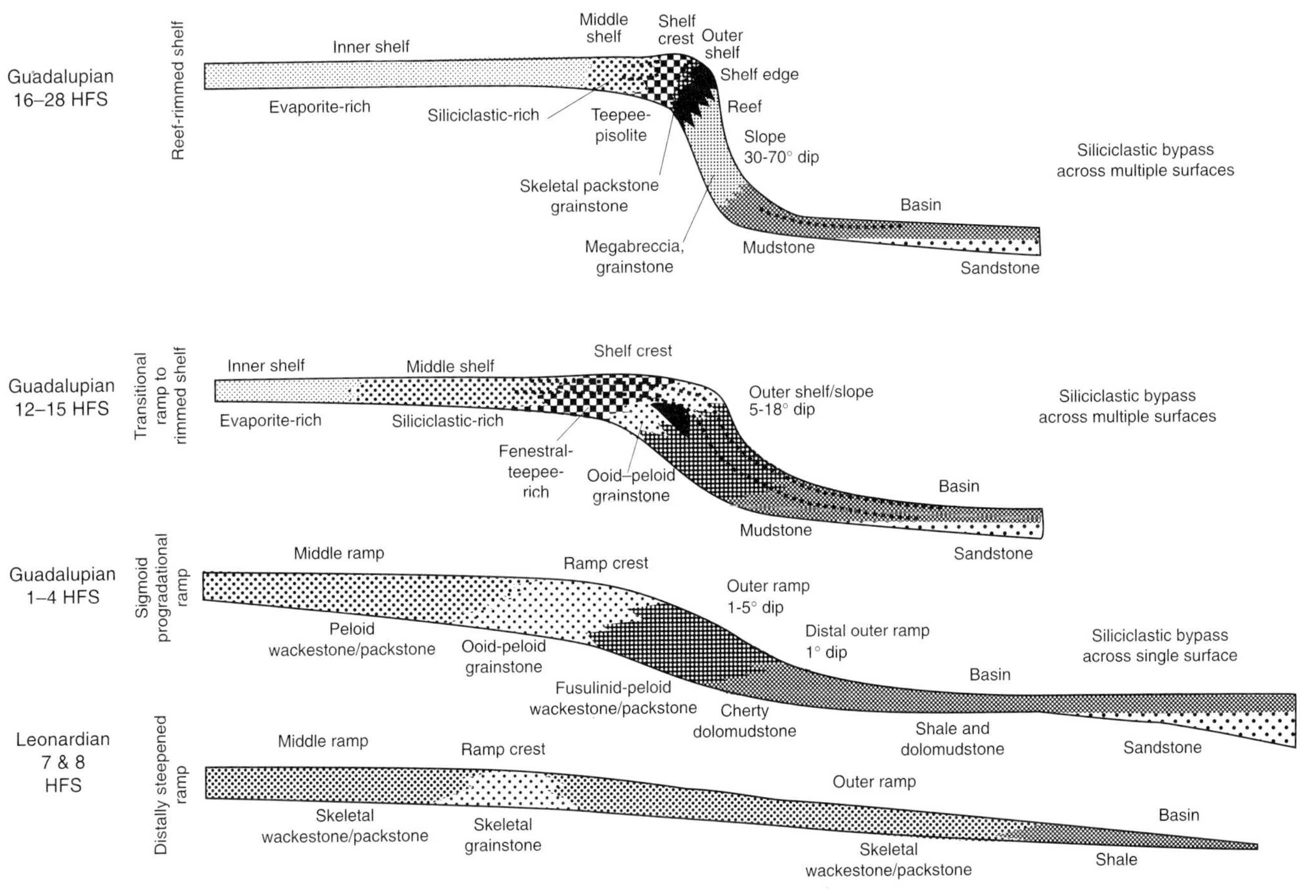

FIG. 6.—Simplified depositional models for the main stages of platform evolution in the latest Leonardian through Guadalupian sequences of the Guadalupe Mountains (modified from Kerans and Fitchen, 1995).

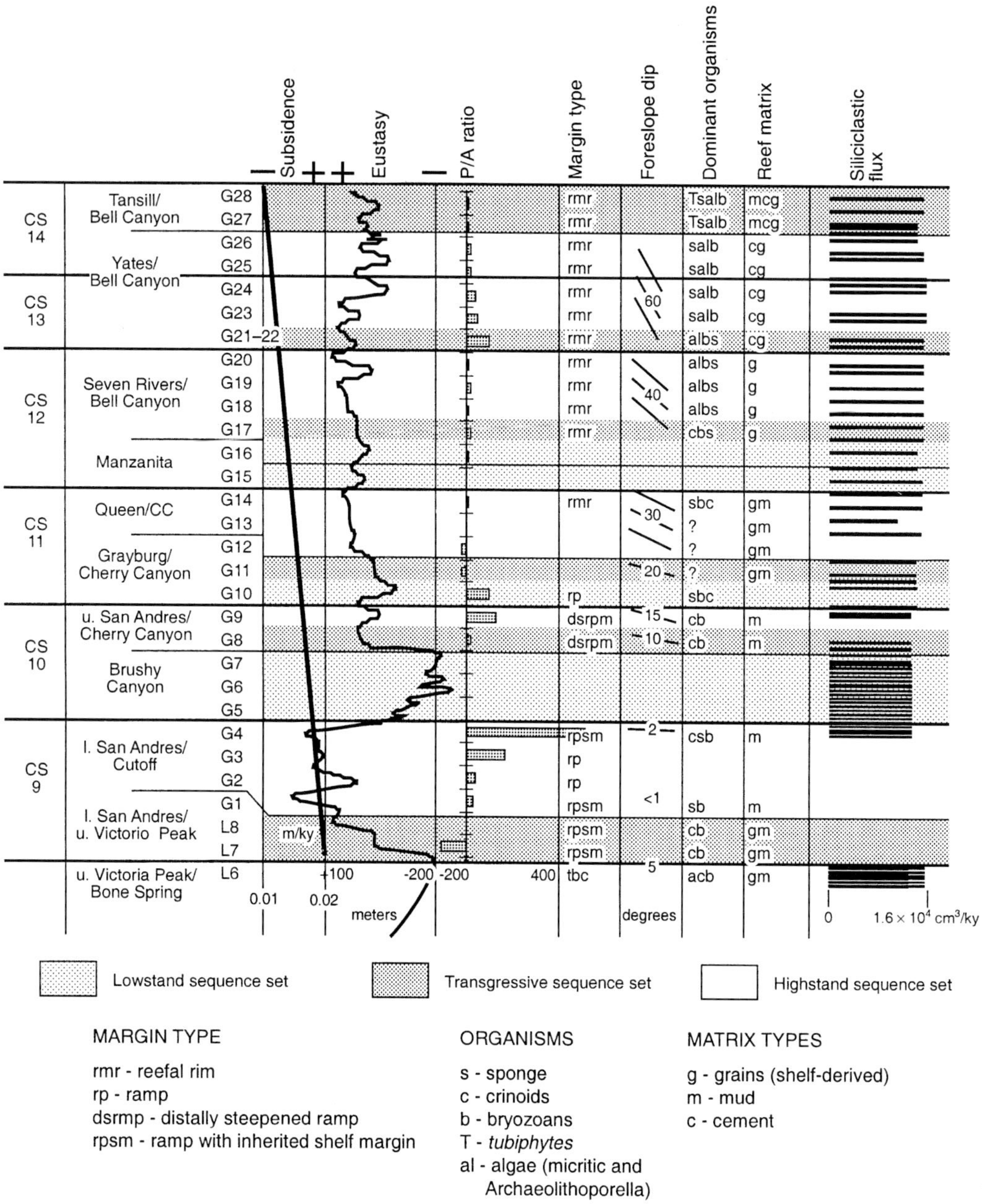

FIG. 7.—Relationship of major extrinsic parameters controlling platform and reef development. Subsidence, eustatic curve, and siliciclastic influx (here given as a rate ranging from 0–1.6 x 10^4 cm^3 per ky) are from Ye and Kerans (1996). Background patterns for LSS, TSS, and HSS are from Fig. 5.

Oceanographic/climatologic effects.—

Oceanographic and climatic factors are critical to reef development, particularly when considering time spans of several million or tens of millions of years during which plate drift, basin evolution, and Milankovitch climate effects may have combined to produce a range of favorable and unfavorable conditions for reef growth (Davies et al., 1989; Golanka et al, 1995; Walker et al., 1995). For the purpose of this analysis, a position approximately 5° north of the paleoequator in a hot, arid climate is consistent with the evaporite facies that are present in inner-shelf facies tracts of all sequences, from late Leonardian through Guadalupian (Fig. 1). The shift in latest Leonardian time (L7-L8 HFS) to a coral-bryozoan-brachiopod-foram assemblage may indicate cooler overall climates.

Deposition of the Capitan-equivalent Bell Canyon Formation is thought to have been influenced by trade winds out of the present-day north, or Permian northeasterlies (Fischer and Sarnthein, 1988; Walker et al., 1995). This would make the Guadalupe margins oblique windward, favoring reef development on this margin relative to the eastern margin of the Delaware Basin, but this has yet to be evaluated with core data.

Eustasy and tectonic subsidence.—

Eustasy and tectonically controlled subsidence for the CS9–14 section are taken from a 2-D backstripping analysis carried out by Ye and Kerans (1996; Fig. 8). Tectonic subsidence is assumed to be a simple quadratic function with a decrease in rate from 2 cm/ky in latest Leonardian time to 1 cm/ky during late Guadalupian time (Figs. 7 and 8) (Ye and Kerans, 1996). Because the Guadalupe Mountain outcrops are mostly from the same tectonic block within the Northwestern Shelf (Fig. 1), subsidence rates are assumed to apply throughout the study area (Figs. 2 and 3). The relatively stable subsidence history, particularly in relation to the rapidly changing eustatic signal, suggests that this factor can be considered a constant for the late Leonardian-Guadalupian.

The eustatic curve was constructed by picking highstand and lowstand shorelines for each of the HFS using lithologic data to remove effects of compaction and isostasy (Ye and Kerans, 1996). The resulting eustatic curve shows several major and relatively rapid rises and falls with amplitudes of as much as 160 m. In general, Leonardian CS-scale eustatic cycles have greater amplitudes than Guadalupian (Capitan) cycles (25 m versus 10 m), reflecting the overall transition from Early Permian icehouse conditions to Triassic greenhouse conditions (Figs. 8 and 9; Ye and Kerans, 1996).

Eustasy can be the dominant variable controlling reef growth and initiation in carbonate platforms. Too fast a rate of eustatic rise will result in platform drowning, retrogradation of the shelf, and deposition of ramps with minimal buildups or reefs. In the Guadalupe Mountains, the Leonardian 7 and 8 HFS of CS9 illustrate the impact of this rapid rise (Fig. 6). The latest Leonardian sea-level rise, which is thought to be a global event (Kerans and Fitchen, 1995), has a rate of rise of 15 cm/ky for 160 m. This eustatic rise caused shelf-margin back-stepping of 32 km (20 mi) and changed the platform profile from a flat-topped platform with 10° foreslope dips and 5–10% shelf-margin boundstone (Leonardian 6 HFS; Fig. 4) to a bioclastic ramp with less than 0.1° of depositional dip and 0.1% buildup facies (Leonardian 7–8 HFS, Fig. 9).

More moderate eustatic rises in the CS11, CS12, and CS13 are associated with initiation or reoccupation of shelf-margin locations by reefal shelf-margin facies (Fig. 9). Reef growth during or near the end of a significant eustatic rise is expected because of the differential topography created by the high aggradation during such a rise. The exception to this rule is where the rate of rise outstrips the ability of the platform to keep pace with the rise. This relationship has implications for the timing of formation of the Goat Seep reef within CS11. Examination of the eustatic signal for CS11 suggests that conditions would have been favorable for nucleation of shelf margin reef facies in transgressive-early highstand sequences (Guadalupian 11-13 HFS). Stacking of shelf-crest and outer-shelf facies within the Guadalupian 11–13 HFS shows initial back-stepping followed by aggradation (Fig. 9). The current interpretation of this CS is that the Goat Seep shelf-margin reef is contained entirely within the Guadalupian 14 HFS (uppermost Queen). This correlation for the Goat Seep was originally proposed by Franseen et al. (1989) by recognizing an erosional escarpment separating massive-reef and bedded-shelf facies of the earlier Queen and Grayburg (G11–13 HFS) strata. Sharp, near-vertical contacts between massive shelf-margin reef facies and bedded shelf strata, however, do not necessarily imply single event collapses as illustrated by similar platform margin "unconformities" at Nadji Cave and at Windjana Gorge in the Frasnian of the Canning Basin (Playford, 1980; Playford et al., 1989). We interpret these scarps not as single-event collapse features but, rather, as sharp facies transitions that form through periodic and repeated erosion and recolonization of the scarp as a natural process in platform-margin deposition (Playford, 1980; Playford et al., 1989). It is also possible to interpret the abrupt transition separating Goat Seep and bedded-shelf facies of Guadalupian 12 HFS and, particularly, the Guadalupian 13 HFS, as depositional contacts. In this interpretation, lower Goat Seep facies would be facies equivalents of the highly aggradational Guadalupian 12 and/or 13 HFS.

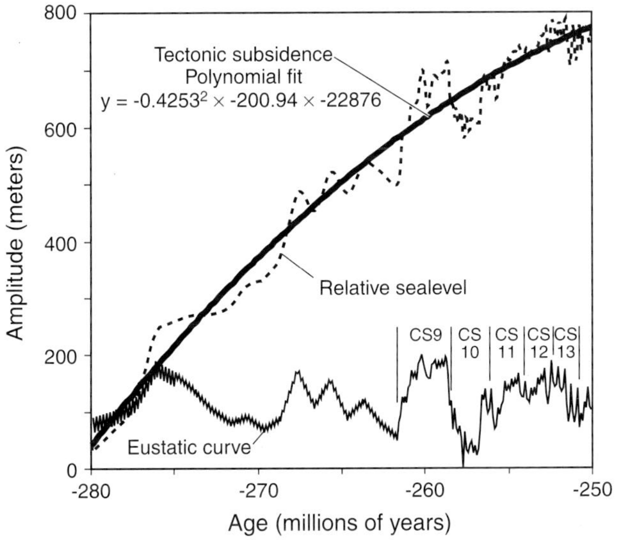

FIG. 8.—Composite (subsidence plus eustasy) and eustatic curves for the Permian as calculated by Ye and Kerans (1996). Time scale is based on Ross and Ross (1987).

Smaller-scale eustatic shifts and associated aggradation-progradation pulses at the HFS scale within the CS12 and CS13 (Seven Rivers/Capitan and Yates/Capitan) affect smaller scale modifications of the shelf-margin reef (see more detailed discussion below).

Antecedent depositional topography.—

Given the appropriate depth window and current regime, antecedent depositional topography becomes a critical factor. Positive relief relative to the surrounding sea floor, in combination with substrate stabilization by early lithification, controls buildup/reef development. This topographic relief provides stable substrates for sessile benthos and improved access to currents that bring nutrients and remove wastes. For shelf-margin reefs, a hard substrate open to the basin is essential for growth/attachment sites and for the required open-ocean circulation. Antecedent topography is not a primary extrinsic parameter but is instead a function of both the global process of eustatic change and local processes of tectonic subsidence and sediment supply. Platform-interior buildups are characterized by low relief with an absence of debris aprons and mud-rich matrix.

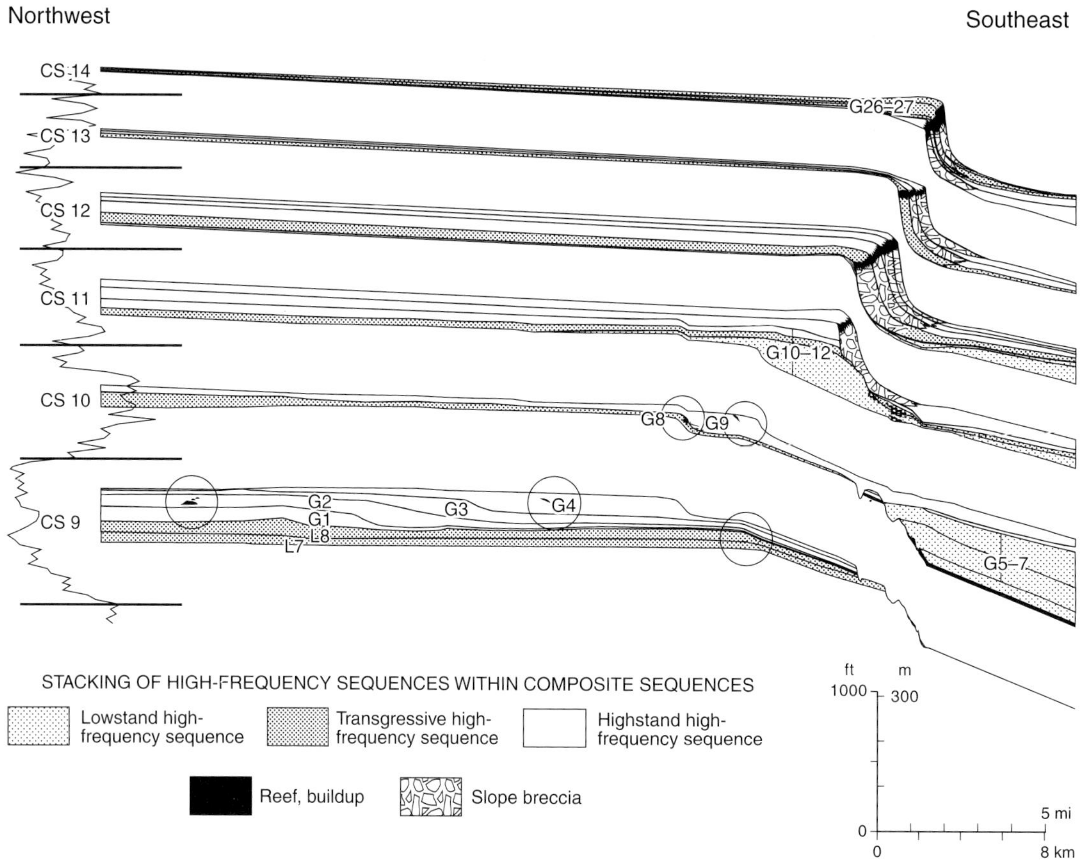

FIG. 9.—Relationship between eustatic curve, composite sequences, and reef/buildup development. Note that the vertical scale for the eustatic curve is time whereas the vertical scale for the composite sequences is thickness, so no exact match occurs.

Platform-margin buildups are either mud-dominated or frame-dominated and typically display well-developed debris aprons (Fig. 10). The G9 and G10 HFS have characteristics transitional between ramps and rimmed platforms and display localized shelf-margin reefs and minor slope breccias within 5°–25°-dipping clinoforms (Fig. 10).

The distribution of buildups within the CS9-14 section provides a variety of examples of the interrelationship of depositional topography with buildup topography (Fig. 10A). Differential compaction of the shelf strata of L7–8 HFS over underlying L2–6 Victorio Peak shelf margins appears to be the main control on development of the bryozoan-pelmatozoan mounds in the L7 and L8 HFS (Fig. 10A). The preferential position of these mounds at the flexural hinge of an otherwise relatively featureless shelf indicates that subtle differences in relief can be sufficient to localize buildups. This has long been recognized in the Western Canada Basin, where the Leduc and other reef trends that have strong linear aspects have been difficult to predict from subtle topographic or basement structural features.

The second occurrence of buildups within the Guadalupian 2 HFS illustrates preferential growth focused by constructional depositional topography (Fig. 10B). These buildups are deeper subtidal, mud-dominated, sponge-bryozoan floatstones with aprons of pelmatozoan debris and no macroscopic early cements. The four buildups, ranging from 2 m to 14 m (8 ft to 45 ft) in thickness, are found near the Guadalupian 2 maximum-flooding surface (MFS) in the early highstand portion of CS9. This position, where the ramp system is in transition from aggradation to strong progradation, approximates maximum aggradational depositional topography for this composite sequence. The most landward encroachment of sponge-bryozoan mud mounds within the San Andres Formation can be found more than 160 km (100 mi) to the east of the Guadalupe Mountains in up-dip portions of the Wasson Field (Willard Unit), some 50 km north of the terminal San Andres shelf margin.

The third buildup occurrence, in Guadalupian 4 HFS, is the first in a series of buildups that are positioned in shallow water depths (approximately 9–12 m; 30–40 ft) on measurable clinoform slopes. The G4 buildups formed on 2°-dipping clinoforms in a position approximating the MFS. The clinoform slope provided differential topography for buildup growth and also favored the concomitant decrease in sediment influx.

The sponge-bryozoan-crinoid buildup in the G8 HFS of CS10 represents the first occupation of inherited shelf-margin topography (Fitchen, 1992, 1993) (Fig. 10D). This small buildup in the transgressive sequence of CS10 is mud-dominated, was formed in 6–15 m (20–50 ft) water depths, and displays 15° foreslope strata composed of pelmatozoan wackestones and packstones, rather than the steeply dipping megabreccias common in the younger shelf-edge reefs. Inherited relief for this mound was the stranded G4 HFS margin that was exposed and lithified during the major CS10 lowstand (Figs. 7 and 8).

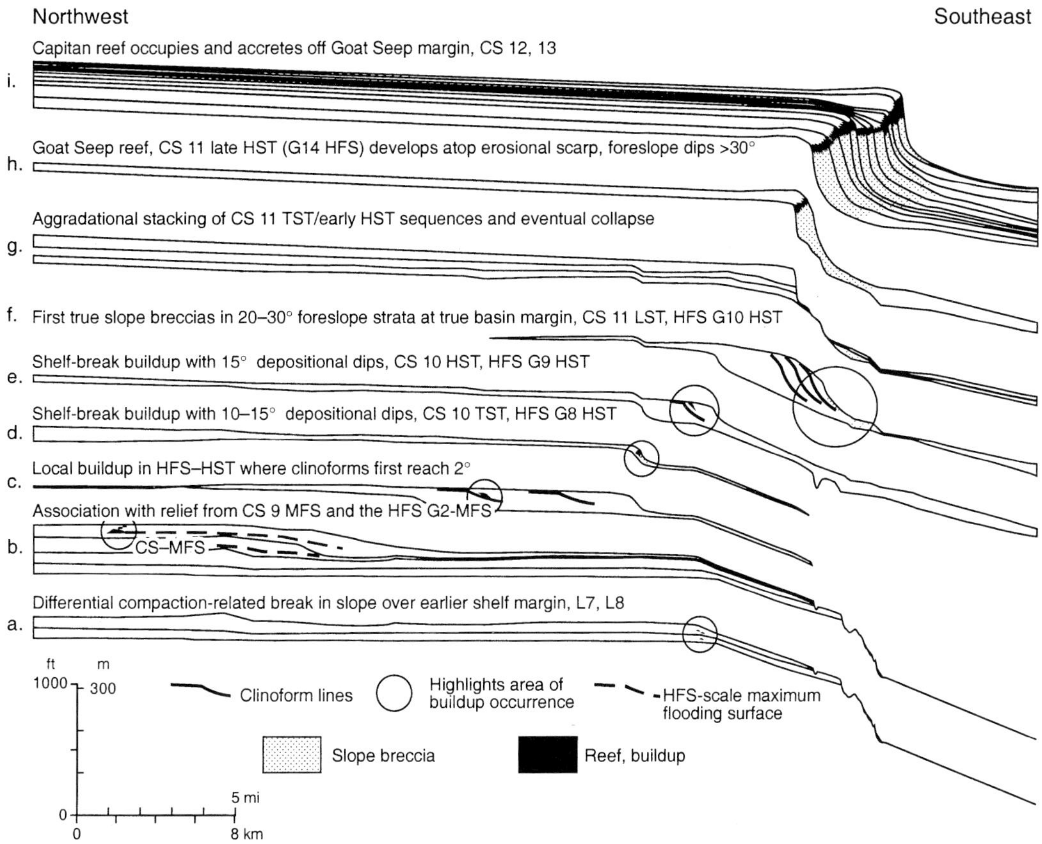

FIG. 10.—Location of shelf buildups and shelf-margin reefs within the CS9-14 Leonardian-Guadalupian section of the Guadalupe Mountains. See text for discussion.

Sonnenfeld and Cross (1993) have documented buildups in the uppermost San Andres (G9 HFS) of Last Chance Canyon. These buildups occur in the early and late highstand cycles of this uppermost sequence in water depths of 3–12 m (10–40 ft) and are otherwise comparable in style and origin to those discussed in the G8 HFS. The best developed of these buildups is as much as 7.6 m (25 ft) thick and has 17° seaward-dipping flank beds (Fig. 10E). Differential relief set up by 5°–15°-dipping clinoforms with up to 90 m (300 ft) of toplap-to-downlap topography provides sufficient relief for buildup development throughout the highstand of the G9 HFS. The paucity of buildups in this sequence may be related to rapid siliciclastic sediment flux across this front associated with the high rate of progradation during base-level fall. These G8 and G9 buildups are viewed as the beginning of a transition to true shelf-margin reef, but they still lack the slope breccias and macroscopic early cementation associated with the fully developed reef-rimmed margins such as the Miocene of the Mediterranean and Indonesian islands (Esteban, 1996), the Triassic of the Dolomite Alps (Bosellini, 1989), the Devonian of the Canning Basin (Playford, 1980) and Canada (Wendte and Campbell, 1992), and the Cambrian of the Appalachian margins (Read, 1989).

Shelf-margin progradation in earliest Grayburg time (G10 HFS) took place seaward of the terminal San Andres margin as a shelf-margin lowstand wedge (Fig. 10F). This wedge prograded seaward across a basinward-sloping sea floor caused by flexure over deeper (L1-3 HFS) shelf margins (Fig. 10F). As the lowstand wedge prograded seaward, each progradation increment required substantially greater sediment volume because of the increased shelf-to-basin relief seaward. The result of this change in basin accommodation was a decreased progradation distance, gradual steepening of foreslope dips (as much as 25°), increased early lithification resulting from the greater pore-volume circulation through any given amount of sediment as a result of the slower progradation, and the first appearance of true slope and toe-of-slope megabreccias (Crawford, 1989; Fekette, 1986). Although no *in-situ* reef facies have been identified from the G10 HFS, the presence of boundstone blocks in the toe-of-slope breccias suggests that this is more a problem of access to exposures in this rugged outcrop area along the Western Escarpment (Fig. 3) than a lack of shelf-margin reef development.

Late CS11 highstand deposition (G14 HFS Goat Seep Formation) marks the final transition to well-developed reefal margins. Antecedent topography for development of this shelf-edge reef was set up during deposition of the transgressive sequences G11, G12, and G13 HFS. The G11–13 HFS forms a retrogradational to aggradational platform with > 150 m (500 ft) of vertical relief above the G10 HFS top. This relief may or may not have been subsequently enhanced by margin collapse, with the ultimate result being a near-vertical escarpment for nucleation and progradation of the Goat Seep and subsequent Capitan reef margins.

The Guadalupe Mountains data set demonstrates that antecedent topography, in concert with even the most dramatic eustatic shifts, is a key factor controlling the development and location of buildup facies. The evolution of carbonate platforms from ramps to rimmed shelves is a well-known pattern (Read, 1985). Variations in antecedent topography within the Permian Basin associated with along-strike changes in subsidence and sediment supply must have resulted in different timing for ramp-to-rim transition around the basin. Documentation of latest San Andres or earliest Grayburg-age reefs at North McElroy Unit (Longacre, 1980), however, supports the contention that timing for development of shelf-margin reefs varied. Rigorous testing of this hypothesis requires core data that are rarely available in shelf-margin settings for the Permian.

Progradation/aggradation Ratios

Carbonate platforms reflect changes in eustasy, subsidence, and sediment supply through changes in platform architecture and facies composition. One measure of these architectural changes is the relative proportion of platform aggradation versus progradation expressed as a ratio, P/A (Fig. 11). P/A is the progradation vector from the initial shelf-slope break of a sequence to the terminal shelf-slope break of that sequence divided by the stratigraphic thickness of the unit at the landward edge of the ramp- or shelf-crest facies tract (Fig. 11A). This value can be calculated for any scale of stratigraphic element for which margins and facies tracts are mapped. Because steep ramps and rimmed platforms are less sensitive to P/A than broader ramps, a better measure for describing geometric response to changing accommodation on these platforms is the shelf-margin P/A (SMP/A, Fig. 11A). SMP/A compares the amount of stratigraphic climb from one shelf-slope break to the next against the distance that the margin progrades in a given stratigraphic time step. Thus, this measure uses the same progradation vector as P/A, but substitutes aggradation of the shelf margin for the aggradation vector in place of aggradation measured in back of the shelf crest (used for P/A). In the following discussion, we use P/A to contrast platform evolution at the HFS and composite sequence scale across all of the Leonardian-Guadalupian sequences, and SMP/A to investigate platform evolution at the HFS scale within the CS12–14 reef-rimmed platforms.

P/A describes the net sediment accumulated on the platform top (aggradation vector) versus the distance that the shelf margin moves laterally, either basinward (progradation vector) or landward (retrogradation vector), during the deposition of the given sequence. P/A is useful for describing the relative proportion of sediment moved off the shelf versus that amount preserved on the shelf. It is not a measure of net sediment volume production. P/A reflects the compound interaction of shelf accommodation, basin accommodation, and sediment supply. Shelf accommodation is a function of eustatic changes, changes in regional subsidence, and local effects such as compaction. Basin accommodation is a reflection of preexisting topography; in other words, how deep is the hole to be filled?

Sediment supply in carbonate systems is generally dominated by the area of the carbonate shelf that occurs between 0.3 m and 10 m (1 ft and 30 ft) of sea level (Schlager, 1992). In this depth range light penetration is greatest, and photosynthetically driven carbonate fixation combines with the production of skeletal invertebrates to generate abundant sediment. This sediment is then exported from the "factory" landward toward the tidal-flat repository and seaward into slope and basin repositories. In siliciclastic systems, sediment is brought into the shallow shelf system by alluvial or eolian processes

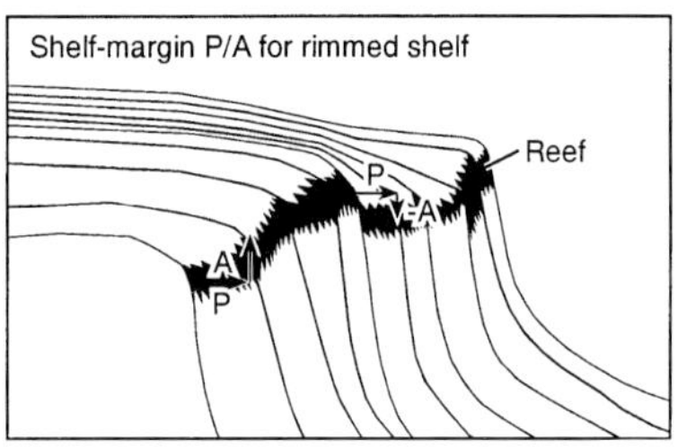

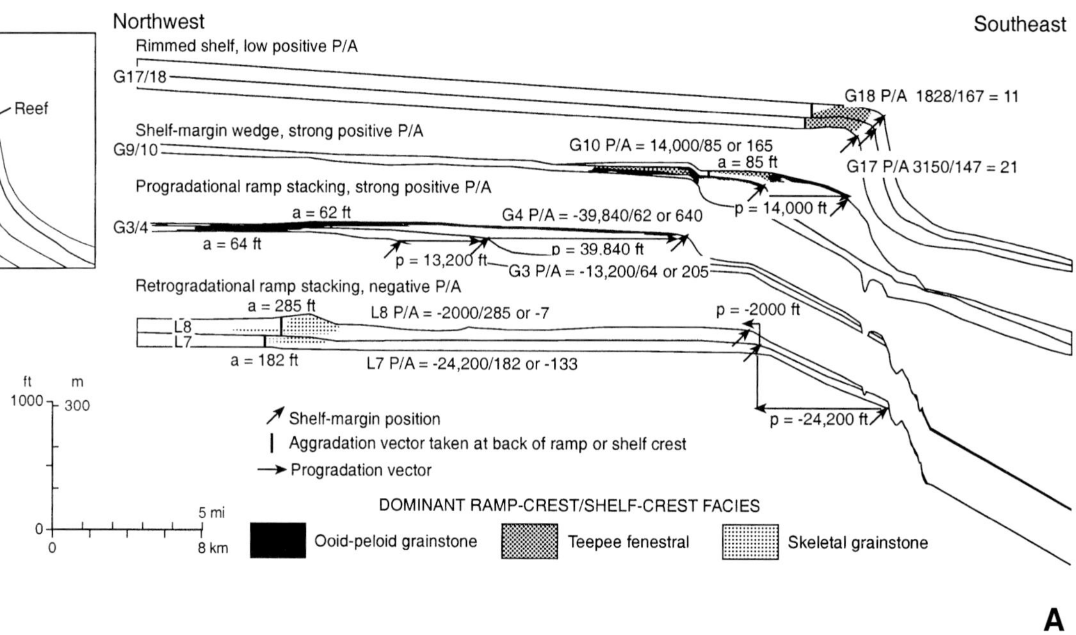

FIG. 11.—Illustration of how P/A and shelf margin P/A values are calculated for different styles of platforms. (A) The shelf margin P/A values include both a positive P/A and a negative P/A. (B) Progradation-aggradation ratios for the Leonardian-Guadalupian sequences of the Guadalupe Mountains.

and then distributed by marine traction and sediment gravity flow processes. Mixed systems such as CS10–13 include a high-frequency reciprocal relationship between both sediment supply systems. Sediment supply is important in considering P/A, as it is the driving mechanism for both progradation and aggradation, and the evolution of the platform profile is a feedback system between rate of eustatic rise or fall, width of the carbonate factory, and therefore the rate of sediment supply and progradation. This critical link is what drives carbonate systems and what differentiates principles of carbonate sequence stratigraphy from those of siliciclastic sequence stratigraphy (Schlager, 1992).

Progradation/aggradation styles for ramps and rimmed shelves are illustrated in Fig. 12. TSS and early HSS ramps with relatively high shelf accommodation but minor bathymetry, such as those in early highstand settings of a CS, yield P/A values of between –50 (retrogradational) and 50. HSS ramps forming in low shelf accommodation settings, or forced regressive settings yield P/A values of as much as 1,000 (Fig. 12). TSS-rimmed platforms may either backstep or prograde gradually in response to base-level rise (Fig. 12). This balance is determined by the relative rate of rise and the increased production potential resulting from an increase in carbonate factory area (Tinker, 1998; Garber et al., 1989). HSS-rimmed shelves with low shelf accommodation typically display an initial increase in P/A that

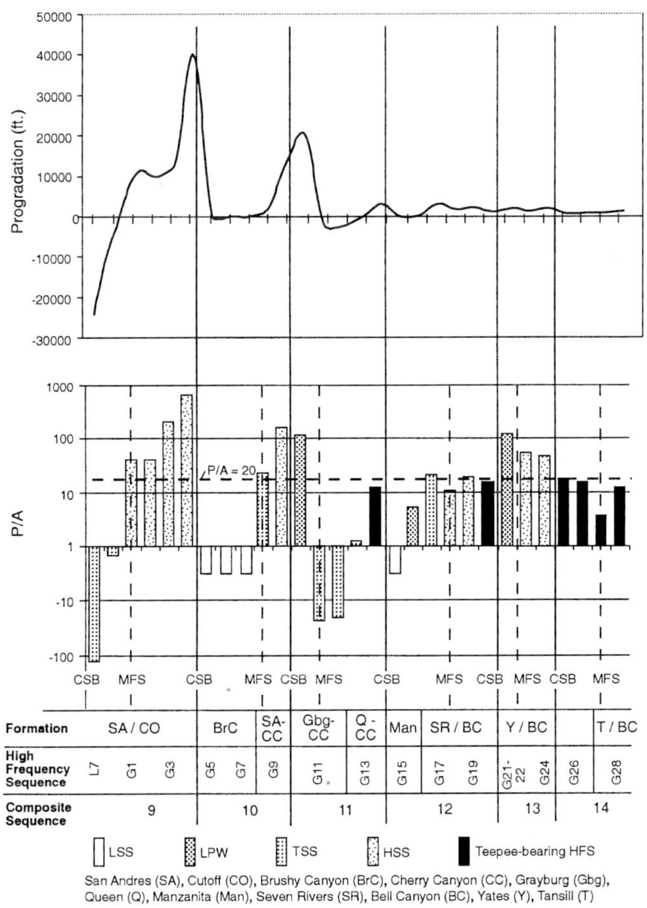

B

FIG. 11.—Continued.

is later offset by the narrowing of the carbonate factory and a decrease in sediment supply from the shelf (Fig. 12).

Siliciclastic influx at the CS scale also tracks with P/A. Times of high accommodation and low or negative P/A such as CS9 and CS12 correspond to a decrease in siliciclastics on the platforms (Figs. 7 and 11B). In contrast, sequences characterized by high P/A display significant sand tongues preserved on the shelf (for example Yates G23-24 HFS).

Scaled illustrations of ramps and rimmed shelves from selected HFS of CS9, CS10, CS12, and CS13 show the marked contrasts in these different styles of carbonate platforms (Fig. 13). The G2 in the early HSS of the CS9 is characteristic of a moderately progradational ramp with relatively high aggradation potential (P/A = 42). The G4 HFS in the late highstand of CS9 displays a P/A of 642, the highest in the Guadalupe Mountains. This high P/A is a function of three factors. First, G4 HFS was developed on the falling limb of the CS9 eustatic curve, and thus had low shelf accommodation, transporting all sediment produced on the shelf seaward. Second, the low basin accommodation on this shallowly dipping ramp, with only 130 m (430 ft) of top-set to toe-set bathymetry, fosters rapid progradation. Third, the high rate of progradation in these two sequences is enhanced by the width of the carbonate factory. Both the G2 and G4 have factory widths exceeding 45,000 ft.

The G9 HFS in the late HSS-forced regressive portion of CS10 has a P/A of 161 (Fig. 13). Factors that increase progradation of this ramp are low shelf accommodation (forced regressive, oblique progradational style) and increased siliciclastic input during both lowstand and highstand. Together these processes produce a platform across which the ramp can prograde, decreasing the basin accommodation from 137 m to 82 m (450 ft to 270 ft). Factors that temper progradation include limited carbonate factory (4.5 km or 15,000 ft in dip-width) and increasing basin accommodation (>130 m or 500 ft and increasing rapidly near the end of progradation).

Geometry and P/A values for the G17 from the TSS of CS12, and the G24 from the HSS of CS13, are representative of the Guadalupian rimmed platforms (Fig. 13). Both of these sequences display substantially lower P/A values than for the ramp systems, regardless of their positions within the composite sequence. This can be related to the increased basin accommodation of each of these sequences (470 m and 400 m or 1,550 ft and 1,310 ft, respectively) and to the much reduced width of the carbonate factories. Average carbonate factory width for the rimmed margins is 700 m (2,250 ft) versus between 500 m and 15,000 m and 16,666 m (50,000 ft) for the ramp systems. It is important to note that, even though the P/A ratios of CS12–14 are significantly lower than for CS9–11, the relative change in SMP/A between TSS (lower SMP/A) and HSS (higher SMP/A) is a well-documented indicator of similar responses at longer term stratigraphic scales.

The CS12 (Seven Rivers and lower Capitan) and CS13 (Yates and middle Capitan) are excellent examples of the complex response of rimmed margins to changes in base level (Fig. 14). The response of the platform during CS12 TSS deposition (G17) is one of back-stepping of the shelf crest driven by sea-level rise and progradation of the margin driven by increased production from a broader outer-shelf carbonate factory (Tinker, 1998; Fig. 14). Though not reported in all margins, this dual response of shelf-crest retrogradation and shelf-margin progradation has been reported from other settings (e.g., Sonnenfeld and Cross, 1993) and is a reflection of the unique autochthonous sediment supply system in carbonate margins. It is interesting to note that the shelf-margin Capitan facies remained deep during this progradation, suggesting that progradation is driven by the outer-shelf carbonate factory instead of by high rates of reef production, which would yield a raised rim profile characteristic of many platform margins (Kendall and Schlager, 1981).

The G20 in the HSS of C12 illustrates the response of the Capitan margin system during falling base level (Fig. 14). The increased progradation expected as a response to the decrease in shelf accommodation is offset by the decrease in total sediment produced by narrowing the carbonate factory. Associated with this change in base level, the shelf-margin reef becomes shallower. Variable depth of the reef margin will be discussed below.

Progradation/aggradation ratios were used by Sonnenfeld and Cross (1993) to describe the San Andres margin, and by Tinker (1998) and Osleger and Tinker (1998) for the Seven Rivers-Yates margins, respectively. The measure used by these workers focuses on the vertical versus horizontal migration of the shelf margin. To avoid confusion with P/A, Osleger and Tinker (1998) referred to this parameter as reef P/A. We prefer shelf-margin progradation/aggradation ratio or SMP/A (Fig. 11A) as this measure is equally applicable to nonreefal clinoforming shelf-margin complexes (e.g., Sonnenfeld and Cross, 1993). The SMP/A uses the same progradation value as P/A above, but substitutes for the aggradation value the vertical change from one shelf margin to the next youngest shelf margin. This SMP/A ratio is more sensitive to shelf-margin changes but is also more affected by local compactional effects near the shelf margin and less representative of overall shelf subsidence patterns.

Tinker (1998) calculated SMP/A for CS12 and 13 (Seven Rivers through Yates, Guadalupian HFS 17–25; Fig. 14) and has shown that SMP/A ratios track the transgressive-regressive history of these HFS. In all but HFS 18 (Tinker's SR2), the transgressive component of the HFS has a more aggradational (lower SMP/A ratio) style, whereas the highstand portion steps seaward more strongly (Fig. 14).

On the composite sequence scale, SMP/A is 12.9 for the Seven Rivers and 12.4 for the Yates, suggesting little difference in the style of margin development. Average P/A for CS12 is 17 versus 52 for CS13. This, coupled with the significant increase in siliciclastics in the Yates, suggests an overall lower accommoda-

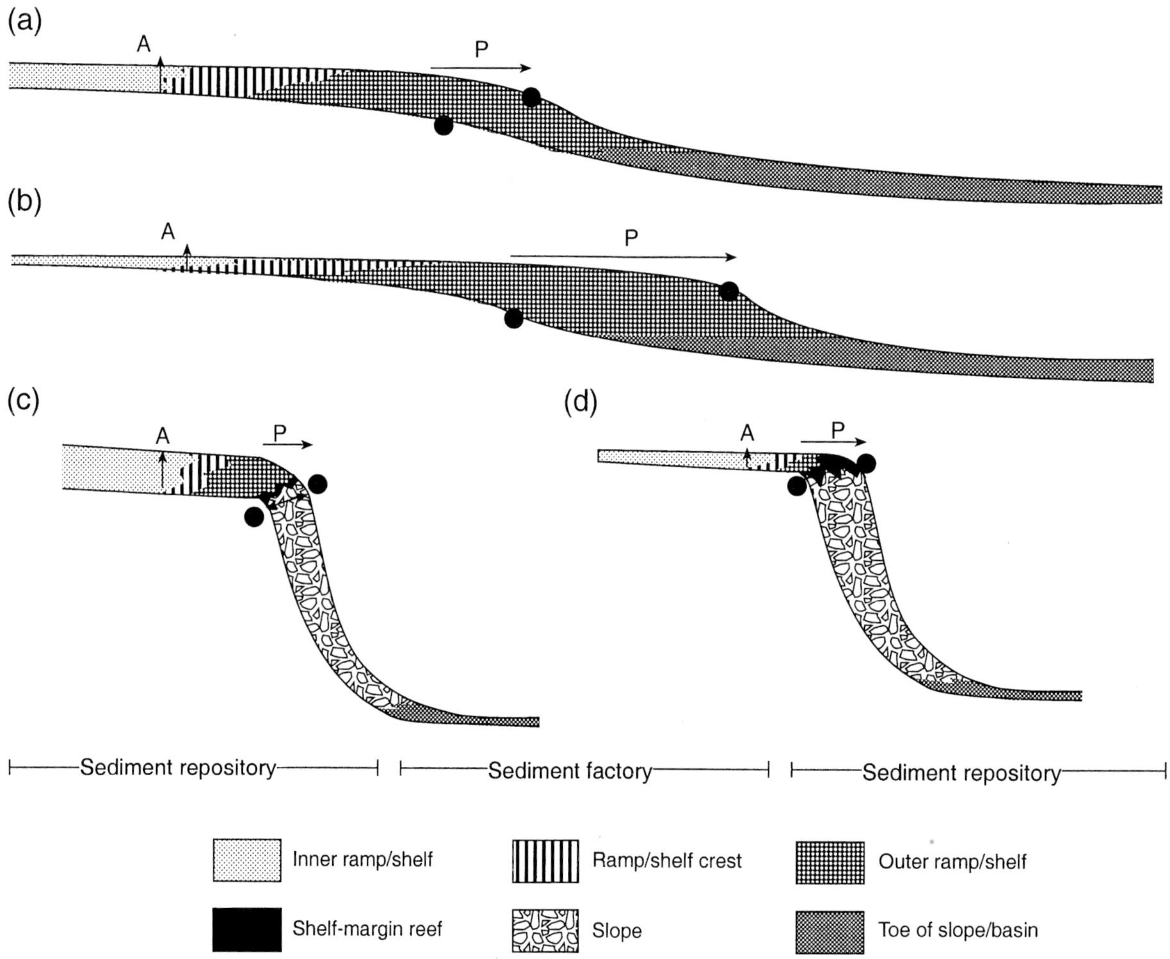

FIG. 12.—Comparison of progradation/aggradation styles for ramps and rimmed shelves, including changes in carbonate factory (= outer portion of shelf crest, outer shelf, and shelf-margin reef).

tion for the shelf. These differences are important, as both the estimated water depth for the Seven Rivers shelf-margin reef and the fabric of the reef are significantly different from the Yates.

Using P/A to analyze platform geometry allows for the normalization of data from sections of uncertain duration. Garber et al. (1989) pointed out that in both the subsurface and outcrop it appeared that the Seven Rivers-equivalent "lower" Capitan margin recorded two-thirds of the total Capitan progradation. Tinker's (1996) analysis suggests the two units record near-equal total progradation. Using P/A, it can be seen that in relative terms, the Seven Rivers and Yates sections are relatively similar in P/A and, therefore, in relative degree of progradation for a given number of cycles on the platform.

Evolving Profiles: Facies Tract Substitution and Facies Proportions

Facies-tract substitution and facies proportions are important stratigraphic tools that look at the changing make-up of facies tracts and individual facies as a function of their controlling parameters. Comparison can be carried out at any scale of stratigraphic hierarchy (high-frequency cycle or cycle set, high-frequency sequence, composite sequence) but should be made within a consistent level of hierarchy and between similar systems tracts. In addition, documenting the evolution of a depositional profile, the process of comparing facies tracts across systems tracts in a HFS, or across HFS in a composite sequence, allows leveraging of data and interpretations from extensively studied and understood units to those that are less well understood.

Facies tract substitution of shelf-margin reefs.—
The ramp-to-rimmed shelf evolution illustrated in this paper is a good example of the use of facies-tract substitution. Comparison of the outer-ramp facies tract of CS9–11 ramps (San Andres and lower Grayburg formations) and the outer shelf, shelf margin, and forereef slope of CS12–14 (Seven Rivers through Tansill/Capitan) illustrates this principal (Fig. 15).

In the ramp model, fusulinid-peloid packstones and wackestones with minor dasycladacean algae and pelmatozoan debris dominate the outer-ramp facies tract, which occupied a 0.5–2° seaward-dipping profile from just below fair-weather wave base (FWWB) at approximately 10 m (30 ft) deep, to depths of 100 m (300 ft) (Kerans and Fitchen, 1995). This model, based on San Andres CS9 ramps, has been developed from detailed sedimen-

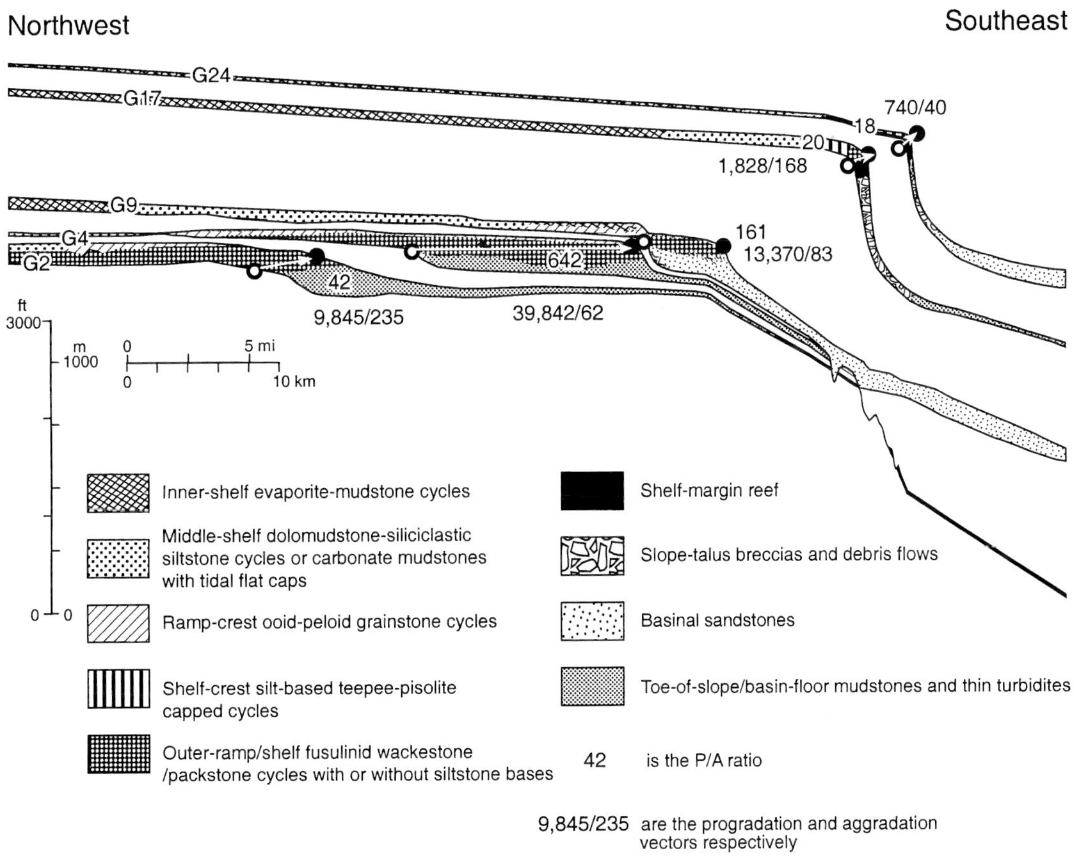

FIG. 13.—Relative rates of P/A for contrasting styles of ramps and rimmed shelves. The carbonate ramps all display P/A > 100, whereas the two rimmed margin profiles illustrated from the lower CS12 (Seven Rivers) and upper CS13 (Yates) are characteristic of the average P/A 28. Change in P/A between G2 and G4 reflects a decrease in shelf accommodation associated with CS9 sea-level fall. The G9 P/A reflects a balance between low shelf accommodation (forced regressive offlap), increased shelf-to-toe-of-slope bathymetry, and increased input of siliciclastics, which increase total progradation by helping build a slope platform for carbonates to prograde across. Similar ratios of P/A for the G 17 and 24 reflect changes in shelf accommodation that are offset by changes in carbonate productivity. See discussion in text and Fig.13.

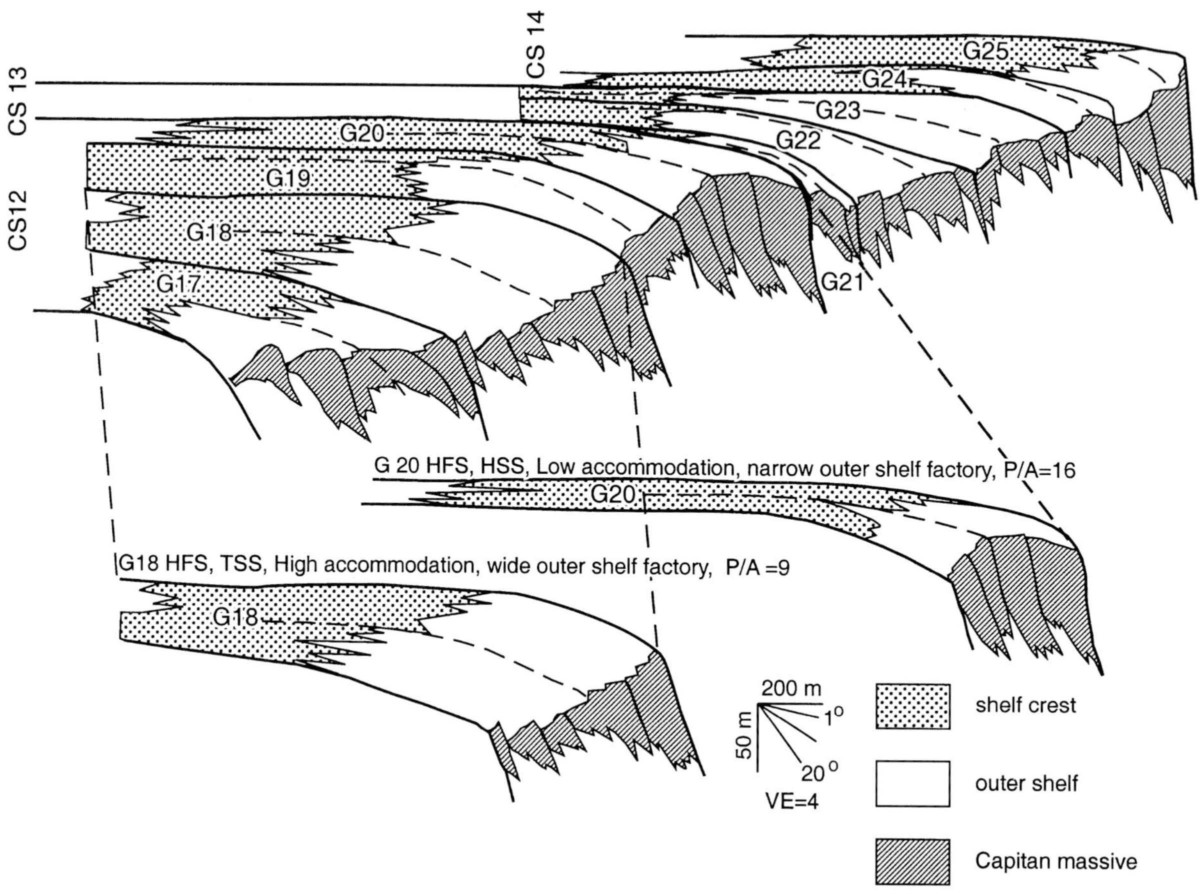

FIG. 14.—Changing styles of progradation and carbonate shelf margin development in composite sequences 12 and 13. In the TSS the shelf crest backsteps, increasing the area of the outer shelf. Because the main carbonate factory for CS12/13 is in the outer shelf, this widening of the factory allows progradation of the margin to continue even as the shelf crest backsteps. During HSS deposition, shelf crest progradation causes a narrowing of the outer shelf factory. This decrease in sediment production is offset by a decrease in shelf accommodation that causes the available sediment produced to be pushed basinward and thus a moderate P/A is maintained. This relationship also applies to the comparison of the Seven Rivers (CS12) versus the Yates (CS13). Progradation rates are nearly identical for the two sequences despite the decrease in accommodation in the CS13.

tologic analysis within a high-frequency cycle-scale framework with well-controlled datums and unit tracing (Kerans and Fitchen, 1995). Some shallowing and lateral compression of this facies tract was noted in the forced regressive G9 HFS at Last Chance Canyon (Sonnenfeld and Cross, 1993). However, a comparable sub-FWWB setting in 10 m to > 70 m (30 ft to > 200 ft) in water depth is also observed in the CS10 and the CS11 Grayburg ramps (Kerans and Nance, 1991; Barnaby and Ward, 1995).

The nature of the depositional profile of the Capitan-equivalent reef-rimmed margins has remained controversial since it was first studied (see discussions in Yurewicz, 1977; Garber et al., 1989; Hurley, 1989; Saller, 1996). Uncertainty still exists regarding whether the basinward dips of the outer shelf immediately landward of the shelf-margin reef of between 1° and 10° represent syndepositional relief or post-depositional compaction (Borer and Harris, 1991; Tinker, 1996; Hurley, 1989; Osleger, 1998; Saller, 1996). Using facies-tract substitution between the CS9 and CS12–14 profiles it is possible to show the following: (1) the outer-ramp facies tract is replaced on its down-dip end by the shelf-margin reef and forereef-slope facies tracts; (2) the outer-ramp facies tract has contracted substantially and changes from a fusulinid-dominated assemblage to a dasycladacean-pelmatozoan-fusulinid assemblage; and (3) the ramp-/shelf-crest has evolved from ooid-peloid-dominated cycles to those dominated by pisolites, teepees, and fenestral-laminite-capped cycles (Fig. 15).

The facies-tract substitution (Fig. 15) implies that the Capitan shelf-margin reef evolved through substitution with the lower portion of the outer-ramp facies tract, making a sub-fair-weather wave-base origin in water depths of between 30 m and 70 m (100 ft and 200 ft) likely. If the Capitan were truly a barrier complex developed at or near wave base, the facies substitution relative to the CS9 should have occurred close to the transition between the ramp-crest grainstones and the outer ramp rather than on the seaward end of the outer ramp. This facies substitution analysis at the CS scale corroborates abundant other evidence that favors a deeper water origin for the Seven Rivers-equivalent Capitan (Yurewicz, 1977; Hurley, 1989; Tinker, 1996) and a progressively shallowing Capitan reef during Seven Rivers through Tansill deposition (Babcock, 1977; Babcock and Yurewicz, 1989; Osleger and Tinker, 1998; Tinker, 1998).

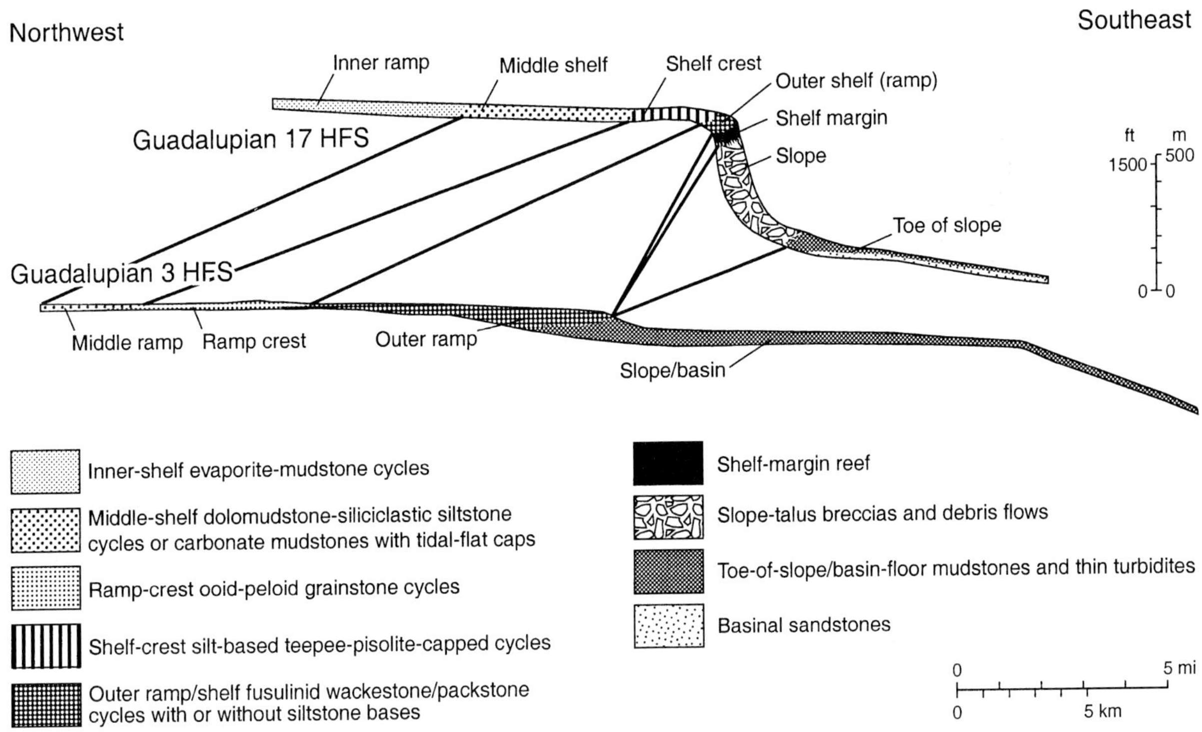

FIG. 15.—Example of facies tract substitution from Guadalupian 3 HFS ramp to Guadalupian 17 HFS rimmed shelf. Shelf-margin facies tract is substituted for the lower portion of the outer ramp facies tract. This location indicates a deeper water (>100 ft) water depth for this reef margin. See discussion in text.

The teepee-reef link.—

Placing facies data within the context of the high-frequency-sequence framework is also extremely useful for establishing links between facies and facies tracts. Concomitant development of teepee-dominated shelf-crest facies tracts and shallow reefal shelf margins is discussed here as an example of a linked facies relationship that might otherwise go unnoticed. To appreciate the interrelationship of these apparently unrelated facies, it is necessary to understand the role of platform architecture in controlling teepee development.

Kerans and Fowler (1995) demonstrated that well-developed teepee complexes like those occurring in the Yates and Tansill at Walnut Canyon (Esteban and Pray, 1977; Neese, 1989) or the entrance to Dark Canyon occur only on shelf margins with P/A of < 20 (Fig. 11B). Understanding the link between teepee development and P/A requires an understanding of the processes driving teepee development. Teepees form through multiple cycles of exposure, desiccation, flooding, sediment infill, and/or marine cementation. Well-developed teepees are not found on rapidly prograding shelf or ramp margins because the steady migration of the shoreline position from one cycle to the next does not permit repeated cycles of desiccation and flooding. This is because after only one or two exposure and flooding events, the shoreline steps sufficiently basinward such that the ensuing transgression does not reclaim this area, leaving it instead in a condition of subaerial weathering.

On platform margins where progradation is limited to a P/A of < 20 either by high shelf accommodation (late TST or early HST settings) or by high basin accommodation (either TST or HST) or both, the shoreline position of each successive cycle or sequence is offset only slightly. This reoccupation of the shoreline forces the repeated desiccation/cementation cycle required for teepee formation. High basin accommodation also fosters steep-rimmed margins and narrow facies-tract widths (compare widths of outer ramp and outer-shelf facies tracts of ramp and steep-rimmed shelf sequences in Fig. 16). This is important for teepee formation because it brings the shelf-crest complex closer to open ocean circulation and thus enhances early marine cementation, a process most likely responsible for expansive buckling of crusts (Dunham, 1972). Indeed, many of the marine cements in teepee complexes are as impressive as those in shelf-margin facies.

Teepee complexes and reefal shelf margins are favored by steep platform margin topography for different reasons. Reefs occur preferentially at such margins because of the high nutrient supply, oxygenation, early cementation, and hard substrates provided by this setting. Teepee facies are favored along steep margins because of the generally stable shoreline position that allows repeated exposure and desiccation, and the proximity to high pore-fluid flux of marine waters at the basin margin for enhanced cementation. Thus, a good relationship exists between teepee formation and reefal shelf-margin development (Fig. 16).

The link between teepee complexes and reef development can be taken farther by comparing teepee occurrence and water depth to reefal shelf margin for the Goat Seep/Capitan during CS11-14 deposition. Many authors have noted that the Goat Seep/Capitan reef facies do not occur at a constant water depth (Babcock, 1977; Yurewicz, 1977; Babcock and Yurewicz, 1989;

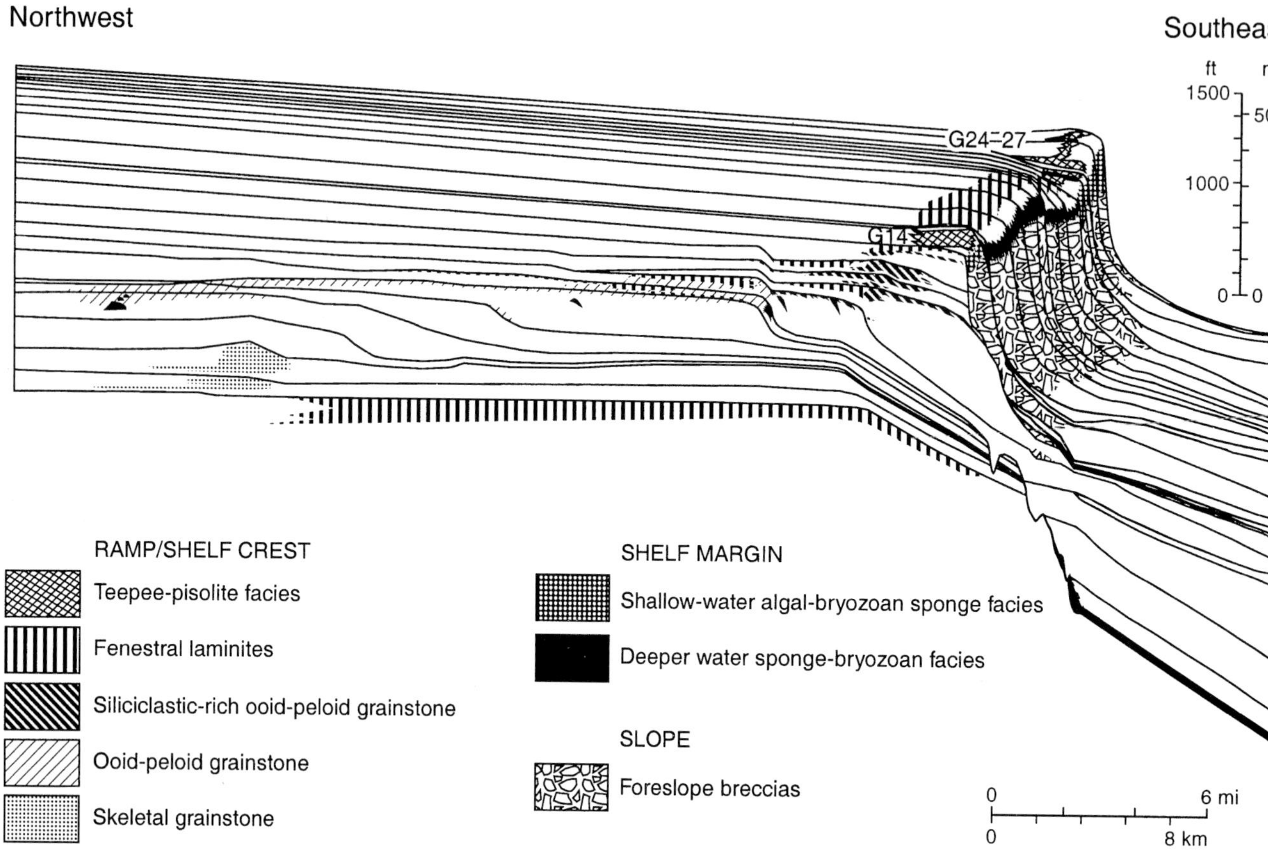

FIG. 16.—Distribution of ramp crest and shelf crest dominant facies and relationship to reef development.

Crawford, 1989; Hurley, 1989; Tinker, 1996). At the CS scale, Yurewicz (1977) pointed out that the older lower Capitan (CS12) occurred at consistently deeper water depths (30–100 m; 100–300 ft) and had a higher matrix sediment content than the younger middle and upper Capitan (CS13 and 14; Fig. 16). Closer examination of the link between teepee complexes and water depth of reef formation illustrates a good correspondence between well-developed teepee complexes and shallow reefs (< 3 m or 40 ft) (Fig. 16). Sequences with shallow (7–13 m; 20–40 ft water depth) reef margins and well-developed teepee complexes include the G14 (upper Queen/Goat Seep), G24 and G25 (upper Yates/Capitan), and G26-27 (Tansill/upper Capitan). Sequences with deep reefs are predominantly G17–19 (Seven Rivers/lower Capitan) and these display a notable absence of well-developed teepee complexes.

A possible explanation for this phenomenon may come from examining the distribution of modern reefs in south Florida and the Great Bahama Bank. P. M. Harris (pers. comm., 1998) noted from observations of Ginsburg, Enos, Gebelein, and others, that modern Florida Keys and Great Bahama Bank reefs show preferential development of reef communities along portions of the shelf forming just seaward of Pleistocene islands and barriers (the Florida Keys or Andros Island, respectively). The indication is that reef communities living on the seaward flank of narrow shelves with Pleistocene islands are protected from influx of platform interior waters that might be detrimental to reef growth as a result of elevated salinity and/or nutrient levels. Reef communities are far less developed or are absent in front of the passes between the Keys or parts of the Great Bahama Bank that are not protected by these barriers.

By analogy, it is proposed that the teepee complexes acted as topographic highs or barriers on the shelf profile (Dunham, 1972; Esteban and Pray, 1977; Kerans and Harris, 1993; Tinker, 1998), protecting the shelf-margin reefs from outflow of evaporitic inner-shelf waters. In the transgressive-dominated Seven Rivers composite sequence, where the shelf-crest facies tract migrated first substantially landward and then seaward, teepee complexes were never established. Reef faunas in these G17–19 HFS were restricted to water depths of >30 m (100 ft) and were dominated by sponge-bryozoan faunas. As teepee complexes developed in the aggradational upper Yates and Tansill G24–27 sequences, reef faunas developed in increasingly shallow water depths and displayed a substantial diversification of faunas with substantial green algal flora not observed in earlier deeper reefs (Babcock, 1977; Yurewicz, 1977).

Through sedimentologic analysis within the HFS and CS framework provided by this study, it is possible to propose a model that synthesizes the relationships between facies, facies tracts, P/A, and inherited depositional topography that can be tested by additional field studies. This model shows a three-stage evolution that combines aspects of Goat Seep through Capitan margins (Fig. 17). The model shows that progradation may be high during initial

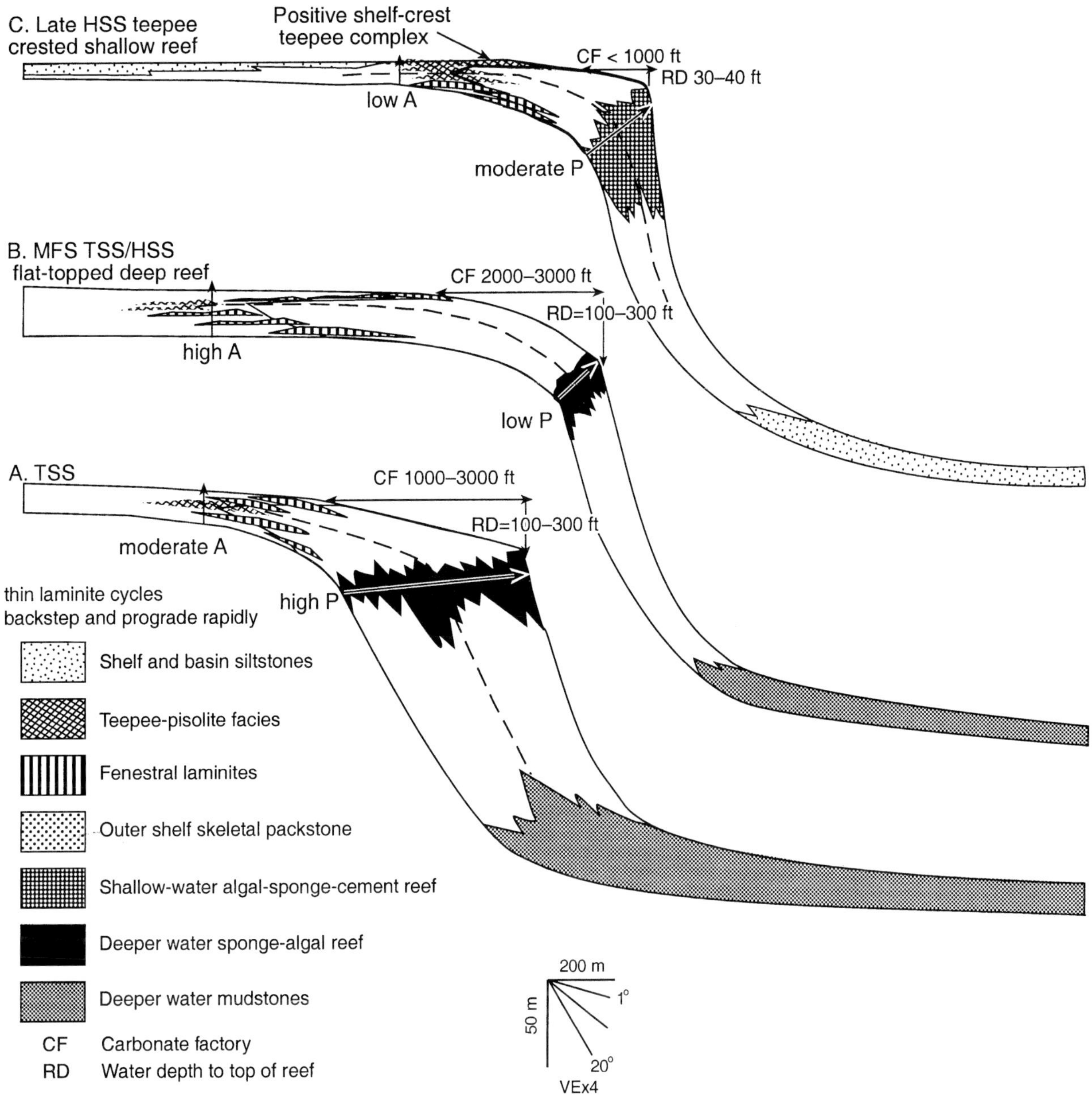

FIG 17 —Three-stage simplified model of CS11-14 rimmed margin development emphasizing the links between base-level setting, P/A, carbonate factory area, shelf-crest teepee development, and depth to reef. (A) During early TSS development, a moderate shelf aggradation occurs with strong progradation due to increased production in the enlarged outer-shelf carbonate factory. Rapid landward/seaward migration of the shelf-crest and the greater distance to shelf margin allows formation of laminites as the dominant exposure facies, with small teepees restricted to HFS-MFS/early HST. Lacking the shelf-crest barrier, reef development is restricted to deep sponge-bryozoan communities with as much as 80% allocthonous sediment. (B) At MFS of CS, base-level rise forces increased backstepping of shelf-crest and development of thick aggradational shelf succession. Reef margin is still deep, lacking the shelf-crest barrier. Stacking of the shelf margin is more aggradational than progradational as less sediment is exported from the outer-shelf factory. (C) HSS conditions include a lower P/A despite only minor shelf aggradation. This is driven by lower progradation rates associated with decreased outer-shelf factory area. Decreased P/A and narrowed outer-shelf facies tract width coupled with steady shelf-crest progradation allows well-developed teepees to develop in MFS and early HST portions of the HFS. The barrier represented by the teepee complex allows the reef margin to evolve into a shallow (30-40 ft) reef margin. Minor carbonate is exported to the toe-of-slope and basin, which is dominated by siliciclastics forced across the shelf by low shelf accommodation.

transgression as a result of increased production potential driven by an increased carbonate factory area on the shelf. This dual shoreline retrogradation and shelf-margin progradation is a characteristic signature of carbonate systems, where base-level rise does not necessarily imply decreased sediment supply. If the rate of transgression is slow and the platform aggrades rather than retrogrades, teepee complexes will form, but in the example of Fig. 17A, modeled after the CS12 TSS, shoreline retrogradation is significant and teepee complexes are restricted to the TSS-HSS turnaround.

During maximum transgression in Fig. 17B, the rate of rise outstrips production and the margin aggrades while the shoreline continues to retrograde. Finally, in the HSS, decreased shelf accommodation causes initial aggradation and stacking of the shoreline (as opposed to retrogradation in TSS and MFS settings) with the result that teepee complexes begin to develop. The aggradational and ultimately progradational style of the shelf crest constrains the outer-shelf carbonate factory to a relatively narrow belt and restricts sediment supply and associated production to a moderate rate. Slowed production is also shown by the change in toe-of-slope facies that becomes dominated by bypassing siliciclastics in the HSS. In response to development of a raised teepee-pisolite complex, reef faunas build to shallow depths and evolve into more diverse green-algal/sponge/bryozoan facies that make up a high proportion of the total volume of the shelf-margin facies tract (Fig. 17).

This model explains both the link between teepees and deep-versus-shallow reef faunas, and the connection between teepee complexes and P/A, thus ultimately linking reef development and style of platform development as controlled in this case largely by inherited topography and accommodation. It is important to recognize that this depositional model acts on at least two, and probably three scales: the HFS, the CS, and to some degree the second-order sequence scales.

DISCUSSION

This review of build-up facies in the late Leonardian through Guadalupian of the Guadalupe Mountains has attempted to place the Capitan and other important reef/build-up occurrences in the late Leonardian/Guadalupian succession within a common high-frequency sequence framework. Within the context of this sequence framework, selected stratigraphic controls/attributes that relate to build-up/reef formation have been evaluated in terms of their impact on occurrence and facies composition of the build-ups/reefs. The focus is on extrinsic controls, including climatic/oceanographic setting, eustasy, tectonic subsidence, antecedent depositional topography, shelf accommodation as represented by P/A, and facies-tract substitution. The discussion that follows touches on some of the interrelationships between these attributes.

Climatic/oceanographic controls are of first-order importance but remain an area that is not constrained by detailed data. New paleomagnetic data and more precise geochronology analyzed in the context of a high-frequency sequence framework will make these independent data quite valuable. Oceanographic considerations must be combined with antecedent topography, in that open-ocean circulation is one of the key criteria that affects the selection of preferential buildup nucleation sites on a profile.

Eustatic effects have an overriding influence when the rate of rise or fall is great (Figs. 7 and 9). Platform back-stepping in latest Leonardian time associated with a global rise changed platform style from a mixed buildup/grainstone rim (Leonardian 6 HFS) to one of carbonate ramps with <1% build-up development. It is interesting to note that no evidence of keep-up type pinnacles such as those described from the Silurian of the Michigan Basin (Sears and Lucia, 1980), the Devonian of the Western Canada Basin (Wendte and Campbell, 1992; Tinker and Kirkby, 1992), or the Jurassic Cotton Valley of the East Texas Basin (Goldhammer, 1998) are observed in outcrop. Keep-up style buildups of comparable age are known from the Leonardian 7 through 8 HFS from the east side of the Midland Basin, which suggests that the relative rise (eustatic plus tectonic effect) rather than the eustatic effect alone affected this development. Small buildups in the Leonardian 7–8 HFS that were localized by differential compaction show the potential for greater development elsewhere. Eustatic rise may also have affected the final ramp-to-rim transition during CS11 deposition. The retrogradational-to-aggradational stacking of the G11–13 HFS likely helped create the differential topography that favored reef-rim development.

Major eustatic sea-level falls such as those associated with the termination of CS9 or at the end of CS11 (Fig. 9), where base-level shifts to well below the previous shelf margin (that is, not gradual forced regressive events), appear to have little effect on subsequent buildup/reef development. Once differential topography is constructed and that topography is transgressed gradually, then buildup/reef growth will ensue in close to the same shelf-margin position.

The impact of antecedent topography on location and style of buildups is dominant in the late Leonardian-Guadalupian data (Fig. 10). Ramp build-ups are focused on areas of maximum differential topography. Whether it is constructional topography associated with high-accommodation maximum-flooding surfaces, or positive areas associated with differential compaction or draping over an older shelf margin, build-up distribution is not random (Fig. 10). Topography is also critical for development of shelf-margin reefs. Regardless of shelf-margin position, extensive reef rims do not develop on distally steepened ramps with < 30° of depositional dip (Figs. 7 and 10). This slope angle in the Guadalupe data defines a break-over for development of a bypass surface for unconsolidated sediment, favoring formation of the lithified substrates required for attachment of encrusting fauna/flora. Slope angle in carbonate platform margins is a complex variable that is dependent on grain size and sorting of material as well as degree of lithification (Grammer et al., 1993), and thus, this slope angle may not be the same in other platform systems.

One of the more predictable patterns of carbonate platform evolution is that from broad carbonate ramps to flat-topped, steep reef-rimmed shelves (Read, 1985). The CS9–14 ramp-to-rim transition illustrates a clear progression from the following:

1. ramps with mud-dominated mounds lacking macroscopic evidence of early lithification or debris and localized near MFS surfaces (L7-G3); to
2. discontinuous mud-dominated mounds within 5–15°-dipping highstand clinoforms lacking slope debris (G4-9); to
3. steep 20–25° clinoform slopes with slope and toe-of-slope breccias containing reef blocks indicating shelf-margin reef development (G10); to
4. continuous reef-rimmed margins with well-developed slope

debris and abundant evidence of early lithification (see Figs. 10 and 11).

The dominant factor controlling the transition from ramps to rimmed shelves in this example is antecedent topography, driven by eustacy, subsidence, and sediment supply. Following platform-margin back-stepping, CS9–10 ramps prograded rapidly across the relatively shallow flooded platform (P/A values of 20–600) to regain the earlier Victorio Peak (L6 HFS) shelf margin. Once this margin was reached, each increment of seaward progradation required a greater volume of shelf/shelf-margin sediment as the platform built out over ever-greater water depths. The effect was to slow progradation and develop steeper clinoform slopes near the angle of repose. The slower progradation rate of the margin, combined with increased pore-fluid exchange at the high-energy margin, favored increased early lithification. Early lithification, steeper slopes, and slower sedimentation rate all favor formation of a stable substrate, and eventually this hard, stable substrate at the shelf margin became a favored site for reef development.

If depositional topography is a dominant factor controlling development of reef-rimmed margins in the Permian, then it is possible to predict that initiation of reef margins should vary considerably around the basin depending on variations in tectonic subsidence and structural setting. Hence, the occurrence of shelf-margin reefs in the Grayburg at the North McElroy Unit on the east side of the Central Basin Platform (Longacre, 1990) probably reflects development of sufficient localized differential topography by drape over the underlying Leonardian (Clear Fork) platform margin to allow nucleation of the reefs. In areas where San Andres margins did not backstep, but instead perched above the Clear Fork shelf margin, such as on the west side and in the southeast corner of the Central Basin Platform (Fig. 1), reef facies may also have been developed more widely throughout the San Andres Formation.

P/A provides one measure of the response of the platform to complex changes in eustasy, subsidence, and sediment supply. Ramps have higher P/A values than rimmed margins for a given shelf accommodation because there is generally less basin accommodation and a larger carbonate factory. P/A patterns within the classic Capitan rimmed margins are not a simple linear response to eustatic change. Initial rise generally increases the carbonate factory area, thus increasing progradation rate temporarily until it becomes outstripped by the rise. Subsequent progradation becomes somewhat self-limiting when the shelf-crest facies tract progrades more rapidly than the shelf margin, causing a reduction in carbonate factory area. P/A and SMP/A are useful descriptors of platform evolution when used in the context of other key parameters such as carbonate-factory size and shelf-to-basin bathymetry, but cannot be used as simple indicators of changing base level.

Another method of stratigraphic analysis available once the shelf-margin Capitan is placed into a high-frequency sequence context is facies-tract substitution. The evolution of facies tracts from earlier ramp profiles to the reef-rimmed Capitan model provides a line of reasoning that indicates a deeper water origin of the lower Capitan that is independent yet supportive of data used in other studies, thereby strengthening the existing interpretation. The linkage of facies and facies tracts within a sequence is critical to improved interpretation of the origin and distribution of these facies. The genetic link between shelf-crest subaerial teepee-pisolite complexes and marine shelf-margin reefs is a good example of the facies interdependence. The complex relationship between P/A values, teepee development, and the deep-versus-shallow setting of shelf-margin reef growth illustrates how data from one facies can be translated into a better model of the origin of another facies that is not contiguous or directly linked genetically.

CONCLUSIONS

The high-frequency sequence framework presented here for the Guadalupe Mountain section provides a valuable context for studying the distribution and internal make-up of build-ups and shelf-margin reefs, of which the Capitan is the dominant component. The framework includes 30 high-frequency sequences (HFS) grouped into lowstand, transgressive, and highstand elements of six composite sequences (CS). Capitan reef growth spans 12 high-frequency sequences in composite sequences 12–14 equivalent to the Seven Rivers, Yates, and Tansill formations on the shelf and the Bell Canyon in the basin.

The Leonardian-Guadalupian CS9 was initiated by a major eustatic rise that resulted in flooding of the precursor Leonardian 6 flat-topped platform. Widespread development of ramp systems characterized this CS9 phase. Ramp buildups consisting of mud-dominated sponge-bryozoan-pelmatozoan assemblages occur in water depths of 10–30 m (30–100 ft) and make up <1% of most of these HFS. Though small in volume, the position of these buildups is predicted from antecedent topography associated with underlying constructional topography near maximum-flooding surfaces (low P/A) and associated with compactionally related drape over the underlying Leonardian margin.

The transition from ramps with minor buildups to flat-topped, reef-rimmed platforms in the Guadalupe Mountains' Guadalupian examples is controlled by antecedent topography. As the CS10 and early CS11 ramps prograded out across the more strongly basinward-dipping Leonardian platform margins, progradation rate was slowed by the increased accommodation potential. This sets off a positive feedback system of clinoform oversteepening, increased sediment bypass, early lithification, and habitation of these firm bypass slopes by encrusting reef dwellers. Reef-rimmed margins may have begun to develop as early as the CS10 lowstand (lower Grayburg) and became well established as a combination of antecedent topography and CS10 eustatic rise constructed a vertically aggrading, oversteepened margin.

The goal of this analysis is to use the high-frequency sequence framework to capture the dynamic aspects of platform evolution and to use this information to better understand the evolving facies patterns. The use of P/A values at the CS and HFS scale across ramps and rimmed margins helps to point out the fundamental differences in these two systems. The high rates of progradation of ramp systems, with P/A values greater than 100, are clearly poor settings for shelf-margin build-up development as the high net sediment input, the lack of a lithified stable substrate, and the limited topographic relief to assist in focusing current energy combine to minimize development of this facies in ramp settings. Rimmed margins, with their low P/A, abundant early lithification and stable substrate, and steep topography form ideal sites for build-up development. P/A value captures these aspects of the evolving platform style.

Analysis of platform-margin geometry within the CS12–14 is best characterized using SMP/A ratios. From this analysis (Osleger and Tinker, 1998; Tinker, 1998) it is possible to define a systematic trend from moderate SMP/A in the TSS to low SMP/A at MFS, to high SMP/A within the HSS. This trend is coupled with the evolution of the shelf margin reef facies as well as with the development of teepee complexes. Thus, the SMP/A values can be used to link shelf-margin development with base-level signal.

Within the high-frequency sequence framework, facies-tract substitution and facies proportions can be used to better constrain interpretations of the dynamic water-depth setting of the Capitan margin and factors controlling its position on the profile. The substitution of Capitan and Goat Seep margin facies in a position between the outer shelf and slope, when compared to earlier ramp systems, is clear evidence for a deeper water origin for at least part of the Capitan. The link between teepees and shelf-margin reef development is a good example of the utility of the sequence framework in facies analysis. Teepees and shelf-margin reef facies, facies that on first examination would not be assumed as genetically related, are shown here to be linked by a common dependence on platform morphology and P/A or SMP/A. Using this link we are able to propose for the first time a model for the evolution of the Capitan that provides a causal mechanism for reef depth and at the same time permits prediction of style of reef development from facies composition of the shelf crest.

ACKNOWLEDGMENTS

The authors acknowledge the support of their respective organizations, the Bureau of Economic Geology, The University of Texas at Austin, and Marathon Oil Company, for permission to publish results of this ongoing research. Kerans also acknowledges the support of RCRL sponsor companies Altura, Amoco, Aramco, ARCO, British Petroleum International, Elf Aquitaine, Exxon, Japan National Oil Corporation, Marathon Oil Company, Oxy USA, Oxy International, Pennzoil, Petroleum Development Oman, Phillips, Texaco, TOTAL, Unocal, Geoquest, Landmark, Terra Science, and Geomath. Many individuals spent considerable time contributing to and revising the stratigraphic framework that we have presented for the Guadalupe Mountains, including Bill Fitchen, Mike Gardner, Mark Sonnenfeld, Mitch Harris, Jerry Lucia, Willis Tyrell, Lloyd Pray, Don Bebout, and many others. To them we offer special thanks.

Mike Grammer, Mitch Harris, and Dave Osleger reviewed the manuscript and offered significant constructive criticism. Figures were drafted by Pat Alfano and Chris Conly under the direction of Joel Lardon, Bureau of Economic Geology. Published with permission of the Director, Bureau of Economic Geology, The University of Texas at Austin.

REFERENCES

BABCOCK, J. A., 1977, Calcareous algae, organic boundstones, and the genesis of the Upper Capitan Limestone (Permian, Guadalupian), Guadalupe Mts., West Texas and New Mexico, *in* Hileman, M. E., and Mazzullo, S. J., eds., Upper Guadalupian Facies, Permian Reef Complex, Guadalupe Mountains, New Mexico and West Texas, 1977 Field Conference Guidebook: Midland, Society of Economic Paleontologists and Mineralogists, Permian Basin Section, Publication 77-16, p. 3-44.

BABCOCK, J A., and YUREWICZ, D. A., 1989, The massive facies of the Capitan Limestone, Guadalupe Mountains, Texas and New Mexico, *in* Harris, P. M., and Grover, G. A., eds., Subsurface and outcrop examination of the Capitan shelf margin, Northern Delaware Basin: Tulsa, Society of Economic Paleontologists and Mineralogists Core Workshop 13, p. 365-372.

BARNABY, R. J., and WARD, W. B., 1995, Sequence stratigraphic framework, high-frequency cyclicity, and three-dimensional heterogeneity: Grayburg Formation, Brokeoff Mountains, New Mexico, *in* Pause, P. H., and Candelaria, M. P., eds., Carbonate Facies and Sequence Stratigraphy: Practical Applications of Carbonate Models: Midland, Society of Economic Paleontologists and Mineralogists, Permian Basin Section, Publication 95-36, p. 37-50.

BEBOUT, D. G., and KERANS, C., eds., 1993, Guide to the Permian Reef Geology Trail, McKittrick Canyon, Guadalupe Mountains National Park, West Texas, Guidebook 26: Austin, Bureau of Economic Geology, University of Texas, 48 p.

BORER, J. M., and HARRIS, P. M., 1991, Depositional facies and cyclicity in the Yates Formation, Permian Basin—Implications for reservoir heterogeneity: American Association of Petroleum Geologists Bulletin, v. 75, p. 726–779.

BOSELLINI, A., 1989, Dynamics of Tethyan carbonate platforms, *in* Crevello, P. D., Wilson, J. L., Sarg, J. F., and Read, J. F., eds., Controls on Carbonate Platform and Basin Development: Tulsa, Society of Economic Paleontologists and Mineralogists, Special Publication 44, p. 3-14.

CRAWFORD, G. A., 1989, Goat Seep–precursor to the Capitan, *in* Harris, P. M., and Grover, G. A., eds., Subsurface and outcrop examination of the Capitan shelf margin, Northern Delaware Basin: Tulsa, Society of Economic Paleontologists and Mineralogists Core Workshop 13, p. 373-378.

DAVIES, P. J., SYMONDS, P. A., FEARY, D. A., and PIGRAM, C. J., 1989, The evolution of the carbonate platforms of northeast Australia, *in* Crevello, P. D., Wilson, J. L., Sarg, J. F., and Read, J. F., eds., Controls on Carbonate Platform and Basin Development: Tulsa, Society of Economic Paleontologists and Mineralogists Special Publication 44, p. 233-258.

DUNHAM, R. J., 1972, Capitan Reef, New Mexico and Texas: Facts and Questions to Aid Interpretation and Group Discussion: Midland, Society of Economic Paleontologists and Mineralogists, Permian Basin Section, Publication 72-14, 291 p.

ESTEBAN, M., and PRAY, L. C., 1977, Origin of pisolite facies of the shelf crest, *in* Hileman, M. E., and Mazzullo, S. J., eds, Upper Guadalupian Facies, Permian Reef Complex, Guadalupe Mountains, New Mexico and West Texas, 1977 Field Conference Guidebook: Midland, Society of Economic Paleontologists and Mineralogists, Permian Basin Section, Publication 77-16, p. 479-486.

FEKETTE, T. E., 1986, The sedimentology and stratigraphy of the Grayburg Formation and its associated erosion surface along the high Western Escarpment of the Guadalupe Mountains, Texas: Unpublished M. S. Thesis, University of Wisconsin, Madison, 174 p.

FISCHER, A. G., and SARNTHEIN, M., 1988, Airborne silts and dune-derived sands in the Permian of the Delaware Basin: Journal of Sedimentary Petrology, v. 58, p. 637-643.

FITCHEN, W. M., 1992, Sequence stratigraphy of the upper San Andres Formation and Cherry Canyon Tongue (Permian, Guadalupian), southern

Brokeoff Mountains, New Mexico: Unpublished M.A. Thesis, The University of Texas, Austin, 155 p.

FITCHEN, W. M., 1993, Sequence stratigraphic framework of the Upper San Andres Formation and equivalent basinal strata in the Brokeoff Mountains, Otero County, New Mexico, *in* Love, D.W., et. al., eds., New Mexico Geological Society Guidebook, 44th Field Conference, Carlsbad Region, New Mexico and West Texas, p. 185-193.

FRANSEEN, E. K., FEKETTE, T. E., and PRAY, L. C., 1989, Evolution and destruction of a carbonate bank at the shelf margin: Grayburg Formation (Permian), western escarpment, Guadalupe Mountains, Texas, *in* Crevello, P. D., Wilson, J. L., Sarg, J. F., and Read, J. F., eds., Controls on Carbonate Platform and Basin Development: Tulsa, Society of Economic Paleontologists and Mineralogists Special. Publication 44, p. 289-304.

GARBER, R. A., GROVER, G. A., and HARRIS, P. M., 1989, Geology of the Capitan shelf margin – subsurface data from the northern Delaware Basin, *in* Harris, P. M., and Grover, G. A., eds., Subsurface and outcrop examination of the Capitan shelf margin, Northern Delaware Basin: Tulsa, Society of Economic Paleontologists and Mineralogists Core Workshop 13, p. 3-272.

GOLONKA, J., ROSS, M. I., and SCOTESE, C. R., 1995, Phanerozoic paleogeographic and paleoclimatic modeling maps, *in* Embry, A. F., Beauchamp, S., and Glass, D. J., eds., Pangea: Global Environments and Resources: Calgary, Canadian Society of Petroleum Geologists Memoir 17, p. 1-47.

GOLDHAMMER, R. K., 1998, Second-order accommodation cycles and points of "stratigraphic turn-around": Implications for high-resolution sequence stratigraphy and facies architecture of the Haynesville and Cotton Valley Lime "Pinnacle reefs" of the East Texas Salt Basin: West Texas Geological Society Bulletin, v. 37, p. 5-9.

GRAMMER, G. M., GINSBURG, R. N., and HARRIS, P. M., 1993, Timing of deposition, diagenesis, and failure of steep carbonate slopes in response to a high-amplitude/high-frequency fluctuation in sea level, Tongue of the Ocean, Bahamas, *in* Loucks, R. G., and Sarg, J. F., eds., Carbonate Sequence Stratigraphy, Recent Developments and Applications: Tulsa, American Association of Petroleum Geologists Memoir 57, p. 107-131.

HAYES, P. T., 1964, Geology of the Guadalupe Mountains, New Mexico: Washington, D.C., U.S. Geological Survey Professional Paper 446, 69 p.

HUNT, D., and TUCKER, M. E., 1992, Stranded parasequences and the forced regressive wedge systems tract; deposition during base-level fall: Sedimentary Geology, v. 81, p. 1-9.

HURLEY, N. F., 1989, Facies mosaic of the lower Seven Rivers Formation, McKittrick Canyon, New Mexico, *in* Harris, P. M., and Grover, G. A., eds., Subsurface and outcrop examination of the Capitan shelf margin, Northern Delaware Basin: Tulsa, Society of Economic Paleontologists and Mineralogists Core Workshop 13, p. 325-346.

KENDALL, C. G. St. C., and SCHLAGER, W., 1981, Carbonates and relative changes in sea level: Marine Geology, v. 44. p. 181-212.

KERANS, C., FITCHEN, W. M., GARDNER, M. H., SONNENFELD, M. D., TINKER, S. W., and WARDLAW, B. R., 1992, Styles of sequence development within uppermost Leonardian through Guadalupian strata of the Guadalupe Mountains, Texas and New Mexico, *in* Mruk, D. H., and Curran, B. C., eds., Permian Basin Exploration and Production Strategies: Applications of Sequence Stratigraphic and Reservoir Characterization Concepts: Midland, West Texas Geological Society Symposium, Publication 92-91, p. 1–7.

KERANS, C., and FITCHEN, W. M., 1995, Sequence Hierarchy and Facies Architecture of a Carbonate-Ramp System: San Andres Formation of Algerita Escarpment and Western Guadalupe Mountains, West Texas and New Mexico: Austin, Bureau of Economic Geology, The University of Texas Report of Investigations 235, 86 p.

KERANS, C., and FOWLER, D., 1995, Role of high-frequency cycles in analysis of ancient facies: An example from shelf-crest teepee-pisolite facies of the Guadalupian section, Guadalupe Mountains and subsurface of the Central Basin Platform (abs.): Tulsa, American Association of Petroleum Geologists Annual Convention, p. 49a.

KERANS, C., and HARRIS, P. M., 1993, Outer shelf and shelf crest, *in* Bebout, D. G., and Kerans, C., eds., Guide to the Permian Reef Geology Trail, McKittrick Canyon, Guadalupe Mountains National Park, West Texas, Guidebook 26: Austin, Bureau of Economic Geology, University of Texas, p. 32-42.

KERANS, C., and NANCE, H. S., 1991, High-frequency cyclicity and regional depositional patterns of the Grayburg Formation, Guadalupe Mountains, New Mexico, *in* Meader-Roberts, S., Candelaria, M. P., and Moore, G. E., eds., Sequence stratigraphy, facies and reservoir geometries of the San Andres, Grayburg and Queen Formations, Guadalupe Mountains, New Mexico and Texas: Midland, Society of Economic Paleontologists and Mineralogists, Permian Basin Section, Publication 91-32, p. 53-96.

KING, P. B., 1948, Geology of the Southern Guadalupe Mountains, Texas: Washington, D.C., U.S. Geological Survey Professional Paper 215, 183 p.

LONGACRE, S. A., 1980, Dolomite reservoirs from Permian biomicrites, *in* Halley, R. B., and Loucks, R. G., eds., Carbonate reservoir rocks, Notes for SEPM Core Workshop 1: Tulsa, Society of Economic Paleontologists and Mineralogists, p. 105-117.

LONGACRE, S. A., 1990, The Grayburg reservoir, North McElroy Unit, Crane County, Texas, *in* Bebout, D. G., and Harris, P. M., Geologic and Engineering Approaches in Evaluation of San Andres/Grayburg Hydrocarbon Reservoirs-Permian Basin: Austin, Bureau of Economic Geology, University of Texas, p. 239-274.

MITCHUM, R. M., Jr., and VAN WAGONER, J. C., 1991, High-frequency sequences and their stacking patterns: sequence-stratigraphic evidence of high-frequency eustatic cycles, *in* Biddle, K.T., and Schlager, W., eds., The Record of Sea-Level Fluctuations: Sedimentary Geology, v. 70, p. 131–160.

NEESE, D. G., 1989, Peritidal facies and cycles in Yates Formation outcrops, Guadalupe Mountains, New Mexico, *in* Harris, P. M., and Grover, G. A., eds., Subsurface and outcrop examination of the Capitan shelf margin, Northern Delaware Basin: Tulsa, Society of Economic Paleontologists and Mineralogists Core Workshop 13, p. 295-304.

OSLEGER, D. A., 1998, Sequence architecture and sea-level dynamics of Late Permian shelfal facies, Guadalupe Mountains, southern New Mexico: Journal of Sedimentary Research, v. 68, p. 327-346.

OSLEGER, D. A., and S. W. TINKER, 1999, Three-dimensional architecture of Upper Permian high-frequency sequences, Yates-Capitan Shelf Margin, Permian Basin, USA, *in* Harris, P. M., Saller, A. H., and Simo, J. A., (eds.), Advances in Carbonate Sequence Stratigraphy: Application to Reservoirs, Outcrops, and Models: Tulsa, SEPM Special Publication (in press).

PLAYFORD, P. E., 1980, Devonian "Great Barrier Reef" of Canning Basin, Western Australia: American Association of Petroleum Geologists Bulletin, v. 64, p. 814-840.

PLAYFORD, P. E., HURLEY, N. H., KERANS, C., and MIDDLETON, M. F., 1989, Reefal platform development, Devonian of the Canning Basin, *in* Crevello, P. D., Wilson, J. L., Sarg, J. F., and Read, J. F., eds., Controls on Carbonate Platform and Basin Development: Tulsa, Society of Economic Paleontologists and Mineralogists Special Publication 44, p. 187-202.

READ, J. F., 1985, Carbonate platform facies models: American Association of Petroleum Geologists Bulletin, v. 69, p. 1-21.

READ, J. F., 1989, Controls on evolution of Cambrian-Ordovician passive margin, U. S. Appalachians, *in* Crevello, P. D., Wilson, J. L., Sarg, J. F., and Read, J. F., eds., Controls on Carbonate Platform and Basin Development: Tulsa, Society of Economic Paleontologists and Mineralogists Special Publication 44, p. 147-166.

ROSS, C. A., and ROSS, J. R. P., 1987, Late Paleozoic sea levels and depositional sequences, *in* ROSS, C. A., and Haman, D., eds., Timing and depositional history of eustatic sequences: Constraints on seismic stratigraphy: Cushman Foundation for Foraminiferal Research, Special Publication 24, p. 137-168.

SALLER, A., 1996, Differential compaction and basinward tilting of the prograding Capitan reef complex, Permian, west Texas and southeastern New Mexico: Sedimentary Geology, v. 101, p. 21-30.

SCHLAGER, W., 1992 Sedimentology and sequence stratigraphy of reefs and carbonate platforms: Tulsa, American Association of Petroleum Geologists Continuing Education Course Note Series 34, 135 p.

SEARS, S. O., and LUCIA, F. J., 1980, Dolomitization of northern Michigan Niagara reefs by brine refluxion and freshwater/seawater mixing, *in* Zenger, D. H., Dunham, J. B., and Ethington, R. L., eds., Concepts and Models of Dolomitization: Tulsa, Society of Economic Paleontologists and Mineralogists Special Publication 28, p. 215-235.

SONNENFELD, M. D., and CROSS, T. A., 1993, Volumetric partitioning and facies differentiation within the Permian upper San Andres Formation of Last Chance Canyon, Guadalupe Mountains, New Mexico, *in* Loucks, R. G., and Sarg, J. F., eds., Carbonate Sequence Stratigraphy, Recent Developments and Applications: Tulsa, American Association of Petroleum Geologists Memoir

57, p. 435-474.

TINKER, S. W., 1996, Reservoir-scale sequence stratigraphy: McKittrick Canyon and 3D subsurface examples, West Texas and New Mexico: Unpublished Ph.D. Dissertation, University of Colorado, Boulder, 245 p.

TINKER, S.W., 1998, Shelf-to-basin facies distributions and sequence stratigraphy of a steep-rimmed carbonate margin: Capitan depositional system, Mckittrick Canyon, New Mexico and Texas: Journal of Sedimentary Research, v. 68, p. 1146-1174.

TINKER, S. W., and KIRKBY, K. C., 1992, The Keg River/Winnipegosis petroleum system in Northeastern Alberta (abs.): Tulsa, American Association of Petroleum Geologists 1993 Annual Convention Abstracts, p. 66.

VAIL, P. R. 1987, Seismic stratigraphy interpretation procedure, *in* Bally, A.W., ed., Atlas of Seismic Stratigraphy, v. 1: Tulsa, American Association of Petroleum Geologists Studies in Geology 27, p. 1–10.

VAN WAGONER, J. C., POSAMENTIER, H. W., MITCHUM, R. W., VAIL, P. R., SARG, J. F., LOUTIT, T. S., and HARDENBOL, J., 1988, An overview of the fundamentals of sequence stratigraphy and key definitions, *in* Wilgus, C. K., Hastings, B. J., Posamentier, H., Van Wagoner, J. C., Ross, C. A., and Kendall, C. G. St. C., eds., Sea-Level Change: An Integrated Approach: Tulsa, Society of Economic Paleontologists and Mineralogists Special Publication 42, p. 39–46.

WALKER, D. A., GOLANKA, J, REID, A., and REID, S., 1995, The effects of paleolatitude and paleogeography on carbonate sedimentation in the late Paleozoic, *in* Huc, A., ed., Paleogeography, Paleoclimate, and Source Rocks: Tulsa, American Association of Petroleum Geologists Studies in Geology 40, p. 133 – 155.

WENDTE, J., d. F., and CAMPBELL, C., 1992, Devonian-Early Mississippian Carbonates of the Western Canadian Sedimentary Basin: Tulsa, Society of Economic Paleontologists and Mineralogists Short Course Notes 28, 268 p.

YE, Q., and KERANS, C., 1996, Reconstructing Permian eustasy from 2-D the Brushy Canyon play in outcrop and subsurface: Concepts and examples, *in* DeMis, W. D., and Cole, A. G., eds., Guidebook: Backstripping and its use in forward: Midland, Society of Economic Paleontologists and Mineralogists, Permian Basin Section, Publication 96-38, p. 69-74.

YUREWICZ, D. A., 1977, The origin of the massive facies of the Lower and Middle Capitan Limestone (Permian), Guadalupe Mountains, New Mexico and West Texas, *in* Hileman, M. E., and Mazzullo, S. J., eds., Upper Guadalupian Facies, Permian Reef Complex, Guadalupe Mountains, New Mexico and West Texas, 1977 Field Conference Guidebook: Midland, Society of Economic Paleontologists and Mineralogists, Permian Basin Section, Publication 77-16, p. 45-92.

SUBSURFACE EXPRESSION OF THE CAPITAN DEPOSITIONAL SYSTEM AND IMPLICATIONS FOR HYDROCARBON RESERVOIRS, NORTHEASTERN DELAWARE BASIN

PAUL M. (MITCH) HARRIS

Chevron Petroleum Technology Company, 2811 Hayes Road, Houston, TX 77082

AND

ARTHUR H. SALLER

Unocal Corporation, Sugar Land, Texas 77478

ABSTRACT: The Capitan depositional system was studied in the subsurface using seismic and well data from the northeastern Delaware basin. Seismic data of the Capitan depositional system show characteristics that include (1) a massive prograding reef/slope, (2) back-reef/shelf reflectors that dip and diverge basinward before disappearing into the massive reef, and (3) layered bottomset beds that thicken basinward by the addition of younger reflectors. A wireline log cross-section of nearby wells illustrates the stratigraphy in more detail than the seismic line. Basinward-dipping shelf strata are interbedded sandstones and carbonates that diverge and pass basinward into massive carbonate of the reef. Correlative markers within the massive reef are difficult to find. Slope carbonate beds thin and basinal siliciclastics thicken toward the basin. Bottomset beds in the basin consist of interbedded sandstones/siltstones and low-porosity carbonates. This subsurface stratigraphy is very similar to outcrop stratigraphy described in the Guadalupe Mountains.

Lithologic differences between outcrops and their subsurface equivalents are due largely to variations in dolomitization and evaporite dissolution on outcrops. Distribution of porosity in the Capitan depositional system is closely related to depositional facies. Back-reef sandstones and some shelf carbonates adjacent to the reef have good porosity and moderate permeability, but porosity and permeability in those strata generally decrease landward. The subsurface Capitan reef has moderate porosity and high permeability and is a regional aquifer. Carbonate beds in the basin are generally not porous, but some basinal sandstones filling elongate channels have good porosity and moderate permeability.

Hydrocarbons are not present in the Capitan reef because it does not occur in a setting that allows structural or stratigraphic closure and/or isolation from active meteoric aquifers. Many oil fields (10-400 million barrels recoverable) occur in back-reef equivalents of the Capitan reef, primarily the Seven Rivers and Yates formations, on the Northwestern Shelf and western edge of the Central Basin Platform. Those reservoirs are generally in stratigraphic or combination stratigraphic-structural traps, where porous and permeable sandstones pass up-dip into impermeable sandstones/siltstones, carbonates, and/or evaporites. Oil also occurs in channelized basinal sandstones equivalent to the reef, but the basinal fields have <5–30 million barrels of oil recoverable, and hence are generally smaller than those of the back-reef.

INTRODUCTION

The Capitan shelf margin is the youngest of a series of Permian shelf-margin complexes developed around the Delaware Basin (King, 1948; Newell et al., 1953). The trend of three of these margins, the Abo, Goat Seep, and Capitan, around the northern end of the basin, illustrates the long-term progradational history of the basin margin (Fig. 1). Outcrops of the Capitan depositional system (Capitan shelf margin and its shelf and basin equivalents) in the Guadalupe and Delaware Mountains of west Texas and southeastern New Mexico are some of the most studied and most frequently visited outcrops in the world. The allure of these outcrops is in part due to their exquisite nature, and also to their close proximity to a major hydrocarbon province, the Permian Basin. The purposes of this paper are: (1) to describe the subsurface expression of the Capitan depositional system using seismic and well data; (2) compare that subsurface expression to outcrop data from the Guadalupe Mountains; and (3) summarize hydrocarbon production from the Capitan depositional system.

Formation names applied to the rock units of the Capitan depositional system are shown in Fig. 2. Those stratigraphic units are based on the work of King (1948), Newell et al. (1953), and Hayes (1964). The Capitan Formation includes carbonates deposited in reef, forereef, and slope environments. Shelfal equivalents of the Capitan are bedded carbonates, siliciclastics, and evaporites of the Seven Rivers, Yates, and Tansill (youngest) formations. The basinal equivalent of the Capitan Formation is the Bell Canyon Formation, which is dominantly siliciclastic. The Bell Canyon Formation contains carbonate interbeds (Hegler, Pinery, Rader, McCombs, and Lamar, in ascending order) along the basin edge.

PREVIOUS STUDIES

Because the Capitan reef is not a hydrocarbon reservoir, subsurface studies of it are scarce. In contrast, there are several subsurface studies of the shelf and basin equivalents of the reef. Hill (1996) published a comprehensive review of outcrop and subsurface data that discusses facies, diagenesis, and several long-standing controversies regarding the Capitan depositional system. Garber et al. (1989) described lithologies, facies, and diagenesis of back-reef, reef, slope, and basinal strata cored in the Gulf PDB-04 well (see Fig. 1 for location). They also developed a gross stratigraphic framework for the Capitan shelf margin in the subsurface that included maps, seismic lines, and well log cross-sections.

Other subsurface studies of the Capitan depositional system generally concentrated on the lithologies and reservoir quality of either back-reef or basinal strata. Back-reef and shelf interior strata equivalent to the Capitan reef have been described in the subsurface by Crawford and Dunham (1982), Ordonez (1984), Ward et al. (1986), Garber et al. (1989), Borer and Harris (1991a, b), Andreason (1992), and Broadhead (1993a). The basin equivalent of the Capitan reef is described in the subsurface by Hull

Geologic Framework of the Capitan Reef, SEPM Special Publication No. 65

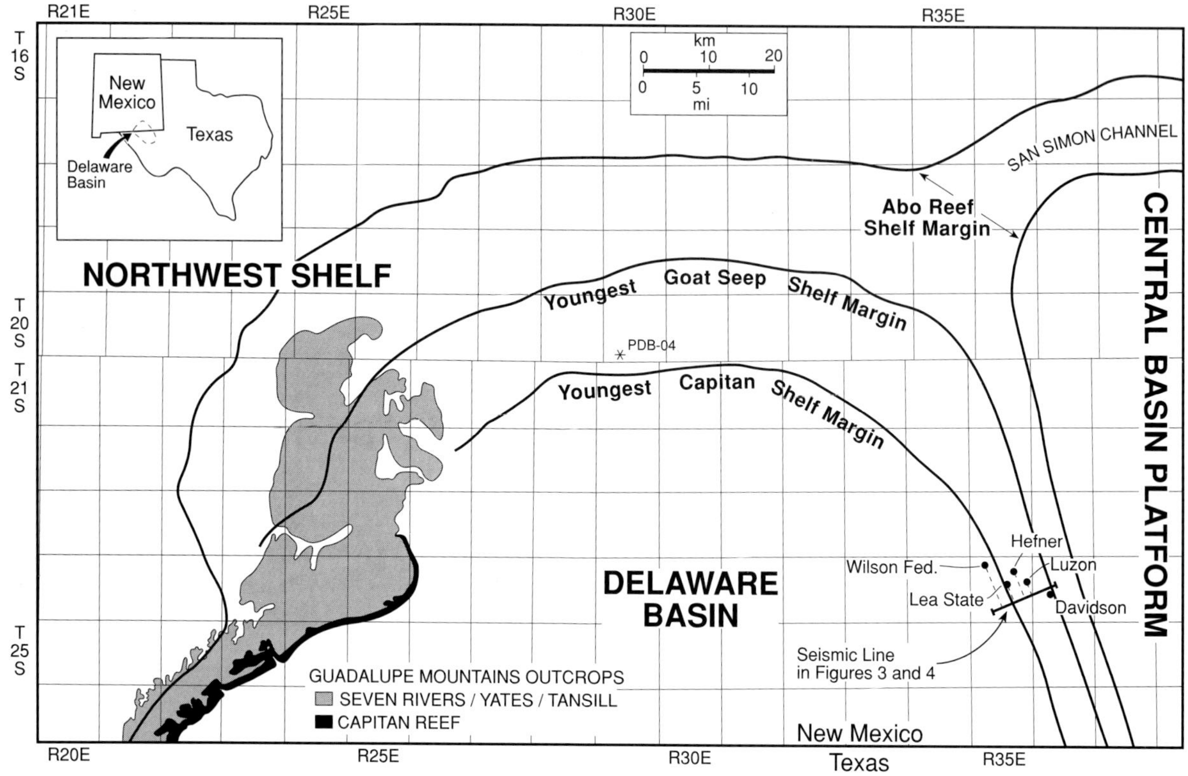

FIG. 1.—Map of northern Delaware Basin showing location of seismic line and well data included in this paper. Location of Guadalupe Mountains outcrops and Gulf PDB-04 well are also shown. Map indicates the long-term progradational history by showing the positions of the youngest shelf margins of the Abo, Goat Seep, and Capitan. Modified after Garber et al. (1989).

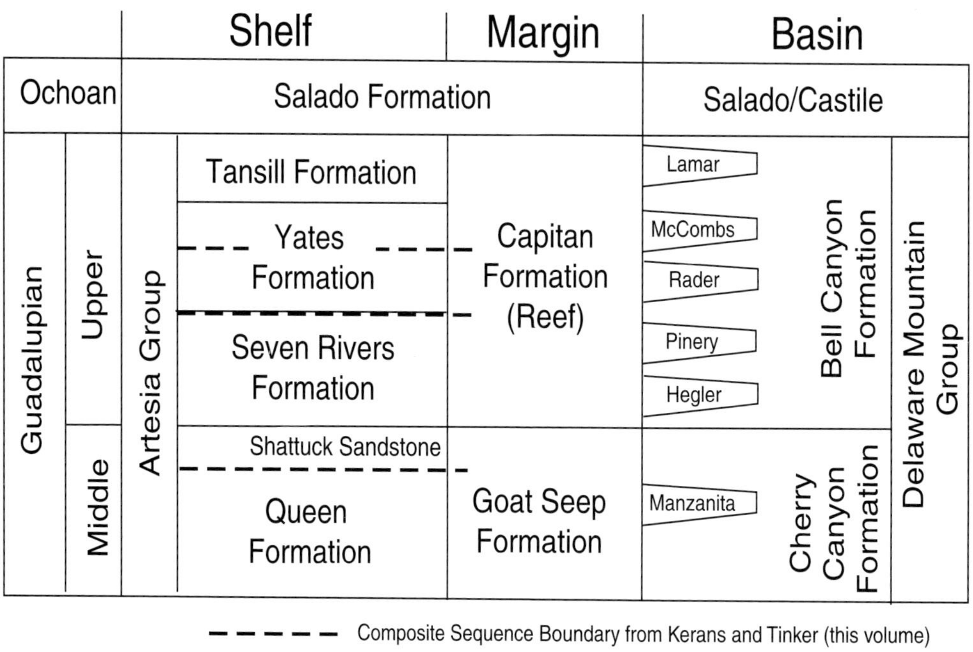

FIG. 2.—Stratigraphic nomenclature for Capitan depositional system and earlier Goat Seep margin. The Seven Rivers, Yates, and Tansill formations on the shelf correlate with the Capitan Formation of the reef margin and slope, and with basin fill of the Bell Canyon Formation. Five limestone members of the Bell Canyon Formation are regionally recognized at the basin edge. Composite sequence boundaries of C. Kerans and S. W. Tinker (this volume) are also shown.

(1957), Payne (1976), Bozanich (1979), Broadhead (1993b), Borer and Harris (1995), Basham (1996), and many others.

INTERPRETATION OF SUBSURFACE DATA

The foundation of this paper is a reprocessed seismic line that images the Capitan depositional system much better than previously published seismic lines. A better understanding of stratal geometries from this seismic data led to improved correlations between wells with wireline logs. Our wireline log cross-section also includes porosity logs that were not presented in previously published subsurface cross-sections in Garber et al. (1989). Fig. 1 shows the location of the seismic line and wells in cross-sections relative to the outcrops in the Guadalupe Mountains.

Seismic Data

Although seismic data lack the resolution of large-scale canyon outcrops in the Guadalupe Mountains, some large-scale correlations between the shelf and basin can be better understood with the seismic data because the seismic data contain the entire stratigraphic section. In contrast, all outcrop sections have either substantial amounts of the Capitan and depositionally equivalent strata eroded from the top of the section or covered below the outcrop section. The seismic line shown in Figs. 3 and 4 is a 1991, 160-fold Vibroseis high resolution line with an orientation almost perpendicular to depositional strike. Data were acquired from 6–125 Hz, and a time invariant filter of 8–90 Hz was used during processing. Fig. 3 (top) shows an uninterpreted line with little vertical exaggeration; whereas Fig. 3 (bottom) is a lightly annotated version of the same line that is hung stratigraphically from the base of the Salado Formation. Fig. 4 shows the same line in uninterpreted (top) and interpreted (bottom) versions, again flattened on the base of the Salado Formation, but the line has an expanded vertical scale.

The seismic data (Fig. 4) show many stratigraphic characteristics of the Capitan depositional system, including (1) a massive prograding reef/slope, (2) back-reef/shelf interior reflectors that dip and diverge basinward before disappearing into the

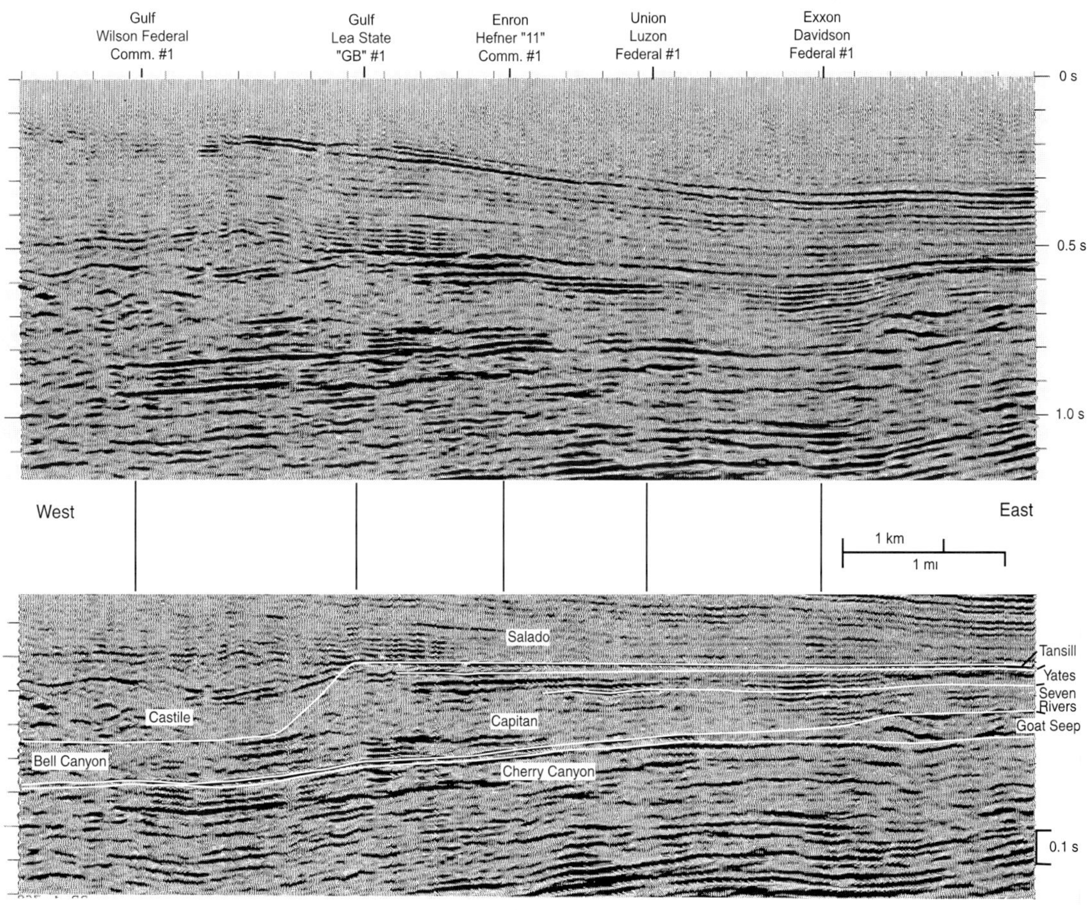

FIG. 3.—A 1991, 160-fold Vibroseis high-resolution seismic line showing shelf-to-basin reflection characteristics for the Capitan depositional system. (Top) Uninterpreted form. (Bottom) Annotated version of the same seismic line with stratigraphic control provided from nearby wells. The line is flattened on the base of the Salado Formation. Location of the line is shown in Fig. 1. Positions of wells are extrapolated onto the seismic line. None of the wells actually occur on the seismic line (see Fig. 1). The vertical scale is two-way-travel time. At these scales, the horizontal scale is similar to the vertical scale (little vertical exaggeration). Seismic data were provided by Seismic Exchange, Inc.; annotation is that of the authors. Data were used with the permission of Chevron North America Production Company.

massive reef, and (3) layered bottomset beds that thicken basinward through the addition of younger beds. The massive Capitan reef/slope seismic facies is bounded on the up-dip (east) end by shelf reflectors and on the down-dip (west) end by the lower slope reflectors. The long-term progradational history of the seismic facies that equates with the reef/slope is quite evident in Fig. 4. Greater than 3 mi (5 km) of basinward progradation is apparent for the reef/slope facies in this area. Back-reef/shelf interior reflectors indicate that shelf strata dip and also prograde basinward.

Comparison of the synthetic seismogram shown in Fig. 5 with the log response shown in Figs. 6 and 7 indicates that reflectors in back-reef/shelf interior strata are caused largely by acoustic impedence contrasts associated with interbedded sandstones and carbonates. In contrast, the lack of strong reflectors in the reef/slope (Figs. 4 and 5) indicates a lack of density or velocity changes related to interbedded sandstones or other lithologic variations of sufficient thickness to produce a contrast in acoustic impedence. The reef/slope generally lacks reflectors indicative of bedding or timelines. Many of the vague seismic reflectors present in the reef/slope interval in Fig. 4 (bottom) are thought to be the result of multiples.

Distinct reflection patterns are present in the lower slope and basin beds (Fig. 4 bottom, 5). In general, those reflections are nearly parallel to the top of the Tansill and are the result of interbedded slope carbonates and onlapping basinal sandstones. The number of slope to basin reflectors of Capitan age increase in a basinward direction with reflectors being added to the upper part of the section. Figs. 6 and 7 show that this situation apparently is due to the addition of younger basin sandstone and carbonate beds in more basinward locations during progradation.

Well Log Cross-section

Wireline logs, principally gamma ray, acoustic, and neutron porosity, indicate features of lithologic and stratigraphic importance to the Capitan depositional system (Fig. 6). Three main lithologies (carbonates, sandstones/siltstones, and evaporites) are present locally in the subsurface and can be identified from modern wireline logs; no cuttings or core samples were examined from wells shown in Fig. 6. Permian sandstones in the Delaware basin are generally feldspathic, and hence give off

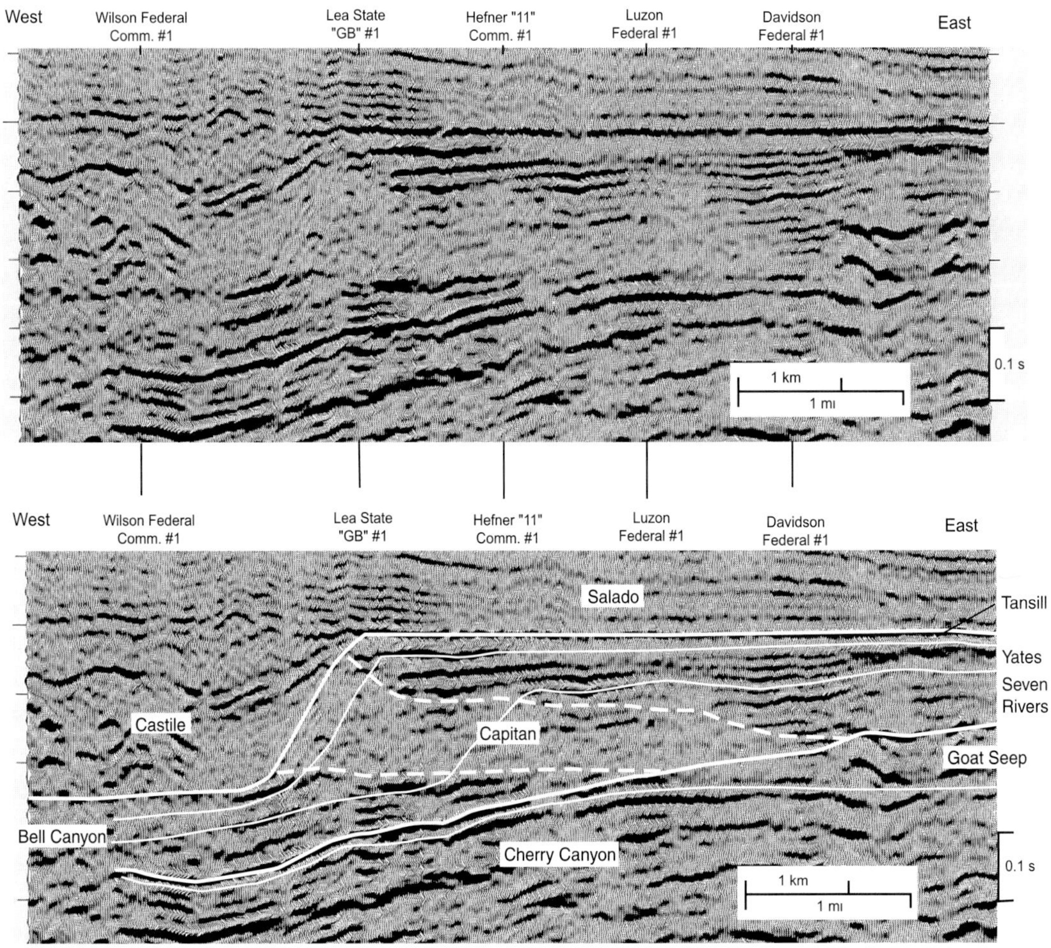

FIG. 4.—Uninterpreted (top) and interpreted (bottom) seismic line of the Capitan depositional system with vertical scale enlarged from Fig. 3. See Fig. 6 for details of well data. Fig. 1 shows location of seismic and well data. Note that positions of wells are extrapolated onto the seismic lines, and none actually occur on the line. The line is flattened on the base of the Salado Formation. The vertical scale is two-way-travel time. The vertical exaggeration is approximately 2:1, although it varies with the velocity of the strata. Seismic data were provided by Seismic Exchange, Inc.; interpretation is that of the authors. Data were used with the permission of Chevron North America Production Company.

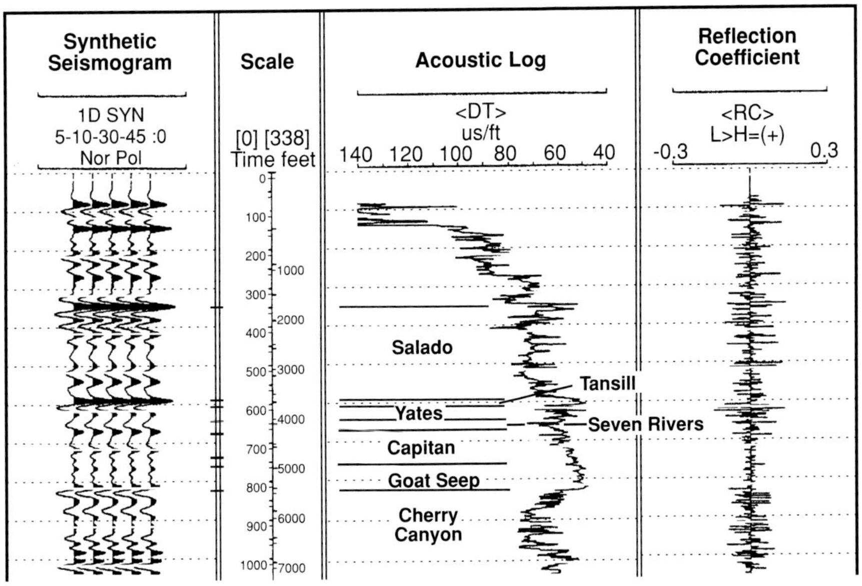

FIG. 5.—Synthetic seismogram for Humble Oil and Refining (Exxon) Davidson-Federal No. 1 well showing the relationship between acoustic log, seismic reflections, and stratigraphy of the Capitan depositional system.

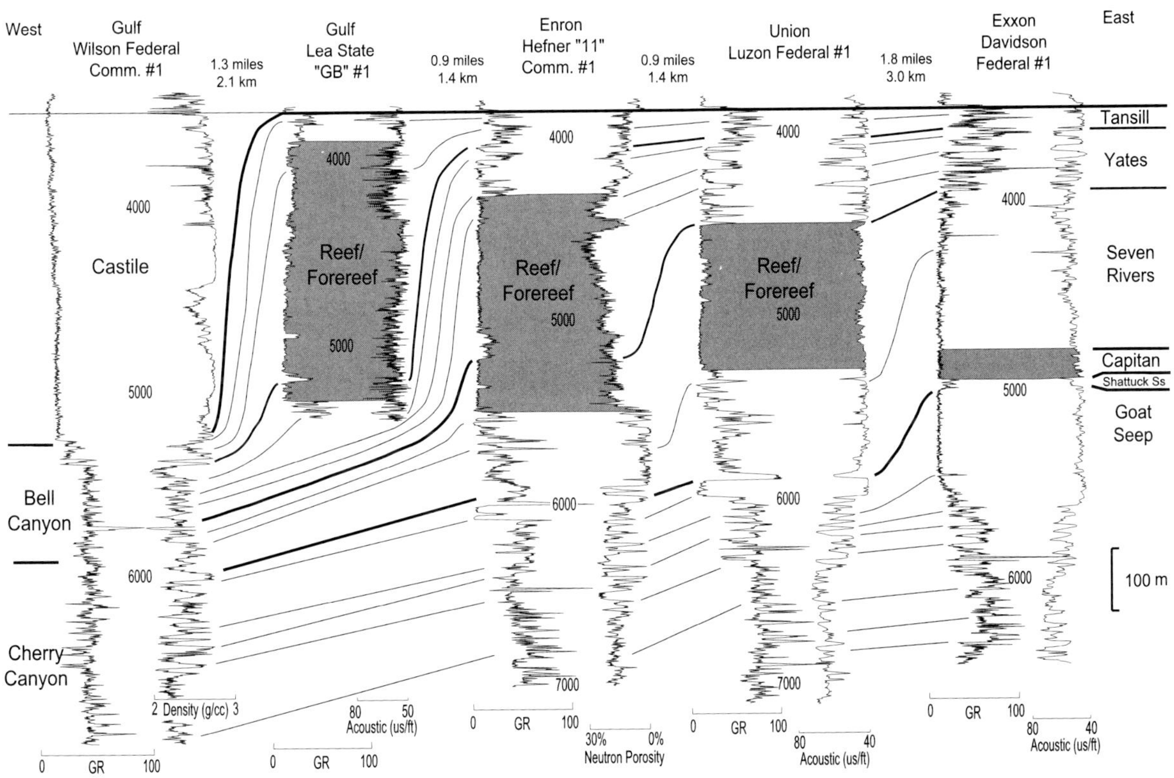

FIG. 6.—Wireline log cross-section using wells located near the seismic line shown in Figs. 3 and 4. Stratigraphy and shelf-to-basin correlations for the Capitan depositional system are shown. Section is approximately dip-oriented. Correlations follow seismic geometries as well as log patterns. Section is stratigraphic; datum is top Tansill Formation except for basin well where top Castile is used. Well locations are shown in Fig. 1.

substantial gamma-ray radiation. In contrast, strata with low gamma-ray radiation are generally carbonates or evaporites. Carbonates are mainly limestone and dolomite, and evaporites include anhydrite and halite. On wireline logs, distinctly higher density and acoustic velocity distinguish anhydrite from carbonate and halite, whereas distinctly lower density and acoustic velocity distinguish halite from carbonate. Some carbonates have substantial gamma-ray response due to uranium enrichment; these radioactive carbonates are generally not porous due to pervasive cementation (Garber et al., 1990) and hence can be differentiated from sandstones by their higher density and/or acoustic velocity.

The steepness of the Capitan shelf margin and great difference in paleodepth between shelf and basin complicate well-to-

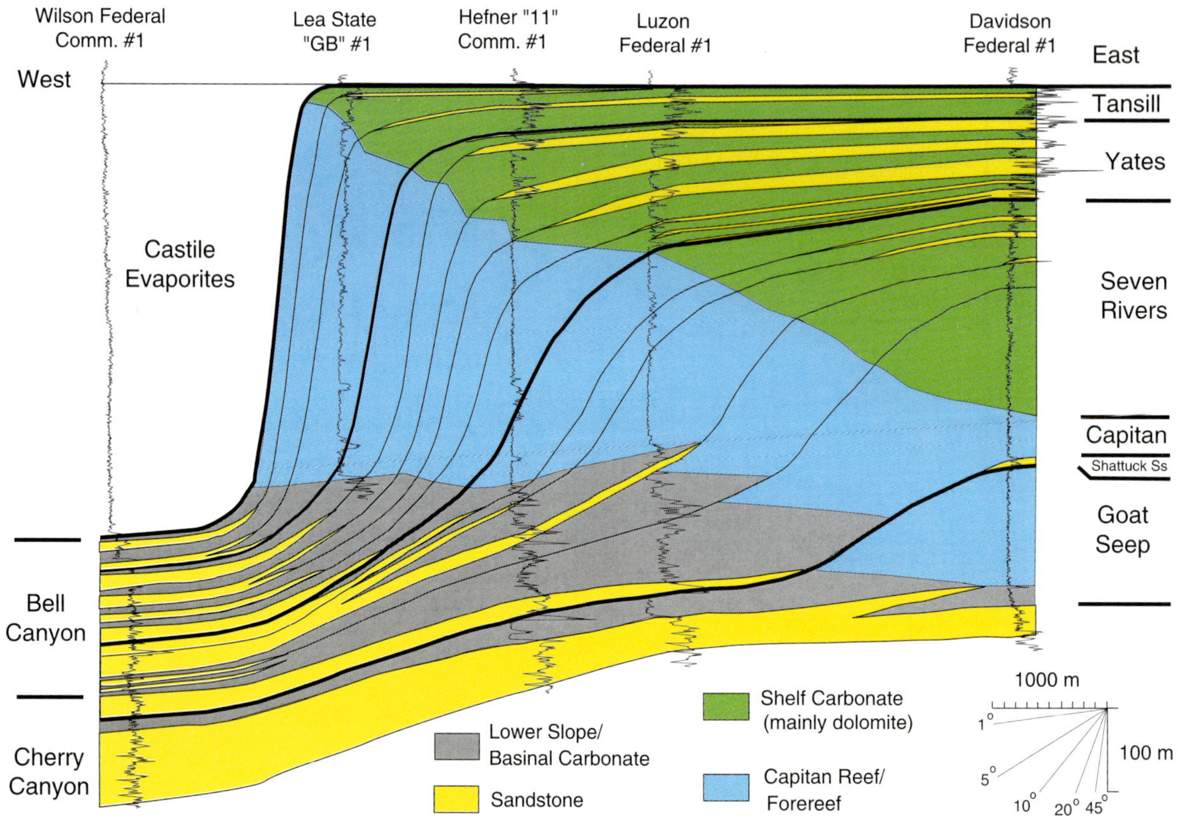

FIG. 7.—Stratigraphic cross-section of the Capitan depositional system showing gamma-ray logs, subsurface stratigraphy, lithologies and depositional environments interpreted for wells in Fig. 6.

well correlations. A wireline log cross-section of wells in the vicinity of the seismic line (Fig. 6) illustrates the stratigraphy of the shelf-to-basin profile of the Capitan depositional system in more detail than possible from the seismic data alone. The correlations shown in Figs. 6 and 7 are guided by the stratal geometries interpreted in Fig. 4 (bottom) and from the dip of the Capitan reef/slope on outcrops. Dip of the Capitan reef/slope is commonly 20° to >45°, as seen in McKittrick Canyon (Bebout and Kerans, 1993; Tinker, 1996, 1998). Equally steep dips are indicated by the seismic correlations shown in Fig. 4 (bottom); dips of the Capitan reef/slope steepen and height of the margin increases from Seven Rivers to Tansill times. Although the well-to-seismic ties for Fig. 4 (bottom) are not exact, they are close enough that the well log correlations between the shelf and basin in Fig. 6 are those suggested by the seismic data.

Seismic reflectors in the Seven Rivers, Yates, and Tansill formations clearly show basinward divergence, and hence back-reef sands should be correlated accordingly (Fig. 7). As a result, the Yates time-stratigraphic unit thickens basinward from 104 m (340 ft) in the Davidson well to 143 m (470 ft) in the Luzon well before merging into the Capitan reef (Figs. 6 and 7). The Tansill displays a similar basinward thickening from 40 m (130 ft) in the Davidson well to 50 m (165 ft) in the Luzon well and 61 m (200 ft) in the Hefner well. If only wireline logs were available, then one might (incorrectly) correlate back-reef sandstones parallel to the top of the Tansill.

Distinct lithologic patterns are apparent in subsurface shelf strata. Sandstones are most abundant in the Yates Formation, where they constitute approximately half the section in shelfward wells such as the Davidson (Fig. 7). Sandstone beds range in thickness from <1 m (3 ft) to approximately 15 m (50 ft), with several sandstones in the Hefner, Luzon, and Davidson wells that are approximately 10 m (30 ft) thick (Fig. 7). In the Yates interval, shelf sandstones decrease in number and thickness basinward and eventually pinch out behind the reef as carbonates increase. Sandstones are less abundant in the Seven Rivers Formation due in part to the basinward location of our wells. Sandstones are rare and thin in the subsurface Tansill Formation.

Although interbedded with sandstones, shelf carbonates contain little sand, as indicated by gamma-ray logs (Fig. 6). Porosity logs (acoustic, density, neutron) indicate that most of the carbonate has low porosity, though some more basinward shelf carbonates have porosity >5%. Shelf interior strata of the Tansill Formation are characterized by substantial gamma-ray radiation, high density, and high acoustic velocity, which suggests that the Tansill in these wells is similar to the dense carbonate with evaporites described by Garber et al. (1989) from the PDB-04 cores. It is not clear where the radioactive elements producing the gamma-ray response occur, but it may be in magnesite, as shown by Garber et al. (1990).

Gamma-ray response is not uniform throughout the reef-dominated section in all wells. Siliciclastic-rich intervals are indicated within reef section equivalent to the basal Yates in the Hefner well, in the reef equivalent to the upper Seven Rivers in the Luzon well, and at the Capitan–Goat Seep boundary, that is, the Shattuck Sandstone, in the Davidson well. These carbonate-clastic alternations may be responsible for some of the low-amplitude, discontinuous reflectors that are evident within the Capitan reef as interpreted in Fig. 4 (bottom). The siliciclastics are thin and laterally discontinuous enough, however, that significant amplitude variations and continuous seismic reflectors generally do not occur. The boundary between the Capitan and Goat Seep reefs does not coincide with a strong seismic reflector (Fig. 5), even though the boundary is marked by the prominent Shattuck Sandstone on outcrop and its likely equivalent between 4900–5000 ft (1500–1525 m) in the Davidson well (Figs. 6 and 7).

Fig. 7 also shows that additional (higher) sandstone beds occur on the slope and basin edge immediately basinward of where shelf sandstones pinch-out against the reef. This observation supports the idea that sandstones in the basin and shelf occur immediately above the same stratigraphic boundary and is consistent with the depositional model shown in Fig. 8.

DISCUSSION OF SUBSURFACE AND OUTCROP DATA

Depositional Model

The stratigraphic and depositional framework interpreted for the subsurface Capitan system (Figs. 4, bottom, and 7) is generally consistent with models proposed from outcrop equivalents (Smith, 1974; Borer and Harris, 1995; Tinker, 1996, 1998; Osleger, 1998; D. A. Osleger and S. W. Tinker, in press, 1999; Kerans and Tinker, this volume). These models follow the concepts of Meissner (1972), Mazzullo et al. (1985), and Fischer and Sarnthein (1988) regarding the relationship between shelf and basin strata. During highstands, the shelf was flooded and carbonates were deposited on the outer 10–20 km (6–12 mi) of the shelf as well as the shelf margin and slope (Fig. 8, top). Down-slope carbonate debris beds accumulated repeatedly (Garber et al., 1989; Brown and Loucks, 1993 a, b). The shelf margin and slope prograded basinward as the shelf aggraded during highstands. During sea-level fall, the shelf was apparently subaerially exposed, allowing sands and silts to be transported across the shelf and into the basin (Fig. 8, middle). Carbonate debris beds generated during lowstand conditions contain a siliciclastic matrix (Garber et al., 1989). The relative lack of sand in the reef and upper slope (Figs. 6 and 7) indicates that they were generally bypass zones during times of low sea level. Though the reef and upper slope remained subtidal during base-level fall, they were too steep to accumulate significant amounts of siliciclastic material. Some sand did fill internal cavities, vugs, and fracture porosity in the reef (Garber et al., 1989; Kirkland et al., 1993), and minor amounts of sandstone/siltstone occur on the upper slope (Mruk and Bebout, 1993). Although depositional models for the basinal sandstones are still being debated (Hill, 1996), several studies, e.g. Borer and Harris (1995), suggest that the time of maximum siliciclastic deposition in the basin was during lowstands. Sand was not trapped on the shelf until the subsequent transgression, when the shelf was reflooded (Fig. 8, bottom) (e.g., Fischer and Sarnthein, 1988; Borer and Harris, 1995; Osleger, 1998).

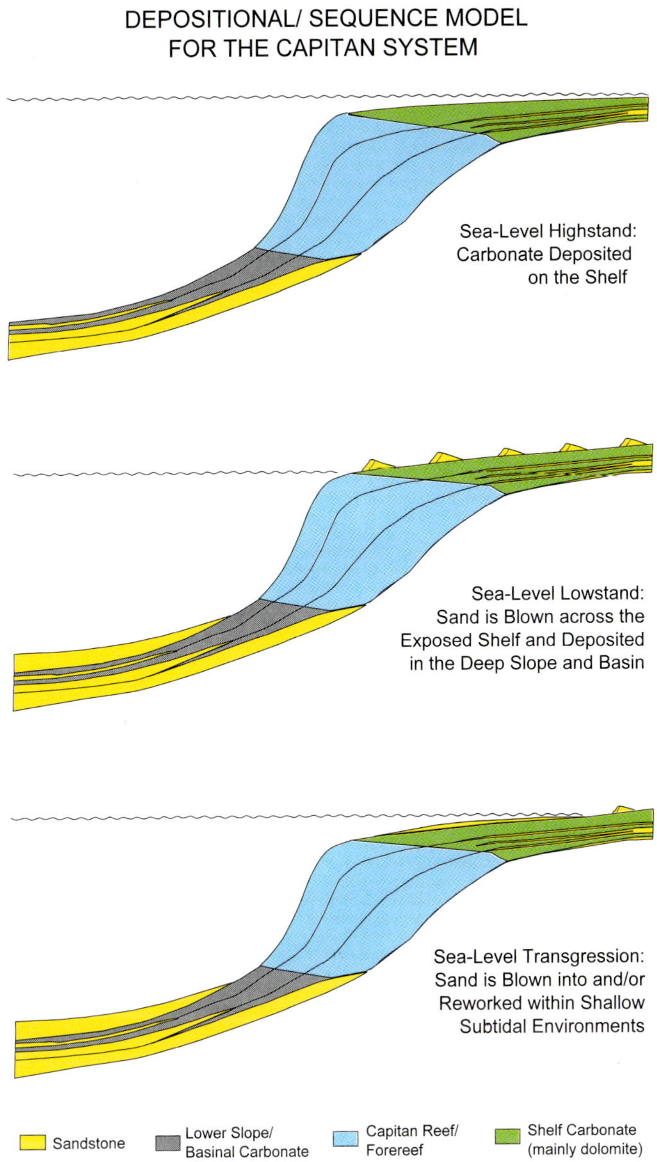

FIG. 8.—Simplified model applicable to the Capitan depositional system. The model is based on concepts of Meissner (1972), Borer and Harris (1995), Rankey and Lehrmann (1996), and Osleger (1998) that stress changes related to fluctuations in sea level and related shelf, shelf margin, and basin stratigraphy.

Shelf cycles produced by deposition portrayed in Fig. 8 are generally asymmetric regressive hemicycles that vary systematically across the shelf and within the context of smaller-scale sequences. Kerans and Harris (1993), Rankey and Lehrmann (1996), Tinker (1996, 1998), Osleger (1998), and D. A. Osleger and S. W. Tinker (in press, 1999) examined the lateral variation and stacking patterns of cycles outcropping in McKittrick and Slaughter canyons. Their work shows that there is substantial variation in the nature of a cycle depending on its position along the depositional profile and within a sequence. A good indication of this lateral variation within a depositional cycle/sequence is shown by the changes in log character within any one of the units correlated in Fig. 6. The lateral variation is also likely responsible for changes in seismic reflectors within the shelf strata, e.g., the number and amplitude of reflectors within the Yates Formation varies laterally between the Davidson and Hefner wells in Fig. 4 (bottom).

Sequence Stratigraphy

Recent analysis of the Permian in the Guadalupe Mountains has emphasized the application of sequence stratigraphic concepts and the analysis of high-frequency (Milankovitch-band) cyclicity to guide facies analysis and refine shelf-to-basin correlation (Sarg and Lehmann, 1986; Sonnenfeld, 1991; Kerans et al., 1992, 1993, 1994; Kerans and Fitchen, 1995). Only recently have the studies of Kerans and Harris (1993), Borer and Harris (1995), Rankey and Lehrmann (1996), Tinker (1996, 1998), Osleger (1998), D. A. Osleger and S. W. Tinker (in press, 1999), and C. Kerans and S. W. Tinker (this volume) begun to place the Capitan depositional system into a high-resolution stratigraphic context. Meissner (1972), Kerans et al. (1992, 1993), Borer and Harris (1995), and C. Kerans and S. W. Tinker (this volume) extrapolated shelf cycles and sequences to the time-equivalent basin deposits of the Bell Canyon Formation. Borer and Harris (1995) suggested that siliciclastics bypassed the shelf and were delivered to the basin during high-frequency lowstands of sea level during deposition of the Yates Formation. The record of high-frequency bypass is also evident in slope equivalents of the youngest portion of the Yates Formation and the Tansill Formation in the McKittrick Canyon outcrops (Brown and Loucks, 1993a, b; Mruk and Bebout, 1993; Brown, 1996).

Shelf-to-basin correlations discussed in Garber et al. (1989) suggest that deposition of 50% to possibly 70% of the Bell Canyon siliciclastics occurred in the basin while a correspondingly thick sequence of very sand-poor Seven Rivers carbonates accumulated on the shelf. In contrast, Borer and Harris (1995) suggest that over 50% of the Bell Canyon Formation equates to the Yates shelf deposits. Figs. 4 (bottom) and 6 suggest that approximately 40% of the Bell Canyon siliciclastic section is equivalent to the Seven Rivers Formation, approximately 40% is equivalent to the Yates Formation, and approximately 20% is equivalent to the Tansill Formation.

Kerans and Tinker (this volume) interpret three composite sequences within the Capitan depositional system, basing their analysis on the large-scale stratigraphic framework developed for the Guadalupe Mountains by Kerans et al. (1992, 1993) and Kerans and Fitchen (1995) and the detailed work within McKittrick Canyon by Tinker (1996, 1998). Their correlations between the shelf and basin edge carbonates include Seven Rivers composite sequence to Manzanita, Hegler, and Pinery members of the Cherry Canyon and Bell Canyon formations, Yates composite sequence to Rader and McCombs members of the Bell Canyon Formation, and Tansill composite sequence to the Lamar member of the Bell Canyon Formation. The correlation of the Manzanita member of the Cherry Canyon to the lowest Seven Rivers Formation remains controversial and is not supported by our subsurface data; however, our correlations in the basin are equivocal and hampered because the basin-margin carbonate members of the Bell Canyon Formation could not be picked in the Wilson well with any certainty. The Tansill/Yates sequence boundary of Tinker (1998) and Kerans and Tinker (this volume) is below the traditional subsurface and outcrop pick for the Yates and Tansill formations (King, 1948; Newell et al., 1953; Hayes, 1964; Garber et al., 1989). Our Tansill-Yates boundary is picked higher than the sequence boundary of Kerans and Tinker (this volume), but is approximately at the traditional subsurface pick for the boundary as shown by Garber et al. (1989).

Progradation of the Capitan

Maximum progradation of the Capitan shelf margin occurred in the north-central portion of the Delaware basin, with substantially less progradation in the northeastern and northwestern portions of the Delaware basin (Silver and Todd, 1969; cross-sections by West Texas and Roswell Geological Societies; Garber et al., 1989). Fig. 1 illustrates the variable amount of progradation by comparing the position of the youngest Goat Seep margin, which is essentially the starting point for Capitan reef growth, with the position of the youngest Capitan reef margin during Tansill time. Figs. 4 (bottom) and 7 indicate approximately 5 km (3 mi) of Capitan reef margin progradation in the northeastern Delaware basin with slightly over half of the progradation occurring below the first Yates sandstone, or during Seven Rivers time. The model of Garber et al. (1989) also indicates maximum progradation during Seven Rivers time. This pattern is similar to that recognized in McKittrick Canyon outcrops from the northwestern portion of the basin. Tinker (1996, 1998) shows approximately 3 km (1.8 mi) of progradation during Seven Rivers deposition in McKittrick Canyon. The reconstruction of eroded Yates and Tansill strata in McKittrick Canyon by Bebout et al. (1993), Borer and Harris (1995), and Tinker (1996, 1998) shows approximately 2 km (1.2 mi) of additional progradation, for a total of approximately 5 km (3 mi) of Capitan progradation. Less progradation occurred during Yates and Tansill time as the margin steepened such that slopes into the basin approached 30° (King, 1948), and water depths increased in the basin to over 600 m (> 2000 ft) in Tansill time. These relationships are also supported by correlations shown in Fig. 4 (bottom).

Recent outcrop studies (e.g., Kerans and Harris, 1993; Tinker, 1996, 1998; Osleger, 1998; D. A. Osleger and S. W. Tinker, in press, 1999) focus on trends in down-dip thickness changes, lateral extent and aspect ratios of facies tracts, progradation:aggradation ratios, and derived offlap angles to better document the details of outer-shelf and margin progradation. The outcrop studies of Tinker (1966, 1998) and the computer modeling of Borer and Harris (1995) show how progradation is expressed by the episodic but progressive seaward step-out of the shelf margin within individual small-scale sequences. This same step-out style of progradation is apparent in Figure 4 (bottom), but because of less seismic resolution, the nature of the progradation shown in Figs. 4 (bottom) and 7 does not have the same detail as outcrop studies.

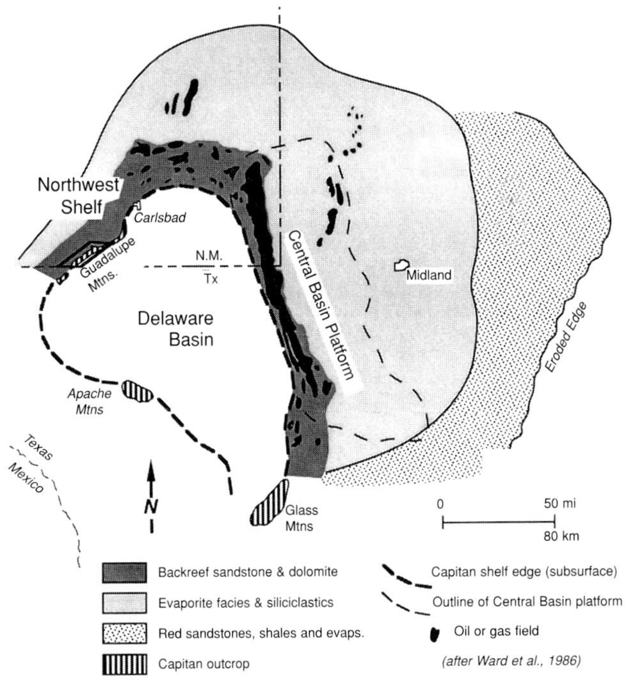

FIG. 9.—Simplified geologic map of depositional environments inferred for Capitan time, which shows the location of hydrocarbon production from shelf reservoirs (modified from Ward et al., 1986). See Table 1 for information on some of the fields.

Basinward Thickening of Shelf Strata

The subsurface correlations of shelf strata (Figs. 6 and 7) show a basinward thickening that is consistent with the more detailed outcrop correlations of Tyrrell (1969), Yurewicz (1977), Hurley (1978, 1989), Borer and Harris (1991a), Longley and Harwood (1992), Kerans and Harris (1993), Tinker (1996, 1998), Osleger (1998), D. A. Osleger and S. W. Tinker (in press, 1999), and C. Kerans and S. W. Tinker (this volume). All of these studies show a basinward thickening and increase in dip of shelf (back-reef) strata into the Capitan reef. King (1948) and Newell et al. (1953) recognized the basinward tilting of outer-shelf deposits, and Pray informally designated these inclined beds as "fall-in" beds (Pray and Esteban, 1977). If the Seven Rivers, Yates, and Tansill formations are defined as time stratigraphic units, then they should also thicken toward the reef. If instead these formations are defined by lithology (that is, Yates Formation is interbedded sand and carbonate), then back-reef formations may have more tabular geometries, as shown by Toomey and Babcock (1983) and Garber et al. (1989). However, those formation boundaries will then cut across timelines.

Thick accumulations of outer-shelf facies accentuate the basinward progradation that occurs within successive small-scale sequences (Tinker, 1996, 1998; Osleger, 1998; D. A. Osleger and S. W. Tinker, in press, 1999). Studies by Yurewicz (1977), Hurley (1978, 1989), and Longley and Harwood (1992) indicated that the basinward dip of the outer-shelf beds could be due to either primary depositional dip or synsedimentary differential subsidence, but not tectonic deformation as thought by Newell et al. (1953) and Kelley (1972). Borer and Harris (1995) interpreted the shift in the locus of sedimentation to be a response to base-level fall along small-scale sequence boundaries, where the abrupt expansion in thickness in each sequence occurs directly above the reef margin of the preceding sequence. Syndepositional differential compaction of slope and basinal strata is interpreted to have generated substantial accommodation space on the prograding Capitan shelf margin (Hunt et al., 1995; Saller, 1996; A. Longley, this volume); however, Osleger (1998) indicates that not all seaward thickening within the Yates sequences can be attributed to compaction. The basinward thickening of shallow shelf and peritidal strata of the Yates and Tansill formations (Figs. 4, 6, and 7) supports basinward tilting and accommodation due to differential compaction of underlying slope and basinal strata.

Evaporites

An important mineralogical difference between outcrops and subsurface equivalents of the Capitan depositional system is the small amount of evaporites in the outcrops. Anhydrite and gypsum are common as interbeds, nodules, and cements in back-reef/shelf interior sandstones and dolomites of the subsurface (Crawford and Dunham, 1982; Ordonez, 1984; Garber et al., 1989; Borer and Harris, 1991a, b; Andreason, 1992), however they are very rare in similar facies in outcrops. In many cases, the anhydrite/gypsum originally present in outcropping strata has simply been dissolved; hence porosity apparent on outcrop may not be representative of the subsurface. In some cases, like that of the Capitan Formation, anhydrite may have, in part, been altered to calcite (Scholle et al., 1992). Evaporites play a major role in the subsurface in trapping of hydrocarbons in shelf reservoirs of the Capitan depositional system.

HYDROCARBON DISTRIBUTION

Porosity in the Capitan depositional system is closely related to facies. The subsurface Capitan reef/slope that rims the Delaware basin has moderate porosity and high permeability. Shelf sandstones and some shelf carbonates (especially grainstones) adjacent to the reef have good porosity and moderate permeability, but porosity and permeability in those beds generally decrease lagoonward, away from the reef margin. Basin carbonate beds are generally not porous, but some basin sandstones in elongate channels have good porosity and moderate permeability.

The upper 122 m (402 ft) of Capitan reef in the PDB-04 well has porosity of 5–25% (average 10%) and permeability of up to 2 darcies (average 256 mD), whereas the lower Capitan is less porous (less than 5%) and permeable (less than 1 mD) (Garber et al., 1989). Wireline logs of Fig. 6 also show that the Capitan reef (especially the upper portion) has substantial porosity (>5%), and hydrologic data (Motts, 1968; Hill, 1996) indicate that the reef has excellent permeability regionally. Although porous and permeable, hydrocarbons do not occur in the Capitan reef because the reef does not have structural or stratigraphic closure. Hydrocarbons migrating out of the Delaware basin apparently moved through the Capitan reef/slope and into permeable shelf deposits up-dip from the reef (Ward et al., 1986). The Capitan Formation is a high permeability, fresh water aquifer around the margins of the basin.

Hydrocarbon reservoirs are present in shelf and basin equivalents to the Capitan Formation (Galloway et al., 1983; Ward et al., 1986; Broadhead, 1993a, b). Hydrocarbon production from shelf portions of the Capitan depositional system generally occurs in stratigraphic and combination stratigraphic-structural traps (Fig. 9; Galloway et al., 1983; Ward et al., 1986; Broadhead, 1993a). Individual siliciclastic reservoir zones show complex interfingering with carbonates in a down-dip direction and evaporites in an up-dip direction (Borer and Harris, 1991a). In addition to these stratigraphic traps caused by facies changes and evaporite cementation, low-relief anticlines caused by compaction and draping over buried structures serve as structural traps (Ward et al., 1986).

Hydrocarbon production on the shelf is primarily from sandstone beds of the Yates and Seven Rivers formations, with minor production from dolomites (Table 1; Galloway et al., 1983; Ward et al., 1986; Borer and Harris, 1991a, b). The most widespread porosity and hydrocarbon reservoirs occur in relatively well-sorted sandstones. Reservoir-grade sandstones have porosities of 15% to 30% and permeabilities of 10 mD to 100 mD (Borer and Harris, 1991a, b). Some porosity also occurs in carbonate beds, especially grainstones in the upper part of beds near the reef (Ordonez, 1984). Porosity in carbonates deteriorates toward the platform interior. The evaporites always have low porosity and permeability.

Many fields along the northern margin of the Delaware basin produce oil from either the Seven Rivers or Yates formation, but not both. In contrast, fields along the western edge of the Central Basin platform produce from stacked reservoirs in both the Seven Rivers and Yates formations (Fig. 9; Ward et al., 1986). The Queen, Seven Rivers, and Yates formations produce with only minor breaks for approximately 145 km along strike along the western edge of the Central Basin platform (Ward et al., 1986). The official field boundaries are generally industrial and not geological. Five fields along this trend have produced over 100 million barrels (Mbbl) of oil, with the North Ward-Estes field (Ward and Winkler Counties, Texas) being the largest, with cumulative production exceeding 350 million barrels (Mbbl) of oil (Table 1). The Tansill Formation produces erratically along the Northwestern shelf and the western edge of the Central Basin platform from small fields in back-reef dolomites (Ordonez, 1984).

A number of small oil fields occur in basin sandstones that were deposited during Capitan time (Galloway et al., 1983; Ward et al., 1986; Williamson, 1977; Broadhead, 1993b). Cumulative production from these fields is generally small, less than 30 million barrels of oil (Table 2). The fields tend to be very elongate (1.5–19 km long by <1–6 km wide; 0.9–11 mi long by <0.6–3.6 mi wide), apparently reflecting accumulation of reservoir sands in deep-water channels (Bozanich, 1979;

TABLE 1.—SOME HYDROCARBON FIELDS PRODUCING FROM CAPITAN SHELF EQUIVALENTS

Name	County-State	Discovery Date	Depth (ft)	Type	Cum. Prod. (million barrels or billion cubic feet)*	Reservoir (Lithology)	Trap Type
Eumont	Lea, NM		5000	Oil	69.100	Yates-Seven Rivers (Ss-St)	Strat (evaporites updip)
Eumont	Lea, NM		4000	Gas	1,451.865	Yates-Seven Rivers-Queen (Ss-St)	Strat (evaporites updip)
Eunice	Lea, NM	1930	5000	Oil	28.409	Seven Rivers-Queen (Ss-St)	Strat (evaporites updip)
GMK S.	Gaines, TX	1957	5598	Gas	15.806	Yates (Ss-St)	Strat
Halley	Winkler, TX	1939	3150	Oil	17.695	Yates-Seven Rivers (Ss-St)	Strat (evaporites updip)
Hendrick	Winkler, TX	1926	3100	Oil	257.685	Yates-Seven Rivers (Ss-St, Dol)	Strat (evaporites updip)
Homann	Gaines, TX	1977	5328	Gas	9.320	Yates (Ss-St)	Strat
Jalmat	Lea, NM		5000	Oil	64.264	Tansill-Yates-Seven Rivers (Ss-St)	Strat (evaporites updip)
Jalmat	Lea, NM		4000	Gas	1,640.896	Tansill-Yates-Seven Rivers (Ss-St)	Strat (evaporites updip)
Kermit	Winkler, TX	1928	2800	Oil	105.576	Yates-Seven Rivers (Ss-St)	Strat (evaporites updip)
Rhodes	Lea, NM	1927	5000	Oil	11.116	Yates-Seven Rivers (Ss-St)	Strat (evaporites updip)
Rhodes	Lea, NM	1927	5000	Gas	197.237	Yates-Seven Rivers (Ss-St)	Strat (evaporites updip)
Scharbrough	Lea, NM	1965	3200	Oil	16.236	Yates-Seven Rivers (Ss-St)	Strat (evaporites updip)
Scharbrough	Winkler, TX**	1927	3200	Oil	27.062	Yates-Seven Rivers (Ss-St)	Strat (evaporites updip)
Seminole	Gaines, TX	1936	5032	Gas	22.085	Yates (Ss-St)	Strat
Shafter Lake	Andrews, TX	1952	3054	Gas	25.067	Yates (Ss-St)	Strat
Ward-Estes N.	Ward/Winkler, TX	1929	3000	Oil	347.958	Yates-Seven Rivers-Queen (Ss-St)	Strat (evaporites updip)
Ward S.	Ward, TX	1938	2700	Oil	101.434	Yates-Seven Rivers (Ss-St)	Strat (evaporites updip)
Wilson	Lea, NM	1928	5000	Oil	9.045	Yates-Seven Rivers (Ss-St)	Strat (evaporites updip)
Shugart	Eddy, NM	1937	2656	Oil	20.179	Yates-Seven Rivers-Queen	Strat

*As of 1984

**Eaves and Leck fields also included

TABLE 2.—SOME HYDROCARBON FIELDS PRODUCING FROM CAPITAN BASINAL EQUIVALENTS

Name	County-State	Discovery Date	Depth (ft)	Type	Cum. Prod. (million barrels or billion cubic feet)*	Reservoir (Lithology)	Trap Type
El Mar	Loving, TX	1959	4532	Oil	17.374	Bell Canyon (Ss-St)	Strat
Grice	Loving, TX	1956	4510	Oil	8.433	Bell Canyon (Ss-St)	Strat
Poquito	Ward, TX			Gas	19.392	Bell Canyon (Ss-St)	Struc
Two Freds	Reeves, Ward, and Loving, TX	1957	4895	Oil	11.290	Bell Canyon (Ss)	Strat
Wheat	Loving, TX	1925	4300	Oil	21.218	Bell Canyon (Ss)	Strat
El Mar	Loving, TX	1959	4532	Oil	17.374	Bell Canyon (Ss-St)	Strat
Geraldine	Culberson & Reeves, TX	1982	3454	Oil	25.043	Bell Canyon (Ss-St)	Strat
Mason N.	Loving, TX	1952	4055	Oil	6.244	Bell Canyon (Ss-St)	Strat
Paduca	Lea, NM	1961	4000	Oil	12.846	Bell Canyon (Ss-St)	Strat
Tunstill	Reeves, TX	1947	3270	Oil	10.615	Bell Canyon (Ss-St)	Strat

*As of 1984

Williamson, 1977; Bashman, 1996). Average porosity and permeability in three Bell Canyon fields were estimated at 24–25% and 10–80 mD, respectively, by Payne (1976). Basin carbonates that are interbedded with sandstones in the Bell Canyon Formation are generally not porous.

CONCLUSIONS

Improved seismic imaging and integrated well data illustrate the similarity of the subsurface Capitan depositional system to outcrops in the Guadalupe Mountains. Shelf strata thicken basinward approaching the reef and shelf sandstones pinchout before they reach the reef. Massive reef/slope facies prograde basinward with complex step-out patterns. Slope carbonates thin and basinal siliciclastics thicken toward the basin. Although the Capitan reef has good reservoir properties, it is not in a stratigraphic and/or structural position that allows for the trapping of hydrocarbons. Instead, large oil fields occur in sandstones and a few carbonate beds in stratigraphically equivalent back-reef (shelf) strata that pinch out up-dip into impermeable sandstones/siltstones, carbonates, and evaporites of the platform interior. Smaller oil fields also occur in deep-water sandstones in the basin.

ACKNOWLEDGMENTS

We thank Jim Medlin of Chevron for help in acquiring and displaying seismic data and Glenn Priess of Chevron for help in acquiring log data. Scott Tinker provided comments that were helpful for comparisons with outcrop data. Reviews by Sal Mazzullo, Gene Rankey, and Chris Kendall helped us improve the manuscript.

REFERENCES

ANDREASON, M. W., 1992, Coastal siliciclastic sabkhas and related evaporative environments of the Permian Yates formation, North Ward-Estes field, Ward County, Texas: American Association of Petroleum Geologists Bulletin, v. 76, p. 1735-1759.

BASHMAN, W. L., 1996, Delaware Mountain group sandstone channel orientations: implications for sediment source and deposition, *in* DeMis, W. D., and Cole, A. G., eds., The Brushy Canyon Play in Outcrop and Subsurface: Concepts and Examples: Midland, Society of Economic Paleontologists and Mineralogists, Permian Basin Section, Publication 96-38, p. 91-102.

BEBOUT, D. G., and KERANS, C., eds., 1993, Guide to the Permian Reef Geology Trail, McKittrick Canyon, Guadalupe Mountains National Park, West Texas, Guidebook 26: Austin, Bureau of Economic Geology, University of Texas, 48 p.

BEBOUT, D. G., KERANS, C., and HARRIS, P. M., 1993, Introduction, *in* Bebout, D. G., and Kerans, C., eds., Guide to the Permian Reef Geology Trail, McKittrick Canyon, Guadalupe Mountains National Park, West Texas, Guidebook 26: Austin, Bureau of Economic Geology, University of Texas, p. 1-4.

BORER, J. M., and HARRIS, P. M., 1991a, Lithofacies and cyclicity of the Yates formation, Permian Basin—implications for reservoir heterogeneity: American Association of Petroleum Geologists Bulletin, v. 75, p. 726-779.

BORER, J. M., and HARRIS, P. M., 1991b, Depositional facies and model for mixed siliciclastics and carbonates of the Yates formation, Permian basin, *in* Lomando, A. J., and Harris, P. M., eds., Mixed carbonate—siliciclastic sequences: Tulsa, Society of Economic Paleontologists and Mineralogists Core Workshop 15, p. 1-133.

BORER, J. M., and HARRIS, P. M., 1995, Computer simulation of the Yates formation (Permian, Delaware basin) - sequence stratigraphy and shelf-to-basin correlation implications, *in* Martin, R. L., ed., In Search of New Permian Basin Oil and Gas Fields: Using Today's Technologies and Tomorrow's Ideas for Exploration, Development, and 3D Seismic in a Mature Basin: Midland, West Texas Geological Society Symposium, Publication 95-98, p. 111-132.

BOZANICH, R. G., 1979, The Bell Canyon and Cherry Canyon formations, eastern Delaware basin, Texas—lithology, environments, and mechanisms of deposition, *in* Sullivan, N. M., ed., Guadalupian Delaware Mountain Group: Midland, Society of Economic Paleontologists and Mineralogists, Permian Basin Section, Publication 79-18, p. 121-141.

BROADHEAD, R. F., 1993a, PB-1, Upper Guadalupian platform, *in* Atlas of major Rocky Mountain gas reservoirs: New Mexico Bureau Mines and Mineral Resources, p. 139-140.

BROADHEAD, R. F., 1993b, PB-2, Delaware Mountain basinal sandstones, *in* Atlas of Major Rocky Mountain Gas Reservoirs: New Mexico Bureau Mines and Mineral Resources, p. 141.

BROWN, A. A., 1996, Position of the *Polydiexodina* last-occurrence datum *in* Guadalupian strata: Revised correlations at McKittrick Canyon, *in* DeMis, W. D., and Cole, A. G., eds., The Brushy Canyon Play in Outcrop and Subsurface: Concepts and Examples: Midland, Society of Economic Paleontologists and Mineralogists, Permian Basin Section, Publication 96-38, p. 75-83.

BROWN, A. A., and LOUCKS, R. G., 1993a, Influence of sediment type and depositional processes on stratal patterns in the Permian basin-margin Lamar limestone, McKittrick Canyon, Texas, *in* Loucks, R. G., and Sarg, J. F., eds.,

Carbonate Sequence Stratigraphy, Recent Developments and Applications: Tulsa, American Association of Petroleum Geologists Memoir 57, p. 133-156.

BROWN, A. A., and LOUCKS, R. G., 1993b, Toe of slope, *in* Bebout, D. G., and Kerans, C., eds., Guide to the Permian Reef Geology Trail, McKittrick Canyon, Guadalupe Mountains National Park, West Texas, Guidebook 26: Austin, Bureau of Economic Geology, University of Texas, p. 5-13.

CRAWFORD, G. A., and DUNHAM, J. B., 1982, Evaporite sedimentation in the Permian Yates Formation, Central Basin platform, Andrews County, west Texas, *in* Handford, C. R., Loucks, R. G., and Davies, G. R., eds., Depositional and diagenetic spectra of evaporites—a core workshop: Tulsa, Society of Economic Paleontologists and Mineralogists Core Workshop 3, p. 238-275.

FISCHER, A. G., and SARNTHEIN, M., 1988, Airborne silts and dune-derived sands from the Permian of the Delaware basin: Journal of Sedimentary Petrology, v. 58, p. 637-643.

GALLOWAY, W. E., EWING, T. E., GARRETT, C. M., TYLER, N., and BEBOUT, D. G., eds., 1983, Atlas of Major Texas Oil Reservoirs: Austin, Bureau of Economic Geology, University of Texas, 139 p.

GARBER, R. A., GROVER, G. A., and HARRIS, P. M., 1989, Geology of the Capitan shelf margin—subsurface data from the northern Delaware basin, *in* Harris, P. M., and Grover, G. A. eds., Subsurface and outcrop examination of the Capitan shelf margin—a core workshop: Tulsa, Society of Economic Paleontologists and Mineralogists Core Workshop 13, p. 3-269.

GARBER, R. A., HARRIS, P. M., and BORER, J. M., 1990, Occurrence and significance of magnesite in upper Permian (Guadalupian) Tansill and Yates formations, Delaware basin, New Mexico: American Association of Petroleum Geologists Bulletin, v. 74, p. 119-134.

HAYES, P. T., 1964, Geology of the Guadalupe Mountains, New Mexico: Washington, D.C., U. S. Geological Survey Professional Paper 446, 69 p.

HILL, C. A., 1996, Geology of the Delaware Basin, Guadalupe, Apache, and Glass Mountains, New Mexico and West Texas: Midland, Society of Economic Paleontologists and Mineralogists, Permian Basin Section, Publication 96-39, 480 p.

HULL, J. P. D., Jr., 1957, Petrogenesis of Permian Delaware Mountain sandstone, Texas and New Mexico: American Association of Petroleum Geologists Bulletin, v. 41, p. 278-307.

HUNT, D., ALLSOP, T., and SWARBRICK, R., 1995, Compaction as a primary control on the architecture and development of depositional sequences: Conceptual framework, applications and implications, *in* Howell, J. A., and Aitken, J. F., eds., High resolution sequence stratigraphy: Innovations and applications: London, Geological Society of London Special Publication 104, p. 321-345.

HURLEY, N. F., 1978, Facies mosaic of the lower Seven Rivers formation (Permian, Guadalupian), Guadalupe Mountains, New Mexico: Unpublished Master's Thesis, University of Wisconsin, Madison, 198 p.

HURLEY, N. F., 1989, Facies mosaic of the lower Seven Rivers formation, McKittrick Canyon, New Mexico, *in* Harris, P. M. and Grover, G. A., eds., Subsurface and outcrop examination of the Capitan shelf margin, northern Delaware basin: Tulsa, Society of Economic Paleontologists and Mineralogists Core Workshop 13, p. 325-346.

KELLEY, V. C., 1972, Geometry and correlations along the Permian Capitan escarpment, New Mexico and Texas: American Association of Petroleum Geologists Bulletin, v. 56, p. 2192-2211.

KERANS, C., and FITCHEN, W. M., 1995, Sequence Hierarchy and Facies Architecture of a Carbonate-ramp System: San Andres Formation of Algerita Escarpment and Western Guadalupe Mountains, West Texas and New Mexico: Austin, Bureau of Economic Geology Report of Investigations 235, The University of Texas, 86p.

KERANS, C., FITCHEN, W. M., GARDNER, M. H., SONNENFELD, M. D., TINKER, S. W., and WARDLAW, B. R., 1992, Styles of sequence development within uppermost Leonardian through Guadalupian strata of the Guadalupe Mountains, Texas and New Mexico, *in* Mruk, D. H., and Curran, B. C., eds., Permian Basin Exploration and Production Strategies: Applications of Sequence Stratigraphic and Reservoir Characterization Concepts: Midland, West Texas Geological Society Publication 92-91, p. 1-6.

KERANS, C., FITCHEN, W. M., GARDNER, M. H., and WARDLAW, B. R., 1993, A contribution to the evolving stratigraphic framework of Middle Permian strata of the Delaware Basin, Texas and New Mexico, *in* Love, D. W., Hawley, J. W., Kues, B. S., Adams, J. W., Austin, G. S., and Barker, J. M., eds., Carlsbad region, New Mexico and West Texas: New Mexico Geological Society, Forty-fourth Annual Field Conference, p. 175-184.

KERANS, C., and HARRIS, P. M., 1993, Outer shelf and shelf crest, *in* Bebout, D. G., and Kerans, C., eds., Guide to the Permian Reef Geology Trail, McKittrick Canyon, Guadalupe Mountains National Park, West Texas, Guidebook 26: Austin, Bureau of Economic Geology, University of Texas, p. 32-43.

KING, P. B., 1948, Geology of the Southern Guadalupe Mountains, Texas: Washington, D.C., U.S. Geological Survey Professional Paper 215, 183 p.

KIRKLAND, B. L., LONGACRE, S. A., and STOUDT, E. L., 1993, Reef, *in* Bebout, D. G., and Kerans, C., eds., Guide to the Permian Reef Geology Trail, McKittrick Canyon, Guadalupe Mountains National Park, West Texas, Guidebook 26: Austin, Bureau of Economic Geology, University of Texas, p. 23-31.

LONGLEY, A., and HARWOOD, G., 1992, Field evidence for differential synsedimentary subsidence of the basin margin, Permian reef complex, Guadalupe Mountains, New Mexico, USA (abs.): Carbonate Platform Margin Conference, France, p. 1.

MAZZULLO, S. J., MAZZULLO, J., and HARRIS, P. M., 1985, Eolian origin of quartzose sheet sands in Permian shelf facies, Guadalupe Mountains, (abs.), *in* Cunningham, B. K., and Hedrick, C. L., eds., Permian Carbonate/clastic Sedimentology, Guadalupe Mountains: Analogs for Shelf and Basin Reservoirs: Midland, Society of Economic Paleontologists and Mineralogists, Permian Basin Section, Publication 85-24, p. 71.

MEISSNER, F. F., 1972, Cyclic sedimentation in middle Permian strata of the Permian basin, *in* Elam, J. G., and Chuber, S., eds., Cyclic Sedimentation in the Permian Basin (2nd Edition): Midland, West Texas Geological Society Publication 72-60, p. 203-232.

MOTTS, W. S., 1968, The control of ground-water occurrence by lithofacies in the Guadalupian reef complex near Carlsbad, New Mexico: Geological Society of America Bulletin, v. 79, p. 283-298.

MRUK, D., and BEBOUT, D. G., 1993, Slope, *in* Bebout, D. G., and Kerans, C., eds., Guide to the Permian Reef Geology Trail, McKittrick Canyon, Guadalupe Mountains National Park, West Texas, Guidebook 26: Austin, Bureau of Economic Geology, University of Texas, p. 14-22.

NEWELL, N. D., RIGBY, J. K., FISCHER, A. G., WHITEMAN, A. J., HICKOX, J. E., and BRADLEY, J. S., 1953, The Permian reef complex of the Guadalupe Mountains region, Texas and New Mexico, a study in paleoecology: San Francisco, W. H. Freeman and Company, 236 p.

ORDONEZ, S. R., 1984, Permian (Guadalupian) shelf deposition and diagenesis: The Tansill Formation of the Cheyenne field, Winkler County, Texas, *in* Moore, G. E., and Wilde, G., eds., Transactions, Southwest Section, American Association of Petrolum Geologists 1984 Convention: Midland, West Texas Geological Society Publication SWS-84-78, p. 61-79.

OSLEGER, D. A., 1998, Sequence architecture and sea-level dynamics of upper Permian shelfal facies, Guadalupe Mountains, southern New Mexico: Journal of Sedimentary Research, v. 68, p. 327-346.

PAYNE, M. W., 1976, Basinal sandstone facies, Delaware basin, west Texas and southeast New Mexico: American Association of Petroleum Geologists Bulletin, v. 60, p. 517-527.

PRAY, L. C., and ESTEBAN, M., 1977, Upper Guadalupian Facies, Permian Reef Complex, Guadalupe Mountains, New Mexico and West Texas, 1977 Field Conference Roadlog: Midland, Society of Economic Paleontologists and Mineralogists, Permian Basin Section, Publication 77-16, 194 p.

RANKEY, E. C., and LEHRMANN, D. J., 1996, Anatomy and origin of toplap in a mixed carbonate-clastic system, Seven Rivers formation (Permian, Guadalupian), Guadalupe Mountains, New Mexico, USA: Sedimentology, v. 43, p. 807-826.

SALLER, A. H., 1996, Differential compaction and basinward tilting of the prograding Capitan reef complex, Permian, west Texas and southeast New Mexico, USA: Sedimentary Geology, v. 101, p. 21-30.

SARG, J. F., and LEHMANN, P. J., 1986, Lower-middle Guadalupian facies and stratigraphy, San Andres-Grayburg formations, Permian Basin, Guadalupe Mountains, New Mexico, *in* Moore, G. E., and Wilde, G. L., eds., Lower and Middle Guadalupian Facies, Stratigraphy, and Reservoir Geometries, San Andres-Grayburg Formations, Guadalupe Mountains, New Mexico and Texas: Midland, Society of Economic Paleontologists and Mineralogists, Permian Basin Section, Special Publication 86-25, p. 1-36.

SCHOLLE, P. A., ULMER, D. S., and MELIM, L. A., 1992, Late-stage calcites in the Permian Capitan formation and its equivalents, Delaware basin margin, west Texas and New Mexico - evidence for replacement of precursor evaporites: Sedimentology, v. 39, p. 207-234.

SILVER, B. A., and TODD, R. G., 1969, Permian cyclic strata, northern Midland and Delaware basins, West Texas and Southeastern New Mexico: American Association of Petroleum Geologists Bulletin, v. 53, p. 2223-2251.

SMITH, D. B., 1974, Sedimentation of upper Artesia (Guadalupian) cyclic shelf deposits of northern Guadalupe Mountains, New Mexico: American Association of Petroleum Geologists Bulletin, v. 58, p. 1699-1730.

SONNENFELD, M. D., 1991, High-frequency cyclicity within shelf-margin and slope strata of the upper San Andres sequence, Last Chance Canyon, *in* Meador-Roberts, S., Candelaria, M. P., and Moore, G. E., eds., Sequence Stratigraphy, Facies, and Reservoir Geometries of the San Andres, Grayburg, and Queen Formations, Guadalupe Mountains, New Mexico and Texas: Midland, Society of Economic Paleontologists and Mineralogists, Permian Basin Section, Publication 91-32, p. 11-51.

TINKER, S. W., 1996, Reservoir-scale sequence stratigraphy: McKittrick canyon and 3-D subsurface examples, west Texas and New Mexico: Unpublished Ph. D. Dissertation, University of Colorado, Boulder, 245 p.

TOOMEY, D. F., and BABCOCK, J. A., 1983, Precambrian and Paleozoic algal carbonates, West Texas-southern New Mexico: Colorado Schol of Mines Professional Contributions, v. 11, 345 p.

TYRRELL, W. W., Jr., 1969, Criteria useful in interpreting environments of unlike but time-equivalent carbonate units (Tansill-Capitan-Lamar), Capitan reef complex, West Texas and New Mexico, *in* Friedman, G. M., ed., Depositional Environments in Carbonate Rocks: Tulsa, Society of Economic Paleontologists and Mineralogists, Special Publication 14, p. 80-97.

WARD, R. F., KENDALL, C. G., and HARRIS, P. M., 1986, Upper Permian (Guadalupian) facies and their association with hydrocarbons—Permian basin, west Texas and New Mexico: American Association of Petroleum Geologists Bulletin, v. 70, p. 239-262.

WILLIAMSON, C. R., 1977, Deep-sea channels of the Bell Canyon Formation (Guadalupian), Delaware Basin, Texas-New Mexico, *in* Hileman, M. E., and Mazzullo, S. J., eds., Upper Guadalupian facies, Permian Reef Complex, Guadalupe Mountains, New Mexico and west Texas: Midland, Society of Economic Paleontologists and Mineralogists, Permian Basin Section, Publication 77-16, p. 409-431.

YUREWICZ, D. A., 1977, The origin of the massive facies of the lower and middle Capitan limestone (Permian), Guadalupe Mountains, New Mexico and west Texas, *in* Hileman, M. E., and Mazzullo, S. J., eds., Upper Guadalupian Facies, Permian Reef Complex, Guadalupe Mountains, New Mexico and West Texas: Midland, Society of Economic Paleontologists and Mineralogists, Permian Basin Section, Publication 77-16, p. 45-92.

DO METER-SCALE CYCLES EXIST? A STATISTICAL EVALUATION FROM VERTICAL (1-D) AND LATERAL (2-D) PATTERNS IN SHALLOW-MARINE CARBONATES-SILICICLASTICS OF THE "FALL IN" STRATA OF THE CAPITAN REEF, SEVEN RIVERS FORMATION, SLAUGHTER CANYON, NEW MEXICO

DANIEL J. LEHRMANN
University of Wisconsin-Oshkosh, Oshkosh, Wisconsin, 54901
AND
EUGENE C. RANKEY
Exxon Production Research, Houston, Texas, 77252

ABSTRACT: The apparent prevalence of meter-scale lithologic cyclicity in shallow-marine carbonate and mixed carbonate-siliciclastic platform strata has led to the widely accepted paradigm that such meter-scale cycles are the fundamental building blocks of platforms. This view commonly is challenged, however, on the grounds that lithologic cyclicity or "ideal lithologic cycles" are difficult to demonstrate from statistical evaluations (for example, Markov chain analysis) of facies successions. This issue is of major importance in determining the accuracy of high-resolution chronostratigraphic correlations using present sequence stratigraphic techniques.

In this study we have used traditional and modified Markov chain analysis to evaluate the presence of cyclicity in backreef strata, the Seven Rivers Formation, of the Capitan reef. Traditional Markov analysis, similar to that used in previous studies, fails to verify the presence of cyclicity. Through the addition of stratigraphic data on the distribution of significant surfaces (subaerial exposure and hiatal) and the two-dimensional distribution of facies bodies, however, Markov analysis verifies the presence of cyclicity and even allows the statistical identification of potential "ideal cycles" in the system. Thus, meter-scale cyclicity does exist in this system, with the implication that previous studies may have failed to incorporate stratigraphic information necessary for an adequate evaluation of the existence of cyclicity.

INTRODUCTION

Lithologic cycles have been widely suggested as the fundamental building blocks of shallow-marine carbonate and mixed carbonate-siliciclastic platforms (e.g., James, 1984). This concept is based on numerous studies of platforms of all ages in which the stratigraphic architecture has been characterized in the context of meter- to decameter-thick lithologic cycles that may be composed of both asymmetrical (shallowing) and symmetrical (deepening then shallowing) facies motifs (e.g., Read, 1973; Wilson, 1975; Enos and Perkins, 1977; Heckel, 1986; Grotzinger, 1989; Watney et al., 1989; Goldhammer et al., 1990, 1991a, 1991b, 1993; Osleger and Read, 1991; Kerans et al., 1992; Montanez and Osleger, 1993; McLean and Mountjoy, 1994; Barnaby and Ward, 1995; Kerans and Fitchen, 1995; Harris and Sheehan, 1996; Sonnenfeld, 1996; Lehrmann and Goldhammer, 1997).

The prevalence of lithologic cyclicity has been challenged, however, based on analyses of whether "ideal cycles," consistently repeated facies motifs, can be statistically verified (Duff and Walton, 1962; papers in Merriam, 1964; Gingerich, 1969; Schwartzacher, 1969, 1975; Doveton 1971, 1996; Carr, 1982). The results of most of these studies have suggested that an ideal cycle is generally not statistically definable, but illustrate that the ideal cycle provides a useful framework for summarizing observations and for evaluating the controlling processes. In recent years, a resurrection of this concept has led some workers to suggest that because an ideal cycle is not present or definable, then shallow-marine carbonate facies successions are predominantly random (Wilkinson et al., 1996a, 1996b, 1997). If true, this notion would suggest that much of sequence stratigraphy is a subjective exercise, because lithologic cycles (high-frequency cycles, meter-scale cycles, or parasequences), the fundamental building blocks of sequences (cf., Van Wagoner et al., 1990), are artificially defined. The question of whether shallow-marine platform stratigraphy is predominantly ordered and cyclic or random has significant implications for how platform strata are described and characterized, how the driving mechanisms of shallow-marine sedimentation are interpreted, and how accurately stratigraphic patterns can be used to make chronostratigraphic correlations (cf., Lehrmann and Goldhammer, 1997).

One of the drawbacks of the previous statistical studies is that they have been based primarily on Markov chain analysis of facies successions in vertical (one dimensional; 1-D) sections. Fundamental to the concepts of stratigraphic cyclicity in general and sequence stratigraphy in particular are (1) the nature and distribution of stratigraphic surfaces (e.g., hiatus, subaerial exposure, or erosion surfaces) and (2) lateral, two- or three-dimensional, facies distribution (cf., Goldhammer et al., 1990; Van Wagoner et al., 1990). Thus, an obvious course for improvement of the statistical evaluation of cyclicity would be to incorporate these parameters in the analysis.

The purpose of this study is to critically evaluate the existence of cyclicity and the presence of "ideal cycles" in a succession where detailed vertical (1-D) and lateral (2-D) stratigraphic data are available and tied by physical correlation. Our approach is to begin with a traditional application of embedded Markov chain analysis, similar to that used in previous studies, using only the vertical facies transitions to evaluate the presence of stratigraphic order and cyclicity. We will then modify the analyses by progressively adding 2-D stratigraphic information on the distribution of subaerial exposure surfaces and 2-D facies geometries into the analysis, again with the goal of evaluating cyclicity and how Markov chain analysis responds to the incorporation of 2-D

Geologic Framework of the Capitan Reef, SEPM Special Publication No. 65
 ISBN 1-56576-063-8, p. 51-62

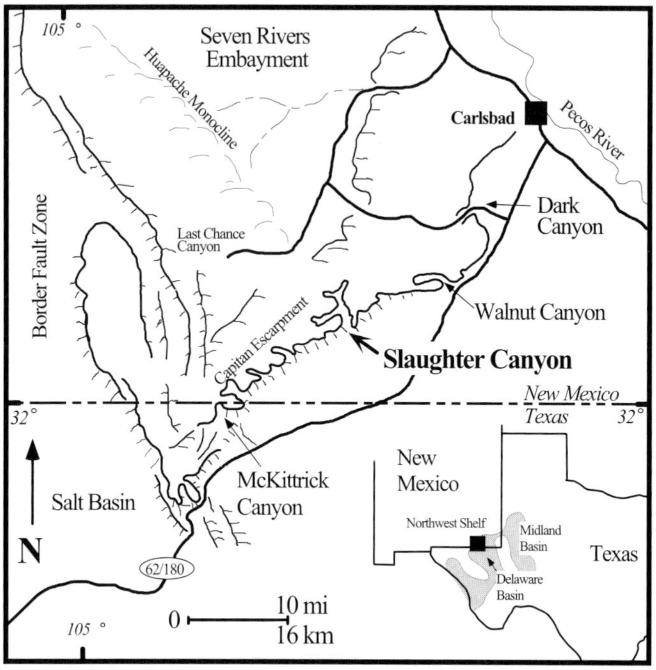

FIG. 1.—Location map, indicating location of Slaughter Canyon along Capitan escarpment, Guadalupe Mountains, New Mexico and Texas, U.S.A.

stratigraphic data. The results of this study illustrate that whereas traditional Markov analysis of data from individual vertical sections suggests the absence of stratigraphic order, the incorporation of 2-D surface and facies information reveals that the system is actually statistically ordered and cyclic.

SETTING

Slaughter Canyon, Guadalupe Mountains New Mexico, exposes a dip-oriented cross-section through the Upper Permian (Guadalupian) reef, back reef and shelf strata of the Permian reef complex in southeast New Mexico (Figs. 1 and 2). The Seven Rivers, Yates, Tansill, and Capitan formations exposed in the canyon contain five depositional sequences, identified on the basis of toplapping and onlapping stratal geometries, which comprise the progradational and aggradational evolution of the Capitan shelf margin (Fig. 3). This study focuses on the upper part of the first sequence, which forms the top of the Seven Rivers Formation (Figs. 3 and 4). The Seven Rivers Formation contains steeply dipping clinoforms that change facies down-dip into the Capitan reef (Figs. 3 and 4). The backreef clinoforms have basinward dips of 1–15° and have been called fall-in beds by Pray (as noted in Hurley, 1989). Details of the stratigraphic architecture, facies distribution, and depositional history have been documented in previous studies (Rankey and Lehrmann, 1996; Lehrmann et al., 1997; Osleger, 1998).

The Seven Rivers Formation in Slaughter Canyon occurs below a pronounced toplap surface (SB-2) and consists of a series of sandstone-bounded carbonate "shingles" or clinoforms, each of which merges updip into the toplap surface and grades downdip into the Capitan Formation (Fig. 3, 4). The Seven Rivers Formation includes eight distinct facies (Table 1). Facies deposited in a peritidal environment include: (1) massive to cryptalgally laminated sandstone/siltstone with evaporite molds, (2) tepee-pisolite packstone, and (3) fenestral and cryptalgal laminite mudstone to packstone. Facies deposited in a subtidal environment include: (4) peloid-skeletal packstone, (5) ooid-skeletal packstone to grainstone, (6) dasyclad-mollusk-

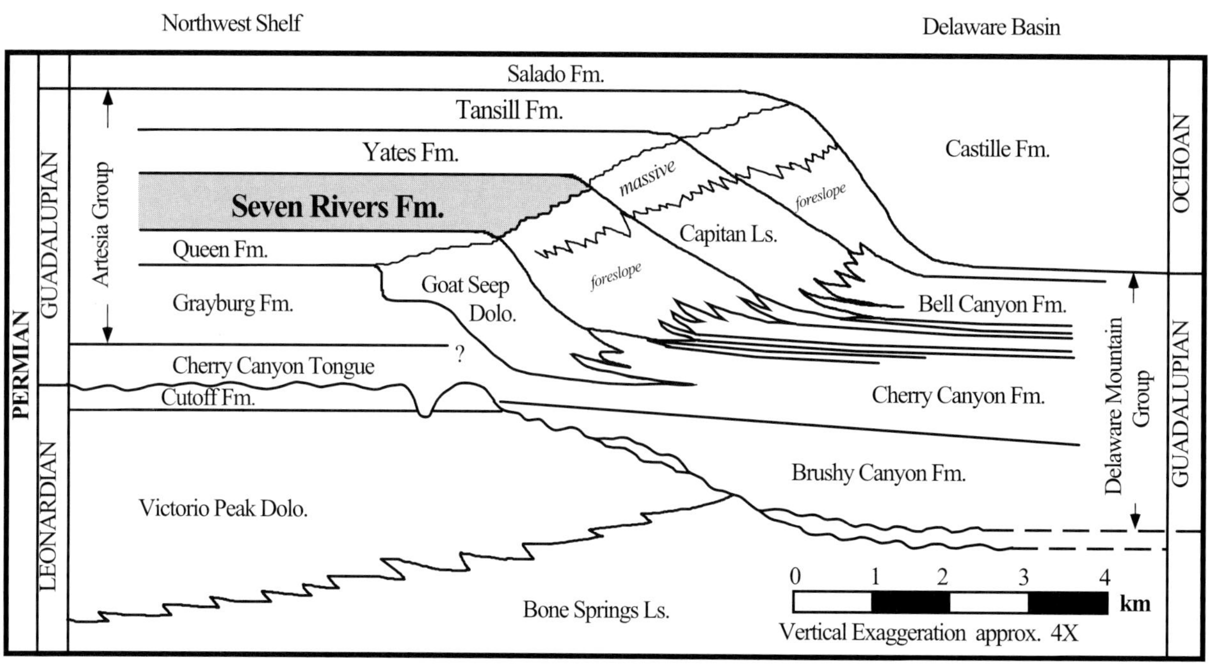

FIG. 2.—Regional stratigraphic framework for Upper Permian strata of the Guadalupe Mountains, west Texas/southeastern New Mexico. The Seven Rivers Formation, the focus of this study, represents shelfal deposits temporally equivalent to the lower Capitan Formation. Note the significant progradation of the Guadalupian shelf edge. Modified from King (1942) and Newell et al. (1953).

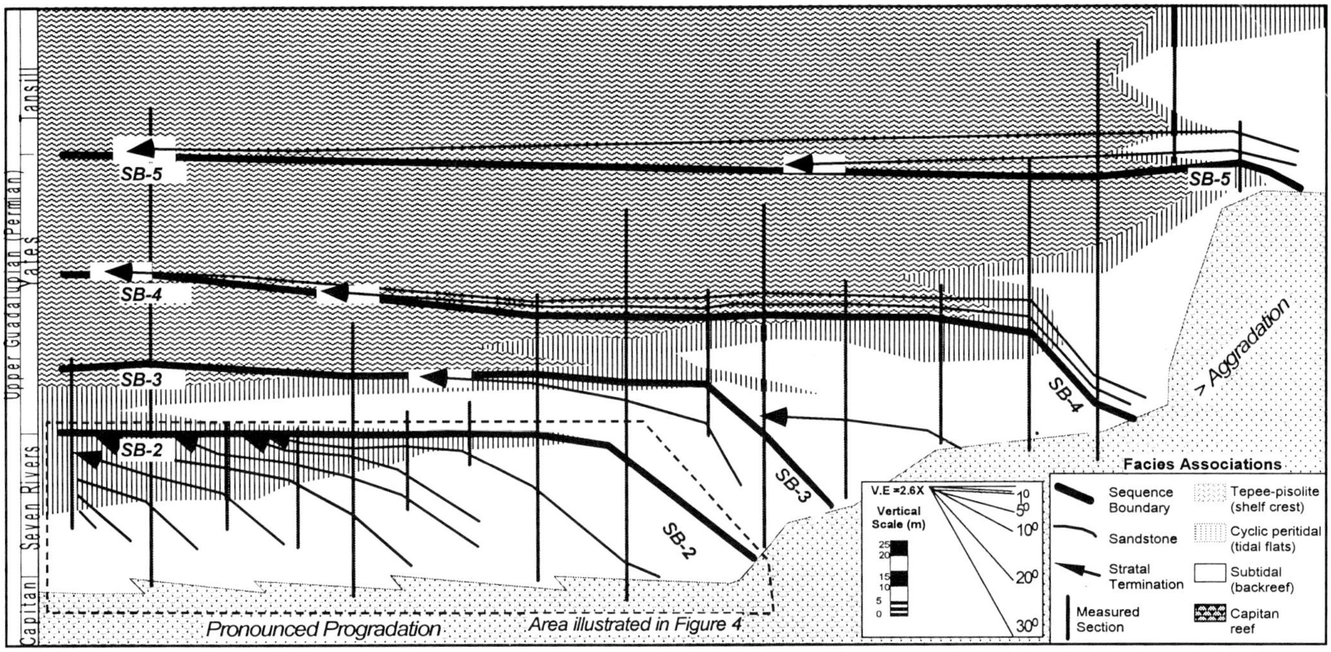

FIG. 3.—Simplified stratigraphic cross-section of the Upper Permian (Guadalupian; Seven Rivers, Yates, Tansill and Capitan formations) exposed on the northeast wall of Slaughter Canyon. Width of the entire cross-section is 2 km. Note depositional sequences (1 through 5) are defined on the basis of toplapping and onlapping stratal geometries. Sequence boundaries are named after the base of the sequence in which they occur. For example, sequence boundary 2 (SB-2) occurs at the base of sequence 2. Each sequence contains a lower retrogradational (landward migrating) and an upper progradational (basinward migrating) facies pattern.

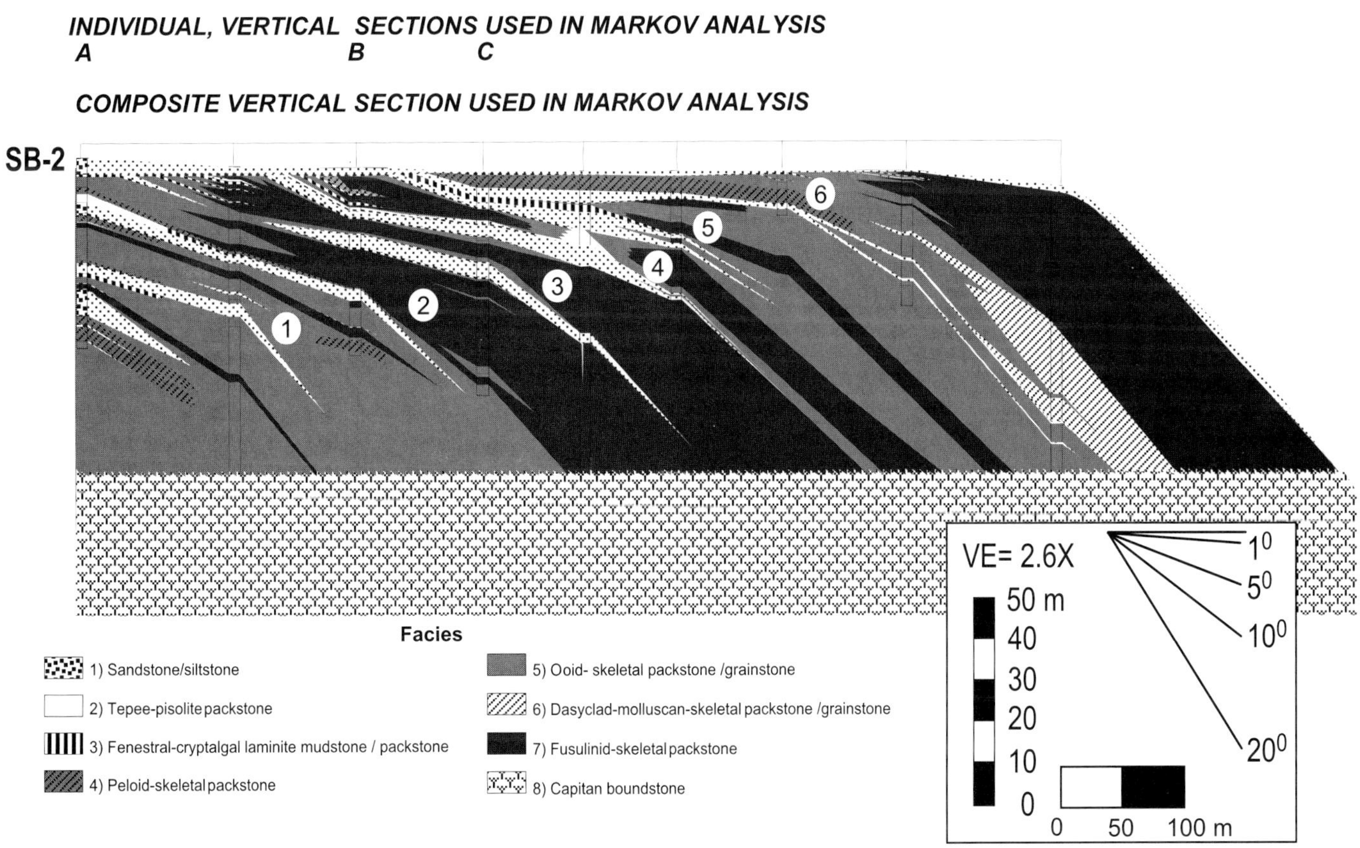

FIG. 4.—Detailed facies patterns and stratal geometries of sequence 1 (top of Seven Rivers Formation; cf. Fig. 3). Stratal geometries were determined by field mapping of the units. Facies interpretations between the closely spaced sections were constrained by observations made while mapping individual stratal units. Note the updip amalgamation of sandstone units into the toplap surface (SB-2) at the top of the depositional sequence-1. Circled numbers correspond to cycle designations indicated in Fig. 5. The facies successions in individual vertical sections A, B, and C and in the composite vertical section (indicated along top of cross-section) were used in Markov chain analysis (see Fig. 6).

TABLE 1.—LITHOFACIES CHARACTERISTICS AND ENVIRONMENTAL INTERPRETATIONS

Facies	1) Massive to fenestral/or cross laminated, fossiliferous silt/sandstone	2) Tepee-Pisolite packstone	3) Fenestral/-cryptalgal laminate mudstone/packstone	4) Peloid-skeletal packstone/grainstone	5) Ooid-skeletal packstone/grainstone	6) Dasyclad-mollusc-skeletal packstone/grainstone	7) Fusulinid-skeletal packstone/grainstone
Component grains	fine sand-coarse silt/ooids, molluscs, fusu.	bioclasts, pisoids, intraclasts	peloids, bioclasts up to 40%, rare silt	peloids, some ooids and bioclasts	ooids, some fusu., peloids and molluscs	biv., gast., peloids, dasyclads, minor fusu., ooids	fusu., peloids, ooids molluscs, dasyclads
Sedimentary structures	evaporite molds, cryptalgal lamination, irregular fenestrae/or ripple x-lam., trough x-bed., planar-tabular x-bed	tepee structures, sheet cracks, irregular and laminar fenestrae; fining and coarsening upward; horiz. and vert. cement filled fracture breccia	irregular and laminar fenestrae, LLH-cryptalgal lamination, evaporite molds, sheet cracks, circum-granular cracks	planar tabular x-bed., trough cross bedding; some units fine upward	planar tabular x-bed.; coarsening-upwards	planar-tabular x-bed., coarse-fine alternations	coarsening-upwards, coarse-fine laminations, planar tabular x-bed, trough x-bed
Environmental interpretation	eolian-derived silt/sand (reworked in peritidal or marine environment)	supratidal	peritidal to subaerial exposure	subtidal restricted to open marine	subtidal open marine	subtidal open marine to restricted	subtidal open marine to slightly restricted

Abbreviations: avg. = average; biv. = bivalve; fenes. = fenestral; fusu. = fusulinid; gast. = gastropod; grst. = grainstone; horiz. = horizontal; lam. = lamination; LLH = laterally linked hemispheroids; mudst. = mudstone; pack. = packstone; silt/sandst. = siltstone to sandstone; seq. = sequence; x-bed. = cross-bedding; x-lam. = cross-lamination; vert. = vertical.

NOTE: Massive to fenestral silt/sandstone and cross-laminated, fossiliferous silt/sandstone facies have been "lumped" into one general facies type here and in statistical analyses because diagnostic sedimentary structures are only sporadically preserved.

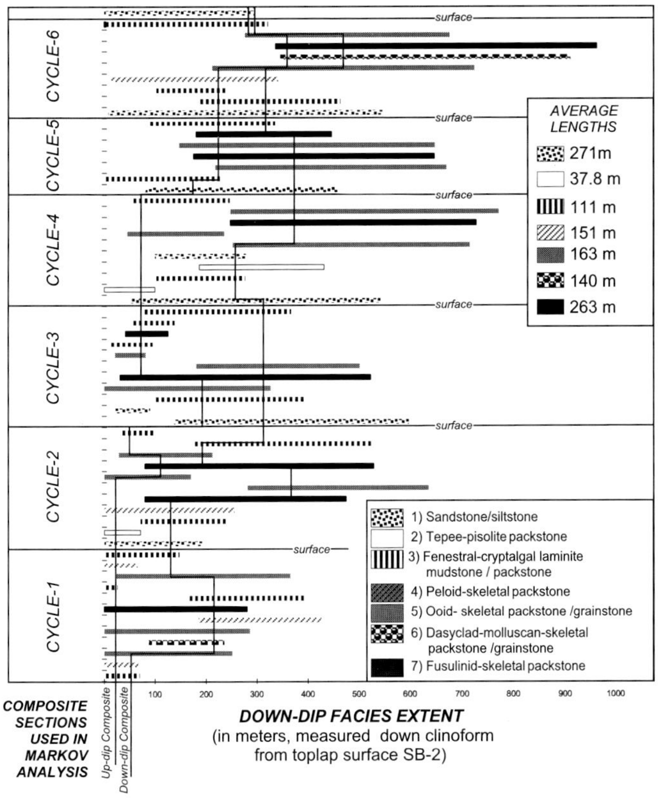

FIG. 5.—Bar chart illustrating facies succession and lateral extent of facies bodies of the Seven Rivers Formation. Lengths of facies bodies were measured as the downdip lengths from surface SB-2 as shown in Fig. 4. Cycle designation numbers are the same as those indicated in Fig. 4. Inset, upper right, shows the average lengths for all strata of a given facies type. Lines labeled "surface" indicate stratal surfaces (subaerial exposure and hiatal) defined on the basis of lateral tracing of stratal units and entered into Markov chain analysis (see Fig. 6). Vertical lines drawn through the succession labeled updip and downdip paths indicate composite facies successions used in Markov chain analysis (see Fig. 6).

skeletal packstone to grainstone, (7) fusulinid-skeletal packstone, and (8) cross-bedded sandstone/siltstone. Capitan age facies and their depositional environments have been extensively discussed in previous studies (Newell et al.,1953; Ward et al., 1986; Garber et al., 1989; Borer and Harris, 1989, 1991, 1995; Rankey and Lehrmann, 1996; Osleger, 1998).

Based on a qualitative recognition of a repeated facies motif and on interpreted depositional environments, we have proposed that the facies arrangements within the sandstone-bounded shingles reflect symmetrical (deepening/shallowing) cycles (Rankey and Lehrmann, 1996). Furthermore, based on physical tracing of subaerial exposure surfaces down preserved paleoslope we interpreted the cyclicity to have resulted from relative changes in sea level (Rankey and Lehrmann, 1996).

To aid in visualization of the facies succession, we have compiled a simplified chart of data of 2-D facies successions and lateral facies extents, measured downdip along clinoform surfaces (Fig. 5). In the updip parts of clinoforms, the strata are bounded along their bases and tops by fenestral facies representing subaerial exposure. Because these subaerial exposure surfaces extend down paleoslope, each is a demonstrated depositional hiatus and a logical break in facies succession for the interpretation of a cycle boundary. The subaerial exposure surface at the base of each cycle is overlain by a sandstone followed by a deepening-upward succession of subtidal facies (e.g., ooid-skeletal packstone to grainstone, peloid-skeletal packstone, fusulinid-skeletal packstone). Maximum deepening is interpreted to be represented by the fusulinid facies near the middle of cycles. This facies is overlain by a shallowing-upward succession of subtidal facies (e.g., fusulinid-skeletal packstone, peloid-skeletal packstone) followed by another fenestral laminite unit (Fig. 5). The subaerial exposure surfaces that bound the cycles are important. These surfaces were physically traced downdip along the clinoforms into entirely subtidal units (Fig. 4). Downdip tracing of surfaces qualitatively reveals a "cycle" motif in facies succession similar to the one above, but lacking subaerially exposed facies (Fig. 5).

ANALYSIS

Traditional Markov Analysis: Embedded Markov Chains

Markov chain analysis is a technique that can be used to determine whether the vertical succession of facies types within a stratigraphic section exhibits non-random behavior (Gingerich, 1969; Krumbein and Dacey, 1969; Carr, 1982; Powers and Easterling, 1982). Specifically, Markov chain analysis addresses the question of whether the frequency distributions of observed facies transitions differs significantly from that of a strictly random succession. In the analysis, vertical facies transitions (facies 1 to 2, 1 to 3, 2 to 3, etc.) are tabulated in a facies transition matrix. Through a series of steps, the facies transition matrix is compared against a matrix of expected facies transitions for a random succession in which the occurrence of each facies type is independent of the occurrence of underlying and overlying types. In the traditional analysis the expected (random) matrix is generated by assuming that the probability of transition from one facies to another is a function only of the proportion of occurrences of those facies in a vertical section. The null hypothesis of randomness or independence is evaluated by using a chi-square test to compare the actual facies transition matrix against the expected matrix for a random succession.

Embedded Markov chain analysis (Gingerich, 1969; Krumbein and Dacey, 1969; Doveton, 1971; Schwartzacher, 1969, 1975; Carr, 1982) determines if a system contains non-random transitions between mutually exclusive states. It differs from conventional Markov chain analysis in that transitions from one state to itself are not counted (cf., Davis, 1986). As used in this study, embedded Markov chain analysis provides a measure of whether vertical transitions between facies in a succession are non-random.

Several authors have noted that Markov chain analysis will generally yield an ordered verdict for long stratigraphic sections in which there is a significant non-stationarity in facies proportions (Doveton, 1996; Wilkinson et al., 1997; Lehrmann and Goldhammer, 1997). This non-stationarity is represented in carbonate platform strata as long-term deepening and shallowing trends, usually encompassing tens or hundreds of meters of vertical section. Thus, the Markov chain analysis of a non-stationary suc-

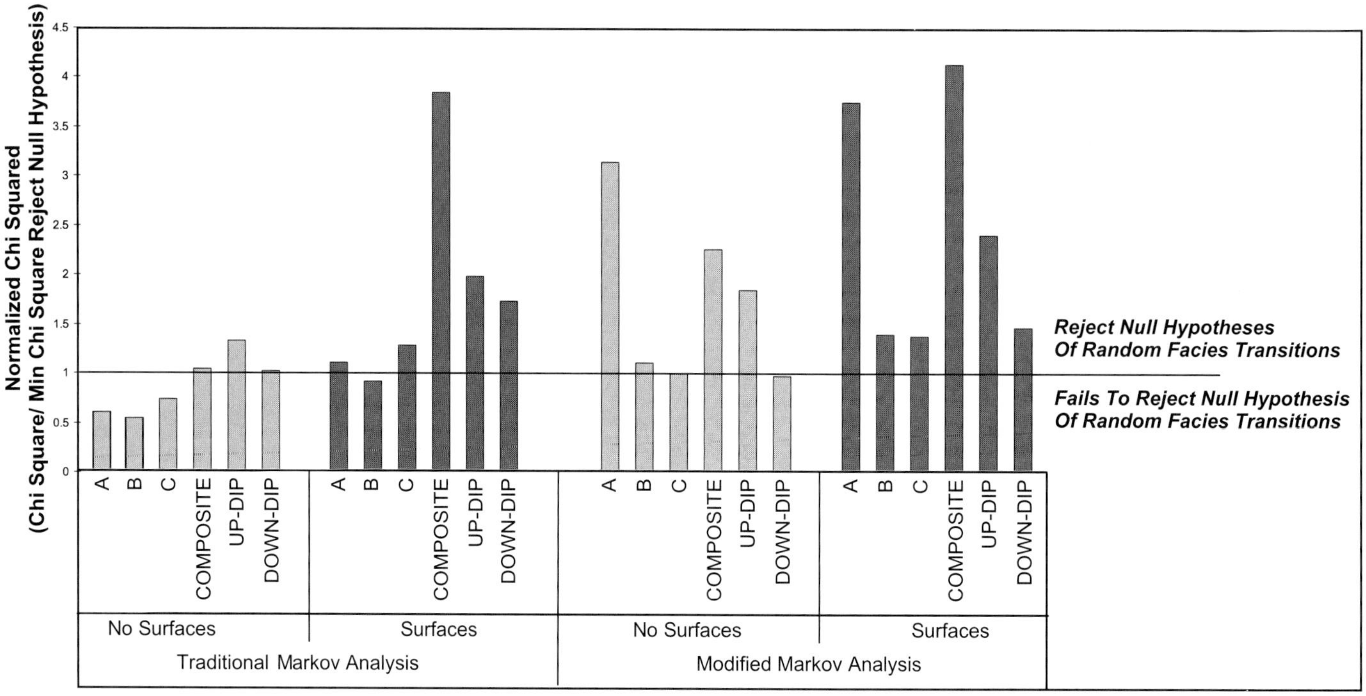

FIG 6.—Comparison of results from Markov chain analyses. Calculated chi-square values (y-axis) are normalized against the minimum chi-square required to reject null hypothesis of random facies transitions at the 95% C. I. Therefore, a value less than 1 indicates that the null hypothesis of random facies transitions cannot be rejected whereas a value greater than 1 allows rejection of the null hypothesis of a random succession. Traditional Markov analysis refers to standard approach of calculation of chi-square by comparison of actual transition frequencies against expected (random) transition frequencies calculated from the proportion of facies occurrences in vertical sections (see text for details). In modified Markov chain analysis, facies dimensions are factored into the expected (random) facies distribution using the technique shown in Fig. 7. "No surfaces" categories indicate analyses calculated from facies transitions only. "Surface" categories are for analyses in which physically mapped stratigraphic surfaces (subaerial exposure and hiatal surfaces) were assigned a "facies" code entered into the analysis.

cession exhibiting *disorder* at the scale of meter-thick cycles has commonly yielded an ordered Markov result because there is a larger-scale, systematic grouping of subtidal-prone facies in transgressive systems tracts and peritidal-prone facies in highstand systems tracts. In Slaughter Canyon, large-scale non-stationarity is manifested as the progradational/ retrogradational facies patterns of depositional sequences (Fig. 3). In order to eliminate the influence of non-stationarity from our analysis we have chosen a narrow stratigraphic window (only 6 cycles thick) that is contained entirely within the highstand systems tract of sequence 1 (Figs. 3 and 4).

The results of our analyses are presented in Table 2 and Fig. 6 under the heading "traditional Markov analysis—no surfaces." Using this approach, results from none of the individual stratigraphic sections through the interval allowed rejection of the null hypothesis of a random facies arrangement at the 95% confidence interval (C.I.). The composite vertical section, an analysis of all of the sections stacked vertically above one another, barely rejected the null hypothesis of a random succession at the 95% C.I. The other "updip" and "downdip" composite sections likewise barely rejected the null hypothesis of random transitions at the 95% C.I. These composite sections were generated from the succession of facies through the updip and downdip parts of the two dimensional cross-section as shown in Fig. 5.

Traditional Markov Chain Analysis Incorporating Stratigraphic Surfaces

As mentioned earlier, the occurrence and geometry of stratigraphic surfaces, such as diastems, subaerial exposure surfaces, and erosion surfaces, are important in any evaluation of stratigraphic cyclicity or sequence stratigraphy. In Slaughter Canyon, these surfaces are especially important because (1) subaerial exposure surfaces extend down paleoslope, each with a demonstrated depositional hiatus and genetic break in facies succession; (2) subaerial exposure surfaces extend down paleoslope, demonstrating relative falls in sea level (Rankey and Lehrmann, 1996); and (3) the surfaces were physically traced down depositional dip across facies tracts through the areas of peritidal to subtidal cycles (Fig. 4). Because of their importance in the stratigraphic framework we have included them in the Markov chain analysis by assigning the surfaces a facies code and incorporating them in the traditional Markov chain analysis.

The results of our analyses are presented in Table 2 and Fig. 6 under the heading "traditional Markov analysis—surfaces." The addition of surfaces increased the chi-squared values by an average of 200% over the traditional analysis with no surfaces. Two of the three analyses of individual vertical sections allow rejection of the null hypothesis of a random succession and analyses of all of the composite sections strongly allow rejection of the null hypothesis of a random succession at the 95% C.I. Relative

TABLE 2.—RESULTS OF MARKOV CHAIN ANALYSES

		SECTION	# FACIES TYPES	# TRANSITIONS	CHI-SQUARED	MIN. CHI-SQUARED (@95% CI)
TRADITIONAL ANALYSIS	NO SURFACES					
		A	7	51	27.56	42.56
		B	6	26	16.41	30.14
		C	6	34	22.25	30.14
		Composite	7	163	44.53	42.56
		Up-dip	7	40	56.7	42.56
		Down-dip	7	44	43.64	42.56
TRADITIONAL ANALYSIS	SURFACES					
		A	8	57	63.18	56.94
		B	7	29	39.1	42.56
		C	7	39	54.78	42.56
		Composite	8	193	220.4	56.94
		Up-dip	8	46	113.6	56.94
		Down-dip	8	50	99.01	56.94
MODIFIED ANALYSIS	NO SURFACES					
		A	7	51	167.1	42.56
		B	6	26	39.01	30.14
		C	6	34	34.94	30.14
		Composite	7	163	110.6	42.56
		Up-dip	7	40	97.14	42.56
		Down-dip	7	44	48.69	42.56
MODIFIED ANALYSIS	SURFACES					
		A	8	57	257.3	46.94
		B	7	29	69.81	42.56
		C	7	39	69.13	42.56
		Composite	8	193	277.7	56.94
		Up-dip	8	46	164	56.94
		Down-dip	8	50	98.27	56.94

to the Markov analysis without surfaces, the chi-squared values increased 370% for the composite vertical section, 150% for the updip composite section, and 170% for the downdip composite section.

Modified Markov Chain Analysis Incorporating Lateral Facies Dimensions

Markov chain analysis is based on a chi-square comparison of the observed facies transition frequencies (probabilities) with the frequencies (probabilities) "expected" in a random succession. In traditional analysis, the expected (random) transition probabilities (frequencies) are determined from data in the actual vertical section. In a strictly random succession, in which there is no preference of a given facies to be followed by another, the probability of transition from a given facies to another is determined exclusively by the proportions of occurrences of those facies in the vertical section. In other words, in a randomly "reshuffled" version of the actual section, two facies with greater abundance will likely have a higher number of transitions to each other simply due to their abundance.

In the case of the Slaughter Canyon data we can improve the chi-square analysis by incorporating lateral (2-D) facies dimensions and their abundances (Fig. 7). With 2-D data, it is obvious that in a randomly reshuffled succession the likelihood (probability) that a given facies will be in transition with another specific facies is dependent not only on the number of occurrences of those two facies but also their lateral extents (Figs. 5 and 7). This concept is based on the idea that any transition from one facies to another is a function of the potential contact area between the two facies (Fig. 7). Given similar number of occurrences of facies bodies, the facies with greater lateral extents will have a greater probability of transition (contact) than facies with smaller lateral extent. In our modified Markov chain analysis, we use this philosophy by comparing the actual facies transition matrix from the vertical and composite sections against the

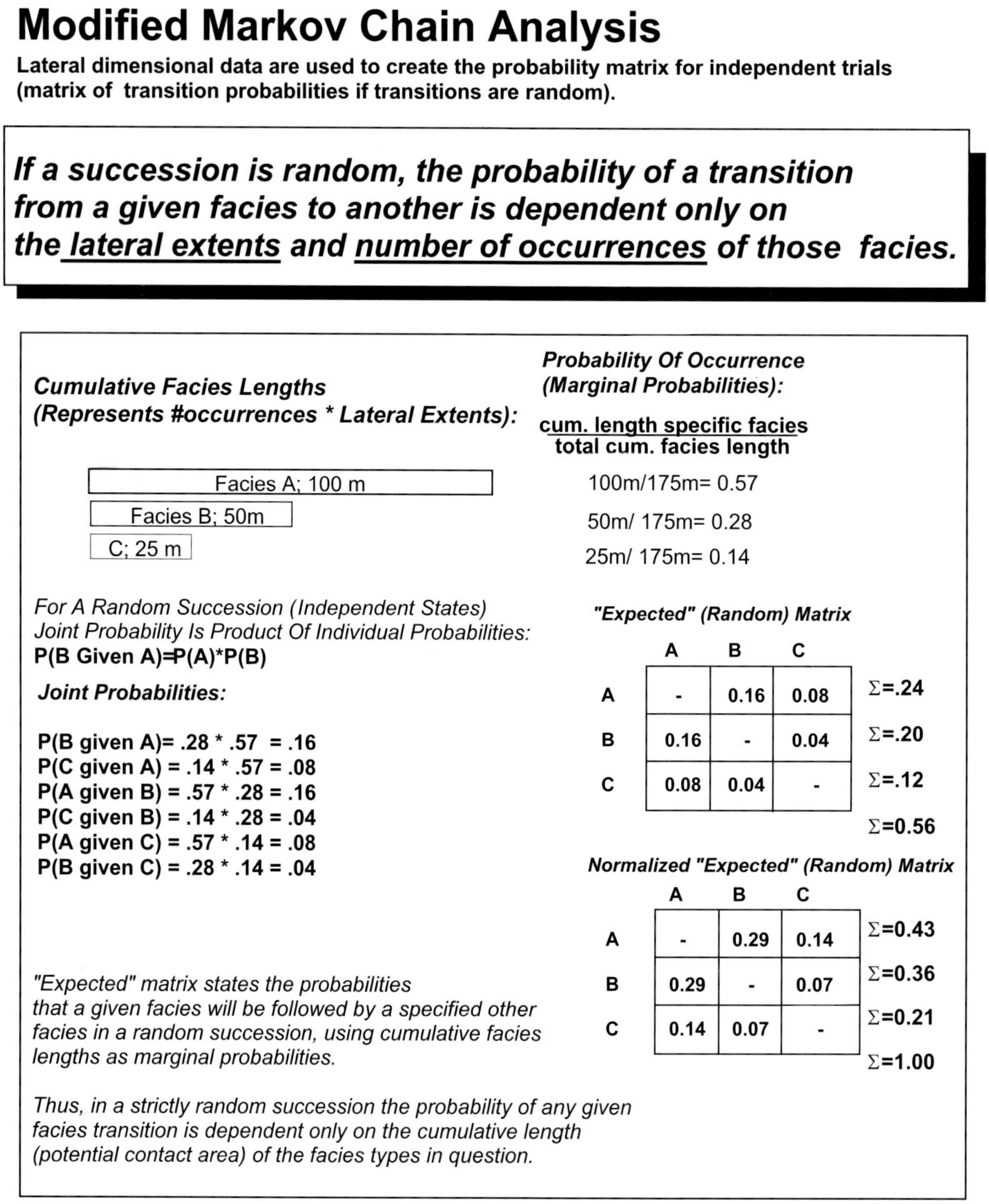

FIG. 7.—Illustration of the modified Markov chain analysis technique in which lateral (2-D) facies dimensions are integrated into the analysis. The modified Markov analysis is performed in exactly the same way as the traditional approach (e.g., Davis, 1986) except that the cumulative lengths of each facies type (which represents the lengths of each facies body and number of facies bodies in Fig. 5) are used to construct the marginal probabilities. The marginal probabilities are then used to construct the expected (random) matrix as shown (for a concise and simple explanation of Markov chain analysis see Davis, 1986).

expected (random) matrix of transitions in which the probabilities of transitions are based only on the number of occurrences and lateral extents of facies bodies (Fig. 7). In this analysis, the matrix of expected (random) probabilities is calculated from the ratio of the cumulative length of a given facies to the cumulative length of all facies. The cumulative length ratio incorporates both the frequency of occurrence of a given facies type as well as its lateral extent (Fig. 7). This philosophy is essentially the inverse of an approach that has been employed to use Markov chains to generate 2-D/3-D forward reservoir models (Doveton, 1996).

The results of our analyses are presented in Table 2 and Fig. 6 under the heading "modified Markov analysis — no surfaces" and "modified Markov analysis — surfaces." Without surfaces, the incorporation of facies dimension data increased the chi-squared values by an average of 213% over the traditional analysis with no surfaces. With surfaces, chi-squared values were increased an average of 150% over traditional analysis with surfaces and by 290% over traditional analysis without surfaces.

Statistical Evaluation of "Ideal Cycles"

To further investigate the dramatic improvement in the level of order resulting from integration of lateral facies extents in the analysis, one can examine the character of individual cells of the facies transition matrices. Individual cells reveal the difference

Observed Facies Transition Matrix

from Facies number	to 1	2	3	4	5	6	7	8
1		1	12	2	3	1	5	1
2				1			1	
3		1		3	13	1	4	15
4			4		2		6	
5	1	1	17	1			10	6
6			1		1		1	
7			4	4	15	1		6
8	29						1	

Transition Frequency matrix expected for Random Succession
(based on number of occurrences and length of facies bodies in 2-d)

from Facies number	to 1	2	3	4	5	6	7	8
1	0	0.38	6.69	0.94	8.74	2.53	7.93	8.7
2	0.38	0	0.35	0.05	0.46	0.13	0.41	0.45
3	6.69	0.35	0	0.87	8.06	2.34	7.31	8.02
4	0.94	0.05	0.87	0	1.14	0.33	1.03	1.13
5	8.74	0.46	8.06	1.14	0	3.05	9.55	10.48
6	2.53	0.13	2.34	0.33	3.05	0	2.77	3.04
7	7.93	0.41	7.31	1.03	9.55	2.77	0	9.51
8	8.7	0.45	8.02	1.13	10.48	3.04	9.51	0

Chi Squared matrix: (Observed-Expected)^2/ expected)

from Facies number	to 1	2	3	4	5	6	7	8
1	0	1.02	***4.22***	1.19	**3.77**	0.93	1.08	**6.82**
2	0.38	0	0.35	***18.37***	0.46	0.13	0.83	0.45
3	**6.69**	1.21	0	***5.23***	**3.03**	0.76	1.5	**6.07**
4	0.94	0.05	***11.29***	0	0.66	0.33	**23.99**	1.13
5	**6.86**	0.65	***9.92***	0.02	0	**3.05**	0.02	1.92
6	**2.53**	0.13	0.76	0.33	1.38	0	1.13	**3.04**
7	**7.93**	0.41	1.5	**8.57**	**3.11**	1.13	0	1.3
8	**47.36**	0.45	**8.02**	1.13	***10.48***	**3.04**	**7.61**	0

"Ideal cycles"
1-3-8-1
1-3-5-3-8-1
1-3-4-7-5-3-8-1
1-3-4-7-4-3-8-1

Bolded cells with chi-squared values >3
Bold italic= # Transitions > expected from random succession
Bold underlined= # Transitions < expected from random succession

Facies codes:
1) Sandstone/ siltst.
2) Tepee/ pisolite
3) Fenestral Laminite
4) Peloid packstone
5) Ooid packstone / grainstone
6) Dasyclad molluscan packstone
7) Fusulinid packstone
8) Surface (exposure surface and downdip equivalent)

FIG. 8.—Facies transition matrix and individual chi-squared values for the modified Markov chain analysis of the vertical composite section (modified, COMPOSITE, with surfaces from Figs. 4 and 6). The chi-squared matrix shows the difference between actual facies transition frequencies and those expected from a random distribution. The "expected" transition frequencies (random distribution) were calculated according to the technique illustrated in Fig. 7. Also shown (lower) are potential "ideal cycles" derived from chi-squared values greater than 3 from the chi-squared matrix (see text for details).

between the number of observed facies transitions from one given facies to another and the number of transitions expected from a random succession (Fig. 8). The high overall chi-squared value results from the fact that certain facies transitions occur far more frequently than expected for a random succession (Fig. 8). The high overall chi-squared is also bolstered, however, by the fact that some facies transitions occur far less frequently than would be expected from a random succession (Fig. 8). For example, the great abundance and lateral extent of two facies such as sandstone and fusulinid-skeletal packstone suggest that in a random succession these facies should frequently be in contact (exhibit transitions) but in the real data these facies are in fact never in contact (Fig. 8).

Using the individual chi-squared values, we can attempt to construct "ideal cycles" based on the facies transitions that occur more frequently than expected from a random succession. Using this approach and the chi-squared values from Fig. 8, potential "ideal cycles" include: (1) sandstone-siltstone, fenestral laminite mudstone-packstone, surface, sandstone-siltstone; (2) sandstone-siltstone, fenestral laminite mudstone-packstone, peloid-skeletal packstone, fenestral laminite mudstone-packstone, surface, sandstone-siltstone; (3) sandstone-siltstone, fenestral laminite mudstone-packstone, peloid-skeletal packstone, fusulinid-skeletal packstone, ooid packstone-grainstone, fenestral laminite mudstone-packstone, surface, sandstone-siltstone; and (4) sandstone-siltstone, fenestral laminite mudstone-packstone, peloid-skeletal packstone, fusulinid-skeletal packstone, peloid-skeletal packstone, fenestral laminite mudstone-packstone, surface, and sandstone-siltstone. Note that some of these statistical reconstructions of "ideal cycles" closely resemble the symmetrical cycles that would be qualitatively inferred from Fig. 5.

DISCUSSION

Through a series of steps in which progressively more stratigraphic data were entered into Markov chain analysis, we have shown that meter-scale lithologic cyclicity and potential "ideal cycles" are statistically definable within a narrow stratigraphic interval in the upper part of the Seven Rivers Formation in Slaughter Canyon. The first analysis, which included only 1-D vertical facies succession data, revealed either disorder or a low level of order. The second set of analyses improved on the first by including stratigraphically significant surfaces. These analyses indicated that most of the sections yielded non-random successions. The third set of analyses, the most geologically robust, added information on lateral (2-D) facies distribution and abundance. These analyses revealed that, when 2-D facies information is added, the succession is decidedly non-random.

We also have found that traditional Markov chain analysis is probably inadequate for evaluating the existence of cyclicity in the Seven Rivers Formation. The progressive addition of geologic data to the analyses made it clear that a traditional quantitative analysis of 1-D (vertical) facies data may not be sufficient to characterize or evaluate the presence of cyclicity. It was only through the incorporation of 2-D distribution of stratigraphic surfaces and facies dimensions that the Markov test yielded an unambiguous result of an orderly succession. These results illustrate that stratigraphic cyclicity should not be viewed as simply a vertical (1-D) attribute of carbonate platforms, but is best viewed in the context of 2-D/3-D stratigraphic information. Conceptually, the 2-D distribution of surfaces of subaerial exposure, erosion, or hiatus and the lateral distribution and geometry of facies bodies should play an important role in the evaluation of cyclicity in any given section.

These findings have significance for the more general question of whether meter-scale lithologic cycles are prevalent in the strata of shallow-marine carbonate and mixed carbonate-siliciclastic platforms. Previous studies have generally found that meter-scale cyclicity and "ideal cycles" are statistically undefinable with traditional Markov chain analysis (e.g., Schwartzacher, 1969, 1975; Doveton, 1971; Wilkinson et al., 1997). Although we recognize that significant levels of randomness exist in the facies successions of shallow-marine platforms (Lehrmann and Goldhammer, 1997), the fact that cyclicity is frequently reported on the basis of qualitative observations suggests that cyclicity indeed may be prevalent in shallow-marine carbonate and carbonate-siliciclastic systems. Previous failed attempts to use Markov chain analysis to demonstrate cyclicity in such systems may stem from limitations of the test and failure to incorporate stratigraphic surface information and 2-D facies relationships into the analysis. It may be that, for many of the previously investigated stratigraphic systems (e.g., Wilkinson et al., 1997), lithologic cyclicity could be statistically validated if the critical 2-D stratigraphic information were factored into the analysis. Geological factors that could complicate the facies variability in a single vertical (1-D) section include finite facies extent (due to lateral facies change or pinch-out), changing position of the shoreline or shelf break, changing platform morphology, and lateral changes in the character of subaerial exposure, erosion, or hiatal surfaces.

Given that numerous detailed 2-D sequence stratigraphic and facies data are now available in areas with continuously exposed platforms such at the Northwest Shelf in the Guadalupe Mountains (Sonnenfeld and Cross, 1993; Kerans and Fitchen, 1995; Rankey and Lehrmann, 1996; Tinker, 1996), quantitative 2-D statistical studies provide a promising approach to further addressing the question of the degree of prevalence of meter-scale lithologic cycles in the stratigraphic record. Such analyses also will allow evaluation of the accuracy of chronostratigraphic correlation, using present sequence stratigraphic techniques, and provide insight into the relationships between vertical and lateral stratigraphic patterns.

ACKNOWLEDGMENTS

We thank Thomas Algeo and Christopher G. St. C. Kendall for providing critical reviews of the manuscript. John Mitchell and William Mode are acknowledged for comments and suggestions on an early draft of the manuscript. We also thank Bruce Wilkinson for stimulating our interest in this study through recent publications and numerous e-mail discussions on the issues of stratigraphic cyclicity.

REFERENCES

BARNABY, R. J., and WARD, W. B., 1995, Sequence stratigraphic framework, high frequency cyclicity and three dimensional heterogeneity: Grayburg Formation, Brokeoff Mountains, New Mexico, *in* Pause, P. H., and Candelaria, M. P., eds., Carbonate Facies and Sequence Stratigraphy: Practical Applications of Carbonate Models: Midland, Society of Economic Paleontologists and Mineralogists, Permian Basin Section, Publication 5-95, p. 37-50.

BORER, J. M., and HARRIS, P. M., 1989, Depositional facies and cycles in Yates Formation outcrops, Guadalupe Mountains, New Mexico, *in* Harris, P. M., and Grover, G. A., eds., Subsurface and outcrop examination of the Capitan shelf margin, northern Delaware Basin: Tulsa, Society of Econonomic Paleontologists and Minereralogists Core Workshop 13, p. 305-318.

BORER, J. M., and HARRIS, P. M. , 1991, Lithofacies and cyclicity of the Yates Formation, Permian Basin: Implications for reservoir heterogeneity: American Association of Petroleum Geologists Bulletin, v. 75, p. 726-779.

BORER, J. M., and HARRIS, P. M., 1995, Computer simulation of the Yates Formation (Permian, Delaware Basin)—sequence stratigraphy and shelf-to-basin correlation implications, *in* Martin, R. L., ed., In Search of New Permian Basin Oil and Gas Fields: Using Today's Technologies and Tomorrow's Ideas for Exploration, Development, and 3D Seismic in a Mature Basin: Midland, West Texas Geological Society Symposium Publication 95-98, p. 111-132.

CARR, T. R., 1982, Log-linear models, Markov chains, and cyclic sedimentation: Journal of Sedimentary Petrology, v. 52, p. 905-912.

DAVIS, J. C., 1986, Statistics and Data Analysis in Geology: New York, John Wiley and Sons, 646 p.

DOVETON, J. H., 1971, An application of Markov chain analysis to the Ayrshire Coal Measures succession: Scottish Journal of Geology, v. 7, p. 11-27.

DOVETON, J. H., 1996, Theory and application of vertical variability measures from Markov chain analysis, *in* Yarus, J. M., and Chamberlain, A. L., eds.: Tulsa, American Association of Petroleum Geologists, Computer Applications in Geology 3, p. 55-64.

DUFF, M. D., and WALTON, E. C., 1962, Statistical basic for cyclothems: A quantitative study of the sedimentary succession in the east Pennine coalfield: Sedimentology, v. 1, p. 235-255.

ENOS, P., and PERKINS, R. D., 1977, Quaternary sedimentation in South Florida: Boulder, Geological Society of America Memoir 147, 198 p.

GARBER, R. A., GROVER, G. A., and HARRIS, P. M., 1989, Geology of the Capitan shelf margin — Subsurface data from the northern Delaware

Basin, *in* Harris, P. M., and Grover, G. A., eds., Subsurface and outcrop examination of the Capitan shelf margin, northern Delaware Basin: Tulsa, Society of Economic Paleontologists and Minereralogists Core Workshop 13, p. 3-272.

GINGERICH, P. D., 1969, Markov analysis of cyclic alluvial sediments: Journal of Sedimentary Petrology, v. 39, p. 330-332.

GOLDHAMMER, R. K., DUNN, P. A., and HARDIE, L. A., 1990, Depositional cycles, composite sea-level changes, cycle stacking patterns, and the hierarchy of stratigraphic forcing: Examples from Alpine Triassic platform carbonates: Geological Society of America Bulletin, v. 102, p. 535-562.

GOLDHAMMER, R. K., LEHMANN, P. J., TODD, R. G., WILSON, J. L., WARD, W. C., and JOHNSON, C. R., 1991a, Sequence Stratigraphy and Cyclostratigraphy of the Mesozoic of the Sierra Madre Oriental, Northeast Mexico, a Field Guidebook: Tulsa, Society of Economic Paleontologists and Mineralogists, Gulf Coast Section, 85 p.

GOLDHAMMER, R. K., OSWALD, E. J., and DUNN, P. A., 1991b, Hierarchy of stratigraphic forcing; Example from Middle Pennsylvanian shelf carbonates of the Paradox Basin, *in* Franseen, E. K., Watney, W. L., Kendall, C. G. St. C., and Ross, W., eds., Sedimentary Modeling; Computer Simulations and Methods for Improved Parameter Definition: Kansas Geological Survey Bulletin 233, p. 361-413.

GOLDHAMMER, R. K., LEHMANN, P. J., and DUNN, P. A., 1993, The origin of high frequency platform carbonate cycles and third-order sequences (Lower Ordovician El Paso Group, West Texas); constraints from outcrop data and stratigraphic modeling: Journal of Sedimentary Research, v. 63, p. 318-359.

GROTZINGER, J. P., 1989, Facies and evolution of Precambrian carbonate depositional systems; emergence of the modern platform archetype, *in* Crevello, P. D., Wilson, J. L., James, J., Sarg, J. F., and Read, J. F., eds., Controls on Carbonate Platform and Basin Development: Tulsa, Society of Economic Paleontologists and Mineralogists Special Publication 44, p. 79-106.

HARRIS, M. T., and SHEEHAN, P. M., 1996, Upper Ordovician-Lower Silurian depositional sequences determined from middle shelf sections, Barn Hills and Lakeside Mountains, eastern Great Basin, *in* Witzke, B. J., Ludvington, G. A., and Day, J., eds., Paleozoic Sequence Stratigraphy; Views from the North American Craton: Boulder, Geological Society of America Special Paper 306, p. 161-176.

HECKEL, P. H., 1986, Sea-level curve for Pennsylvanian eustatic marine transgressive-regressive cycles along midcontinent outcrop belt, North America: Geology, v. 14, p. 330-334.

HURLEY, N. F., 1989, Facies mosaic of the Seven Rivers Formation, McKittrick Canyon, New Mexico, *in* Harris, P. M., and Grover, G. A., eds., Subsurface and Outcrop Examination of the Capitan Shelf Margin, Northern Delaware basin: Tulsa, Society for Economic Paleontologists and Mineralogists Core Workshop 13, p. 325-347.

JAMES, N. P., 1984, Shallowing-upward sequences in carbonates, *in* Walker, R. G., ed., Facies Models, Second Edition: St. John's, Geological Association of Canada, p. 229-244.

KERANS, C., HOLTZ, M., and TINKER, S., 1992, Recognition and correlation of high frequency cyclicity in mixed clastic-carbonate sequences: Reservoir Characterization Research Laboratory: Austin, Bureau of Economic Geology, University of Texas, Short Course Notes, October 11-14, 1992, 54 p.

KERANS, C., and FITCHEN, W. M., 1995, Sequence stratigraphy and facies architecture of a carbonate ramp system: San Andres Formation of Algerita Escarpment and western Guadalupe Mountains, west Texas and New Mexico: Austin, Bureau of Economic Geology, University of Texas, Report of Investigations 235, 86 p.

KING, P. B., 1942, Permian of west Texas and southeastern New Mexico: American Association of Petroleum Geologists Bulletin, v. 26, p. 535-763.

KRUMBEIN, W. C., and, DACEY, M. F., 1969, Markov chains and embedded Markov chains in geology: International Association of Mathematical Geology , v. 1, p. 79-96.

LEHRMANN, D. J., AND GOLDHAMMER, R. K., 1997, Guidelines for application of stacking patterns techniques in platform carbonates from ice-house to green-house periods of Earth History (abs.): American Association of Petroleum Geologists, Annual Meeting Abstracts, v. 6, p. 69.

LEHRMANN, D. J., RANKEY, E. C., and GOLDHAMMER, R. K., 1997, Relationships between parasequence stacking patterns and depositional sequence geometries: Upper Guadalupian (Permian) strata of Slaughter Canyon, New Mexico: Tulsa, American Association of Petroleum Geologists Annual Convention Official Program, p. A68.

MCLEAN, D. J., and MOUNTJOY, E. W., 1994, Allocyclic control on Late Devonian buildup development, southern Canadian Rocky Mountains: Journal of Sedimentary Research, v. B64, p. 326-340.

MERRIAM, D. F., 1964, Symposium on cyclic sedimentation: Kansas Geological Survey Bulletin, v. 169, 640 p.

MONTANEZ, I. P., and OSLEGER, D. A., 1993, Parasequence stacking patterns, third order accommodation events, and sequence stratigraphy of Middle to Upper Cambrian platform carbonates, Bonanza King Formation, southern Great Basin, *in* Loucks, R. G., and Sarg, J. F., eds., Carbonate Sequence Stratigraphy: Recent Developments and Applications: Tulsa, American Association of Petroleum Geologists Memoir 57, p. 305-326.

NEWELL, N. D., RIGBY, J. K., FISCHER, A. G., WHITEMAN, A. J., HICKOX, J. E., and BRADLEY, J. S., 1953, The Permian Reef Complex of the Guadalupe Mountains Region, Texas and New Mexico: San Francisco, W. H. Freeman Company, 236 p.

OSLEGER, D. A., 1998, Sequence architecture and sea-level dynamics of Upper Permian shelfal facies, Guadalupe Mountains, southern New Mexico: Journal of Sedimentary Research, v. 68, p. 327-346.

OSLEGER, D. A., and READ, J. F., 1991, Relation of eustasy to stacking patterns of meter-scale carbonate cycles, Late Cambrian, U.S.A.: Journal of Sedimentary Petrology, v. 61, p. 1225-1252.

POWERS, D. W., and EASTERLING, R. G., 1982, Improved methodology for using embedded Markov chains to describe cyclical sediments: Journal of Sedimentary Petrology, v. 52, p. 913-923.

RANKEY, E. C., and LEHRMANN, D. J., 1996, Anatomy and origin of toplap in a mixed carbonate-clastic system: Seven Rivers Formation (Permian, Guadalupian), Guadalupe Mountains, New Mexico, USA: Sedimentology, v. 43, p. 807-826.

READ, J. F., 1973, Carbonate cycles, Pillara Formation (Devonian), Canning Basin, Western Australia: Bulletin of Canadian Petroleum Geology, v. 21, p. 38-51.

SCHWARTZACHER, W., 1969, The use of Markov chains in the study of sedimentary cycles: Mathematical Geology, v. 1, p. 17-39.

SCHWARTZACHER, W., 1975, Sedimentation Models and Quantitative Stratigraphy: New York, Elsevier, 382 p.

SONNENFELD, M. D., 1996, Sequence evolution and hierarchy within the Lower Mississippian Madison Limestone of Wyoming, *in* Longman, M. W., and Sonnenfeld, M. D., eds., Paleozoic Systems of the Rocky Mountain Region: Tulsa, Society of Economic Paleontologists and Mineralogists, p. 165-192.

SONNENFELD, M. D., and CROSS, T. A., 1993, Volumetric partitioning and facies partitioning within the Permian Upper San Andres Formation of Last Chance Canyon, Guadalupe Mountains, New Mexico, *in* Loucks, R. G., and Sarg, J. F., eds., Carbonate Sequence Stratigraphy: Tulsa, American Association of Petroleum Geologists Memoir 57, p. 434-474.

TINKER, S. W., 1996, Reservoir scale sequence stratigraphy; McKittrick canyon, and three-dimensional subsurface examples, West Texas and New Mexico: Unpublished Ph.D Dissertation, University of Colorado, Boulder, 245 p.

VAN WAGONER, J. C., MITCHUM, R. M., CAMPION, K. M., and RAHMANIAN, V. D., 1990, Siliciclastic sequence stratigraphy in well-logs, cores, and outcrops: Tulsa, American Association of Petroleum Geologists Methods in Exploration Series 7, 55 p.

WARD, R. F., KENDALL, C. G. St. C., and HARRIS, P. M., 1986, Upper Permian (Guadalupian) facies and their association with hydrocarbons—Permian Basin, west Texas and New Mexico: American Association of Petroleum Geologists Bulletin, v. 70, p. 239-262.

WATNEY, L., FRENCH, J. A., and FRANSEEN, E. K., 1989, Sequence stratigraphic interpretations and modeling of cyclothems in the Upper Pennsylvanian (Missourian) Lansing and Kansas City Groups in eastern Kansas, 41st Annual Field Trip Guidebook: Kansas Geological Society, p. 108-114.

WILKINSON, B. H., DIEDRICH, N. W., and DRUMMOND, C. N., 1996a, Facies successions in peritidal carbonate sequences: Journal of Sedimentary Research (B), v. 66, p. 1065-1078.

WILKINSON, B. H., DRUMMOND, C. N., and DIEDRICH, N. W., 1996b, Shoaling cyclicity in carbonate sequences: Tulsa, American Association of Petroleum Geologists, Annual Meeting Abstracts, v. 5, p. 151.

WILKINSON, B. H., DRUMMOND, C. N., ROTHMAN, E. D., and, DIEDRICH, N. W., 1997, Stratal order in peritidal carbonate sequences: Journal of Sedimentary Research, v. 67, p. 1068-1082.

WILSON, J. L., 1975, Carbonate Facies in Geologic History: New York, Springer-Verlag, 471 p.

FORMAL DESIGNATION: REEF TRAIL MEMBER, BELL CANYON FORMATION, AND ITS SIGNIFICANCE FOR RECOGNITION OF THE GUADALUPIAN-LOPINGIAN BOUNDARY

GARNER L. WILDE
GLW International, Midland, Texas 79705
SHANNON F. RUDINE
McDonald Observatory, Ft. Davis, Texas 79734
AND
LANCE L. LAMBERT
Southwest Texas State University, San Marcos, Texas 78666

ABSTRACT: The Guadalupian Series, with its type section in the Guadalupe Mountains, west Texas and New Mexico, has generally become the accepted worldwide standard for the Middle Permian. The Cisuralian of Russia is the accepted standard for the Early Permian; and the Lopingian of China has been accepted as the Late Permian standard. Most recently, the task of securing this threefold worldwide standard has been pursued by a host of workers in many specialties in order that no major gaps or overlaps are left for chronostratigraphic correlation.

Following established international procedures for defining a chronostratigraphic boundary, the base of the Guadalupian Series is recognized at the stratigraphically lowest occurrence of the conodont *Jinogondolella nankingensis* in a section recording continuous deposition in the middle of the El Centro Member of the Cutoff Formation, Roadian Stage, Guadalupe Mountains. This designation is in basic agreement with the ranges of important species of three other major faunal groups: the ammonoids, brachiopods, and fusulinaceans.

Designation of a formal top of the Guadalupian Series has remained more elusive inasmuch as the final establishment of the base of the overlying Lopingian in China has remained in a rather unsettled state until recently. Enough is now known to insure that the Lopingian base will fall chronostratigraphically somewhere between the upper Lamar Limestone, Bell Canyon Formation, and the base of the Castile Formation of the Delaware Basin. This interval includes the so-called "post-Lamar beds" of the upper Bell Canyon Formation.

As a means of equating the lithostratigraphy to the chronostratigraphy, we describe herein the Reef Trail Member of the upper Bell Canyon Formation by establishing its type section at McKittrick Canyon, Guadalupe Mountains, as a formal replacement name for the informal "post-Lamar beds." We also identify the fusulinacean and conodont faunas from composited sections of Lamar Limestone and Reef Trail members, and extend their correlations biostratigraphically from the Guadalupes to the Apache and Glass mountains. The McKittrick Canyon Limestone is a new name given to the "Middle" Limestone (Brown, 1996), following his recognition that this limestone had for too long been misidentified as the McCombs Limestone. Finally, correlation of faunal zones established in the Delaware Basin from the three mountainous areas are extended globally in an attempt to reconcile chronostratigraphic problems.

INTRODUCTION

Application of the Guadalupian as a viable series name for the rocks, faunas, and floras commonly assigned to the middle portion of the Permian System in North America has a long and storied history. Girty, 1908 (1909), described many of the fossils that have since become critical to understanding and defining this body of rock. Indeed, Girty had proposed as early as 1902 that the rocks and fauna exposed in the magnificent outcrops of the Guadalupe Mountains were so significantly unique as to warrant the naming of the Guadalupian Series. Since that time to the present, many workers of various backgrounds have built on Girty's fundamental framework with enlightened detail. Those studies were summarized by Glenister et al. (1992), who formally proposed the Guadalupian for the world standard Middle Permian reference.

The basal boundary of the Guadalupian Series had been explicitly established by Girty (1902) at the base of the "black limestone" near the floor of Guadalupe Pass. Subsequent miscorrelations of the "black limestone" confused the lithic, and as a result, faunal, definition of the basal Guadalupian—which remained unsettled while different criteria for definition were attempted. At the Guadalupian Symposium (convened in Alpine, Texas), Lambert et al. (1991), and in a forthcoming paper (1998) corrected the ingrained miscorrelation and proposed a return to Girty's 1902 definition. They demonstrated that such action would both stabilize the basal Guadalupian boundary and corroborate Wilde's (1975, 1986a, 1986b, 1988, 1990), Wilde and Todd, (1968) fusulinid correlations across the Delaware Basin. This synthesis galvanized the titular members of the Permian subcommission who were in attendance, and since then great strides have been made toward finalizing a world standard Permian time scale (summarized in Jin et al., 1994a, 1994b, 1997). Unfortunately, the proceedings volume from the Guadalupian Symposium (Smithsonian Contributions to the Earth Sciences 32) has languished in press since that time. As a consequence, manuscripts from the proceedings (including resolution of the "black limestone" miscorrelations) have only been circulating unevenly among interested groups, and some have yet to be made aware of their existence. For example, an extensive compilation summarizing Delaware Basin geology (Hill, 1996) was bereft of the advances still to be described in that volume.

Following established international procedures for defining a chronostratigraphic boundary (Cowie et al., 1986; Remane et al., 1996; see also Lambert et al., 1995; Remane, 1996), the base of the Guadalupian Series is recognized at the stratigraphically lowest occcurrence of the conodont *Jinogondolella nankingensis* in a section recording continuous deposition in the middle of the El Centro Member of the Cutoff Formation, Roadian Stage,

Geologic Framework of the Capitan Reef, SEPM Special Publication No. 65

Guadalupe Mountains. Location is at "Stratotype Canyon", approximately 1.0 km south-southeast of Bone Canyon, west side of the Guadalupe Mountain escarpment, Texas (Glenister et al., 1992; Lambert and Wardlaw, 1996; and Rohr, 1996). This designation is in basic agreement with ranges of important species of three other major faunal groups, the ammonoids, brachiopods, and fusulinaceans (Glenister et al., 1992).

The reference boundary is defined at a precise point within the paedomorphocline from which *Jinogondolella nankingensis* evolved from *Mesogondolella idahoensis* (Lambert and Wardlaw, 1992, 1996). A paedomorphocline is a temporal series of populations through which juvenile characteristics of an ancestral species become progressively expressed in increasingly adult stages of the descendant species. Delineation of the complete transition in the basinal "black limestone" demonstrates depositional continuity. The point within the transition recommended to define the base of the Guadalupian can be recognized both qualitatively and quantitatively (Lambert, 1994; Lambert and Wardlaw, 1996).

The formal position of the boundary will be selected by ballot of the titular members of the Guadalupian working group, then ratified by the International Commission on Stratigraphy. Once established, boundary correlations will be made to the fixed position of the resulting "golden spike" (Ager, 1964, 1981).

In Stratotype Canyon a relatively complete section of the Cutoff Formation (Girty's "black limestone") rests on the Bone Spring Formation (with which it has commonly been miscorrelated; Lambert, 1992). The transition takes place within the El Centro Member (M. T. Harris, pers. comm., 1998) of the Cutoff Formation 42.7 m (140 ft) above the base of the formation. There, within monotonous thin-bedded pelagic calcilutites, a sample 28.5 cm (11.25 in) below a prominent shale band contains the earliest known *Jinogondolella nankingensis*. Samples from the 8 underlying beds are currently being evaluated to confirm that this is the best boundary horizon. A postal ballot of the titular members of the Subcommission on Permian Stratigraphy in July 1996 formally adopted the Guadalupian as defined above, along with the preceding Cisuralian and succeeding Lopingian series, to comprise the Permian System (Spinosa, 1996; Jin et al., 1997).

The top of the Guadalupian will become fixed by formal establishment of the basal Lopingian boundary, defined in China by the same procedures. There, a concerted effort to biostratigraphically characterize this interval has been undertaken recently by Chinese scientists. Correlation between the upper type Guadalupian and the lower type Lopingian requires precise correlation to demonstrate both continuity and to insure that no significant gap or overlap compromises the integrity of either series. The beds in the boundary interval of the type Guadalupian have traditionally been cited informally as "post-Lamar beds" (following King, 1948). This nomenclature implies a lack of stratigraphic precision that does not exist. The purpose of this paper is to remove that impediment by formally naming the Reef Trail Member of the Bell Canyon Formation, and by providing initial lithostratigraphic and biostratigraphic characterization and correlation of the unit. Specifically we will: (1) describe the Reef Trail member, Bell Canyon Formation; (2) identify the fusulinacean and conodont faunas of the Lamar and Reef Trail members; (3) correlate equivalent faunas regionally between the Guadalupe, Apache, and Glass mountains of west Texas (Fig. 1); and (4) briefly consider a global extension of the chronostratigraphic boundary between the Guadalupian and Lopingian series.

GUADALUPE MOUNTAINS

"Post-Lamar Beds"

King (1948) described a relatively thin sequence of thinly laminated limestone, siltstone, and anhydrite between the Lamar Limestone and the Castile Formation, which he referred to simply as the "basal beds of the Castile Formation" (p. 58). King noted that out in the basin these beds are well exposed in a 6.0 m (20 ft) section a short distance southeast of U.S. Highway 62, on the north bank of a creek, one-half mile northeast of benchmark 4729. He further commented that these beds are somewhat thicker nearer the Reef Escarpment, with an increase in limestone. In nearby areas, between McKittrick and Big Canyon draws, the same beds are preserved as scattered outliers, forming light-colored knolls in stark contrast to the underlying dark beds of the Lamar Limestone. Probably the best exposed section of these rocks is on the north side of Big Canyon Draw, near the state line, and three-quarters of a mile northwest of the Old Gray Ranch. This section, 6.0 m (20 ft) of limestone with crushed brachiopod shells (King, 1948), is overlain by the Castile Formation.

An interesting alignment of three identical sections is seen in King (1948, pl. 3), including the last two described, above, and another one about one-half mile and almost due south of Old Gray Ranch. Whether this represents the last remnant of a Castile strike line is unknown (Fig. 2).

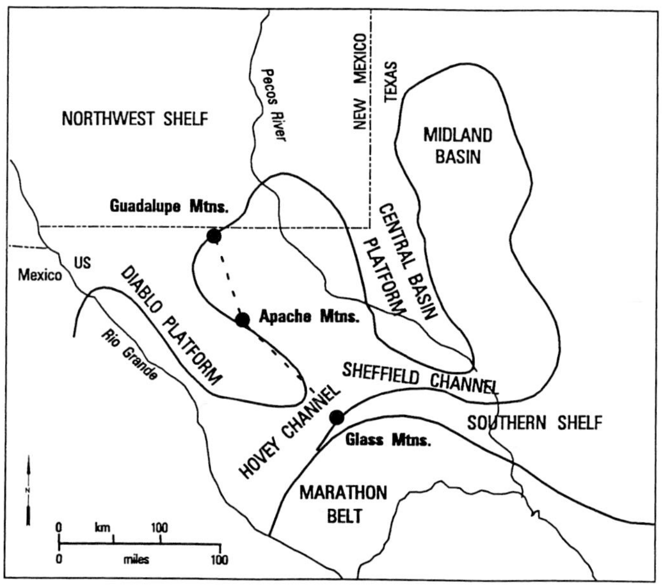

FIG. 1.—General location map for cross-sections (Figs. 3, 3A) and detailed location maps—(Figs. 2, 4, 5) of areas discussed in this paper.

Two other sections in the general area, reported by Wilde (1955), are hills with elevations of 5195 ft (reported erroneously as 5193 ft), and 4913 ft, located on the ridge between Malpais and Six canyons, just north of McKittrick Canyon. All of the sections reported above are on private property, and access in recent years has been limited (Fig. 2).

King (1948) traced the "post-Lamar beds" to the base of the Reef Escarpment and commented, "at McKittrick Canyon and elsewhere, the Lamar Member is overlain by a small thickness of massive dolomitic Capitan limestone. This limestone is evidently equivalent to the sandstones and slabby limestones not far to the southeast, but the actual connection between them has been removed by erosion, so that the change cannot be traced from one outcrop to the other."

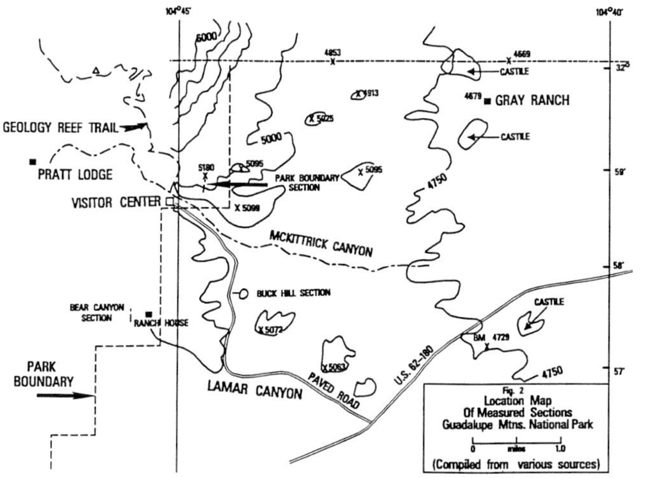

FIG. 2.—Location map of measured sections, Guadalupe Mountains National Park.

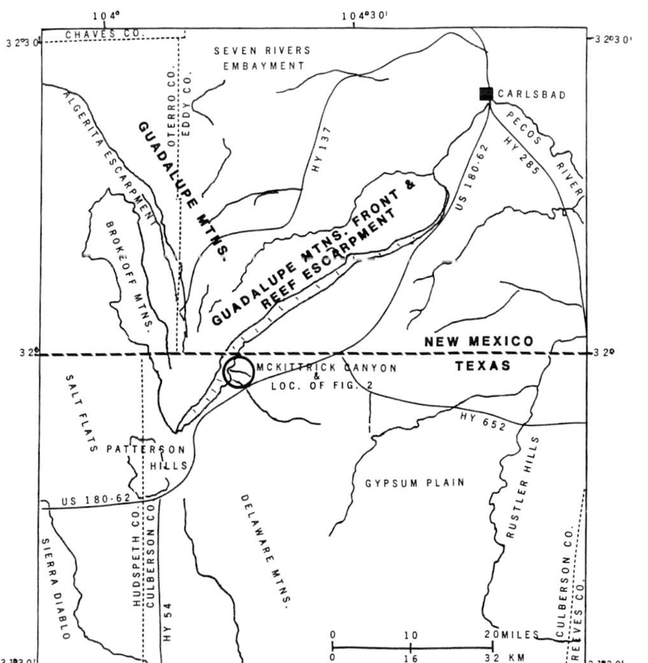

FIG. 2A.—Sketch map of west Texas and southeastern New Mexico, showing location of Fig. 2 (modified after Hill, 1996).

Geologists who recently prepared the Permian Reef Geology Trail Guidebook 26 (Bebout and Kerans, 1993) traced the "post-Lamar beds" from shelf to basin by working along both the north and south walls of McKittrick Canyon (see especially Figs. 16, 20). That work proved what had been maintained for many years based upon fusulinacean studies (Skinner and Wilde, 1954, 1955; Wilde, 1955; Tyrrell, 1962, 1969; and Tyrrell et al., 1978), namely that the Lamar and "post-Lamar beds" are stratigraphic equivalents of the Tansill Formation on the shelf. The Tansill contains an identical fusulinacean fauna in the same sequence as the Lamar and "post-Lamar beds" (Wilde, 1990; G. L. Wilde and S. F. Rudine, in prep.).

At the Guadalupian II Symposium, Wilde and Rudine (1996a) introduced for the first time the name Reef Trail Member for the post-Lamar/pre-Castile beds described above. Later, Wilde and Rudine (1996b) presented this name at the International Field Seminar on the Permian Capitan System, and offered notice of plans to describe the new member formally, in association with Lambert, who is working with the conodont faunas. The present paper is the result of those plans. Subsequent papers will provide more detailed descriptions of the fauna.

Reef Trail Member, Bell Canyon Formation

The name Reef Trail Member derives from the proximity of the type section to the Geology Reef Trail at the Guadalupe National Park Visitors Center, McKittrick Canyon, Culberson County, Texas (location maps, Figures 2, 2A). The section begins at the top of the Lamar Limestone on a small escarpment less than one-half mile northeast of the Visitors Center and continues to the top of hill 5180 west of the Park boundary line. Remainder of the section was projected from earlier collecting to hills 5195 and 4913 to the northeast and farther downdip. The total section thus measures approximately 12.6 m (41.25 ft).

At the type section (Fig. 3B), the lowest unit above the Lamar comprises 2.75 m (9 ft) of yellowish weathering calcareous siltstone, which forms a partially covered slope. At its top is a gray lime mudstone with a maximum thickness of 2.0 cm. (5 in). This is followed upwards by a gray-weathering, talus-covered, platy, silty lime mudstone, 0.91 m (3.0 ft) thick, and capped by another lime mudstone 1.8-2.0 cm (4.5-5.0 in) thick. This, again, is followed upward 1.4 m (4.5 ft) to the top of hill 5180 by a third cycle of gray, platy, silty lime mudstone, capped by a 1.2-cm (3.0-in) thick lime mudstone bed. Total section at this locality from Lamar Limestone to top of hill 5180 is 5.5 m (18.0 ft).

Remainder of the projected section, east of the Park boundary, comprises alternating beds of yellowish weathering siltstone with occasional interbeds of thin, gray lime mudstone. Contact with the overlying light-colored Castile evaporite is always abrupt but gradational. Such is the case wherever the contact has been seen, both at the surface and in the subsurface (King, 1948; Wilde, 1975; Cys, 1978).

Greater details of the lithostratigraphy of the Reef Trail Member type section are offered in Fig. 3B, Plate 3, and

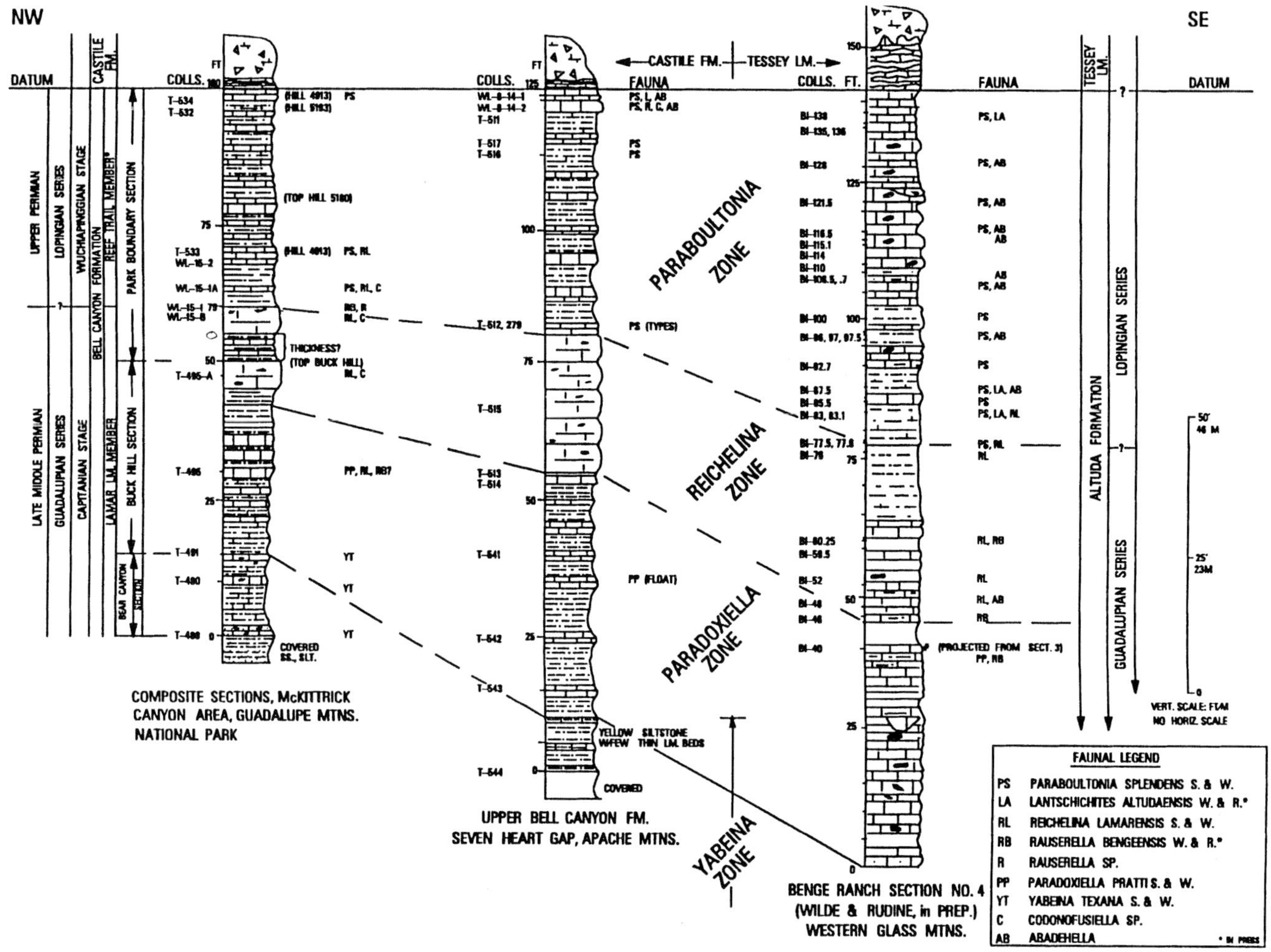

FIG. 3.—Stratigraphic cross-section of Permian beds (Guadalupe-Apache-Glass Mountains), west Texas, showing faunal occurrences and zonation.

Appendix B. The section is divided into seven sedimentary cycles consisting of varying thicknesses of gray, platy, silty lime mudstone, wackestone, and packstone, with numerous thin, silty shale partings. Commonly the capping limestone of each cycle contains a detrital mixture of slope, toe-of-slope, and rare reef-derived faunas and algal fragments. Most likely, the algae, all dasycladacean, were swept through or over the diminished reefs from the shelf.

More than these seven cycles undoubtedly exist in the Reef Trail, but regardless of the number, they appear to reveal a pattern of siliciclastic and carbonate mud deposition, interrupted occasionally by turbidity flow of fine carbonate materials derived from periodic buildups along the shelf margin.

Most of the limestones, regardless of their grain content, are mud-supported, which offers the appearance on weathered surfaces of rather smooth, gray, lime mudstones. Only in thin section in some instances can one discern the great variety of finely comminuted fossil types.

Laurel Babcock (1977) undertook a rather comprehensive paleoecological study of the underlying Lamar Limestone, which bears somewhat upon the sedimentology of the Reef Trail Member, although she did not extend her studies above the Lamar. The main differences between the two members are rather obvious in that abundant reef growth occurred throughout Lamar time, whereas during Reef Trail time, reef growth was somewhat restricted, and indeed, absent much of the time. This allowed a more even distribution of silt and sand material entering the basin.

Nevertheless, as far as the Lamar was concerned, Babcock's (1977) studies revealed that life was quite abundantly represented as far as nine miles out into the basin. Much of the material had been washed in from the basin margin; however, many forms of both benthonic and pelagic life existed far beyond the basin edge.

The same probably was true for the Reef Trail Member also, but in a more restricted way. With abundant life shutting down along the shelf margin, with evaporitic conditions encroaching from the surrounding shelves, and with silt and sand pouring into the basin, life became difficult. By Castile time, practically all life was gone from the Delaware Basin.

One obvious fact remains, however, fresh sea water, although possibly altered somewhat in chemical content from earlier times, continued to flow into the Delaware Basin right up to the beginning of Castile deposition, because, as we shall see, fusulinacean and conodont faunas continued to proliferate, and

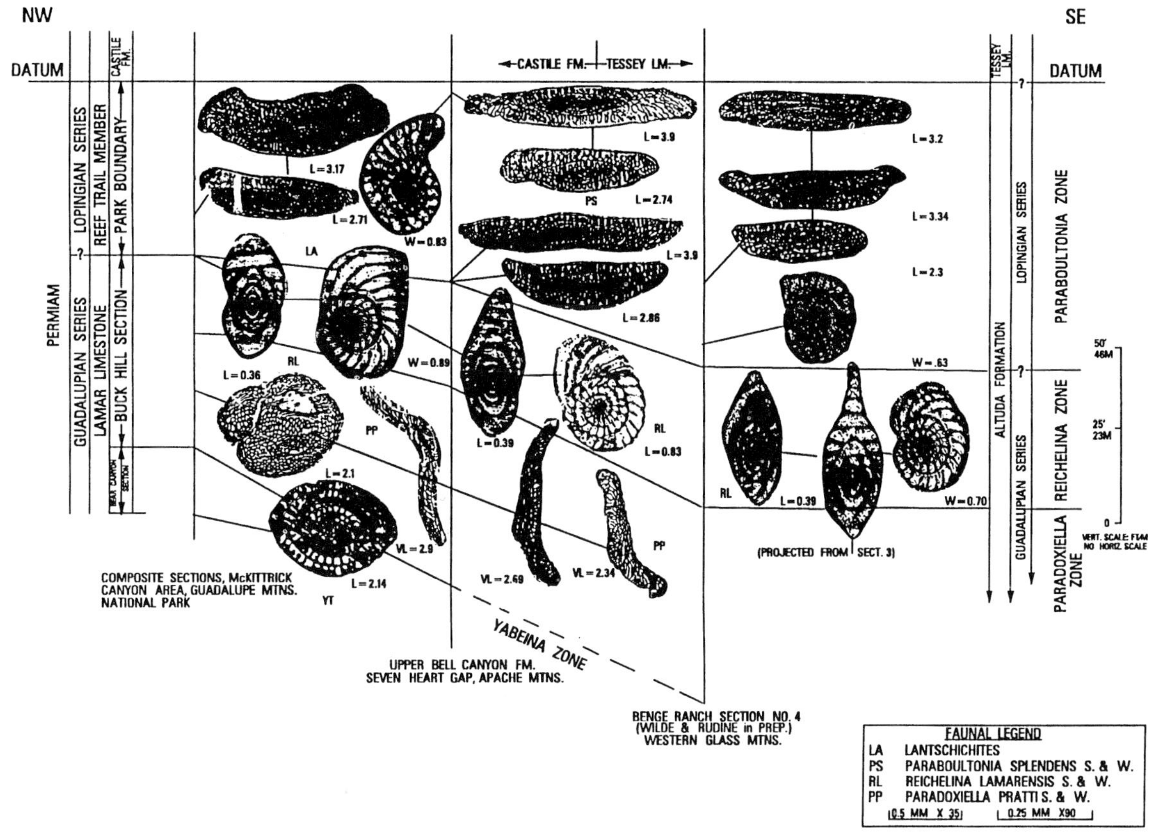

FIG. 3A.—Same as Fig. 3, fauna only.

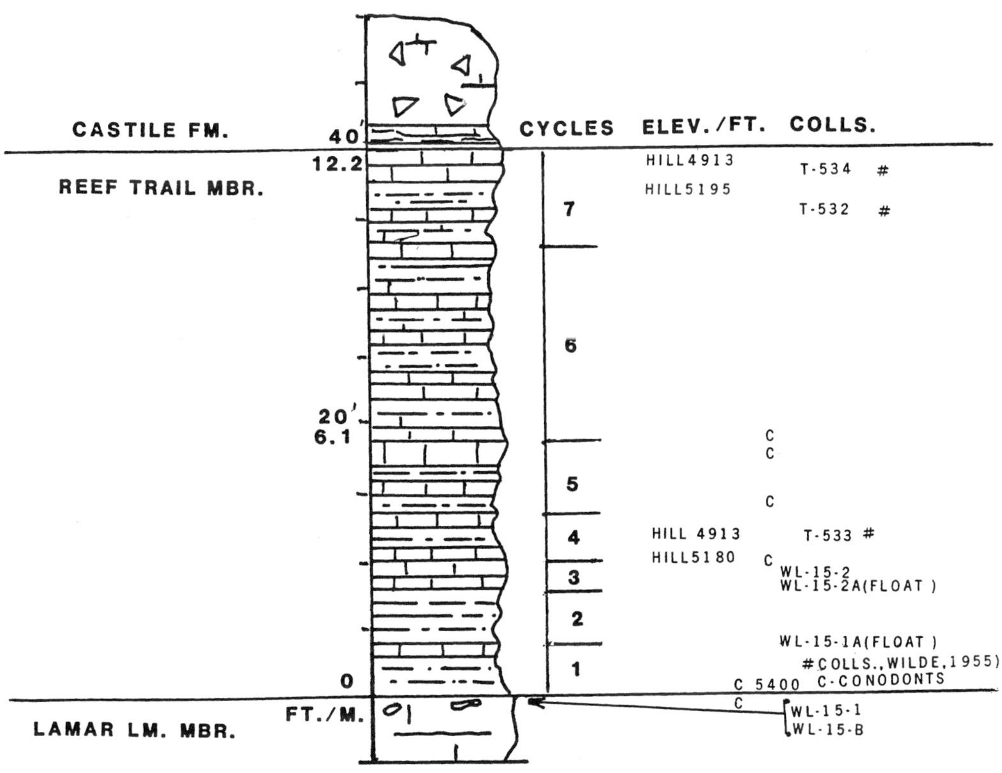

FIG. 3B.—Lithostratigraphy of stratotype section, Reef Trail Member, Bell Canyon Formation, showing colls. described in Appendix B and Pl. 4.

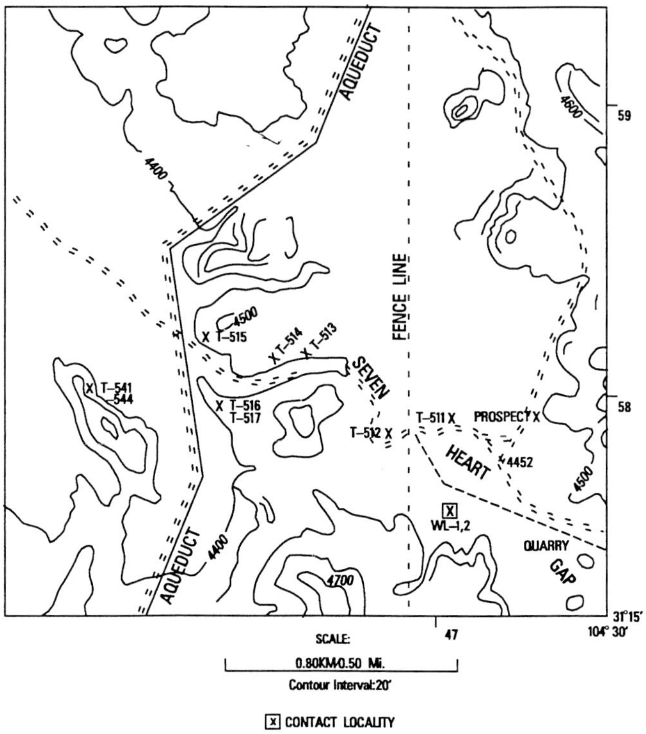

FIG. 4.—Topographic map, Apache Mountains, showing location of Castile-Bell Canyon contact and fusulinid colls. (After Cys, 1978)

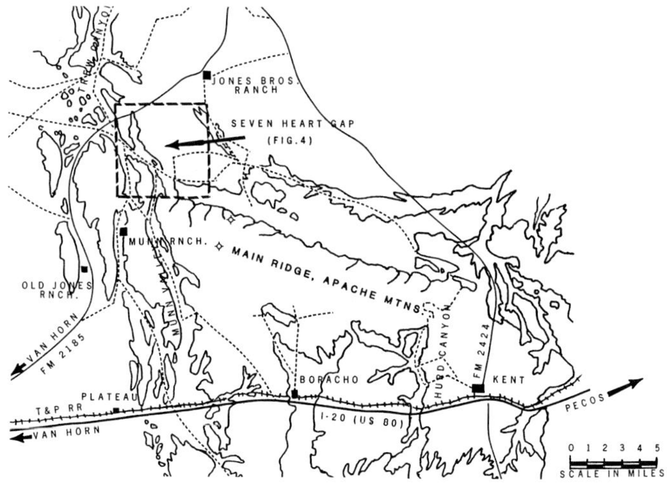

FIG. 4A.—Sketch map, Apache Mountains, showing location of Fig. 4.

evolved in a manner similar to those in other parts of the world. Closing of the Hovey Channel altered deposition in the Delaware Basin for the remainder of Permian time.

More nearly complete sections are possibly available for the Reef Trail Member; however, problems of access give the above section priority at this time. Also, the location of this section near the depositional toe of slope allows an integration of fusulinacean and conodont faunas, whereas a section farther out in the basin might contain conodonts and other pelagic faunas, but fail to encounter useable fusulinaceans.

Lamar – Reef Trail Sequences In the Guadalupe Mountain National Park

It is important to understand both the Lamar and Reef Trail faunas together in order to grasp their importance to chronostratigraphy. As a consequence, the following discussion seeks to provide a background for the composite section shown in Fig. 3 for the area around McKittrick Canyon. Fig. 3 also illustrates the relationship of these sections to those in the Apache and Glass mountains. Fig. 3-A illustrates representative fusulinid markers from these same sections.

The basal unit, which we call the Bear Canyon Section (Skinner and Wilde, 1955), is located near the head of Bear Canyon on a small spur south of the draw over 183 m (about 200 yds) southwest of a water well pumping unit. Locality is one-half mile west of the Wallace Pratt home, which he affectionately called "La Casa Manzanita" (actually a madrone), "The Home of the Little Apple", and which the Park Service now calls "Ship on the Desert", referring to its structure. This name might also have been suggested by Wallace Pratt. From the house the locality is on a bearing of 277°. Here the Lamar Limestone is largely in a siltstone, sandstone and silty detrital limestone phase. Its base is possibly not seen, and there is a covered slope of sandstone and siltstone measuring 1.5 m (5.0 ft) overlain by less than 1.0 m of hard, cherty, silty limestone. The remainder of the Bear Creek Section consists of 4.0 m (13.75 ft) of yellowish, interbedded siltstone and silty lime mudstone with two thin beds of detrital lime mudstone near the top. Total section collected is nearly 6.5 m (21.3 ft). *Yabeina texana* Skinner and Wilde was found at three levels in this section.

The next higher section of Lamar is at Buck Hill (Figs. 2 and 3, Skinner and Wilde, 1955), located just to the east of the paved road into McKittrick Canyon, and little over a mile east-northeast of the Bear Canyon Section. Here the Lamar is in a more toe-of-slope position, with less siltstone and more fossiliferous lime mudstone, and with a total section of 11.0 m (36.25 ft). At its top is a 1.5 m (5.0 ft) section of relatively light-gray colored, cherty, fossiliferous lime mudstone and wackestone. The beds contain abundant *Reichelina lamarensis* Skinner and Wilde. The slope below is the zone of *Paradoxiella pratti* Skinner and Wilde, which occurs mainly in three or four thin limestone beds near the middle of the slope.

Tracing of the Buck Hill Section into the next and highest section (Park Boundary Section, Figs. 2, 3) has not been accomplished with precision, other than by reasonable projection. Therefore, the interval from the top of Buck Hill to the base of the highest Lamar Limestone unit at the Park Boundary Section has only been estimated. This interval appears to consist of thin beds of lime mudstone with silty shale interbeds. This entire portion has been assigned a thickness of 3.0 m (10 ft), but it might be twice that thickness. Total thickness for the three composited sections thus measures approximately 20.5 m (67.25 ft). This thickness is in reasonable agreement with that of King (1948) whose Section 32, near the McCombs ranch house, measured 22.25 m (73.0 ft).

The covered interval at the base of the Bear Creek Section possibly should not be included in the Lamar total thickness, with the 2.0 m thickness of hard, cherty, silty limestone representing the base of the Lamar. This cherty limestone contains the lowest occurrence of *Yabeina texana* Skinner and Wilde thus far found in the Guadalupe Mountains. This same bed also contains *Codonofusiella*.

Lamar – Reef Trail Faunal Zones

Fusulinacean biostratigraphy of the Lamar Limestone has been mentioned in general, above. As a way of comparison to the fusulinaceans of the Reef Trail Member, a summary is in order.

In the past (Skinner and Wilde, 1954, 1955; Wilde, 1955, 1975, 1990), the Lamar and "post-Lamar beds" have been referred to as the Zone of *Paraboultonia splendens* Skinner and Wilde. Recently, however, a review of some of the older data, supplemented with new collections, reveals that *Paraboultonia* is rare to absent in the Upper Lamar. On the other hand, it is common to abundant in some of the "post-Lamar" (Reef Trail) beds.

This zonation is closely similar to that which has been observed in the Tansill Formation on the shelf in the Guadalupe Mountains (Wilde, 1955, 1975; Tyrrell, 1969; Tyrrell et al., 1978). These last listed workers pointed out that *Paraboultonia* is restricted to the Ocotillo Silt Member and younger beds of the Tansill. We believe that the Ocotillo Silt Member correlates to the top of the Lamar in the basin, and marks an important sequence boundary (Kerans and Harris, 1993).

Occurrence of the genus *Abadehella* in the Reef Trail Member and its equivalents is of more than passing interest. This tetrataxid-like form was first described from the Abadeh beds of Iran (Okimura et al., 1975), and has been reported from many areas containing Late Permian rocks (Loeblich and Tappan, 1988). *Abadehella* was observed in a single collection in the Reef Trail Member, in equivalent rocks at Seven Heart Gap, Apache Mountains (Plate 1), and in the Altuda Formation of the Western Glass Mountains.

In the Altuda, however, *Abadehella* appears to have gotten its start at least as early as the main zone of *Reichelina*; in other words, as early as upper Lamar Limestone equivalent time, although it was not seen that low in any of the Guadalupe or Apache Mountain collections.

At present there exists a possible problem with some of the reported occurrences of *Abadehella*. In the Guadalupe and Glass mountains we have never observed *Abadehella* in rocks older than Late Guadalupian. In fact, a single, poor specimen of questionable identity, observed with *Paradoxiella* in the Buck Hill Section of the Lamar Limestone, is the oldest occurrence of the genus that we have seen in the three mountain areas reported herein.

Recently, Stevens et al. (1997, Table 1), reported the occurrence of *Abadehella* sp. in Upper Wordian rocks of the Guadalupe Mountains, but offered no supportive evidence. In the Guadalupe Mountains, to our knowledge, *Abadehella* has never been seen anywhere in the *Polydiexodina* zone, which occupies a major portion of the Late Guadalupian (Capitanian). Thus, the possibility of a Wordian appearance of *Abadehella*, that is, prior to *Polydiexodina* time, seems unlikely there.

In the same Table 1, Stevens et al. (1997) also reported *Abadehella* sp. in Wordian rocks from the Cache Creek terrane of British Columbia and northwestern Washington. No confirming documentation was offered by these authors; therefore, these occurrences are at present unverified. Rui and Nassichuk (1996) published a rather thorough study of the fusulinaceans from the Cache Creek, which they reported to be of Wordian age, but they gave no hint of *Abadehella* occurrences.

The reported occurrence by Stevens et al. (1997) of *Abadehella* sp. from the subject area of their paper, the Chugach terrane, Kenai Peninsula, Alaska, was based upon a single, poorly preserved, but illustrated, specimen of questionable affinity.

Davydov et al. (1996) recently published a foraminiferal study of the Permian of the Koryak terrane, northeastern Russia, in which they identify *Abadehella* from "Early Midian (Upper Wordian)" limestones. The associated fusulinacean faunas appear to represent a mixture of age groups, typical of accreted terranes, and thus difficult to assess. These workers, however, agreed with Leven (1993) as to Tethyan Midian associations.

Finally, Kobayashi (1997a) recently reported a new species of *Abadehella* from the Upper Permian of the Iwai-Kanyo Area, West Tokyo, Japan, which he correlates to the Changxingian type section of South China. In an earlier paper, however, Kobayashi (1996) discussed, from an evolutionary point of view, both Middle to Late Permian occurrences of *Abadehella*. Although we have not seen that paper, we do take note of the fact that Kobayashi recognizes earlier representatives of the genus than its occurrence in the Late Guadalupian of the Delaware Basin.

Concerning assumptions of earlier ages of certain fusulinacean genera, we also take note of a very recent paper by Kobayashi (1997b), in which he assigned *Polydiexodina oregonensis* Bostwick and Nestell (1965) to *Eopolydiexodina* Wilde, and subsequently placed the species in the Roadian Stage. This is clearly a misidentification of the genus and subsequent chronostratigraphic "fit" based on the earlier assumption. *Eopolydiexodina* is of Middle Guadalupian (Wordian) age; *Polydiexodina* is of Late Guadalupian (Capitanian) age. This is only an example, and not meant as critical response; but it does represent the kinds of assignments that engender problems in chronostratigraphy, of which *Abadehella* is one of the many.

We cannot categorically deny the presence of *Abadehella* in rocks older than late Guadalupian; however, we do emphasize that if this scenario is true, the genus must have migrated out of the Permian Tethys by End Middle Guadalupian time prior to entering the Delaware Basin. We also believe that this late arrival must be taken into consideration when correlating the Tethyan realm with the North American realm, particularly in the accreted terranes.

The lenticular fusulinacean genus *Reichelina*, whose earliest known occurrence in the Guadalupe Mountains is with *Polydiexodina*, reached its zenith in the upper Lamar and Reef Trail members. Little difference has been noted in the species *Reichelina lamarensis* Skinner and Wilde in the two members; however, detailed studies have not been completed (see Plate 1). The Reef Trail forms may be somewhat smaller, at least in axial length, than those of the Lamar.

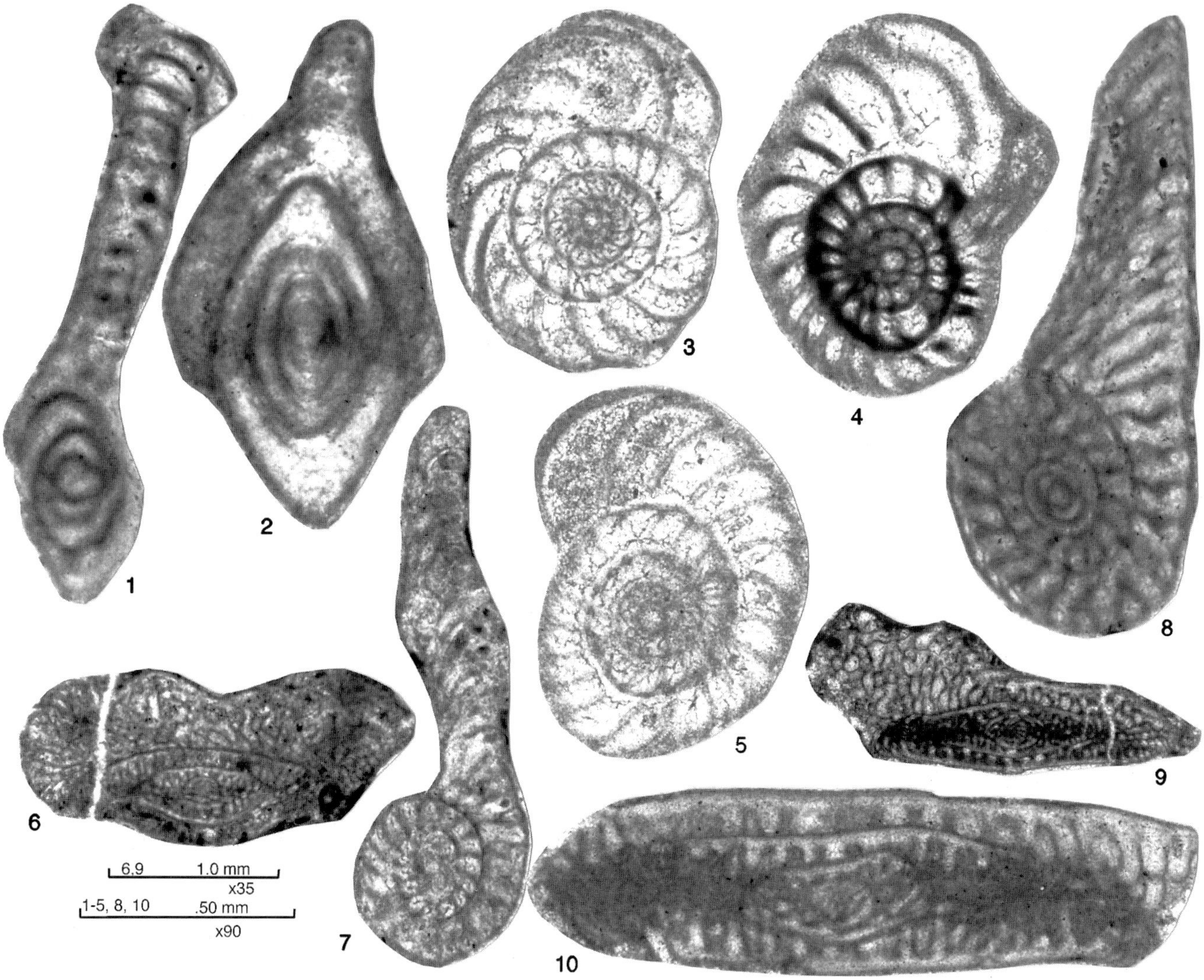

PLATE I.—Examples of Fusulinaceans.*

Fig. 1—*Advenoreichelina apacheensis*, n.gen., n. sp., X80, Coll. 2-AP, top of westernmost hill of the Apache Mtns., east of paved county road, FM 2185, Culberson County, Texas.

Figs. 2–5—*Reichelina* n. sp., X90. Fig. 2, axial; Figs. 3–5, sagittals. Col. HY-POLY, U.S. Hwy. 180, near Patterson Hills, Guadalupe Mtns., Hudspeth County, Texas. We believe this to be a down-faulted section of McCombs Limestone.

Fig. 6—*Codonofusiella* sp. A, axial section, X35. Coll. WL-8-14-2, Upper Bell Canyon Formation, 5.0 ft below Bell Canyon-Castile contact, Seven Heart Gap Section, Apache Mtns.

Fig. 7—*Codonofusiella* sp., sagittal section, X90. Slope of Buck Hill, *Paradoxiella* zone, Guadalupe Mtns.

Fig. 8—*Codonofusiella* sp., sagittal section, X90. Coll. WL-15-2, Park Boundary Section, Guadalupe Mtns.

Fig. 9—*Lantschichites* ? sp., axial section, X35. Coll. WL-8-14-1, 2.0 ft below Bell Canyon-Castile contact, Seven Heart Gap Section, Apache Mtns.

Fig. 10—*Paraboultonia* ? sp., axial section, X90. Same coll. as in Fig. 9.

*Specimens to be reposited at University of Iowa (SUI) upon completion of faunal studies.

By Reef Trail time other exotic, small, but complex fusulinid genera developed. *Paraboultonia*, an elongate boultonid, which became inflated in the final whorl but failed to uncoil, is the dominant form present. Rare specimens may occur in the Lamar Limestone, but the genus dominates the Reef Trail Member (Wilde, 1955). Occurring with *Paraboultonia splendens* Skinner and Wilde are undescribed species of the codonofusiellid subgenus, *Lantschichites* Toumanskaya, 1953, which is like *Paraboultonia* but which uncoil completely in the last whorl (Wilde, 1990; G. L. Wilde and S. F. Rudine, in prep.). A new species occurs abundantly with *P. splendens* Skinner and Wilde in the Glass Mountains, but that species has not been identified in the Reef Trail section with certainty. Other undescribed species of *Lantschichites* may be present, however (see Fig. 3A and Plate 1).

Accompanying the aforementioned forms are several small, undescribed fusulinacean and nonfusulinacean species not observed in the Lamar Limestone below.

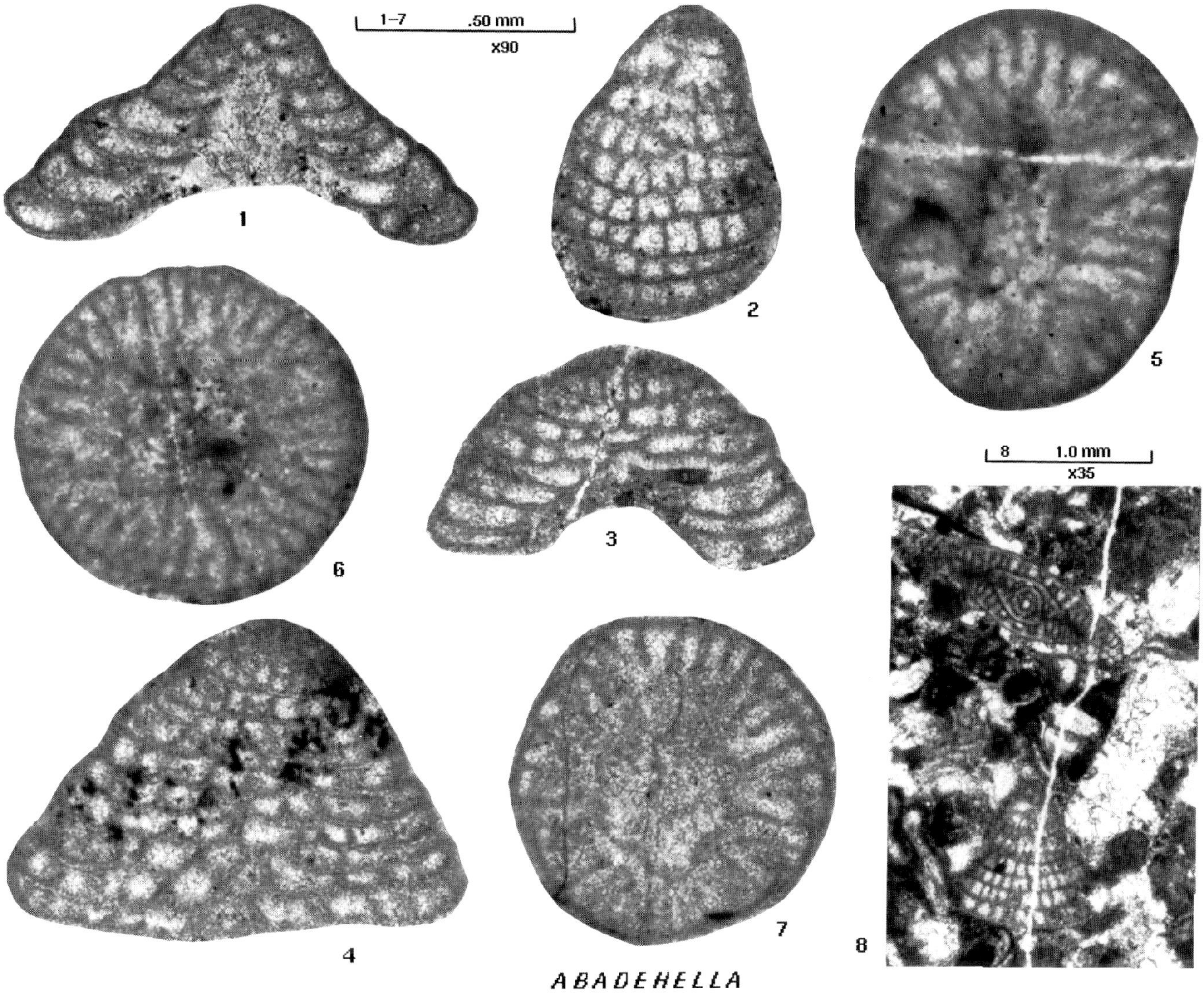

PLATE II.—*Abadehella*.*

Figs. 1–8—*Abadehella* cf. *coniformis* Okimura and Ishii, 1975. Figs. 1–7; Fig. 8, X35, also depicting a coiled axial section of a probable *Codonofusiella* with uncoiled portion missing. Bell Canyon Formation, Seven Heart Gap, Apache Mountains; within 5.0 ft of transitional contact of Bell Canyon and Castile formations.

*See Plate I.

APACHE MOUNTAINS

Seven Heart Gap Section

Approximately 97 km (60 mi.) southeast of McKittrick Canyon, at the confluence of the Delaware escarpment and the Apache Mountains, (Figs. 1 and 4) is Seven Heart Gap, which forms an erosional break between these two entities. The Delaware escarpment exposes the basinal equivalents of the Leonardian, Guadalupian, and Lopingian? (Castile) rocks of the western Delaware Basin. The Apache Mountains expose Capitan reef on the southwest side of the Delaware Basin. Indeed, were it not for Tertiary blockfaulting of Salt Flat Graben, the Capitan reef of the Apaches would probably connect its equivalents in the southern Guadalupe Mountains.

At the east end of the Apache range, however, these same rocks dip into the subsurface, and are not seen again until they emerge in the Glass Mountains, nearly 129 km (80 mi) from the Apache Mountain outcrops. The subsurface trend, however, was interrupted in late Capitan time in the form of a narrow channel, called the Hovey Channel (Kroenlein, 1939). Existence of a "Hovey Channel" has been questioned by Hill (1996); however, most workers have employed its existence to explain a most likely connection to the open sea during Permian time. The later closing of this "channel" sealed off the Delaware Basin from the conditions that had fed the vigorous reef life of the Capitan. Evaporitic conditions thus ensued, viable life shut down, and the Castile rocks began to fill the Delaware Basin. We suggest that these changing conditions might have begun in at least two earlier pulses: (1) at the time of the demise of the huge *Polydiexodina* faunas (Wilde, 1955), and (2) near the close of

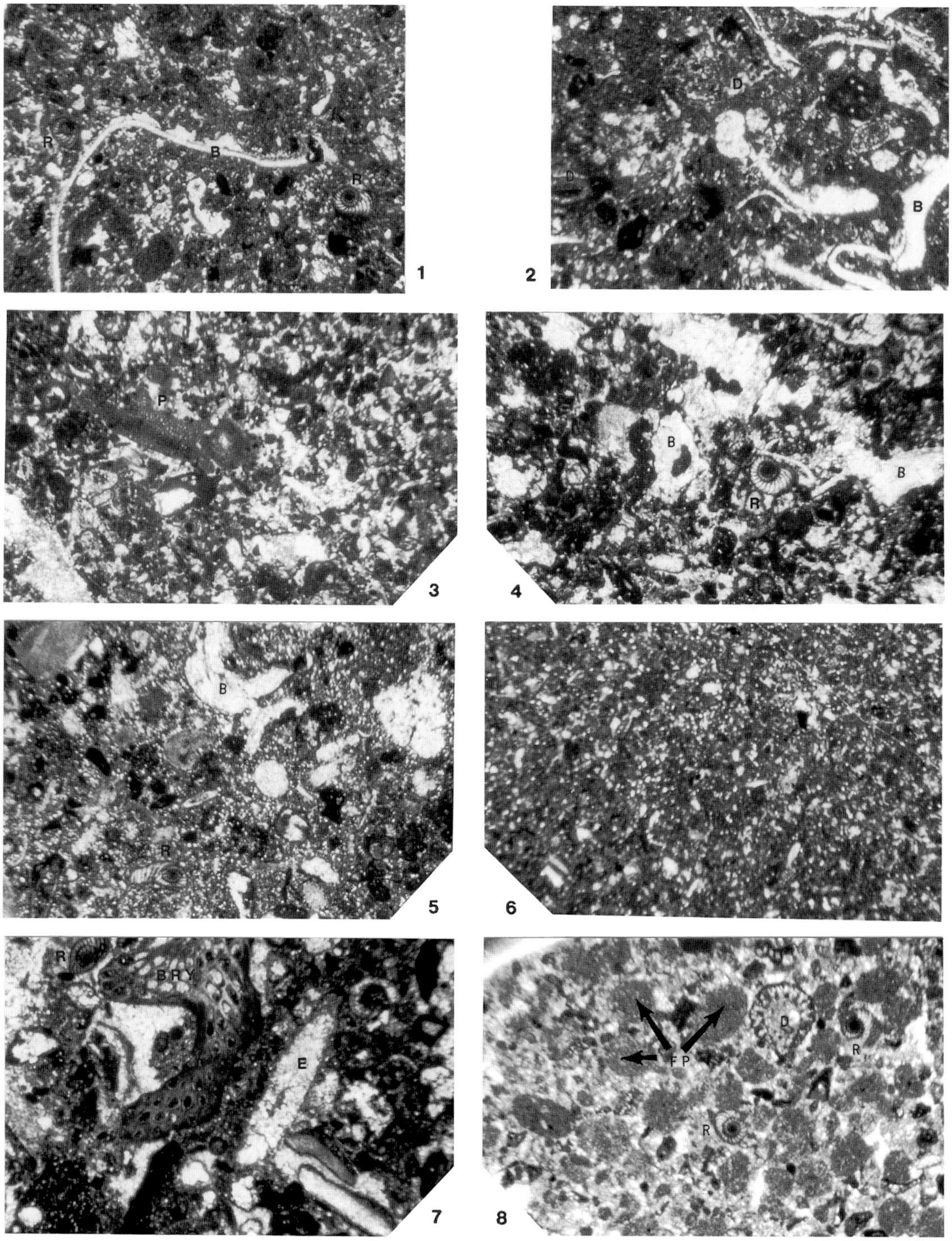

PLATE III.—Thin Section Identifications of Some Lamar-Reef Trail Lithologic Types (all magnifications X10).

Legend for photomicrographs on this plate: R, *Reichelina;* P, *Paraboultonia;* D, Dasyclad algae; B, Brachiopod; BRY, Bryozoan; E, Echinoderm spine; FP, Fecal Pellet.

Figs. 1, 2.—Top Lamar Limestone Member. Coll. WL-8-15-1. Fossiliferous, lime wackestone, with *Reichelina,* brachiopods, and dasyclad algal fragments.

Figs. 3–5.—Reef Trail Member. Float sample, WL-8-15-1A, just above Lamar-Reef Trail contact. Figs. 3, 4, siliceous, fossiliferous lime packstone, and Fig. 5, silty lime packstone. Contains *Reichelina, Paraboultonia,* and *Lantschichites,* bryozoan and echinoderm fragments.

Figs. 6, 7.—Reef Trail Member. Coll. WL-8-15-2, 9.3 ft (2.8 m) above base in Park boundary measured section. Fig. 6, fine-grained, silty, lime mudstone to finely comminuted lime wackestone. Fig. 7, siliceous lime wackestone to packstone with much fragmented fossil material. Contains *Paraboultonia, Lantschichites, Reichelina,* and fragmented brachiopods, gastropods, bryozoans, sponges, echinoderms, and dasyclad algae.

Fig. 8.—Reef Trail Member. Float sample, Coll. WL-8-15-2A. Fecal pellet, slightly fossiliferous, lime wackestone, with scattered *Reichelina* and dasyclad algae *(Mizzia?).*

Lamar Limestone deposition. The last-named may well have coincided with changes of a global magnitude (Wilde, 1998).

Returning to the section at Seven Heart Gap, where access problems have hampered progress for a number of years, a sequence of upper Bell Canyon Formation of approximately 38 m (125 ft) has been pieced together from past work as well as more recent work at the Bell Canyon-Castile boundary (Fig. 3).

Collections from past years, not presently in our possession, are noted on the section. These collections must be examined or recollected in the future. Wilde had a note in his files to the effect that a float sample containing the tiny *Yabeina* had been picked up near the base of the section and later in the laboratory filed as "*Yabeina* float in top miscellaneous drawer, no number." At the time the sample was found, its interval appeared to belong to a yellow siltstone unit containing a few thin limestone beds, partially covered. For practical reasons, and because of its lithologic similarity to the Bear Canyon section in the Guadalupes, the top of the *Yabeina* zone has been drawn at the top of the yellowish siltstone unit. The exposed interval is 3.0 m (10.0 ft) thick.

Some 7.5 m (25 ft) higher in the section another float sample was collected containing *Paradoxiella pratti* Skinner and Wilde. This sample remains in the senior author's possession. Collections T-541, 542, and 543 also contain that species; however, these are not in the author's possession, and must be recollected.

The section representing the *Paradoxiella* zone at Seven Heart Gap consists of 14.0 m (47 ft) of buff gray alternating silty shale and fossiliferous limestone. Its top is drawn at the base of the thickly bedded limestone unit here designated as the *Reichelina* zone.

Above what we presume to represent the *Paradoxiella* zone is a well-developed section of fossiliferous, cherty limestone filled with *Reichelina lamarensis* Skinner and Wilde. The section is 7.6 m (25 ft) thick, and we believe that it represents the same zone of upper Lamar as seen in the McKittrick Canyon area, but perhaps is somewhat thicker.

Immediately above this thickened carbonate unit is a 6-in bed of limestone that is the type locality for *Paraboultonia splendens* Skinner and Wilde. Here the specimens are so abundant that Wilde (1955) was inclined to calculate their numbers in the millions per cubic foot! This bed appears to represent the base of the *Paraboultonia* zone, which is a little more than 13.7 m (45 ft) to the base of the Castile. Interestingly, *Paraboultonia, Lantschichites, Codonofusiella*, and *Rauserella*, accompanied by numerous nonfusulinacean forms, occur to within less than 2.0 ft of the base of the Castile Formation.

GLASS MOUNTAINS

Benge Ranch Section, Upper Altuda Formation

In 1984, Rudine measured and collected a number of sections of the upper Altuda Formation in the western Glass and northern Del Norte mountains (Figs. 1 and 5). Later, Rudine invited the senior author to join in a study of the fusulinaceans, which resulted in a paper describing the fauna (G. L. Wilde and S. F. Rudine, in prep.). We include one of those measured sections (Benge Ranch, Section 4) in the present paper to show the zonal relationships (Fig. 3).

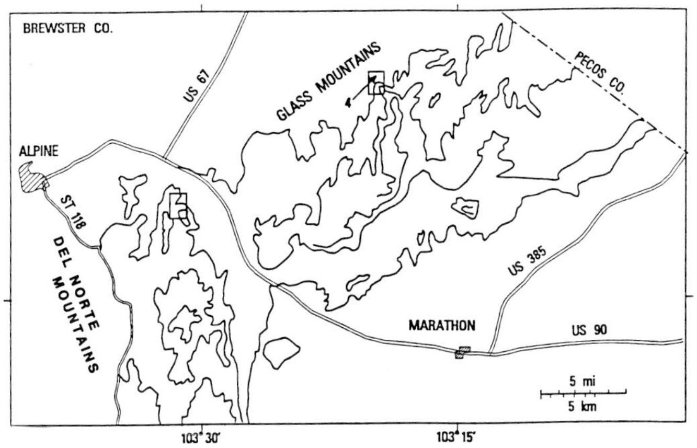

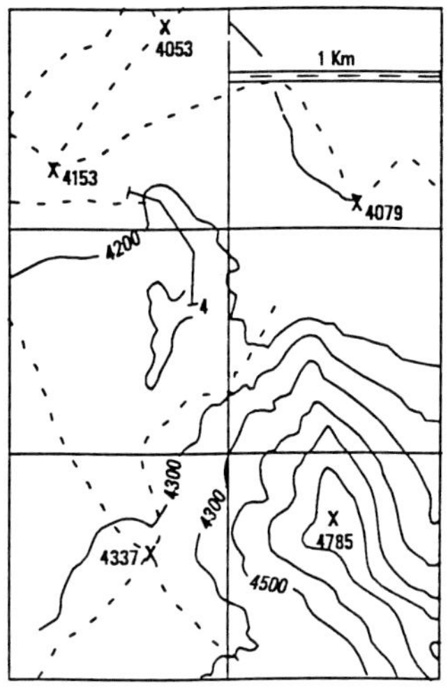

FIG. 5.—Location of Benge Ranch section 4, Western Glass Mountains.

We have no assurance from our collections that the *Yabeina* zone is even present in the Glass Mountains, but there is little reason to suggest its absence. As a consequence we only suggest the possibility of its presence. One way this can be done is by projecting upwards from a known level to the base of the next faunal zone. The highest (last) occurrence datum of *Polydiexodina* is known from collecting points in each of the four measured sections included in a paper by G. L. Wilde and S. F. Rudine (in prep.). Its highest occurrence was normalized to a straight datum for presentation. For the Benge Ranch, Section 4 (Fig. 3), thickness of the section from the highest *Polydiexodina* datum to base of section shown is 39.6 m (130 ft). Above this to the base of the referred *Paradoxiella* zone is possibly no more than 5 or 6 ft, offering a combined total of 41.0 m (135 ft). Inasmuch as very little sand was pouring into the Hovey Channel at that time, and due to the fact that the entire section is composed of partially cherty marine carbonates with numerous zones of sponge spicules and larger faunal components, it seems reasonable to suggest that the *Yabeina* fauna will ultimately be found in the Glass Mountains, if not the genus itself, perhaps an equivalent conodont fauna. Later, we shall discuss briefly the position of the highest *Polydiexodina* datum in the Guadalupe Mountain sequence.

The next higher zone above *Yabeina* is the *Paradoxiella*

zone. Again, there are problems regarding its extent in the Glass Mountains. At the time G. L. Wilde and S. F. Rudine were writing their paper (in prep.), only a single float sample containing this unusual genus was known, from Section 3, on Bird Mountain (formerly Altuda Mountain). We have been unable to return to these sections to date; therefore, we project the assumed level into the section presented herein, and below the earliest *Reichelina lamarensis* Skinner and Wilde of that zone a few meters above. In this instance we know that *Paradoxiella* is present in the Glass Mountains, but its exact level needs to be established in future work. By correlation, the *Paradoxiella* zone at Benge Ranch is given 11.81 m (38.75 ft) of section, but it appears to thicken slightly basinward until at Bird Mountain it is given a maximum thickness of 14.0 m (47 ft).

The overlying *Reichelina* zone is very well controlled in our collections from Benge Ranch. There it is almost 10 m (33 ft) thick, and also contains one of the tiniest species of *Rauserella* yet to be described. We believe that this same *Rauserella* is also present in the Guadalupe Mountains.

Earlier reference was made to the appearance of the tetrataxid genus, *Abadehella*, in the three mountainous areas discussed herein. At Benge Ranch, however, the genus first occurs in the lower part of the *Reichelina* zone, where it is rare but well documented. It appears to be poorly represented throughout the remainder of the *Reichelina* zone, and the individuals are quite small. *Abadehella* increases in abundance and size from about the base of the *Paraboultonia* zone upwards. We believe it is important to make note of this early occurrence of the genus, inasmuch as this might belong to an evolutionary continuum of importance, and should be factored in when considering chronostratigraphic relationships around the world.

Overlying the *Reichelina* zone in proper sequence is the *Paraboultonia* zone. At Benge Ranch it reaches 20.6 m (67.5 ft) in thickness, and contains a surprising addition to its faunal composition. Occurring with *P. splendens* Skinner and Wilde is an undescribed species of *Codonofusiella (Lantschichites)*, one which G. L. Wilde and S. F. Rudine (in prep.) are describing as *L. altudaensis* n. sp.

An interesting aspect of this new discovery is that it is now demonstrable that *P. splendens* never uncoils, but *L. altudaensis* n. sp. always uncoils. Otherwise the two forms have a somewhat similar appearance in the coiled portion of the shell. Toumanskaya (1953) described *Lantschichites maslenikovi* as a subgenus of *Codonofusiella* because of its uncoiling habit. Skinner and Wilde (1954) described *P. splendens* in 1954, emphasizing that the genus never uncoiled, but only inflated somewhat in the final whorl. After that time, however, most workers assumed that the two were the same. Only now can this problem be settled. Both forms appear to be approximately the same age regardless of where they are found, so that no chronostratigraphic problem exists, to our knowledge. In the Tethys, however, *Paraboultonia* and *Lantschichites* appear to have gotten an earlier start.

In summary, it appears that an identical fusulinacean zonation is present in late Guadalupian rocks in at least three mountain ranges along the western and southern boundary of the Delaware Basin. At Benge Ranch, by utilizing the total thickness from the highest *Polydiexodina* datum to the top of the Altuda Formation, the maximum thickness is 84 m (274 ft).

The McKittrick Canyon Limestone
Guadalupe Mountains

In the Guadalupe Mountains, recent observations by Brown (1996) have altered some previous understandings about the last-occurrence *Polydiexodina* datum. King's (1948) "flaggy limestone bed" was determined by Newell et al. (1953) to contain the highest occurrence of *Polydiexodina* in the basin. Recognizing the importance of this datum, Newell later prevailed upon King to join with him in formally naming the flaggy member as the McCombs Limestone Member of the Bell Canyon Formation (King and Newell, 1956). Its type section was designated as bed 15, in section 34 of King (1948).

It has been understood since the work of Newell et al. (1953) that the last occurrence datum of *Polydiexodina* on the shelf of the Delaware Basin is at or near the top of Yates "B" (Garber et al., 1989; Mruk and Bebout, 1993, see especially Fig. 24; Brown, 1996), and in the McCombs Member of the Bell Canyon in the basin (also see Tyrrell et al., 1978; and Wilde, 1990).

Most workers since King's original paper in 1948, including the senior author, have believed that the limestone at the mouth and floor of McKittrick Canyon was the McCombs Member (Skinner and Wilde, 1954; Wilde, 1955; Lawson, 1989). The confusion occurred as the result of assuming that the first prominent limestone below the Lamar Limestone Member would be the McCombs Member, because that is the case farther out in the basin, where the member was defined. However, Brown (1996) found that this "middle" limestone pinches out basinward approximately 3.0 km from the shelf break. Based upon Brown's tracing, the "middle" unit lies 10.6 m (35 ft) below the base of the Lamar at the prominent outcrop in the mouth and floor of McKittrick Canyon. By projection the actual McCombs lies approximately 6.1 m (20 ft) beneath the "middle" limestone at the same location. For more formal recognition of this important limestone unit, we herein designate it to be known as the McKittrick Canyon Limestone, named for its presence near the mouth of McKittrick Canyon.

Recognition that the McCombs and McKittrick Canyon limestones are separate entities is important for many reasons, not the least being that the location at the mouth of the canyon is the type locality for *Codonofusiella extensa* Skinner and Wilde (1954). *Polydiexodina* has not been found in collections taken from this limestone over a period of many years. Further evaluation of the range of *C. extensa* must now be made (Brown, 1996). This might also suggest that *C. extensa* could be found in the Glass Mountains somewhere between the *Polydiexodina* datum and the suggested *Yabeina* zone, which would aid immensely in the further delineation of these important zones.

CHRONOSTRATIGRAPHIC IMPLICATIONS OF THE REEF TRAIL MEMBER

Following many years of wrangling amongst biostratigraphers in meetings, published papers, and *Permophiles* discussion articles, the Permian Subcommission (ICS) has approved names and boundary levels for series and stages of the Permian System based upon marine successions (Fig. 6), as was reported by Jin et al. (1997). It is not the purpose of the present paper to restate all of the arguments as to why these series/stage names were chosen over others. In a separate section, Lambert offers the conodont evidence as it is presently known for establishing chronostratigraphic boundaries for the upper Guadalupian. Because the fusulinacean and conodont evidence for correlation are often difficult to reconcile, particularly during this period of instability in the conodont nomenclature, we present separate correlation charts for the fusulinaceans (Fig. 7) and the conodonts (Fig. 8).

Beginning with what is definitely known regarding the fusulinid evidence, we start at the base of the Lamar Limestone where we encounter the tiny *Yabeina* (*Yabeina texana* Skinner and Wilde). There has been much argument about the status of this tiny species. We believe that most of the contentiousness derives more from a misunderstanding of the chronostratigraphy than from the species itself. The species just happens to be of diminutive size, but it has all the characteristics associated with *Yabeina*.

As has already been discussed, *Yabeina texana* lies well above the highest occurrence of the large *Polydiexodina*, 16.7 m or 54.28 ft in the Bear Creek Section (Figs. 3, 3-A, 3-B), Guadalupe Mountains. Less than 10 m above the highest *Yabeina* is the first *Paradoxiella* (Buck Hill Section, Figs. 3, 3-A, 3-B). In Japan, it is well documented from the Akasaka Limestone that *Paradoxiella* occurs with the important zone marker *Yabeina globosa* (Yabe) (Sada and Skinner, 1977). In

SYSTEM/SERIES		STAGES
TRIASSIC		GRIESBACHIAN
PERMIAN	LOPINGIAN	CHANGHSINGIAN
		WUCHIAPINGIAN
	GUADALUPIAN	CAPITANIAN
		WORDIAN
		ROADIAN
	CISURALIAN	KUNGURIAN
		ARTINSKIAN
		SAKMARIAN
		ASSELIAN
CARBONIFEROUS		GZHELIAN

FIG. 6.—Permian chronostratigraphic subdivisions. (After Jin, Wardlaw, Glenister, and Kotlyar, 1997).

addition, Ishii and Takahashi (1960) had already described *Paradoxiella japonica* from the Kwanto Massif, albeit from conglomerates reworked from older strata. In other parts of the Tethys, Leven (1993) reports *Paradoxiella* with *Lepidolina*, but above *Lantschichites*!

Yabeina globosa (Yabe) is a very large and highly developed species of the genus, an excellent representative of the *Yabeina* zone. *Y. globosa* is often thrown into synonomy with *Y. inouyei* Deprat, although the former species was described some 8 years prior.

Occurring with *Yabeina* in many areas of the world is the more highly advanced genus, *Lepidolina* Lee. *Lepidolina* has often been considered merely an advanced *Yabeina* rather than

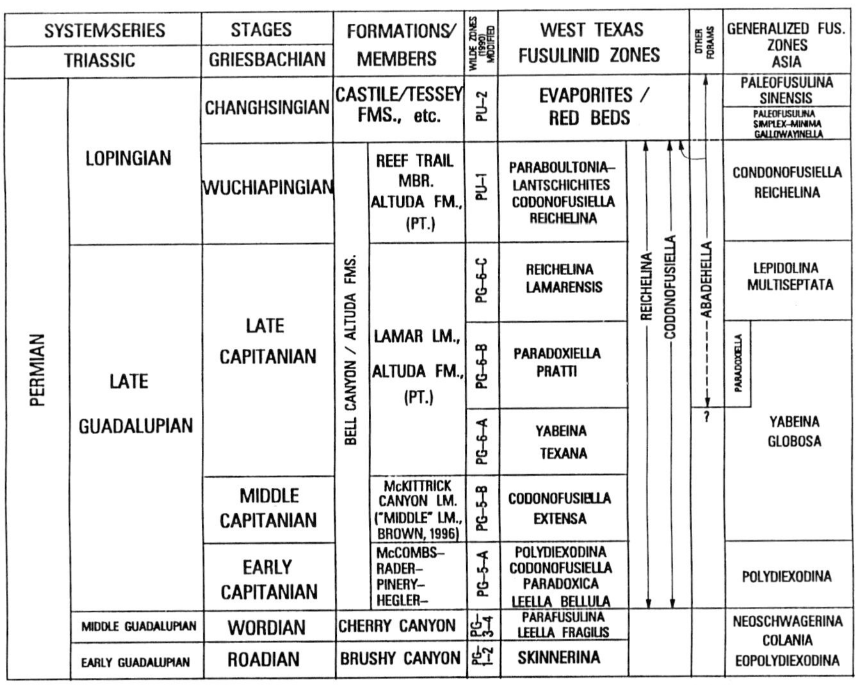

FIG. 7.—Suggested correlation of Fusulinacean biostratigraphic zones, to worldwide chronostratigraphic subdivisions.

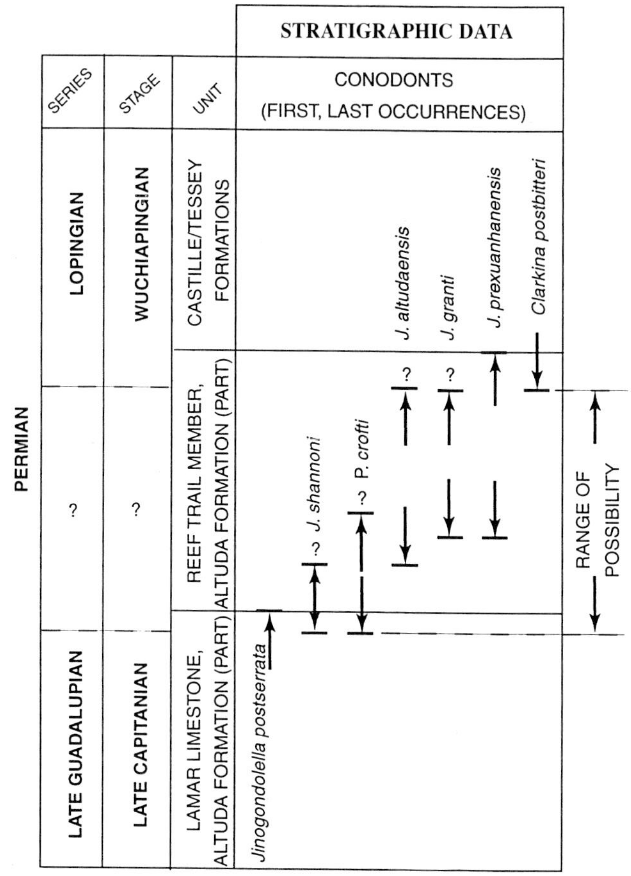

FIG. 8.—Range chart of biostratigraphically important conodonts near the Middle/Upper Permian chronostratigraphic boundary in west Texas. Ranges shown combine our results with those reported by other workers. Range of possibility refers to the interval that brackets the boundary, the precise horizon depending on which conodont species is formally selected for definition. Queries mark the coincident boundary intervals of the Guadalupian/Lopingian and Capitanian/Wuchipingian boundaries. Queries at conodont last occurrence horizons indicate that these species have been reported to range higher elsewhere. Note that Kozur is the only worker to report Protoclarkina crofti from the Lamar Member.

a separate genus (Skinner and Wilde, 1954; Loeblich and Tappan, 1988). We believe that the two genera are separate and distinct. Indeed, there may be some validity in the proposal by Rui and Nassichuk (1996) that *Lepidolina* arose from a separate *Neoschwagerina* stock through *Colania*. *Colania kwangsiana* Lee appears to bear *Neoschwagerina* characteristics in its early volutions, while displaying *Lepidolina* features in later volutions. However, the very same can be said of *Yabeina*. Thus, the point might be rather moot as to whether *Colania* is transitional to *Yabeina* or directly to *Lepidolina*. We must also point out that Minato and Honjo (1965), in their restudy of *Colania kwangsiana* Lee, asserted that *Colania* belongs in the "*Yabeina* globosa zone." Based upon the larger proloculus size in *Colania*, Minato and Honjo further assert that *Colania* is closer to *Lepidolina* than to *Yabeina*, a thesis with which we agree completely.

Regardless, the *Lepidolina* zone continues higher than the youngest *Yabeina* and has been correlated to some part of the Capitanian, to portions of the *Abadehella* beds in Iran, and to the *Reichelina-Codonofusiella* zone (Ingavat et al., 1980; Taraz et al., 1981).

As we mentioned earlier, *Abadehella* is first seen in the *Reichelina* zone at the Benge Ranch Section, Glass Mountains, west Texas; and it continues upward in greater abundance through the entire Reef Trail Member and its equivalents.

In the Kwangsi, Kueichow, and Szechuan provinces of China, Sheng (1963) has recorded species of *Lantschichites* (some are *Paraboultonia*) as occurring in the uppermost *Yabeina* zone. Indeed, it appears that the *Yabeina* zone in most parts of Eastern Asia includes both older elements (*Neoschwagerina*) as well as younger ones. The older elements never seemed to have arrived into the Permian Basin, and the younger ones appear to have arrived later than in Asia overall. More recently, Rui Lin (1981) discussed this very subject by pointing out that during late Maokou limestone deposition, two distinctive fusulinacean biogeographic provinces existed in the South China Sea with two distinctly differing faunas:

(a) Yangzi-Jiangnan Province
Lepidolina
Yabeina
Neomisellina
Neoschwagerina (advanced)
Lantschichites
Kahlerina (non-fusulinid)

(b) Southeast China Province
Neomisellina
Polydiexodina

It is obvious when one compares these faunal lists with those from the Guadalupe Mountains' Lamar and Reef Trail members among those genera that do occur, their arrival into the Permian Basin was somewhat later, which suggests a major faunal boundary in the upper part of the Lamar Limestone. From the standpoint of chronostratigraphy, we believe that this could also mark the Late Guadalupian (Capitanian)—Early Lopingian (Wuchiapingian) boundary. Earlier we (G. L. Wilde and S. F. Rudine, in prep.) suggested that the Reef Trail might be a Dzhulfian equivalent, and while this is also true in part, the Dzhulfian has been left out of the official chronostratigraphic stage names. Kozur (1992a) believes that "the whole Dzhulfian and a part of the Early Changxingian are included in the uppermost part of the Altuda Formation of the Glass Mountains, ...and that the fauna of the Lamar Limestone is...regarded as the topmost Guadalupian..."

The above discussion is important to an understanding of the fusulinacean zonation chart (Fig. 7) developed in the present paper (also see Jin, et al., 1997, Fig. 2). A number of years ago Rui Lin (1979) discussed the need for a refined *Codonofusiella* zone, with *Reichelina*, where the two genera occur above *Polydiexodina*, *Neoschwagerina*, *Yabeina*, and *Lepidolina*, and below *Palaeofusulina*.

This refined zonation of *Codonofusiella* and *Reichelina*, without the older and larger faunas, is precisely the picture of the Reef Trail Member in the Guadalupian Series. The only difference lies in the apparent late arrival of the *Paraboultonia-Lantschichites* faunas to North America, and the failure of the *Palaeofusulina* faunas to arrive at all, due to the closing of the

Delaware Basin.

Palaeofusulina faunas are still not completely understood because of recent splitting (*Nanlingella, Parananlingella*) by Chinese students, in a manner similar to the older *Dunbarula-Paradunbarula* separation. Numerous species have been erected, but geographic distribution is somewhat restricted. Conodonts must supply the necessary link for further chronostratigraphic understanding of this portion of the Lopingian.

SUMMARY OF THE CURRENT CONODONT ZONATION IN THE GUADALUPIAN-LOPINGIAN BOUNDARY INTERVAL

To expedite formal definition of the basal Lopingian boundary, a flurry of recent activity has taken place in China and elsewhere with regard to conodont zonation within that interval. This has resulted in considerable taxonomic flux for the present, because many new species and even genera have been erected without receiving the benefit of peer evaluation. The evolving taxonomic concepts and nomenclature have resulted in as yet unresolved disagreements in the literature, but most conflicts result from differences in systematics rather than lithostratigraphic interpretations (see Kozur and Mostler, 1995 vs. Mei and Wardlaw, 1996). A general consensus is already beginning to emerge from the taxonomic disorder. Because the fundamental work in China has been the focus of collaboration between Mei Shilong, Jin Yugan, and Bruce Wardlaw (Mei, et al., 1994a, b; c; Jin et al., 1994; Mei and Wardlaw, 1996; B. R. Wardlaw and S. Mei, pers. comm., 1998), their taxonomic concepts are followed herein.

In South China, eight conodont zones each have been assigned to the Maokouan (=Wordian + Capitanian) and succeeding Wuchiapingian (early Lopingian) stages. The contact between the Maokou Formation and the Wuchiaping Limestone (and equivalents) is disconformable across most of South China, marking a major low-stand sequence boundary (Mei and Wardlaw, 1996). Each of the conodont zones below that sequence boundary is a first-occurrence interval of the nominate species preceding the appearance of the subsequent zonal name-bearer. Either, or both, species may range into the next overlying zone. *Jinogondolella altudaensis*, for example, ranges through all the strata from the base of its zone to the sequence boundary. Thus the widespread zone of *J. altudaensis* is correlative between South China and North America, whether or not *J. prexuanhanensis* and successors are present to further subdivide that zone in west Texas. The *Clarkina* zonation for the early Lopingian, overlying the Maokouan-Wuchiapingian sequence boundary, was proposed in the same manner.

The Permian pan-tropical distribution of *Jinogondolella postserrata, J. shannoni*, and *J. altudaensis* (in ascending order) provide a worldwide conodont zonation for the Capitanian Stage. The *J. altudaensis* zone can be further subdivided as described above into the (ascending) zones of *J. prexuanhanensis, J. xuanhanensis*, and *J. granti* in South China, with a potential for recognizing these zones elsewhere. Another important faunal element of the late Capitanian is the relatively rare *Protoclarkina crofti. Protoclarkina* crofti is interpreted to be the direct ancestor of *Clarkina postbitteri* (B. R. Wardlaw and S. Mei, pers. comm., 1998), and thereby for the entire *Clarkina* clade. Because the conodont genus *Clarkina* is considered diagnostic for the Lopingian, the first occurrence of *C. postbitteri* has been proposed to denote the basal Lopingian boundary in China (Mei et al., 1994a). Reported appearances of *Protoclarkina crofti* seem to take place low within the zone of *J. shannoni*, and range through the late Capitanian and into the basal Lopingian.

Jinogondolella shannoni occurs commonly in the upper Altuda Formation of the Glass and Del Norte Mountains. That excellent record is reported to continue to the top of the formation, passing through a transitional form to *J. altudaensis* (B. R. Wardlaw and S. Mei, pers. comm., 1998). *Jinogondolella altudaensis* was named from the uppermost Altuda Formation (Kozur, 1992a). The uppermost Altuda Formation has also yielded specimens of *Protoclarkina crofti*, along with some advanced morphotypes of *P. crofti* bearing characters transitional to those of *Clarkina postbitteri* (B. R. Wardlaw and S. Mei, pers. comm., 1998). This requires a very late Capitanian (that is, pre-Lopingian) age for those strata. Some of the youngest specimens may represent *C. postbitteri sensu stricto* (B. R. Wardlaw, pers. comm., 1997), placing the highest Altuda rocks from which they were derived firmly within the Lopingian. Reported occurrences of late Lopingian *Clarkina* species (for example, *C. changxingensis, C. liangshanensis*, etc., Kozur, 1992a, b, c) are attributed to misidentification during this period of unsettled taxonomy (Mei and Wardlaw, 1996; B. R. Wardlaw and S. Mei, pers. comm., 1998).

In the Guadalupe Mountains, the Lamar Limestone has long been assigned to the zone of *Jinogondolella postserrata*, the species most commonly recovered from that unit. Various additional species have been reported to occur in the Lamar, but most appear to be misidentifications. Others, such as *J. denticulata* and *J. babcocki*, are now considered to be poorly circumscribed variants of *J. postserrata*. It should be noted that the holotype of *J. babcocki* is from the lower Lamar (Clark and Behnken, 1979), and its morphological characteristics fall within the broad range exhibited by *J. postserrata*. The morphologic concept widely applied to *J. babcocki*, however, is stable, and is commonly applied to upper Lamar forms now assigned to *J. shannoni* (Kozur and Mostler, 1995). The uppermost Lamar and "post-Lamar" have been reported to produce *J. shannoni, J.altudaensis*, and *Protoclarkina crofti*. No specific stratigraphic assignments are even attempted here for the Guadalupes from the literature, because many workers have assigned these names to different morphotypes and reassigned each other's taxonomic designations several times. Collections cited in those papers have yet to be placed within measured sections, nor have they been placed into any precise stratigraphic context for evaluation.

Although our results at this time are preliminary, they confirm the general succession of conodont zones described from China (Plates 4 and 5). In the slope mileau near the Capitan Reef, the upper Lamar has produced *Jinogondolella postserrata*, followed by the appearance of *J. shannoni* in the uppermost beds. Only within the Reef Trail Member have we recovered *J. altudaensis, J. prexuanhanensis, J. granti*, and *Protoclarkina crofti*. At present, the earliest occurrence of *J. altudaensis* is at 2.9 m (9.5 ft) above the Lamar top. Preliminary examination of a sample at 4.19 m (13.75 ft) above the Lamar top also indicates the presence of this species, plus *J. prexuanhanensis* and *J.*

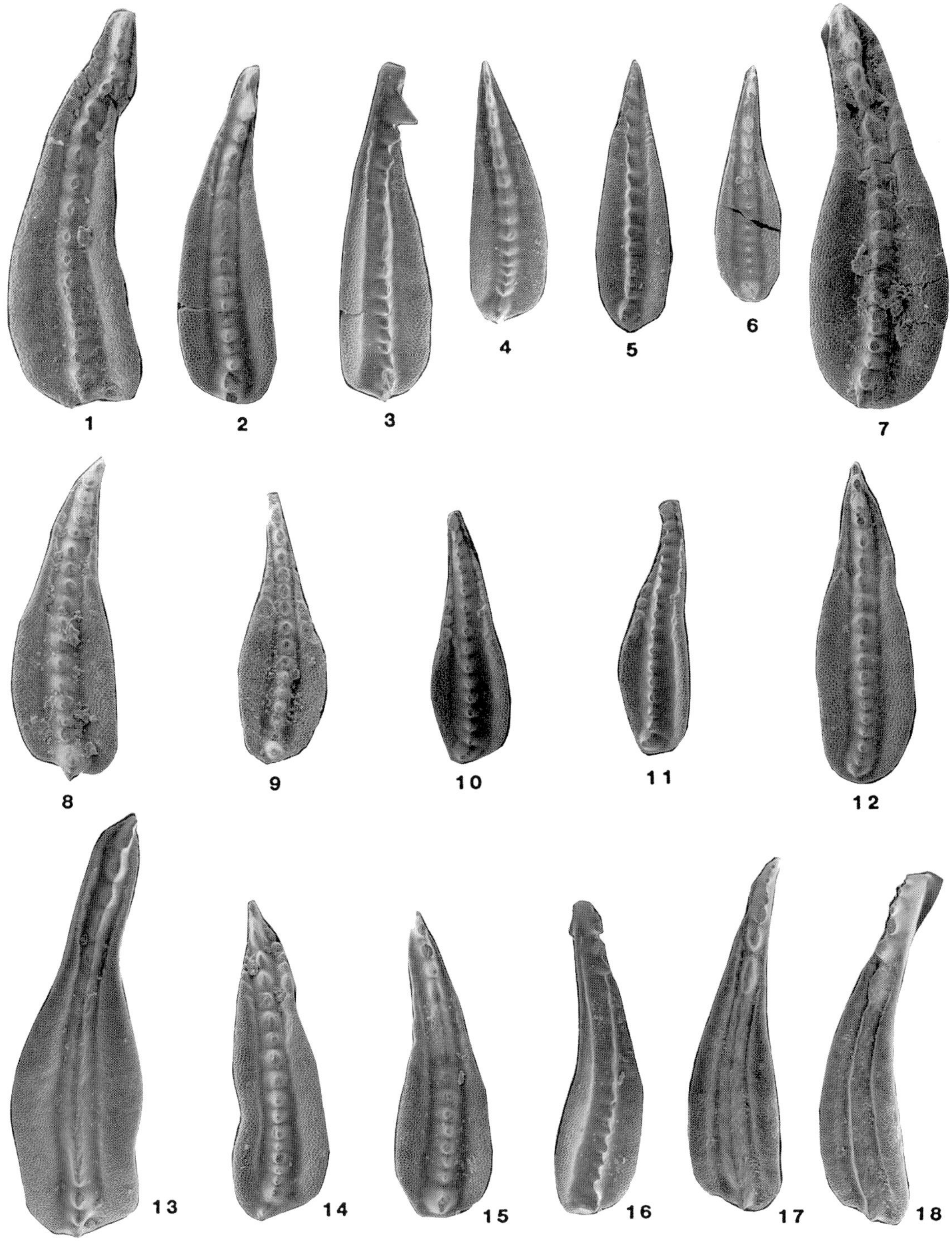

PLATE IV.—Conodonts Recovered During Present Study.

All magnifications X60. All figures are of conodont Pa elements. All views upper unless noted otherwise. Figs. 1–5 from sample 18 ft above top of Lamar; Figs. 6, 7, 10–12 from sample 14.5 ft above top of Lamar; Figs. 8, 9, 13–18 from sample 13.75 ft above top of Lamar. All figured specimens reposited at the Department of Geology, University of Iowa (SUI).

Figs. 1–8, 10–16.—*Jinogondolella prexuanhanensis*. Fig. 1, specimen with characters transitional to *J. xuanhanensis*, SUI 64535; 2, 3, upper and oblique views, SUI 64536; 4, 5, upper and oblique views, SUI 64537; 6, SUI 64538; 7, specimen with characters transitional to *J. xuanhanensis*, SUI 64539; 8, SUI 64540; 10, upper and oblique views of specimen with characters transitional to *Protoclarkina crofti*, SUI 64541; 12, SUI 64542; 13, specimen with characters transitional to *J. granti*, SUI 64543; 14, SUI 64544; 15, 16, upper and oblique views of specimen with characters transitional to *J. granti*, SUI 64545.

Fig. 9.—*Protoclarkina crofti*, SUI 64546.

Figs. 17, 18.—*Jinogondolella granti*, SUI 64547.

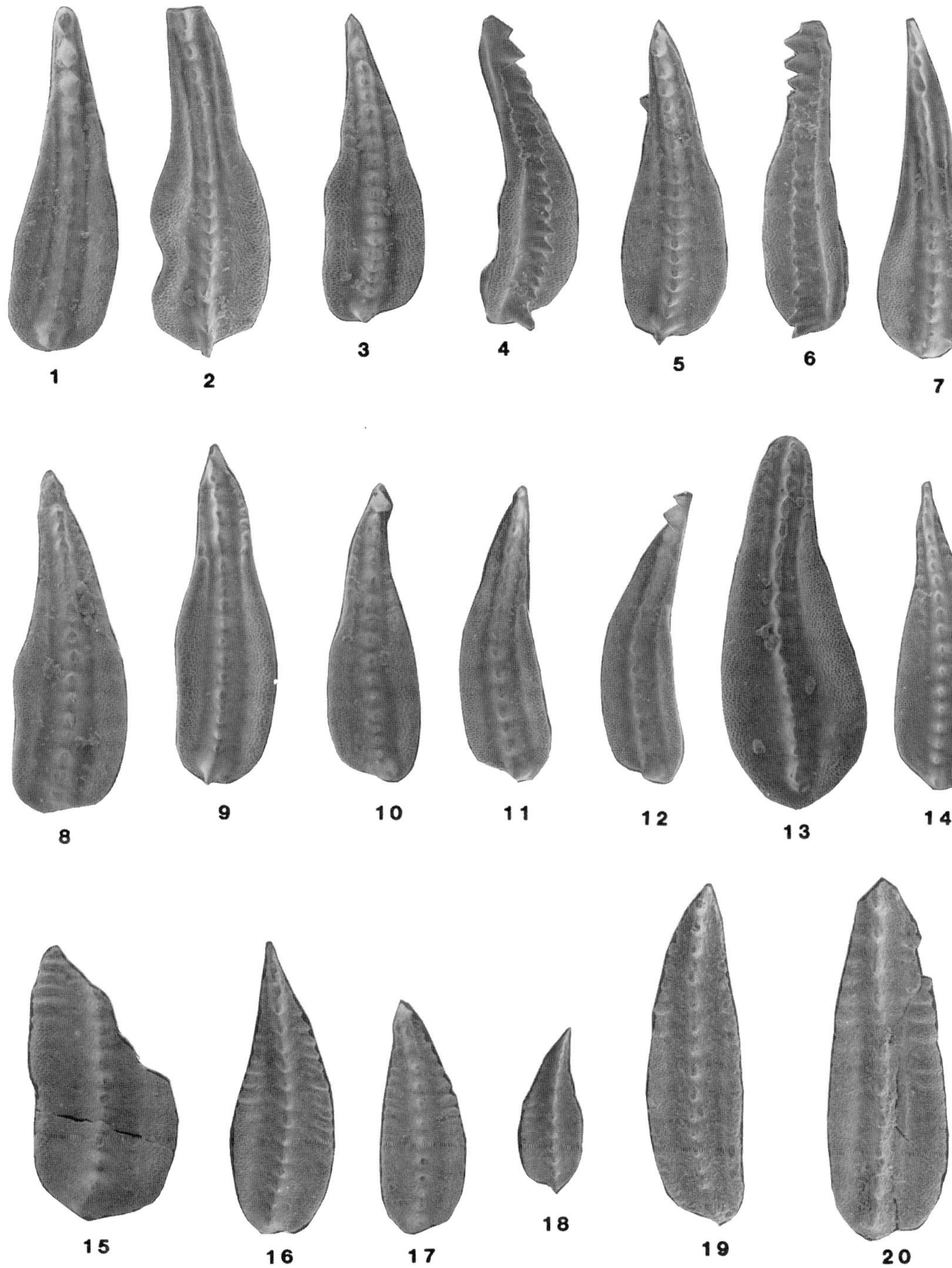

PLATE V.—Conodonts Recovered During This Study.

All magnifications X60. All figures are of conodont Pa elements. All views upper unless noted otherwise. Figs. 1–7 from sample 13.75 ft above top of Lamar; Figs. 8–14 from sample 9.5 ft above top of Lamar; Figs. 15–18 from sample of top bed of Lamar; Figs. 19, 20 from sample of upper Lamar, projected onto the Permian Reef Trail at the 5400 ft elevation marker. All figured specimens reposited at the Department of Geology, University of Iowa (SUI).

Figs. 1–7.—*Jinogondolella prexuanhanensis*. Fig. 1, specimen with characters transitional to *J. granti*, SUI 64548; 2, SUI 64549; 3, 4, upper and oblique views, SUI 64550; 5, 6, upper and oblique views, SUI 64551; 7, SUI 64552.

Figs. 8–14.—*Jinogondolella altudaensis*. Fig. 8, specimen with characters transitional to *J. prexuanhanensis*, SUI 64553; 9, SUI 64554; 10, SUI 64555; 11, 12, upper and oblique views, SUI 64556; 13, SUI 64557; 14, specimen with characters transitional from *J. shannoni*, SUI 64558.

Figs. 15–18.—*Jinogondolella shannoni*. Growth series illustrating the distinctiveness of *J. shannoni* throughout ontogeny. Fig. 15, fragment of very large specimen, SUI 64559; 16, SUI 64560; 17, SUI 64561; 18, SUI 64562.

Figs. 19, 20.—*Jinogondolella postserrata*. Fig. 19, SUI 64563; 10, SUI 64564.

granti, accompanied by *P. crofti*. In our sample from 2.9 m (9.5 ft) above the Lamar top, *J. altudaensis* is recognized without *P. crofti* or others. *J. shannoni* occurs in the top 1.52 m (5.0 ft) of the Lamar Limestone along the reef trail. We have found *J. postserrata* throughout most of the Lamar.

Our preliminary results from the Apache Mountains agree biostratigraphically with the Guadalupe Mountains, in general, but lithostratigraphic thicknesses and characteristics differ somewhat. Details will be provided in a forthcoming paper.

The chronostratigraphic definition of the basal Lopingian boundary will probably be based on conodont transitions because they were rapidly evolving pelagic organisms with rapid dispersal and wide distribution (Lambert et al., 1995). The zonation at the top of the type Guadalupian apparently matches that of the latest Maokouan, in China, demonstrating that no significant gap nor overlap occurs between the type Guadalupian and type Lopingian. The top of the Guadalupian will formally correspond to the basal definition of the Lopingian established in its type area in China. The leading evolutionary event for circumscribing that boundary is the transition from *Protoclarkina crofti* to *Clarkina postbitteri*. A second evolutionary event under consideration is the transition from *Jinogondolella xuanhanensis* to *Protoclarkina crofti* (B. R. Wardlaw, pers. comm., 1997). Defined on one of these two conodont transitions, along with the strong fusulinacean evidence discussed above, the Guadalupian/Lopingian boundary will most likely fall somewhere within the Reef Trail Member (Fig. 8).

ACKNOWLEDGMENTS

We thank Superintendent Larry Henderson, Janice Wobbenhorst, and Fred Armstrong of Guadalupe Mountains National Park for their many kindnesses and permission to measure section and collect material in the Park for our studies. Field expenses for Lambert were defrayed in part by the U. S. Geological Survey Volunteer for Science Program, and by Southwest Texas State University. Thanks are extended to Alton Brown, ARCO Exploration and Production Technology, for sharing in the field with us his data for the recognition of the McCombs Limestone. Appreciation is extended to Mr. and Mrs. Mac Benge, owners of the Benge Ranch, Glass Mountains, for allowing access for collecting and measuring section.

Gratitude is extended to the following peer review editors for their friendly insistence upon correctness and promptness in completing the manuscript: Art Saller, UNOCAL, Houston, Tex.; P. M. Harris, Chevron, LaHabra, Calif.; Brian Glenister, University of Iowa, Iowa City; and Merlynd Nestell, University of Texas, Arlington, who also made arrangements for the use of SEM facilities there.

One of the authors, Shannon Rudine, prepared the more than 250 thin sections of fusulinacean material for the study. Lula Wilde prepared the numerous renditions of manuscript. Without such helpful support, this paper would have been long on arrival.

APPENDIX A

General Description of Faunas Shown On Plates 1 and 2

In order to express the rich variety of fusulinacean and associated material occurring in the Late Guadalupian and Early Lopingian of the Delaware Basin shelf margin, we include herein some examples of that material not shown on the accompanying cross-section (Fig. 3A). A very brief description is given for this material, and one new genus is formally described.

Abadehella cf. *coniformis* Okimura and Ishii
Plate 2, Figs. 1-7

While not identical to this species in every respect, our species is more closely similar to *A. coniformis* than to any other described form. The specimens shown have a maximum width of 1.06 mm, average height of 0.68 mm, and average apical angle of 70°. All of the figured specimens came from the Apache Mountains, Seven Heart Gap, Bell Canyon Formation, within 1.52 m (5.0 ft) of the transitional contact of the Bell Canyon and Castile Formation. At the Park boundary section of the Reef Trail Member, a single grab sample, collected 1.52-2.30 m (5.0 to 7.5 ft) above the base of the member, also contained numerous specimens of *Abadehella*, presumably this same species.

As mentioned earlier, *Abadehella* occurs throughout the *Reichelina* and *Paraboultonia* zones of the Altuda Formation in the Glass Mountains. The earlier occurrences of the genus appear to represent smaller species; however, lineage establishment must await further work.

Advenoreichelina apacheensis, n. gen., n. sp.
Plate 1, Fig. 1

A single axial section of this minute, uncoiled form has been found in the upper Bell Canyon Formation in the Apache Mountains. Although the form is closer in general appearance to the Early Permian *Pseudoreichelina* than to any previously described genus, we do not believe that it belongs to that genus. Most members of that genus are generally poorly preserved like in *Staffella, Nankinella*, and *Pseudoendothyra*, possibly indicative of original aragonitic composition. The present specimen shows no such evidence. Also, the present form exhibits a greatly expanded final septum of the uncoiled last whorl, and is much smaller than *Pseudoreichelina*.

There is also a faint similarity to certain species that are generally referred to *Reichelina*, particularly *R. changhsingensis* Sheng and Chang (1958); however, that species exhibits a sharply angular periphery and lacks the widely expanding final septum.

Although absolute precision in measuring is difficult, this specimen appears to have three to four whorls, the final one exhibiting a greatly exaggerated uncoiling. Axial length is .114 mm, and width of the coiled portion is .23 mm. The uncoiled portion measures .40 mm, nearly twice the coiled width. The form is presumed to be a member of the reichelinid line of succession, but precisely how it fits into that line would call for presumptuous reasoning from this single specimen. We believe that the specimen shown here is so uniquely different as to deserve the new generic name, *Advenoreichelina*, from the Latin *advena*, meaning *stranger*. The holotype is the specimen shown here, and named *A. apacheensis*, n. gen., n. sp. In the same material with *Advenoreichelina apacheensis* we find *Rauserella* sp., *Globivalvulina* sp., and *Monotaxinoides* sp.

Reichelina n. sp.
Plate 1, Figs. 2-5

The four specimens of *Reichelina* illustrated here occur with *Polydiexodina* in what we believe to be a McCombs equivalent. They are similar to, but not identical to *R. lamarensis* Skinner and Wilde, 1954. Adults with three and a half to four and a half whorls average 0.85 mm in sagittal diameter, including the expanded final whorl. Diameters prior to the final inflation average 0.58 mm. Axial length of the specimen shown in Fig. 2 is 0.62 mm.

The types of *R. lamarensis* average five to six whorls, 0.36 to 0.40 mm in axial length and 0.96 to 1.09 mm in sagittal width, including the final inflated whorl. Thus, with an average of one to two more whorls, *R. lamarensis* is larger in sagittal diameter, but smaller by 50% in axial length than the stratigraphically older material.

We have also considered in this study the species described by Dunbar and Skinner (1937) as *Ozawainella delawarensis*. The species occurs with *Polydiexodina shumardi* Dunbar and Skinner at the head of Rader Ridge in the Guadalupe Mountains. We believe that *O. delawarensis* is actually a species of *Reichelina*, but it exhibits an axial diameter of 1.0 mm and a sagittal diameter of 2.0 mm. This seems to indicate a trend of more robust species in the older rocks succeeded by species with shorter axial lengths and much more highly inflated final whorls. This general trend appears to have continued to the top of the Permian.

Codonofusiella sp.
Plate 1, Fig. 6

This axial section from near the Bell Canyon-Castile contact at Seven Heart Gap, Apache Mountains, is closely similar to *C. extensa* Skinner and Wilde, from a limestone between the McCombs and Lamar limestone members in the Guadalupe Mountains, originally thought to be the McCombs Member, but named herein as the McKittrick Canyon Limestone. The coiled portion of the present specimen, however, is more ovate in shape than *C. extensa*, which is diamond-shaped with sharply pointed poles.

The two sagittal sections shown in Figs. 7 and 8, at X90, may or may not be related specifically to the axial section shown here. They do not appear to belong to the same species, either, inasmuch as Fig. 8 represents a larger form at the same magnification. Indeed, Fig. 8 might represent a *Lantschichites*.

Lantschichites sp., X35; *Paraboultonia* sp. X90
Plate 1, Figs. 9, 10, respectively

Accompanying *Paraboultonia splendens* Skinner and Wilde, *Reichelina*, and *Codonofusiella* at the Bell Canyon-Castile contact in Seven Heart Gap, Apache Mountains, we have also discovered two axial sections of minute forms of probable *Lantschichites* and *Paraboultonia*. In one (Fig. 9) the uncoiled flare is present, and the coiled portion has pointed poles, and possible axial filling. In the other (Fig. 10), there is no uncoiled portion present and the shape of the coiled specimen is more elongate, cylindrical. Axial filling is shown to be present. The two specimens might represent two genera, *Lantschichites* and *Paraboultonia*, but with no more material than this, we hesitate to go further.

APENDIX B

Descriptions from Thin Sections and Photomicrographs
Reef Trail Member Collections
(See Fig. 3B and Plate 3)

WL-8-15-1.—Top Lamar. Grayish-brown fossiliferous fecal pellet, and clastic lime wackestone, with *Reichelina, Rauserella*, brachiopod fragments, and dasyclad algal fragments. Small clasts of lime mudstone similar to the large fecal pellets.

WL-8-15-1A.—Float sample, lower slope, just above pink zone at Lamar-Reef Trail contact. This is east of Hill 5180 and just east of Park boundary fence. Grayish-brown, siliceous, fossiliferous lime packstone and silty lime packstone. Both with *Paraboultonia, Lantschichites, Reichelina*, plus bryozoan, and echinoderm fragments.

WL-8-15-2.—9.3 ft (2.8 m) above base Reef Trail Member in measured section. Brown, siliceous lime wackestone to packstone with much fragmented fossil material, and small lime mudstone clasts. *Paraboultonia, Lantschichites, Reichelina* with fragmented brachiopods, gastropods, bryozoans, sponges, and dasyclad algae. Collection also consists of a thin bed of gray to brown fine-grained silty lime mudstone to finely comminuted lime wackestone.

WL-8-15-2A.—(Float) Reef Trail Member, 7.5 ft (2.3 m) above Lamar Reef Trail contact. About 100 yds west of Park boundary fence 5130 ft elevation. Brown, abundantly fecal pellet and small foram lime packstone, containing *Reichelina, Rauserella* ex. gr. *bengeensis* Wilde and Rudine, *Globivalvulina*, and other, unidentifiable small forams. Fecal pellets average 0.5-1.0 mm in size.

REFERENCES

Ager, D. V., 1964, The British Mesozoic committee: Nature, v. 203, p. 1059.
Ager, D. V., 1981, The nature of the stratigraphical record: New York, John Wiley and Sons, 122 p.
Babcock, L. C., 1977, Life in the Delaware Basin: The paleoecology of the Lamar Limestone, *in* Upper Guadalupian Facies, Permian Reef Complex, Guadalupe Mountains, New Mexico and West Texas: Midland, Society of Economic Paleontologists and Mineralogists, Permian Basin Section, Publication 77-16, p. 357-389.

BEBOUT, D. G., and KERANS, C., eds., 1993, Guide to the Permian reef geology trail, McKittrick Canyon, Guadalupe Mountains National Park, west Texas, Guidebook 26: Austin, Bureau of Economic Geology, University of Texas, 48 p.

BOSTWICK, D. A., and NESTELL, M. K., 1965, A new species of *Polydiexodina* from central Oregon: Journal of Paleontology, v. 39, p. 611-614.

BROWN, A., 1996, Position of the *Polydiexodina* last-occurrence datum in Guadalupian strata: Revised correlation in McKittrick Canyon, *in* DeMis, W. D., and Cole, A. G., eds., The Brushy Canyon play in outcrop and subsurface: Concepts and examples, Guidebook: Midland, Society of Economic Paleontologists and Mineralogists, Permian Basin Section, Publication 96-38, p. 75-83.

CLARK, D. L., and BEHNKEN, F. H., 1979, Evolution and taxonomy of the North American Upper Permian *Neogondolella serrata* complex: Journal of Paleontology, v. 53, p. 263-275.

COWIE, J. W., ZIEGLER, W., BOUCOT, A. J., BASSET, M. G., and REMANE, J., 1986, Guidelines and statutes of the International Commission on Stratigraphy (ICS): Courier Forschungsinstitut Senckenberg, v. 83, p. 1-14.

CYS, J. M., 1978, Transitional nature and significance of the Castile-Bell Canyon contact: *in* Austin, G. S., ed., Geology and mineral deposits of Ochoan rocks in Delaware Basin and adjacent areas: New Mexico Bureau of Mines and Mineral Resources, Circular 159, p. 53-56.

DAVYDOV, V. I., BELASKY, P., and KARAVAYEVA, N. I., 1996, Permian fusulinids from the Koryak terrane, northeastern Russia, and their paleobiogeographic affinity: Journal Foraminiferal Research, v. 26, p. 213-243.

DUNBAR, C. O., and SKINNER, J. W., 1937, Permian Fusulinidae of Texas: The University of Texas Bulletin 3701, v. 3, p. 517-825.

GARBER, R. A., GROVER, G. A., and HARRIS, P. M., 1989, Geology of the Capitan shelf marginñsubsurface data from the northern Delaware Basin, *in* Harris, P. M., and Grover, G. A., eds.: Midland, Society of Economic Paleontologists and Mineralogists Core Workshop 13, p. 3-269.

GIRTY, G. H., 1902, The Upper Permian in western Texas: American Journal of Science, 4th Series, v. 14, p. 363-368.

GIRTY, G. H., 1909, (imprint 1908), The Guadalupian fauna: Washington, D.C., U.S. Geological Survey Professional Paper 58, 651 p.

GLENISTER, B. F., BOYD, D. W., FURNISH, W. M., GRANT, R. E., HARRIS, M. T., KOZUR, H., LAMBERT, L. L., NASSICHUK, W. W., NEWELL, N. D., PRAY, L. C., SPINOSA, C., WARDLAW, B. R., WILDE, G. L., and YANCEY, T. E., 1992, The Guadalupian: Proposed international standard for the Middle Permian Series: International Geology Review, v. 34, p. 857-888.

HILL, C. A., 1996, Geology of the Delaware Basin, Guadalupe, Apache, and Glass mountains, New Mexico and West Texas: Midland, Society of Economic Paleontologists and Mineralogists, Permian Basin Section, Publication 96-39, 480 p.

INGAVAT, R., TORIYAMA, R., and PITAKPAIVAN, K., 1980, Fusuline zonation and faunal characteristics of the Ratburi Limestone in Thailand and its equivalents in Malaysia: Geology and Palaeontology of Southeast Asia, v. 13, p. 43-62.

ISHII, A., and TAKAHASHI, T., 1960, Fusulinids from the Upper Permian Ogamata Formation, central part of the Kwanto Massif, Japan: Science Repository, Tokyo Kyoiku Daigaku, section C, v. 66, p. 205-216.

JIN, Y., GLENISTER, B. F., KOTLYAR, G. V., and SHENG, J. C., 1994, An operational scheme of Permian chronostratigraphy: Palaeoworld, v. 4, p. 1-14.

JIN, Y., MEI, S., and ZHU, Z., 1994a, The Maokouan-Lopingian boundary sequences in south China: Palaeoworld, v. 4, p. 119-132.

JIN, Y., SHENG, J. C., GLENISTER, B. F., FURNISH, W. M., KOTLYAR, G. V., WARDLAW, B. R., KOZUR, H., ROSS, C. R., and SPINOSA, C., 1994b, Revised operational scheme of Permian chronostratigraphy: Permophiles (A newsletter of the subcommission on Permian stratigraphy, International Union of Geological Sciences), v. 25, p.1214.

JIN, Y., WARDLAW, B. R., GLENISTER, B. F., and KOTLYAR, G. V., 1997, Permian chronostratigraphic subdivisions: Episodes, v. 20, p. 10-15.

JIN, Y., WARDLAW, B. R., GLENISTER, B. F., and KOTLYAR, G. V., 1997, Permian chronostratigraphic subdivisions: Permophiles (A newsletter of the subcommission on Permian stratigraphy), v. 28, p. 5.

KERANS, C., and HARRIS, P. M., 1993, Outer shelf and shelf crest, *in* Bebout, D. G., and Kerans, C., eds., Guide to the Permian reef geology trail, McKittrick Canyon, Guadalupe Mountains National Park, west Texas, Guidebook 26: Austin, Bureau of Economic Geology, University of Texas, p. 32-43.

KING, P. B., 1948, Geology of the Southern Guadalupe Mountains, Texas: Washington, D.C., U.S. Geological Survey Professional Paper 215, 183 p.

KING, P. B., and NEWELL, N. D., 1956, McCombs Limestone Member of Bell Canyon Formation, Guadalupe Mountains, Texas: American Association of Petroleum Geologists Bulletin, v. 40, p. 386-387.

KOBAYASHI, F., 1996, Morphologic change of *Abadehella* (Foraminiferida) through Middle to Late Permian: Professor Hisayoshi Igo Commemorative Volume on Geology and Paleontology of Japan and Southeast Asia, Gakujyutsu Tosho Insatsu, Tokyo, p. 85-97.

KOBAYASHI, F., 1997a, Upper Permian foraminifers from the Iwai-Kanyo area, west Tokyo, Japan: Journal Foraminiferal Research, v. 27, p. 186-195.

KOBAYASHI, F., 1997b, Middle Permian fusulinacean faunas and paleogeography of exotic terranes in the Circum-Pacific: *in* Ross, C. A., Ross, J. R. P., and Brenckle, P. L., eds., Late Paleozoic Foraminifers; Their Biostratigraphy, Evolution, and Paleoecology and the Mid-Carboniferous Boundary: Cushman Foundation Foraminiferal Research, Special Publication 36, p. 77-80.

KOZUR, H., 1992a, Dzhulfian and Early Changxingian (Late Permian) Tethyan conodonts from the Glass Mountains, west Texas: Neues Jahrbuch fur Geologie und Palaontologie Abhandlungen, v. 187, p. 99-114.

KOZUR, H., 1992b, Boundaries and stage subdivision of the mid-Permian (Guadalupian Series) in the light of new micropaleontological data: International Geology Review, v. 34, p. 907-932.

KOZUR, H., 1992c, Late Permian Tethyan and conodonts from west Texas and their significance for world-wide correlations of the Guadalupian-Dzhulfian boundary: Geol. Palaont. Mitt. Innsbruck, v. 18, p. 179-186.

KOZUR, H., and MOSTLER, H., 1995, Guadalupian (Middle Permian) conodonts of sponge-bearing limestones from the margins of the Delaware Basin, west Texas: Geologica Croatica, v. 48, p. 107-128.

KROENLEIN, G. A., 1939, Salt, potash, and anhydrite in the Castile Formation of southeast New Mexico: American Association of Petroleum Geologists Bulletin, v. 23, p. 1682-1693.

LAMBERT, L. L., 1992, Characterization and correlation of two Upper Paleozoic chronostratigraphic boundaries (Leonardian/Guadalupian; Atokan/ Desmoinesian): Unpublished Ph.D. Dissertation, University of Iowa, Iowa City, 229 p.

LAMBERT, L. L., 1994, Morphometric confirmation of the *Mesogondolella idahoensis* to *M. nankingensis* transition: Permophiles (Newsletter of the Subcommission on Permian Stratigraphy, International Union of Geological Sciences), v. 24, p. 2835.

LAMBERT, L. L., LEHRMANN, D. J., and HARRIS, M. T., 1991, Faunal correlation of the Road Canyon and Cutoff formations, west Texas, and its bearing on the Leonardian/Guadalupian boundary (abs.), *in* Wardlaw, B. R., Grant, R. E., and Rohr, D. M., eds., Proceedings of the Guadalupian Symposium: Alpine, Texas, Sul Ross State University, p. 7.

LAMBERT, L. L., and WARDLAW, B. R., 1992, Morphological transition from *Mesogondolella idahoensis* to *M. serrata*: Basal Guadalupian definition: Appendix II in Glenister et al., *op cit.*: International Geology Review, v. 34, p. 876-880.

LAMBERT, L. L., and WARDLAW, B. R., 1996, Precise boundary definitions for the Guadalupian subseries and its component stages: Analyzing the conodont transitional morphoclines, *in* Wardlaw, B. R., and Rohr, D. M., eds., Abstracts and Proceedings of the Second International Guadalupian Symposium: Alpine, Texas, Sul Ross State University, p. 39-60.

LAMBERT, L. L., WARDLAW, B. R., and GLENISTER, B. F., 1995, West Texas Permian Series as world chronostratigraphic reference standards — A status report, *in* Garber, R. A., and Lindsay, R. F., eds., Wolfcampian-Leonardian shelf margin of the Sierra Diablo: Seismic scale models for subsurface exploration: Midland, West Texas Geological Society Publication 95-97, p. 123-133.

LAWSON, E. C., 1989, Subaqueous gravity flows and associated deposits in the Rader Member, Capitan reef complex (Permian), Delaware Mountains, west Texas: Unpublished Masters Thesis, University of Wisconsin, Madison, 160 p.

LEVEN, E. YA., 1993, Sumatrinid phylogeny and the question of the zonal subdivisions of the Murghabian and Midian stages of the Permian:

Paleontological Journal, v. 3, p. 23-29.

LOEBLICH, A. R., JR., and TAPPAN, H., 1988, Foraminiferal Genera and Their Classification, v. 1: New York, Van Nostrand Reinhold Company.

MEI, S., JIN, Y., and WARDLAW, B. R., 1994a, Succession of Wuchiapingian conodonts from northeastern Sichuan and its worldwide correlation: Acta Micro-palaeontologica Sinica, v. 11, p. 121-139.

MEI, S., JIN, Y., and WARDLAW, B. R., 1994b, Zonation of conodonts from the Maokouan-Wuchiapingian boundary strata, South China: Palaeoworld, v. 4, p. 225-233.

MEI, S., JIN, Y., and WARDLAW, B. R., 1994c, Succession of conodont zones from the Permian "Kufeng" Formation, Xuanhan, Sichuan and its implication in global correlation: Acta Palaeontologica Sinica, v. 33, p. 16-23.

MEI, S., and WARDLAW, B. R., 1996, On the Permian conodont "*liangshanensis-bitteri*" Zone and related problems, *in* Wang, H., and Wang, X., eds., Centennial Memorial Volume of Professor Sun Yunzhu: Paleontology and Stratigraphy: China University Geosciences Press, p. 130-140.

MINATO, M., and HONJO, S., 1965, On *Colania kwangsiana* Lee: Journal of Faculty Science, Hokkaido, Japan, v. 13, p. 1-5.

MRUK, D., and BEBOUT, D. G., 1993, Slope, *in* Bebout, D. G., and Kerans, C., eds., Guide to the Permian Reef Geology Trail, McKittrick Canyon, Guadalupe Mountains National Park, West Texas, Guidebook 26: Austin, Bureau of Economic Geology, University of Texas, p. 14-22.

NEWELL, N. D., RIGBY, J. K., FISCHER, A. G., WHITEMAN, A. J., HICKOX, J. E., and BRADLEY, J. S., 1953, The Permian Reef Complex of the Guadalupe Mountains Region, Texas and New Mexico, a Study in Paleoecology: San Francisco, W. H. Freeman and Company, 236 p.

OKIMURA, Y., ISHII, K., AND NAKAZAWA, K., 1975, *Abadehella*, a New Genus of Tetrataxid Foraminifera from the Late Permian: Memoirs of Faculty Science, Kyoto University, Series of Geology and Mineralogy, v. 41, p. 35-48.

REMANE, J., 1996, Comments to statements opposed to the new subdivision of the Permian System, as adopted by the Subcommission on Permian Stratigraphy of ICS: Permophiles, v. 29, p. 18-19.

REMANE, J., BASSETT, M. G., COWIE, J. W., GOHRBANDT, K. H., LANE, H. R., MICHELSEN, O., and WANT, N., 1996, Guidelines for the establishment of global chronostratigraphic standards by the International Commission on Stratigraphy (ICS) (revised): Permophiles, v. 29, p. 25-30.

ROHR, D. M., 1996, Road logs for the Guadalupian II field conference, *in* Wardlaw, B. R., and Rohr, D. M., eds., Proceedings for the Second International Guadalupian Symposium: Alpine, Texas, Sul Ross State University, p. 24-39.

RUI, L., 1979, Upper Permian fusulinids from western Guizhou: Acta Palaeontologica Sinica, v. 18, p. 271-297, 4 pls.

RUI, L., 1981, Permian fusulinacean zones and their biogeographic provinces in south China: Canadian Journal of Earth Sciences, v. 18, p. 342-349.

RUI, L., and NASSICHUK, W. W., 1996, Upper Permian (Wordian) fusulinaceans from the Cache Creek Terane, northern British Columbia: Canadian Journal of Earth Sciences, v. 33, p. 1022-1036.

SADA, K., and SKINNER, J. W., 1977, *Paradoxiella* from Japan: Journal of Paleontology, v. 51, p. 421.

SHENG, J. C., 1963, Permian fusulinids of Kwangsi, Kueichow, and Szechuan: Palaeontologica Sinica, n. 149, new series B, 247 p. (Chinese and English versions).

SHENG, J. C., and CHANG, L. H., 1958, Fusulinids from the type locality of the Changhsing Limestone: Acta Palaeontologica Sinica, v. 6, 205-214.

SKINNER, J. W., and WILDE, G. L., 1954, The fusulinid subfamily Boultoniinae: Journal of Paleontology, v. 28, p. 434-444.

SKINNER, J. W., and WILDE, G. L., 1955, New fusulinids from the Permian of west Texas: Journal of Paleontology, v. 29, p. 927-940.

SPINOSA, C., 1996, Secretary's notes: Permophiles, v. 29, p. 1-2.

STEVENS, C. H., DAVYDOV, F. I., and BRADLEY, D., 1997, Permian Tethyan fusulinina from the Kenai Peninsula, Alaska: Journal of Paleontology, v. 71, p. 985-994.

TARAZ, H., GOLSHANI, F., NAKAZAWA, K., SHIMIZU, D., BANDO, Y., ISHII, K., MURATA, M., OKIMURA, Y., SAKAGAMI, S., NAKAMURA, K., and TOKUOKA, T., 1981, The Permian and the Lower Triassic Systems in the Abadeh Region, central Iran (Iranian-Japanese Research Group): Memoirs, Faculty of Science, Kyoto University, Series of Geology and Mineralogy, v. 47, p. 61-133.

TOUMANSKAYA, O. G., 1953, Concerning Upper Permian fusulinids of the southern Ussuri territory: Moscow, Trudy, Geological Institute, VSGEI, p. 19-21 (in Russian).

TYRRELL, W. W., JR., 1962, Petrology and stratigraphy of near-reef Tansill-Lamar strata, Guadalupe Mountains, Texas and New Mexico, *in* Permian of the central Guadalupe Mountains, Eddy County, New Mexico, Roswell and Hobbs Geological Society Guidebook 62-48: Midland, West Texas Geological Society, p. 59-69.

TYRRELL, W. W., JR., 1969, Criteria Useful in Interpreting Environments of Unlike but Time-equivalent Carbonate Units (Tansill-Capitan-Lamar), Capitan Reef Complex, West Texas and New Mexico: Midland, Society of Economic Paleontologists and Mineralogists Special Publication 14, p. 80-97.

TYRRELL, W. W., JR., LOKKE, D. H., SANDERSON, G. A., and VERVILLE, G. J., 1978, Late Guadalupian correlations, Permian reef complex, west Texas and New Mexico: in Austin, G. S., ed., Geology and mineral deposits of Ochoan rocks *in* Delaware Basin and adjacent areas: New Mexico Bureau of Mines and Mineral Resources, Circular 159, p. 84-85.

WILDE, G. L., 1955, Permian fusulinids of the Guadalupe Mountains, Guidebook 55-1: Midland, Soceity of Economic Paleontologists and Mineralogists, Permian Basin Section, p. 59-62.

WILDE, G. L., 1975, Fusulinid-defined Permian Stages, *in* Permian Exploration, Boundaries, and Stratigraphy: Tulsa Society of Economic Paleontologists and Mineralogists and West Texas Geological Society, Permian Basin Section, Publication 75-65, p. 67-83.

WILDE, G. L., 1986a, Stratigraphic relationships of the San Andres and Cutoff formations, northern Guadalupe Mountains, New Mexico and Texas, *in* Moore, G. E., and Wilde, G. L., eds., Lower and Middle Guadalupian Facies, Stratigraphy, and Reservoir Geometries, San Andres and Grayburg Formations, Guadalupe Mountains, New Mexico and Texas: Midland, Society of Economic Paleontologists and Mineralogists, Permian Basin Section, Publication 86-25, p. 49-63.

WILDE, G. L., 1986b, An important occurrence of Early Guadalupian (Roadian) fusulinids from the Cutoff Formation, western Guadalupe Mountains, Texas, in Moore, G. E., and Wilde, G. L., eds., Lower and Middle Guadalupian Facies, Stratigraphy, and Reservoir Geometries, San Andres and Grayburg Formations, Guadalupe Mountains, New Mexico and Texas: Midland, Society of Economic Paleontologists and Mineralogists, Permian Basin Section, Publication 86-25, p. 668.

WILDE, G. L., 1988, Fusulinids of the Roadian Stage, *in* Reid, S. T., Bass, R. O., and Welch, P., eds, Guadalupe Mountains Revisited, Texas and New Mexico: Midland, West Texas Geological Society Publication 88-84, p. 143-147.

WILDE, G. L., 1990, Practical fusulinid zonation: The species concept; with Permian Basin emphasis: West Texas Geological Society Bulletin, v. 29, (March), p. 5-13; 15; 28-34.

WILDE, G. L., 1998, A fusulinacean's way of life and death: 25th Anniversary, Guadalupe Mountains National Park, April 22-25, Proceedings and Abstracts, sponsored by Guadalupe Mountains National Park and Texas Tech University.

WILDE, G. L., and RUDINE, S. F., 1996a, Lamar-Post Lamar equivalent fusulinacean sequences (Guadalupe, Apache, and Glass mountains, west Texas) and their relationship to a Guadalupian-Dzhulfian boundary solution (abs.), in Wardlaw, B. R., and Rohr, D. M., eds., Guadalupian II, Second International Guadalupian Symposium and Guidebook: Alpine, Texas, Sul Ross State University, p. 2223.

WILDE, G. L., and RUDINE, S. F., 1996b, Correlation of Late Capitanian fusulinacean sequences in the Guadalupe-Apache-Glass mountains region with emphasis on definition of the Late Guadalupian boundary (abs.), International Field Seminar on the Permian Capitan System: Carlsbad, New Mexico, Oct. 4-6, 1996, sponsored by The University of Texas, Austin, and The University of Cambridge, England.

WILDE, G. L., and TODD, R. G., 1968, Guadalupian biostratigraphic relationships and sedimentation in the Apache Mountain region, west Texas, *in* Silver, B. A., ed., Guadalupian facies, Apache Mountain area, west Texas, Guidebook 68-11: Midland, Society of Economic Paleontologists and Mineralogists, Permian Basin Section, p. 10-31.

DIFFERENTIAL COMPACTION AND ITS EFFECTS ON THE OUTER SHELF OF THE PERMIAN CAPITAN REEF COMPLEX, GUADALUPE MOUNTAINS, NEW MEXICO

ANDREW J. LONGLEY[1]

School of Environmental Sciences, University of East Anglia, Norwich NR4 7TJ, United Kingdom

ABSTRACT: The outer shelf sediments of the Capitan Reef Complex (Permian, Guadalupian) dip towards the basin at variable angles of up to 10°. High-resolution correlation of field logs from several canyons in the Guadalupe Mountains, New Mexico, has revealed a geometry that is most likely to have resulted from differential subsidence with the greatest rate of subsidence occurring at the shelf margin. During deposition of the upper Seven Rivers, Yates, and lower Tansill Formations, the reef accreted in several alternating phases of progradation and aggradation in response to fluctuating sea level. As the reef aggraded, loading of the underlying fore-reef and basin deposits gave rise to episodic compaction-induced differential subsidence, which tilted the outer shelf strata in a basinward direction. This process was controlled by the variable compactability of the fore-reef and basin lithologies as well as by the loading of the aggrading reef. Episodic differential compaction also influenced deposition of outer shelf sediments and induced small-scale relative sea-level changes near the shelf margin that were not experienced by the inner shelf. Mechanical failure of lithified outer shelf sediments occurred as the reef prograded over the compacting fore-reef and basin deposits. Water and liquefied sand from compacting basin sediments appear to have been forcefully injected upwards through mechanically initiated fractures, and the effects of dissolution are apparent along some fracture margins. The importance of differential compaction as a control on platform development must be recognized if the effects of sea level are to be properly understood.

INTRODUCTION

This paper describes in detail the stratal geometry of the outer shelf deposits of the upper Permian (Guadalupian) Capitan Reef Complex of New Mexico and west Texas from outcrop studies in several canyons in the Guadalupe Mountains (Fig. 1). The basinward dipping and thickening of strata on the outer shelf of the Capitan Reef Complex as exposed in the Guadalupe Mountains (Fig. 2) have been described in many of the classic works on the Guadalupe Mountains (e.g., King 1948; Newell et al., 1953). The term "fall-in beds" was first used to describe this feature by Esteban and Pray (1977), who interpreted them as representing a primary depositional dip. This interpretation was supported by geopetal structures in the lower Seven Rivers Formation, which indicate a depositional dip of about 10° (Hurley 1979, 1989). Many later authors have assumed that this was also the case during Yates and Tansill times, although there is no conclusive evidence for this in the upper parts of the stratigraphy (Yurewicz, 1976, 1977; Candelaria, 1982; Babcock and Yurewicz, 1989; Borer and Harris, 1989, 1991; Kerans and Harris, 1993). In this paper it is argued that these geometries are inconsistent with the hypothesis that the fall-in beds represent a depositional dip. Rather, these geometries are most likely to have been caused by episodic syndepositional differential compaction at or near the shelf margin.

The effects of differential compaction on carbonate platform geometry have been described for a variety of geologic settings by Hunt et al. (1995). The possibility that differential compaction may have made a significant impact on the development of the shelf margin of the Capitan Reef Complex has recently attracted particular attention. Studies of geopetal structures in the Capitan Reef suggest that differential compaction of slope and basin sediments tilted the reef and outer shelf sediments basinward shortly after deposition (Saller, 1996). The compaction interpretation of the fall-in beds also features in several earlier works (King, 1942; Hull, 1957; Achauer, 1969).

This paper describes a model for episodic differential compaction of the shelf margin aiming to relate specific features of platform geometry to the different phases of the compaction process. This paper also describes other features in the Capitan Reef Complex that can be attributed to differential compaction, such as small-scale depositional cycles on the outer shelf, which may be the result of compaction-induced sea-level changes at the shelf margin, and extensive fractures and cavities in the reef and outer shelf sediments. This study complements previous publications by other authors advocating differential compaction by proposing a more detailed mechanism for compaction and by providing high-resolution stratigraphic correlations in support of this hypothesis.

REGIONAL SETTING AND STRATIGRAPHY

The Guadalupe Mountains of New Mexico and west Texas are formed by the exhumed topography of the northwestern shelf of the Delaware Basin. During Early Guadalupian time a marine incursion brought carbonate-depositing waters over most of the shelf (Hills, 1972). The Capitan Reef Complex developed as a prograding carbonate platform around most of the margin of the basin, although it is likely to have been broken in several places, particularly along the northern segment (Motts, 1972). There is also evidence that sand-bypass channels breached the reef along the northwest margin of the basin (Harwood, 1989), possibly coinciding with the axes of present-day canyons in the Guadalupe Mountains (Jacka et al., 1985). The paleogeographic evolution of the Delaware Basin is widely acknowledged to have been controlled by differing rates of subsidence between its major depocenters (Horak, 1985), and differential subsidence due to crustal mobility accentuated the shelf-to-basin transition resulting in favorable conditions for the development of prograding carbonate platforms. However, during the Guadalupian, the differences between the subsidence

[1]Present Address: Environment Agency, Isis House, Howbery Park, Wallingford OX10 8BD, UK

Geologic Framework of the Capitan Reef, SEPM Special Publication No. 65
 ISBN 1-56576-063-8, p. 85-105

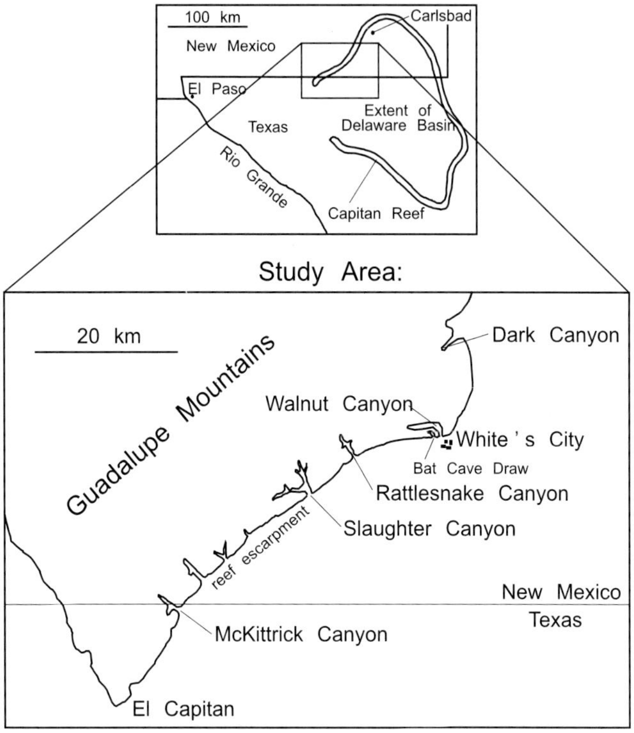

FIG. 1.—Location map of the study area in the Guadalupe Mountains.

rates of the different structural elements are thought to have diminished such that by the end of the Permian, crustal recoupling had occurred (Horak, 1985). There is no evidence of any regional tectonic activity between the rapid subsidence of the Early Permian and the Laramide uplift phase.

The Guadalupian stratigraphy has been subdivided into formations and members according to three regions: the northwest shelf, the shelf margin, and the Delaware Basin (Fig. 3). This study has been limited to the upper Seven Rivers, Yates and lower Tansill Formations and their time equivalents, the Capitan Formation and the Bell Canyon Formation, including the Rader, McCombs, and Lamar Members. This is because high-resolution correlation of shelf strata in the field can be most reliably achieved where sandstones are present and interbedded with carbonates. The Yates Formation is ideal in this respect for a study of this kind. Since this work is dependent on high-resolution correlation, it has been necessary to make further informal stratigraphic subdivisions. Informal subdivisions for the upper Yates and lower Tansill Formations were used by Neese (1979) and Schwartz (1981) and are presented in a field guide by Pray (1985). This study uses their nomenclature, but also extends the subdivisions into the lower Yates Formation. Thus three new subdivisions are proposed: the Mid-Yates Sands, the Lower Yates Dolomite, and the Basal Sands (Fig. 4). Such subdivisions as these have a very limited use due to the lateral variation in the lithologies. The nomenclature used by Neese (1979), Schwartz (1981), and Pray (1985) is based on type localities in Rattlesnake and Walnut Canyons while the type locality for the Mid-Yates Sands, the Lower Yates Dolomites, and the Basal Sands is north Rattlesnake Canyon (Fig. 1). A reference section for the outer shelf facies of the Yates Formation, compiled from these different type localities, is shown in Fig. 4 and compared with the Yates Formation reference section in Dark Canyon proposed by Lanphere (1972). The composite reference section was found to be more useful for correlation since it is more typical of the outer shelf. This is partly because, at the Dark Canyon locality, the lower units of the Yates Formation were deposited close to the shelf margin where the thickness and spacing of beds are not typical of the whole outer shelf, whereas by the time the upper Yates units were deposited at the same locality, the reef had prograded some distance basinward and unit thicknesses are more representative of the rest of the shelf.

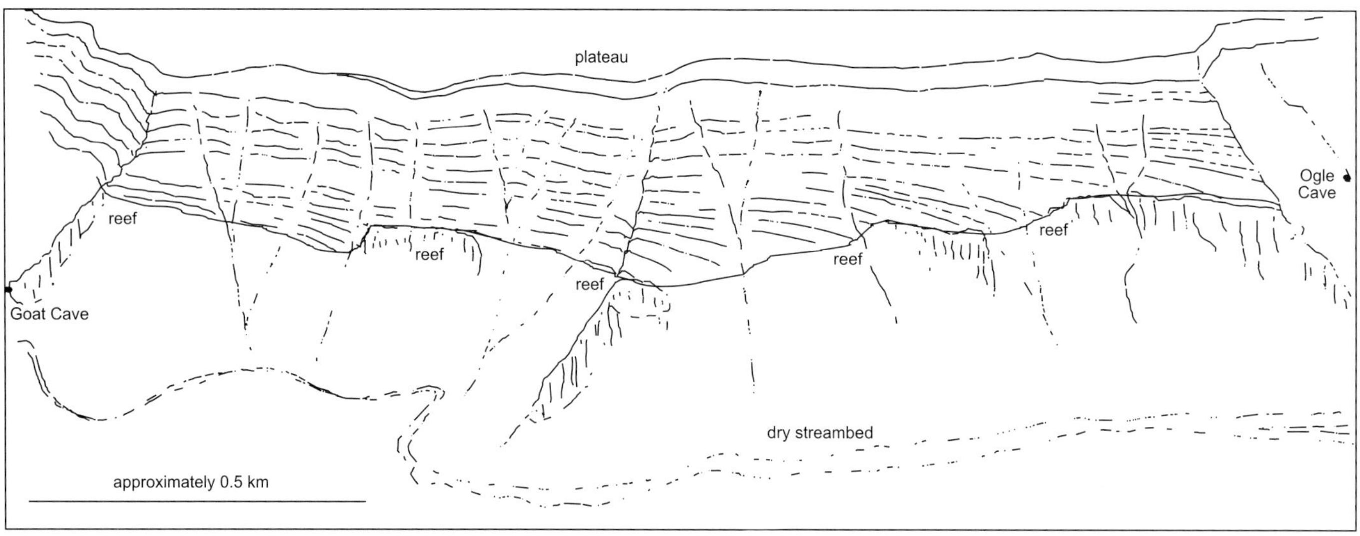

FIG. 2.—Profile of the northeast wall of Slaughter Canyon drawn from oblique aerial photographs demonstrating the occurrence of the fall-in beds. Bedded units belong to the back-reef deposits (Seven Rivers and Yates Formations) while massive units are the fine-grained sediments of the outer shelf (Seven Rivers and Yates Formations) and the boundstone of the reef itself (Capitan Formation).

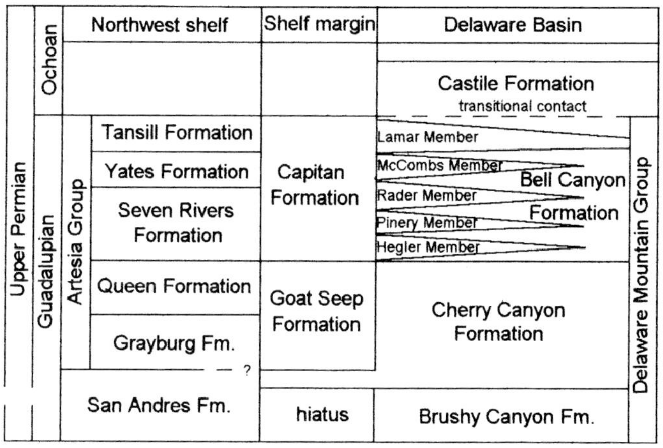

FIG. 3.—Stratigraphic correlations for the Guadalupe Mountains region (not to scale). This correlation is a composite based on works by King (1948), King and Newell (1956), Hayes (1964), Tyrrell (1962, 1969) and Bebout et al. (1993).

SEDIMENTARY FACIES

Many previous papers have attempted to subdivide and describe the different sediments of the shelf (King, 1942, 1948; Newell et al., 1953; Kendall, 1969; Dunham, 1972). A distinction has often been made between outer shelf sediments, from the pisolite facies to the reef, and inner shelf sediments, including calcisphere-rich wackestones and evaporitic dolomites, gypsum, and anhydrite. The outer shelf sediments are the focus of attention in this paper. The subdivision used here presents four broad lateral facies belts with transitional contacts: the reef itself, the grainstone belt, the laminite belt and the pisolite belt, and two more facies representing the temporal variations exhibited in Yates Formation environments, sandstones, and erosion surfaces. It is not the purpose of this paper to give detailed descriptions of each sedimentary facies. However, since these facies belts are used as distinct stratigraphic units in order to describe the geometry of the outer shelf, it is necessary to define exactly what is meant by the divisions used.

The Capitan Reef

The reef itself is defined for the purposes of this study as a massive unit in which some sort of framework is apparent. This predominantly boundstone facies is composed of sponges, bryozoa, algae, inorganic cements, and internal sediments with a wide range of other biotic constituents. There has been much debate about the exact nature of the reef and, in particular, the depth at which it developed (Kirkland et al. 1993). However, recent studies have suggested that the Capitan Reef, at least during Tansill times, was a wave-resistant and highly porous structure consisting of cryptic communities of calcareous sponges in a framework of bryozoans, microbially bound sediment, and extensive cements that developed in relatively shallow water (Wood et al., 1994).

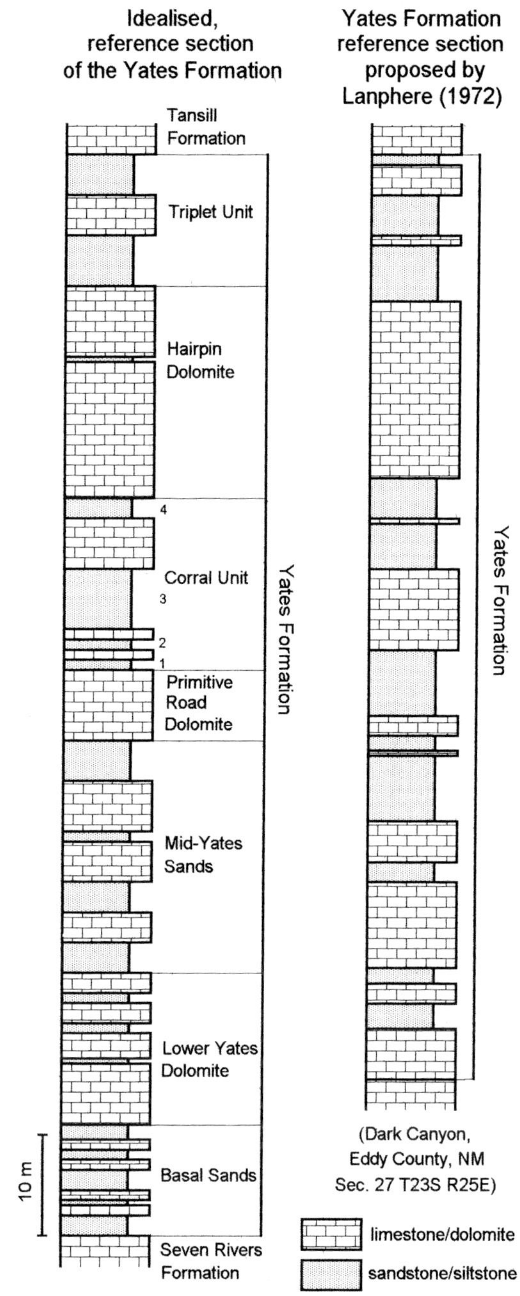

FIG. 4.—Informal subdivisions of the Yates Formation used in this study marked on an idealized section of the Yates Formation and compared with the reference section proposed by Lanphere (1972) from Dark Canyon. The idealised section has taken the most typical thickness of each bed in Slaughter, Rattlesnake, Walnut, and Dark Canyons. Correlations with this section are possible in most of the outcrops studied, whereas the section proposed by Lanphere is only recognizable in the Dark Canyon area.

The Grainstone Facies Belt

The "grainstone" facies is the generalized name given to the next facies in a shelfward direction that typically occupies a belt several kilometers wide. This subdivision includes wackestones and mudstones, which are common in the immediate back-reef, as well as true grainstones. Most sediments of the grainstone facies belt were intertidal or subtidal, but occasional prolonged periods of exposure occurred on the outer shelf. The paleotopography of this lagoonal area was relatively flat with some localized channels and mounds. Immediately shelfward of the reef, the dasyclad *Mizzia* thrived in what has been interpreted as shallow, protected lagoons (Kirkland and Chapman, 1990; Kirkland, 1992).

The Laminite Facies Belt

Laminites are typically mudstones, packstones, or wackestones with distinct millimeter-scale laminae. Fenestrae are very common, though not present in all laminites, and cements occasionally reinforce algal laminations. The laminite facies occupies a belt of usually a few hundred meters or less, immediately shorewards of grainstones, and its contacts with the grainstone belt on its basinward side, and the pisolite belt on its shoreward side, are transitional. However, laminites also commonly grade vertically into grainstones or pisolites within a single carbonate unit and the lateral extent of this facies is therefore often indistinct. These laminites are most likely to have developed in a peritidal environment (Kendall, 1969; Smith, 1974).

The Pisolite Facies Belt

Pisolites, containing pisoids often up to 10 mm in diameter, extend for hundreds of metres to several kilometers from their gradational contact with the laminite facies belt towards the inner shelf, where they merge into wackestones, mudstones, and evaporites. This is one of the most easily recognized and most studied lithologies of the Capitan Reef Complex and it has alternately been interpreted as subaerial (Dunham, 1969), marine vadose (Scholle and Kinsman, 1974) and shallow subtidal (Esteban and Pray, 1983).

Siliciclastics

Sandstones and siltstones are preserved as sheet-like deposits of no more than a few meters thick that appear to have intermittently covered most of the shelf. They do not, therefore, occupy a particular "belt," and the environment in which they were deposited may have varied significantly across the shelf. Opinion is divided amongst previous authors as to whether these beds were deposited by eolian (Kendall, 1969; Mazzullo et al., 1985; Fischer and Sarnthein, 1988; Borer and Harris, 1991) or subaqueous processes (Neese and Schwartz, 1977; Candelaria, 1989).

Erosion Surfaces

Although not strictly regarded as facies, erosion surfaces, like grainstones and siliciclastics, represent particular processes and environments and their temporal and lateral extent can be determined and interpreted. Periods of erosion represent negative rates of accretion, a factor that must be considered in any discussion of periodicity. Erosion surfaces also have important implications for interpretations of paleobathymetry.

STRATAL GEOMETRY OF THE OUTER SHELF

Method

In order to quantify the lateral thickening of back-reef beds towards the reef, and to determine the precise geometry of these beds, 39 field logs were made in canyons in the Guadalupe Mountains, which provided good exposures of the back-reef stratigraphy along a platform-to-basin profile. Since the lateral thickening and corresponding increase in dip towards the basin is a relatively subtle feature (dips are usually less than 10°), it was important not only to draw up closely spaced logs along the platform-to-basin profile, but also to be able to correlate them accurately across large distances laterally (Fig. 5). Sandstone beds were excellent markers for this purpose. Since virtually all the shelf sandstones of the Capitan Reef Complex occur within the Yates Formation, most attention was given to this part of the stratigraphy.

Many of the best exposures of the outer shelf from the lower and middle Yates Formation are found in Slaughter Canyon, the northeast wall of which cuts perpendicular to the reef margin and provides an uninterrupted profile several kilometers long, from the pisolite facies belt to the reef itself. An interpretation of part of this section was presented in a recent study (Rankey and Lehrmann 1996), but marked the base of the Yates Formation at a much higher point in the stratigraphy than can be justified from correlation with the reference section for the Yates Formation (Fig. 4). This study described a toplap geometry in the "Seven Rivers Formation" below a sequence boundary (named as SB-1) that was interpreted as the base of the Yates Formation. However, this surface is in fact in the middle to lower part of the Yates Formation, probably at the boundary of the Mid-Yates Sands and Lower Yates Dolomite (Fig. 4); thick sandstones marked as part of the upper Seven Rivers Formation are only present elsewhere in the Yates Formation.

Though smaller, Rattlesnake Canyon also provides excellent exposures of the outer shelf from the middle Yates Formation along its northeast wall, although here the profile is interrupted by smaller canyons that branch off from the main reef-perpendicular canyon. To study the outer shelf and reef/back-reef transition of the upper Yates Formation it was necessary to examine outcrops in canyons further to the east. Walnut Canyon contains exposures of the uppermost units of the Yates Formation, although the reef of equivalent age is not exposed. Due to the many interruptions in the outer shelf profile, logs were made from both sides of the Walnut Canyon preventing accurate correlation in units where no sandstone is present. In view of these restrictions, only the Triplet sandstone was studied in this canyon. Some logs were also made of the outer shelf from the lower Tansill Formation in Dark Canyon, although correlations between these logs are only approximate due to the lack of reliable marker beds. Logs from each of the canyons studied were then compiled to produce a composite profile (Fig. 5E).

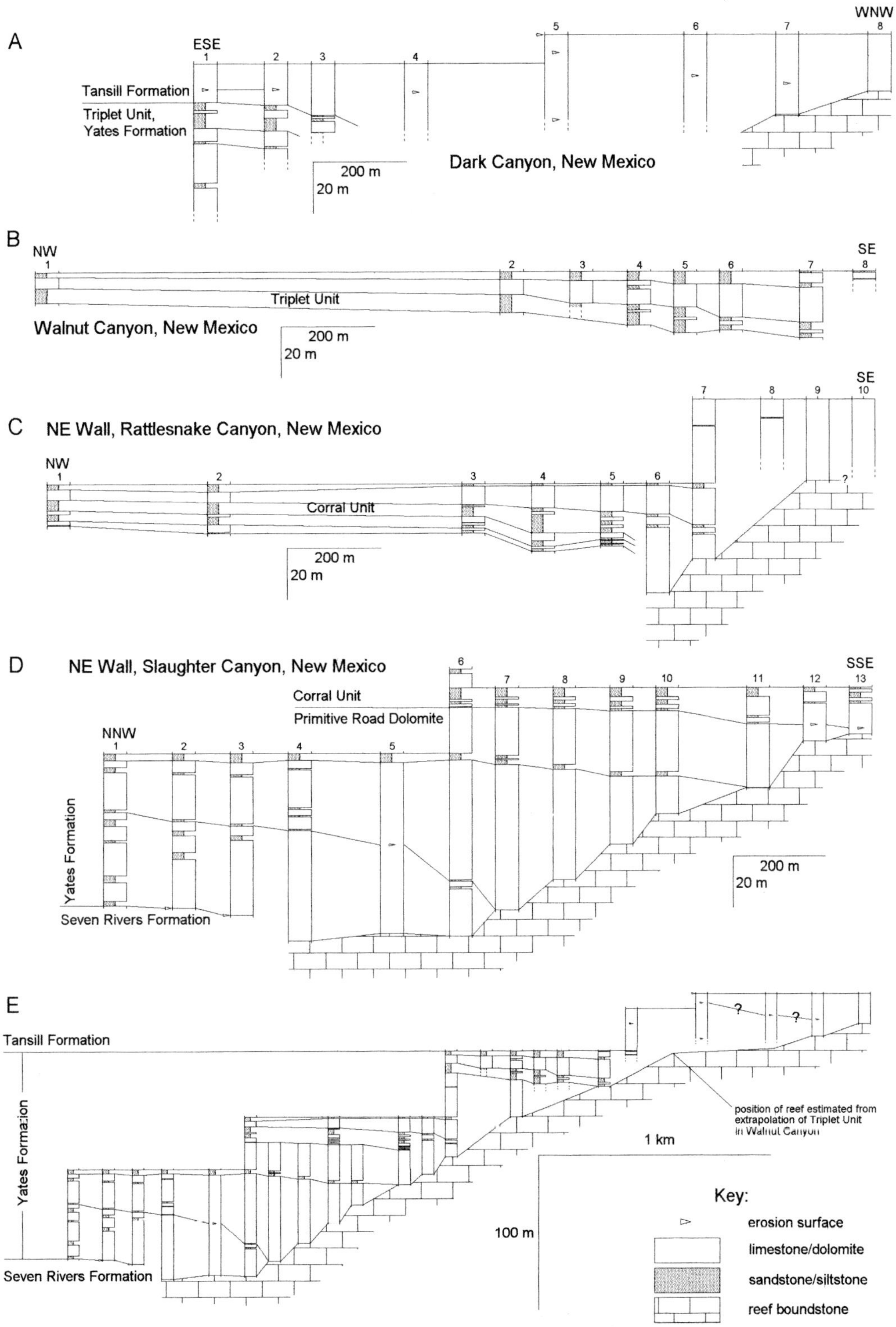

FIG. 5.—Stratigraphic correlations of logged sections from each of the canyons studied. A) Correlation of logged sections, Dark Canyon. Because of the lack of sandstones in the Tansill Formation, correlation was achieved by tracing beds in the field. The datums used are only approximate because of the limited exposure of the outcrop. B) Correlation of logged section, Walnut Canyon. C) Correlation of logged sections, Rattlesnake Canyon. The reason for the shift in datum from the top of the Corral Unit in this section is that the uppermost sandstone beds of the Corral Unit pinch out shorewards of the reef itself and it was therefore necessary to use an alternative marker bed to accurately constrain the position of the reef at its contact with the upper part of the back-reef Corral Unit. D) Correlations of logged sections, Slaughter Canyon. The datum used to correlate sections 1–6 is different from that used to correlate sections 6–13. The purpose of this is to emphasize the lateral thickness changes in each part of the stratigraphy, and by shifting the datum at different points along the platform-to-basin profile it is possible to show the lateral changes in the lower units more clearly. Each datum is therefore "artificial" and the geometry shown may not match the geometry seen in outcrop because, in reality, each datum itself dips basinwards. E) Composite of figures 5A–D showing the outer shelf geometry of the whole Yates Formation.

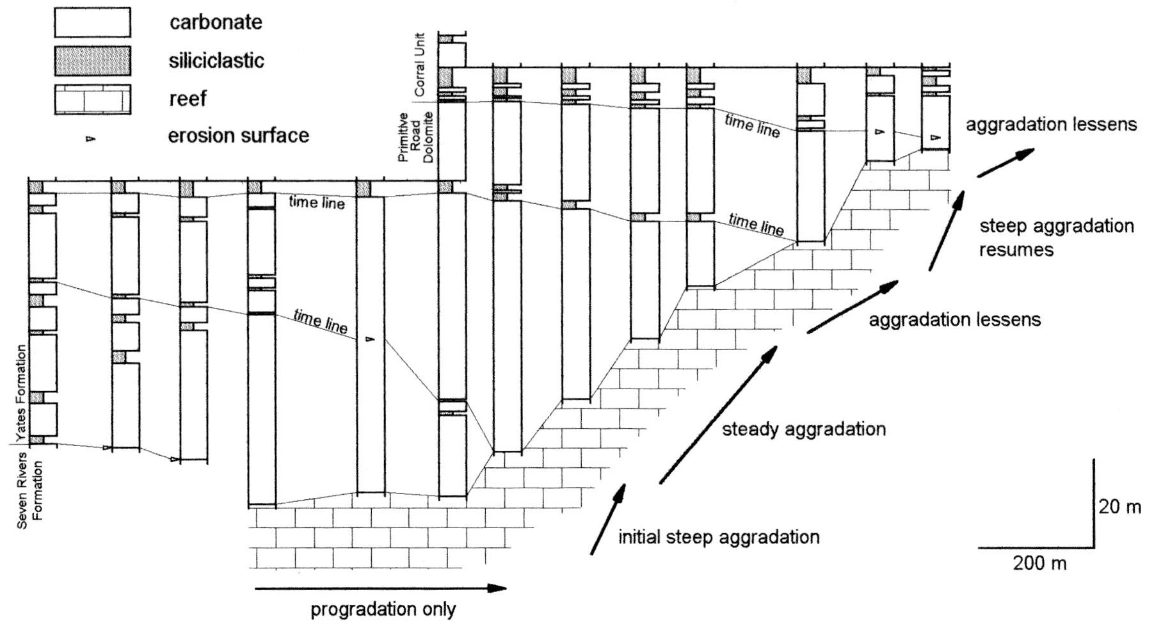

FIG. 6.—Detail of the outer shelf of the lower Yates Formation showing marked change from progradational reef growth to progradation and aggradation at the Yates/Seven Rivers boundary. Other changes in the reef accretion direction are also apparent within the Yates Formation. Note that there is a vertical exaggeration on this profile that emphasizes the aggradational component of reef growth and subdues the progradational element that is dominant throughout most of the stratigraphy.

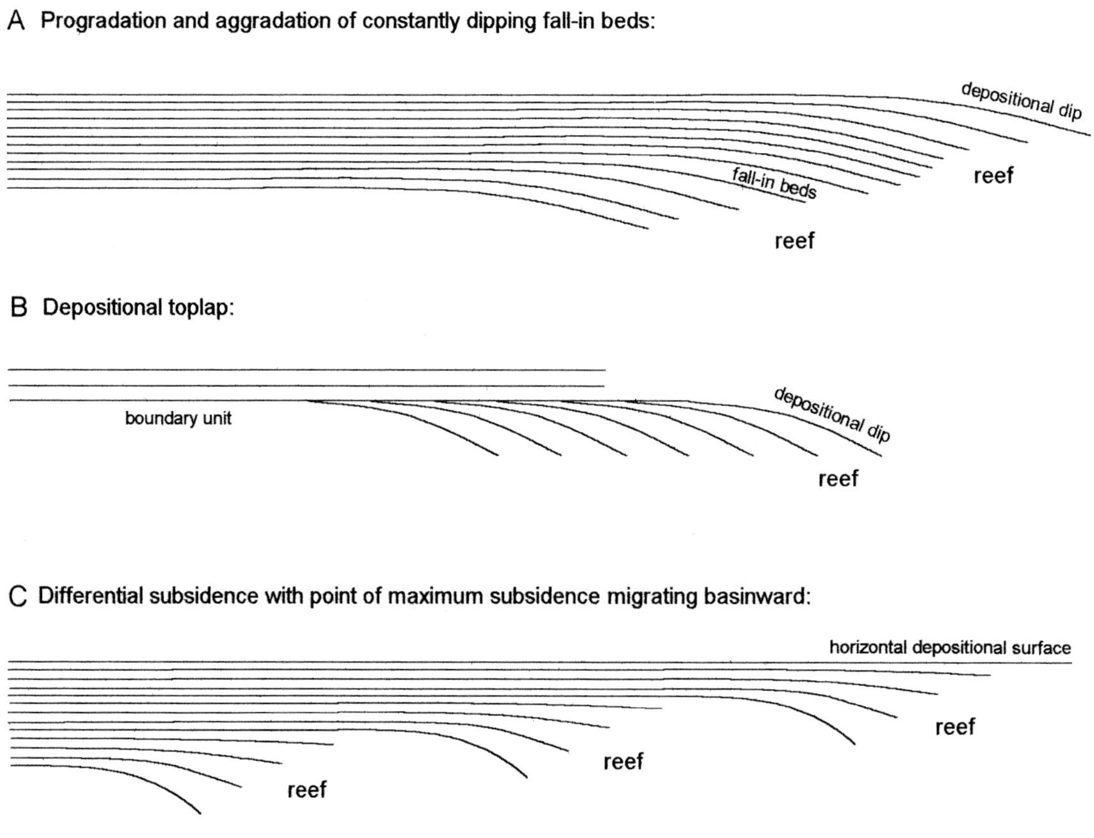

FIG. 7.—Alternative geometries for the fall-in beds (not to scale). A) The result of stacking beds of constant depositional dip in alternating phases of progradation and aggradation. B) Depositional toplap geometry beneath a boundary unit. C) Geometry produced by differential subsidence at the shelf margin with point of maximum subsidence periodically migrating basinward.

Progradation and Aggradation of the Reef

One of the most easily recognized patterns of the geometry of the shelf margin of the Capitan Reef Complex is the characteristic 'stepped' profile of the reef, reflecting its alternating phases of progradation and aggradation. This pattern can be observed in outcrop and is seen on the profile of the northeast wall of Slaughter Canyon shown in Fig. 2. This geometry has also been observed and documented in McKittrick Canyon (Tinker, 1996). On a larger scale, the accretion of the Capitan Reef Complex can be described as progradational during upper Seven Rivers times, alternately progradational and aggradational during Yates times, and dominantly progradational once again during Tansill times. However, a more detailed look reveals that during the Yates Formation, the reef appears to have accreted in several short alternating phases of progradation and aggradation. Fig. 6 shows a detail of the lower Yates Formation from Slaughter Canyon in which some of the smaller scale patterns can be seen. The first aggradational stage of reef growth in the lower Yates Formation represents a marked change from the upper Seven Rivers Formation. Initial aggradation was relatively steep followed by steady upward growth of 30 m or more. Reef growth then appears to have followed a shallower gradient before another very sharp change to relatively steep aggradation at the base of the Primitive Road Dolomite (Fig. 6).

Patterns of progradation and aggradation of the shelf margin are interpreted as responses of the platform to relative sea-level changes (Tinker, 1996; Osleger, 1998), something which is widely recognized in the context of other carbonate platforms (e.g., Bosellini, 1984; Bond et al., 1989). The ratio of progradation to aggradation from measured sections in McKittrick Canyon has been used to examine the response of the shelf margin to relative sea-level changes; reef growth was more aggradational during transgressive phases and more progradational during highstands (Tinker, 1996). It is also possible that a sigmoidal pattern of reef growth may have been responsible for the geometry observed in the Yates Formation in Slaughter Canyon (Fig. 6), where growth during aggradational phases slowed as available accommodation became filled.

The Fall-in Beds

The basinward splay and dip of beds within a kilometer of the reef can be seen from the correlations shown in Fig. 5. Corresponding with increasing dip is a lateral thickening of beds towards the reef. The dip of the fall-in beds is not constant and the steepest angles are associated with the parts of the reef that appear to have aggraded. There is also a distinctive "packaging" of fall-in beds; the dip of the beds decreases upwards through each "package" until the top bed is nearly horizontal, at which point the next package appears several hundred meters in a reefward direction (Fig. 7). Lateral thickening of the outer shelf towards the reef may also involve an increase in the number of beds as wedges of carbonate pinch out into sandstones in a shoreward direction.

It is most unlikely that the geometry of the outer shelf described above resulted from the stacking of beds deposited with a primary basinward dip. Fig. 7 shows various alternative geometries produced by stacking beds with a constant depositional dip (Fig. 7A), depositional toplap (Fig. 7B) as described for the lower part of the stratigraphy (Rankey and Lehrmann, 1996), and differential subsidence with the point of maximum subsidence migrating basinward (Fig. 7C). The stacking of beds

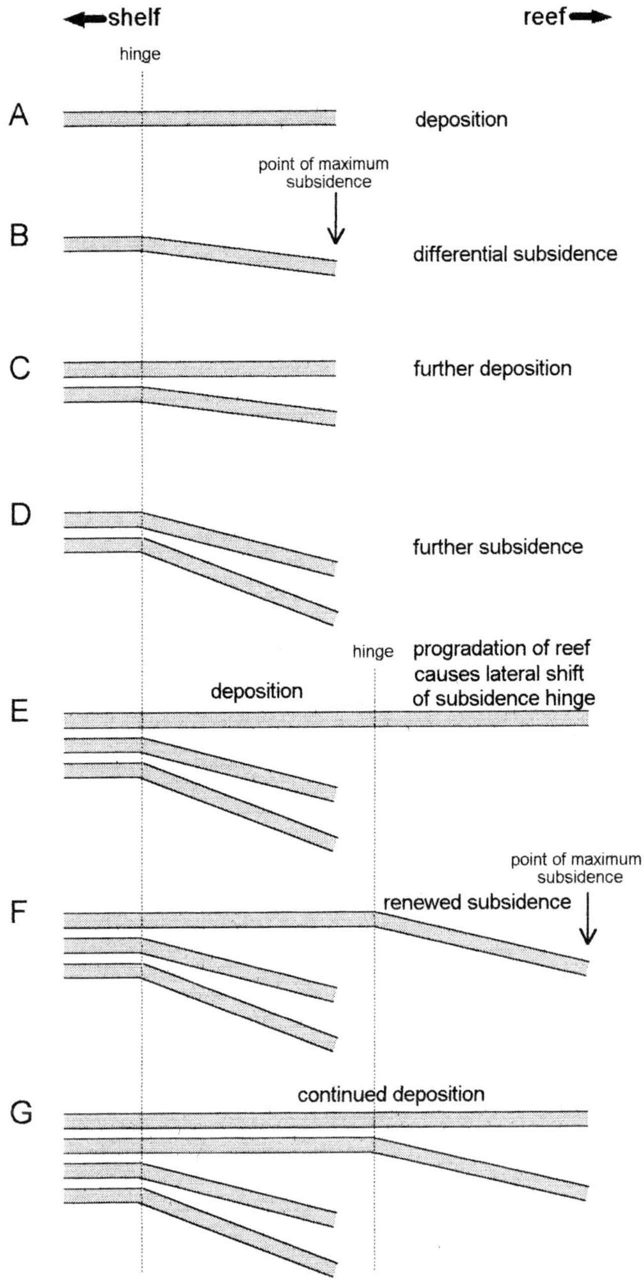

FIG. 8.—The development of the fall-in beds by differential subsidence. A) Deposition of a bed on a flat surface. B) Differential subsidence with point of maximum subsidence at the shelf margin. The hinge represents the furthest point in the shelfward direction to be affected. C) Deposition of further beds. D) Further subsidence to tilt younger beds; original bed tilted further by this process. E) Progradation in a reefward direction increases the depositional area. Deposition takes place over "package" of fall-in beds, which will not be deformed further. F) Point of maximum subsidence migrates basinward as does subsidence hinge. G) Continued deposition creates a second package of fall-in beds, stratigraphically higher and basinward of the first package.

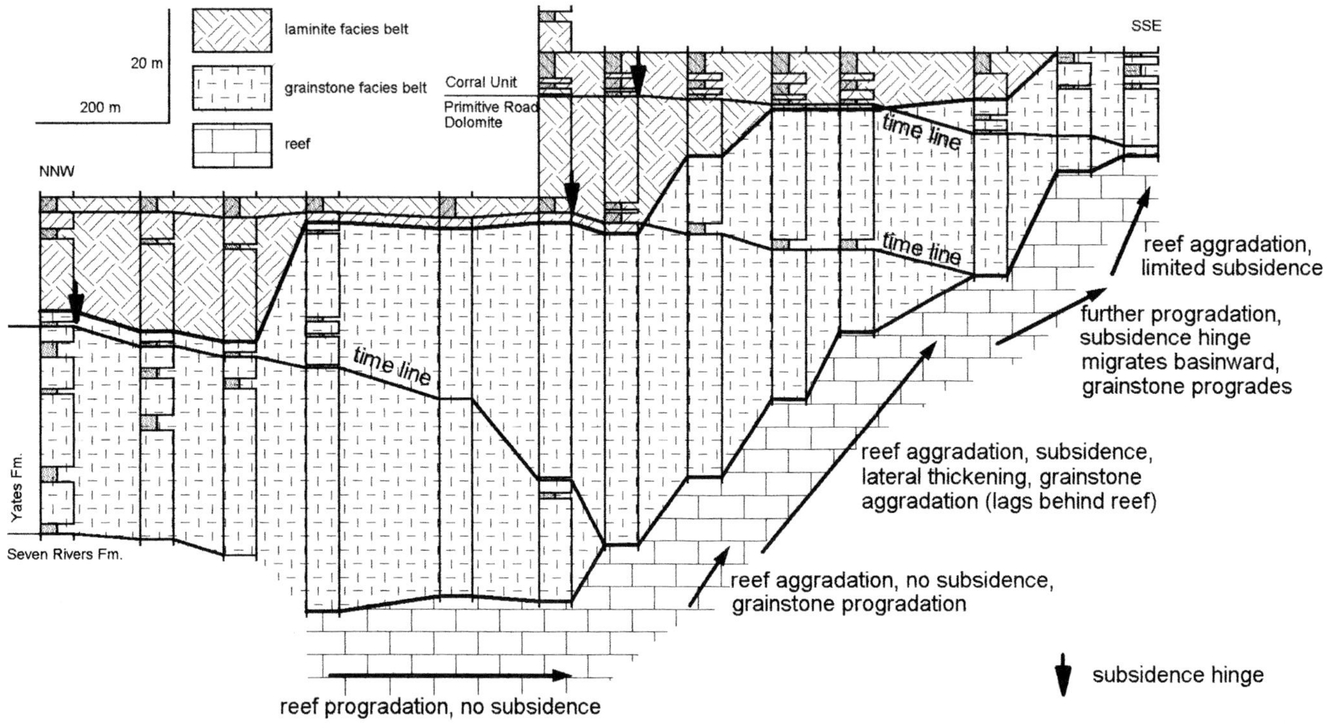

FIG. 9.—Migration of facies belts as observed in Slaughter Canyon. Changes in the reef accretion direction are marked. Timelines are based on units that can be reliably correlated. The oldest timeline indicates that the change from progradation to aggradation of the reef took place before a corresponding change can be seen in the back-reef facies.

with a constant depositional slope would not produce the lateral thickening and splaying of beds that characterizes the outer shelf of the Capitan Reef Complex. Furthermore, there is an apparent point of convergence of the dips of each bed within any one depositional package. This point represents the shelfward extent of the basinward dip, and while it migrates basinward from one stratigraphic package to another, it remains constant within that package. The depositional toplap geometry (Rankey and Lehrmann, 1996) also infers that the dip of the fall-in beds is constant from one bed to another, an interpretation which is not supported by the observations presented in this study.

Recent studies in Slaughter Canyon (Osleger, 1998) and McKittrick Canyon (Tinker, 1996) have also noted the systematic variation in the slope of the fall-in beds within each package, but interpreted this as a variation in depositional slope according to relative sea level. According to this interpretation the depth to the reef and associated outer shelf fall-in was greatest at the maximum flooding surface (Tinker, 1996). Basinward thickening of each package implies that accumulation took place without much accommodation available updip on the platform (Osleger, 1998).

This study proposes that differential subsidence of the outer shelf of the Capitan Reef Complex is the reason for the observed geometry of the fall-in beds. Fig. 8 describes the way in which the observed geometry is most likely to have evolved by differential subsidence. Subsidence at or near the shelf margin produced a shallow gradient in the bedded units of the outer shelf. Continuing subsidence during deposition meant that those beds deposited first were more steeply dipping than those deposited subsequently. The point of maximum subsidence, together with the subsidence "hinge," migrated reefward as the reef prograded. Subsidence was thus episodic and controlled, at least in part, by the loading of the reef, which was greatest during the most aggradational phases of reef growth.

Migration of Facies Belts

The shelf margin geometry produced by progradation and aggradation has already been described in the context of the reef. However, shelfward of the reef there are distinct facies belts, each of which also prograded and aggraded as the platform developed. Fig. 9 shows the distribution of grainstone and laminite facies as well as the reef on the northeast wall of Slaughter Canyon. The geometry of the grainstone/laminite contact is broadly similar to that of the reef/grainstone contact, indicating that the back-reef facies also accumulated in alternating phases of progradation and aggradation. There are, however, several notable differences between these two profiles. By using the correlations based on traced sandstone beds and erosion surfaces (marked as timelines on Fig. 9), the timing of changes in the reef accretion direction can be compared with the timing of equivalent changes in the back-reef. For example, the pronounced change from progradation to aggradation of the reef at the base of the Yates Formation is below the first time line shown on Fig. 9, whereas the equivalent change in the grainstone belt, although just as sharp, postdates this time line. A similar time lag is evident in the next change from aggradation back

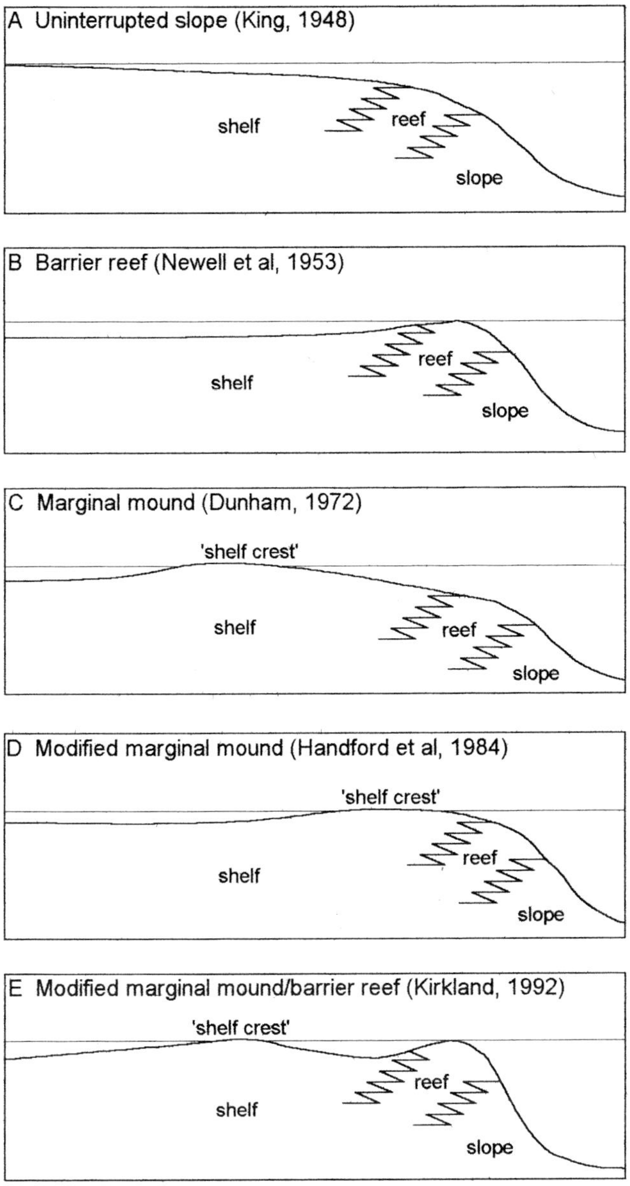

FIG. 10.—Interpretations by previous authors of the paleobathymetric shelf-to-basin profile (not to scale).

to progradation of the grainstone facies belt. From these anomalies it can be inferred that the facies belts migrated as the reef developed, but not always at the same time as the reef.

From the stratigraphic relationships shown in Fig. 9 it is possible to suggest relative timings for changes in reef accretion direction, differential subsidence, and the migration of facies belts. In order to ascertain which of these processes was the more likely cause of the other, the timing of periods of substantial differential subsidence, represented by packages that thicken significantly towards the reef, can be compared with the timing of aggradational phases of reef growth by using the correlated time lines. Although the parts of the stratigraphy that demonstrate the greatest lateral thickening broadly coincide with the most aggradational phases of reef growth, this relationship is not exact. At the base of the Yates Formation the sharp change to reef aggradation predates the period of greatest lateral thickening of the outer shelf which occurred between the first and second time lines on Fig. 9. This implies that the basinward tilting of the outer shelf took place shortly after the reef had started to aggrade, suggesting that the loading of the reef may have caused the outer shelf to be tilted. Differential subsidence appears to have caused a twofold lateral thickening of the outer shelf when reef aggradation was steady. The sharp changes from progradation to aggradation of the grainstone facies, while lagging the equivalent changes in the direction of reef growth, do not appear to be directly related to the onset of differential subsidence.

DIFFERENTIAL COMPACTION

Differential compaction of underlying fore-reef and basin deposits is considered the most likely explanation for the differential subsidence of the outer shelf (Saller, 1996; Longley, 1995), particularly since the upper Guadalupian was a relatively quiescent time tectonically. Since the fore-reef and basin deposits of the Capitan Reef Complex consist of a wide range of lithologies, it is likely that the amount of compaction varied throughout the sedimentary pile. The changing lithologies of the foreslope, "toe of slope" and basin have been described in great detail (Mruk and Bebout, 1993; Brown and Loucks, 1993; Melim and Scholle, 1995).

The upper fore-reef facies contains much material derived from the reef and is made up of intraclastic packstones to rudstones with intraclasts up to 5 m in diameter, with some well-bedded grainstones (Melim and Scholle, 1995). Neither of these facies is likely to have been particularly compactable, especially coarser intraclastic packstones and rudstones which are largely clast-supported. The middle and lower fore-reef facies, however, contain more matrix-supported intraclastic packstones and mudstones that have been interpreted as debris-flow deposits (Melim and Scholle, 1995). The higher matrix content and finer grain size of these deposits suggests that they may have compacted considerably. The most likely units in the basin to have compacted significantly are mud-dominated gravity flow deposits. Calcareous shales of the McCombs Member of the Bell Canyon Formation exposed in the floor of McKittrick Canyon are mostly mudstones and contain dewatering structures, including mud volcanoes and stylolites suggesting that compaction continued after dewatering. These units may be regarded as 'very highly compactable' because of the large amount of fluid that is usually expelled from rocks of this grain size during burial (Shinn and Robbin, 1983).

The compactability of the different lithologies of the fore-reef and basin deposits is highly variable. The facies dependence of the amount of compaction that is possible beneath carbonate build-ups has been demonstrated for the Silurian Pipe Creek Junior complex of east-central Indiana; the most effective compaction occurred by the dewatering of mudstones followed by pressure solution in mudstone-grainstone units (Devaney et al., 1986). It is most likely that differential subsidence in the Capitan Reef Complex originated principally from compaction of basin sediments and mud-dominated debris flows of the forereef. Subsidence of the outer shelf at any one time thus would have partly depended on the characteristics of the sediments directly underneath the reef and outer shelf at that moment. It is possible that the lateral variation of lithologies under the reef would have contributed to the episodic nature of differential compaction, with the rate of aggradation of the reef, which controlled the loading of lithified sediments on the compacting slope and basin deposits.

Implications for the Paleobathymetric Profile of the Outer Shelf

One of the longest-standing controversies about the development of the Capitan Reef Complex has been the interpretation of paleobathymetry across the shelf, from the pisolite facies belt to the reef itself (King, 1948; Newell et al., 1953; Dunham, 1972; Handford et al., 1984; Babcock and Yurewicz, 1989; Kirkland, 1992; Wood et al., 1994). Numerous attempts have been made to reconstruct this paleobathymetric profile (Fig. 10) with arguments principally centered on interpretations of the sedimentology of the outer shelf and reef facies. Although detailed studies of the sedimentology of the outer shelf were made as part of this study, these are presented elsewhere (Longley, 1995). However, the model for differential compaction presented in this paper does have important implications for paleobathymetry as it necessarily implies that the fall-in beds do not represent a depositional topography. Interpretations of paleobathymetry that allow for a relatively shallow reef and outer shelf are thus more consistent with the conclusions of this study. The interpretation given by Kirkland (1992) is considered to be the most consistent with the conclusions of this study (Fig 10E). Low-energy conditions of the outermost shelf are interpreted as an indication of sheltered and restricted conditions as opposed to deeper water. This implies that these beds could have been deposited horizontally, and any depositional dip should therefore be attributed to a secondary mechanism (for example, differential compaction).

Recent studies have shown that the paleobathymetric profile was not constant throughout the development of the Capitan Reef Complex, from Seven Rivers to Tansill times (Tinker, 1996; Osleger, 1998). Furthermore, not only was there a temporal variation, but at any one time there may have also been considerable variations in the profile laterally along the margin of the Delaware Basin. For this reason any "typical" model may not adequately explain all anomalous features.

DIFFERENTIAL COMPACTION-INDUCED CYCLICITY

Like many other shallow water shelf carbonates, the Capitan Reef Complex exhibits rhythms or cycles of several orders. The Permian Basin is comprised of two major subsidence cycles (Sarg, 1988). Within these, formation-scale cycles (3rd order) are widely recognized (Silver and Todd, 1969; Hills, 1972; Harms, 1974; Ross and Ross, 1987; Sarg, 1989). Ross and Ross

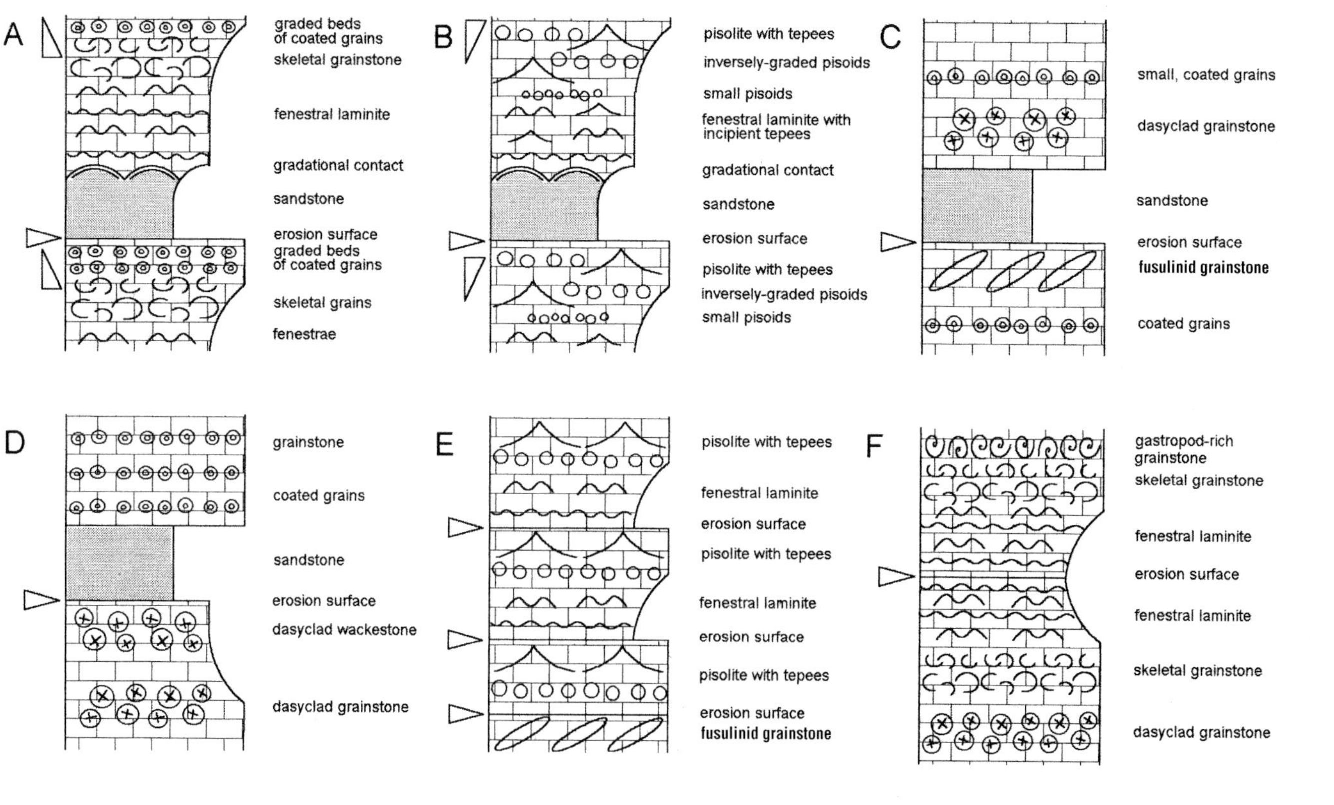

FIG. 11.—Examples of carbonate/clastic cycles and carbonate cycles from the Yates and Tansill Formations. A) Mid-Yates Sands, Log 3, Slaughter Canyon. B) Corral Sandstones, Log 9, Slaughter Canyon. C) Mid-Yates Sands, Log 8, Slaughter Canyon. D) Mid-Yates Sands, Log 2, Slaughter Canyon. E) Primitive Road Dolomite, Log 9, Slaughter Canyon. F) Tansill Formation, Log 5, Dark Canyon.

(1987) considered the Yates and Tansill Formations to have been deposited during a third-order sea-level cycle lasting about 1 m.y. whereas Sarg (1989) proposed that the Yates Formation alone was deposited over two third-order cycles representing about 2 m.y. in total. In the Yates Formation, the most obvious form of cyclicity is the interbedding of siliciclastics and carbonates (fourth-order cycles). For some authors, local (shelf-wide) subsidence was the most likely explanation of these cycles (Hull, 1957; Hills, 1972; Smith, 1974; Pray, 1977; Candelaria, 1985) while others invoked orbitally forced eustatic sea-level changes as the probable cause (Jacka et al., 1985; Meissner, 1985; Borer and Harris, 1989, 1991; Kerans and Harris, 1993). This paper proposes that these fourth-order cycles or parasequences (*sensu stricto*) were modified or partly controlled by differential compaction on the outer shelf of the Capitan Reef Complex during deposition of the Yates Formation.

Cyclicity of Back-Reef Facies

Siliciclastic-carbonate cycles are common in the Yates Formation, although their character varies from one part of the shelf to another. A grainstone-erosion surface-sandstone cycle on the outer shelf may become a pisolite-erosion surface-sandstone-laminite cycle further shelfward. The depositional environment of the sandstone is obviously very important to the interpretation of these cycles. If the sandstones were deposited subaqueously (Candelaria, 1989), then the erosion surfaces are more likely to represent the lowest point of sea level. If, on the other hand, the sandstones were eolian, then it may be possible to use them as marker beds for lowstands (Borer and Harris, 1991).

Common forms of the siliciclastic-carbonate cycle found in the Yates Formation are shown in Fig. 11. For many parts of the shelf, the most common examples are grainstone-erosion surface-sandstone-laminite cycles (Fig. 11A) and pisolite-erosion surface-sandstone-laminite cycles (Fig. 11B). In another case (Fig. 11C) the sandstone lies above the high-energy grainstone and below the deeper water dasyclad grainstone suggesting that it represents a transgressive phase, while the reverse is true in another example (Fig. 11D) where the sandstone represents a regressive phase from the deeper water dasyclad wackestone to the grainstone containing coated grains. Other common cycles do not include sandstones at all. Erosion surface-laminite-pisolite cycles with no sandstones present are very common (Fig. 11E). Fig. 11F shows an unusually symmetrical cycle in which the erosion surface occurs in the middle of fenestral laminites. In this cycle, shallowing upwards occurred from the dasyclad grainstone through the reworked shell horizon to the intertidal/supratidal laminites. Shallowing upwards continued

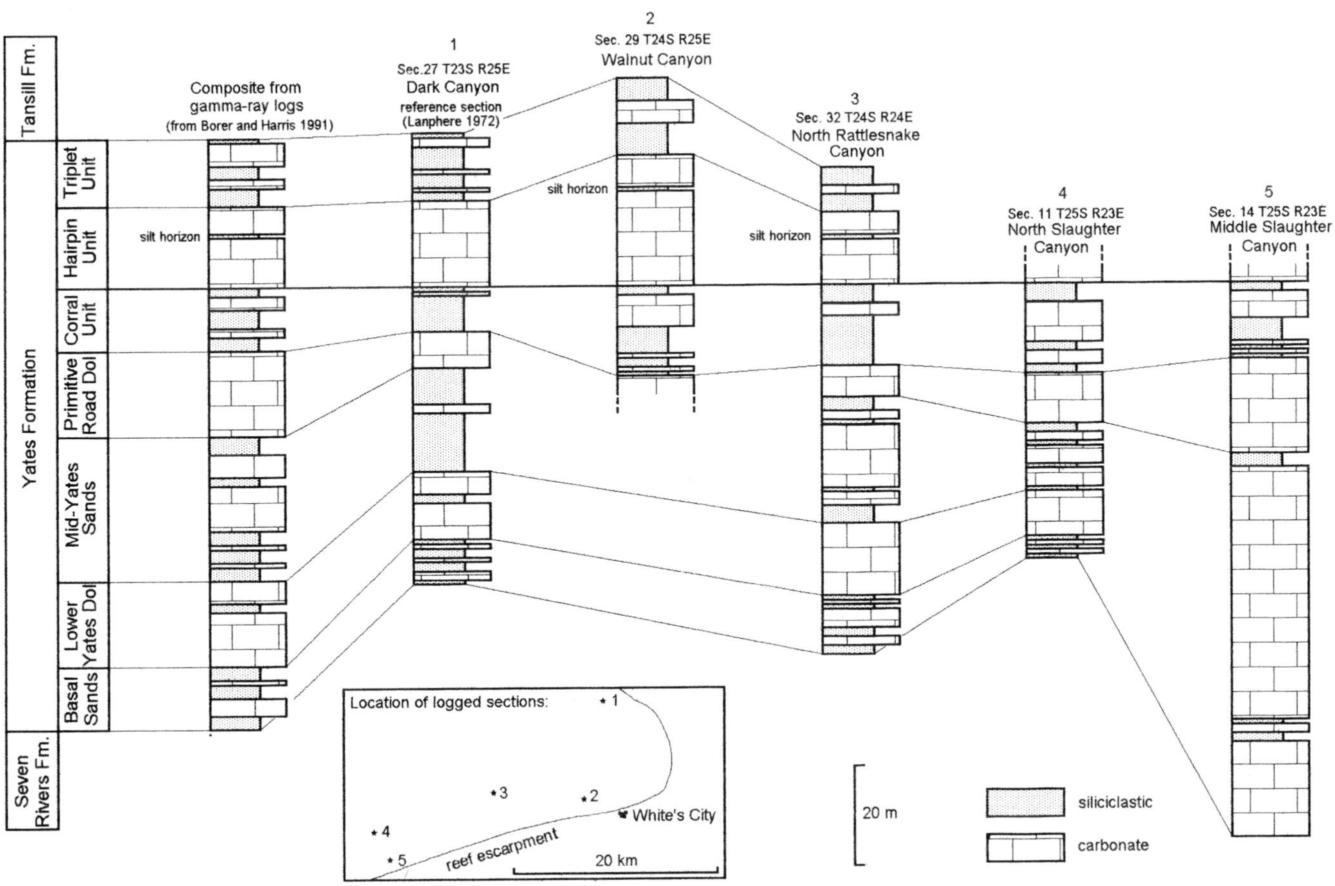

FIG. 12.—Correlation of the Yates Formation between different parts of the Guadalupe Mountains, illustrating the differences in the number, thickness, and spacing of sandstone beds.

still further from this point and subaerial erosion took place before deposition of intertidal laminites resumed, followed by subtidal/intertidal grainstones once again.

Cycles of longer duration than the couplets or "parasequences" described above and shorter duration than the formation-scale (third-order) cycles are difficult to recognize. Although stacked cycles are easily recognized in some intervals, lateral facies changes mean that the entire Yates Formation is not exposed within a single facies belt. Hence in any one section of the Yates Formation, stacked laminite-pisolite cycles at the top of the section may overlie laminite-grainstone cycles below. Borer and Harris (1991) recognized five major groups of siliciclastic/carbonate cycles within the Yates Formation, but these cannot be recognized in the sections presented in this paper.

Although cycles are common in the Yates Formation, they do not conform to an easily interpreted, idealized pattern. Most attention is usually given to sandstone-carbonate cycles, but from the examples shown in Fig. 11 it can be seen that siliciclastic beds may not always be easily interpreted in terms of sea level. Although less obvious in the field than sandstone beds, truncational erosion surfaces may be more useful in interpreting cyclicity in shelf carbonates. There are actually many more erosion surfaces than sandstone beds and a large proportion of these occur within carbonate units rather than at the bases of sandstones.

Lateral Variation of Cycles

Detailed logging of various parts of the Yates Formation at different points on the shelf has revealed that there is considerable lateral variation in the sandstone-carbonate cyclicity of the Yates Formation. While it may be possible to correlate the Yates Formation from some localities to others using groupings of sandstone beds (such as the Triplet and Corral Units), most logged sections show differences between them in the number, thickness, and spacing of siliciclastic intervals.

Fig. 12 shows a correlation of the Yates Formation between different parts of the shelf. Most obvious are the differences between logs in the number of sandstones. Although some differences may be explained by the relative proximity to the reef in some sections compared with others (sandstones interdigitate with carbonates on the outer shelf), there are also differences between outcrops that are the same distance from the coeval part of the reef but are some distance apart laterally along the shelf margin. In Dark Canyon the Triplet Unit contains four sand beds whereas there are only two in the equivalent part of the shelf in Walnut Canyon. Similarly, there are two sandstone beds in the Mid-Yates Sands in Dark Canyon, but four are present in Rattlesnake Canyon.

One of the most significant features of this correlation is that, in some cases, where a sandstone bed appears to be "miss-

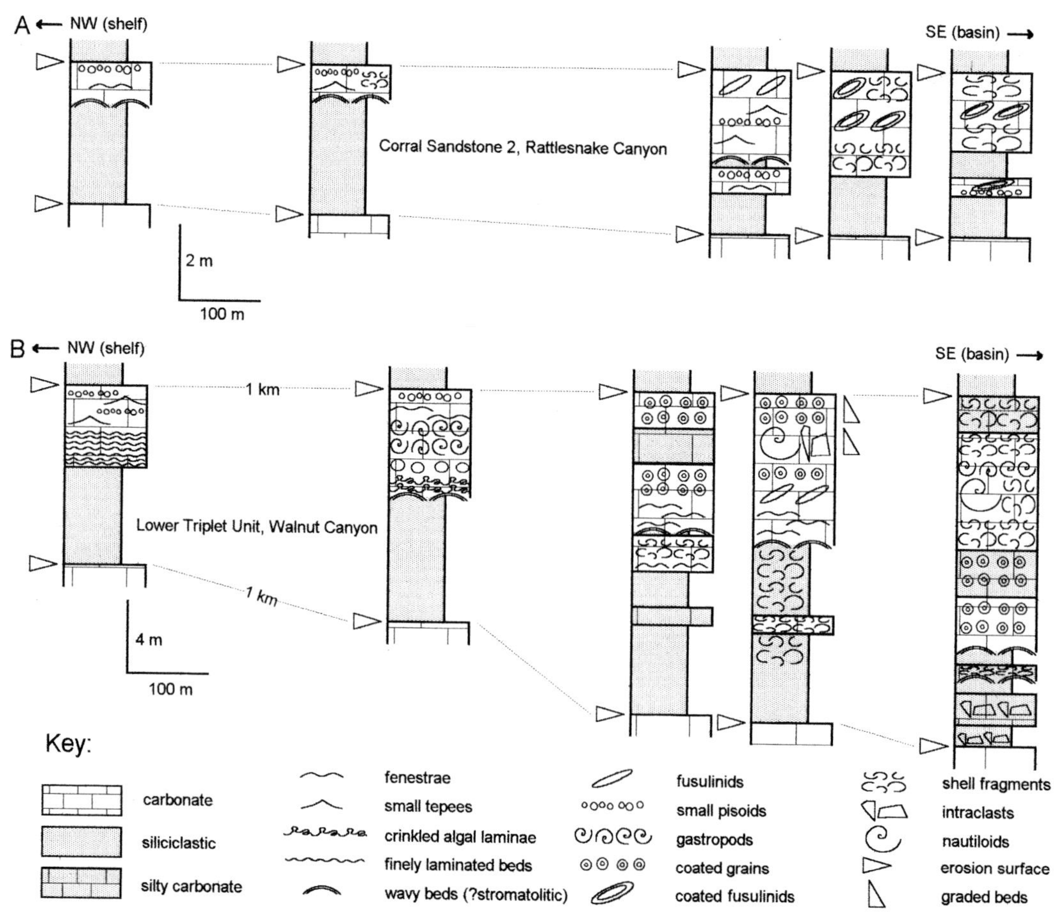

FIG. 13.—Lateral changes in cycles from the inner to outer shelf in the Yates Formation. A) Corral Sandstone 2, Rattlesnake Canyon. B) Lower Triplet Unit, Walnut Canyon.

ing" from an outcrop, a truncational erosion surface is present that correlates with the erosion surface at the base of the sandstone in another outcrop. The significance of this is that the erosion surface is laterally continuous and may be related to a shelfwide lowstand, whereas the sandstone was only deposited locally. This suggests that the deposition of siliciclastics may have been controlled by the position of channels or other topographic features, or else erosion, which may have followed sandstone deposition, removed siliciclastic material from some, but not all, parts of the shelf.

There are also lateral variations in the number and thickness of carbonate cycles, particularly along the platform-to-basin profile. This does not necessarily mean that sea-level variations were different for the outer shelf than for the inner shelf. Shelf-wide sea-level fluctuations may have produced different forms of cycles on different parts of the shelf even though the cause was the same because of differences between facies and the way in which each of them responded differently to sea-level change. However, it is possible that differential subsidence influenced the cyclicity of the outer shelf at certain times while the cyclicity of the inner shelf was dominantly controlled by shelf-wide (and possibly eustatic) sea-level change. Two examples of cycles that appear to have been modified on the shelf edge by differential subsidence are shown in Fig. 13.

Fig. 13A shows a cycle from the Corral Unit in Rattlesnake Canyon. On the shelfward side, this cycle follows a typical erosion surface-sandstone-fenestral laminite-pisolite progression. Further reefwards, however, another smaller cycle is present as the sandstone grades upwards into laminites and pisolites before deposition of an additional thin sandstone bed that is followed by more laminites and pisolites. In effect, at this point on the shelf, the Corral Sandstone 2 cycle has acquired an extra oscillation.

In Fig. 13B a similar transition may be observed. In this case, however, significant lateral thickening of the cycle also occurs. This thickening of shelf strata towards the reef reflects more rapid sedimentation rates on the outer shelf. However, not all beds thicken by the same proportion. While the thickness of the carbonate bed increases from the laminite facies belt to the grainstone facies belt because of the difference between the sedimentation rates of these facies, the thickness of the underlying sandstone bed does not change. Thus, not only is the thickness of the cycle increased reefwards, but its symmetry is also changed.

While some aspects of these facies variations may be due to local anomalies such as channels or dunes, this does not ade-

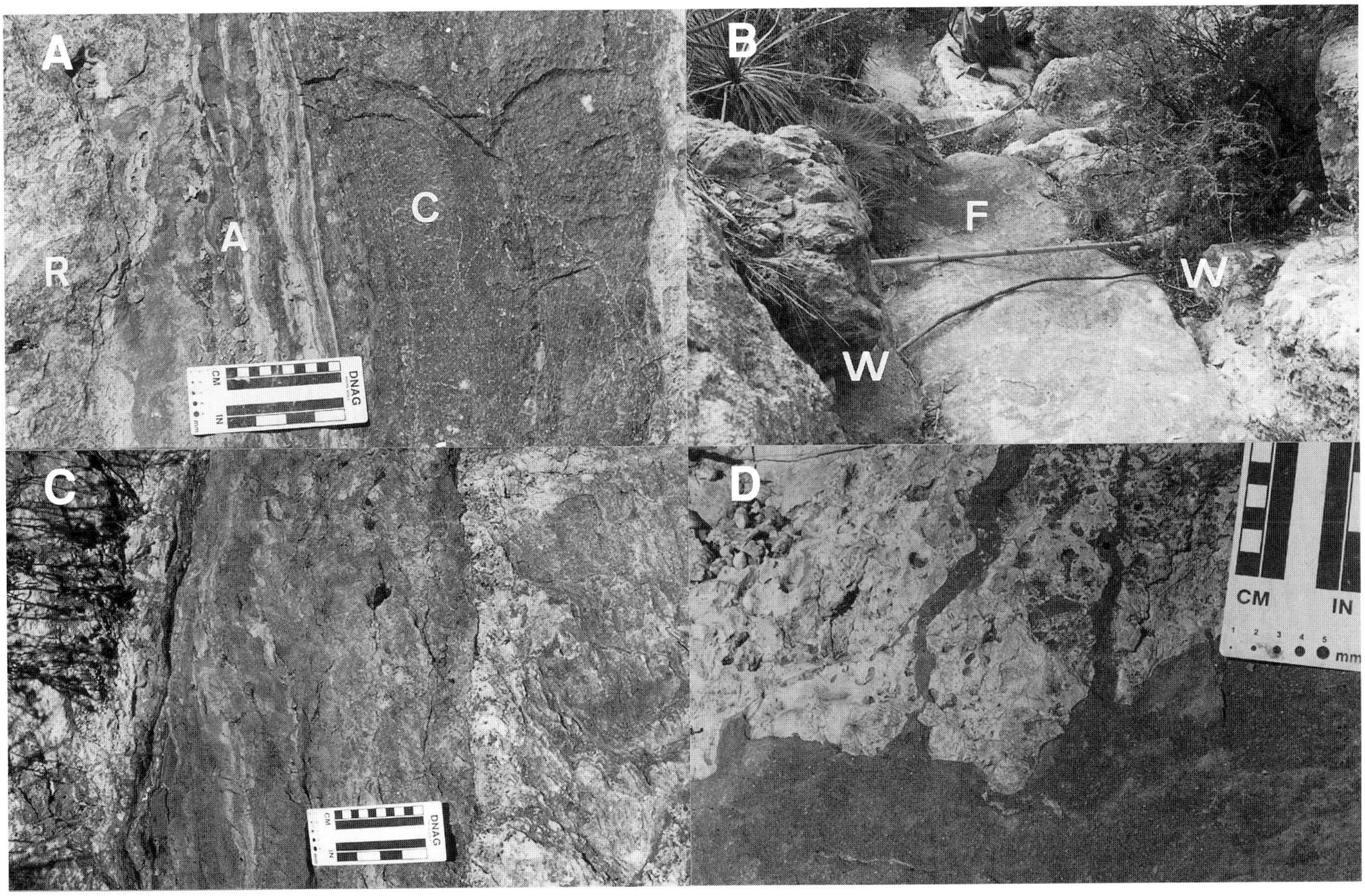

FIG. 14.—Vertical fractures in the Capitan Reef Complex. A) Carbonate-filled fracture from Dark Canyon. Cement (C) and possible encrusting algae (A) fill fracture which truncates reef fabric (R). B) Sand-filled fracture (F) cross-cutting wackestones (W) of the immediate back-reef from Slaughter Canyon (rod is 2 m long). C) Vertical fracture with silt infill showing vertical laminations parallel to fracture walls (wackestones of immediate back-reef), Rattlesnake Canyon. D) Margin of sand-filled fracture from near Ogle Cave, Slaughter Canyon, showing off-shoots of main fracture with sand penetrating into cracks in the host rock.

quately explain why there appears to be a localized shallowing-upward cycle on the outer shelf when no such change is evident further shelfwards. In both of these cases lateral variations in cyclicity from the inner shelf to the outer shelf can be explained by the superposition of relative sea-level change resulting from differential subsidence (the effect of which increases in a reefwards direction) on shelf-wide sea-level change. According to this interpretation, episodic differential subsidence would have created additional accommodation space and increased the depth of the depositional surface on the outer shelf while there was no change at the subsidence hinge point further shelfward, thereby creating the appearance of an "extra" cycle at the shelf margin.

FRACTURE AND CAVITY DEVELOPMENT

Fractures and Cavities in the Capitan Reef Complex

The Capitan Reef Complex as exposed in the Guadalupe Mountains contains a large number of fractures and cavity systems produced by many different mechanisms from the time of deposition to the uplift phase and after (King, 1948; Hayes, 1964; Dunham, 1972; Yurewicz, 1976; Jagnow, 1979; Melim, 1991). In many cases these fractures and cavities are developed in zones already affected by earlier fracturing processes, thus creating a complex pattern of overlapping and overprinting. The number of different fracturing mechanisms, the interaction between them, and, in some cases, the incomplete nature of the exposures make the task of understanding their origins far from simple.

Yurewicz (1976) described three types of "joints": post-Permian gravel-filled joints, Permian joints lined with *Archaeolithoporella* and cements, and silt and sand-filled joints of possible Permian age. Melim (1991) added two further categories: simple joints and "fractures lined with dolomitized clast-supported breccia with blocky calcite cement." Of particular interest to this study are fractures of Permian age, which may be, in some way, related to differential compaction.

Vertical carbonate-filled fractures.—

Vertical carbonate-filled fractures are common in the basinward portion of the massive Capitan reef (Dunham, 1972; Yurewicz, 1976; Babcock, 1977; Pray and Stanton, 1992; Kirkland et al., 1993). These fractures are characteristically between 5 cm and 1 m in width, vertical or near-vertical, have sharp margins and contain vertically banded, fracture-parallel cement and sediment fills. Mechanical failure of the reef initiated fracture formation as indicated by fracture margins that truncate the boundstone fabric of the host rock. The trend of these fractures always approximates the local trend of the shelf margin, which suggests that mechanical failure occurred along the reef front where reef growth had become gravitationally unstable, although there is little evidence to suggest any sort of displacement along fracture surfaces.

The petrology of the fracture fill is highly variable. Most common is a combination of cement and *Archaeolithoporella* and *Tubiphytes*, which lines fracture walls (Fig. 14A). This "lining" would have commenced immediately following mechanical failure and could have effectively "healed" the reef, preventing further movement. In some cases, however, vertical fractures in the reef front are filled with an abundant macroscopic marine biota (Pray and Stanton, 1992; Kirkland et al., 1993). Composition of the infilling material was apparently dependent on the position of the fracture on the reef and the rates at which opening and cementation could occur.

Vertical sand-filled fractures.—

Vertical and near-vertical fractures filled with silt and sand are common in the outer shelf region of the Capitan Reef Complex (Fig. 14B). These fractures are generally between 5 cm and 3 m wide and have sharp margins, usually truncating the fabric of the host rock. Sediment infill varies from fine silts with up to 60% dolomite matrix to coarse gravels with no carbonate content at all. The vertical extent of these fractures can be 100 m or more, passing through both reef and back-reef stratigraphy. The timing of fracture opening can sometimes be constrained by their vertical extent; many are truncated by overlying sediments deposited after fracture formation.

A distinction should be made between sand- or silt-filled fractures of Permian age and gravel-filled fractures of post-Permian age. Two distinct generations have been widely recognized (Hayes, 1964; Dunham, 1972; Yurewicz, 1976; Jagnow, 1979; Melim, 1991) and are usually easy to distinguish on the basis of the vertical extent of the fractures and the composition of the sediment infill, although in some localities (for example, Ogle Cave, Slaughter Canyon, New Mexico) the two phases have become closely linked. Amongst fractures of Permian age, there is also much variation. Some fractures contain orange-brown, medium to fine-grained sandstones, petrographically similar to bedded sandstones of the Yates formation and also similar to Bell Canyon formation sandstones. Others contain fossiliferous dolomitic siltstones that have a distinctive brick red color and are petrographically dissimilar to any siliciclastics of the Yates Formation in composition, grain size, and texture, although they appear from cross-cutting relationships to be syndepositional features.

At least 10 fractures were observed along the northeast wall of Rattlesnake Canyon within a 200 m-wide zone at the reef/back-reef transition. These fractures vary from 20–150 cm in width, although most are between 30 cm and 50 cm. Their strikes are all within 20° of the local trend of the shelf edge, and all dip steeply basinward at angles from 60° to vertical. Many fractures bifurcate upwards and some also coalesce. The sediment infill is laminated, parallel to the fracture margin (Fig. 14C). Petrology of the infill varies from one fracture to another. Most contain iron oxide-rich dolomitic silt, some contain fusulinids, gastropods, or pellets, and brecciated carbonate clasts are also common. Most of these fractures can be traced for tens of meters vertically; one fracture could be traced upwards from the reef, through bedded back-reef sediments for over 50 m until the brick red infill spreads outwards into horizontal sheet cracks. Many fractures seem to pinch out downwards or become progressively less steeply inclined.

In the Devonian reef complexes of the Canning Basin in western Australia, Playford (1984) has observed neptunian

FIG. 15.—Solution-enlarged cavity system, North Slaughter Canyon. A) Margins between the sandstone infill (S) and the grainstone host rock (G) are highly irregular; in some places, the fracture truncates the fabric of the host rock (T) while other margins are smooth and irregular, possibly indicative of dissolution (D). Angular breccia clasts (B) are also common. B) Detail of area shown in Fig. 16A showing truncational margin (T) and possible dissolutional margin (D). C) Thin section of fracture margin shown in Fig. 16B with sharp truncational contact between the grainstone host rock (G) and the sandstone infill (S). Isopachous cement (I) lines fracture wall. (Cross-polarized light; field of view 3 mm.)

dykes up to 6-km long, 20-m wide and 80-m deep (more typically 100-m long, 5-cm wide and 20-m deep) that pinch out at depth or occasionally pass into neptunian sills in underlying marginal slope deposits. It is possible that some of the vertical fractures in the Capitan Reef Complex pass through the reef and become less steep to follow bedding/glide planes in the foreslope deposits, although lack of exposed outcrop has so far prevented conclusive proof of the validity of this analogy. Some care must be taken when studying these fractures, as post-Permian faults have exploited some of them, producing often misleading displacements.

The initiation mechanism of these fractures appears to be the same as for vertical carbonate-filled fractures; sharp, truncational margins suggest mechanical failure of the host rock. Mechanisms for development and infill of the fractures that have been proposed include tectonic opening of joints and washing in of sand from above (Jagnow, 1979; Melim, 1991), intrusion of sandstone into fractures formed during reef growth (Adams and Frenzel, 1950), and forceful emplacement of sandstone dikes from below (Owen, 1984).

Observations made by the author strongly support this latter hypothesis; the vertical banding in many fractures is not diagenetic and the total absence of horizontal bedding suggests injection rather than passive infill from above. Also supporting this explanation is the fact that sand intrudes into narrow crevices in the wall of the host rock, which suggests pressurized infill (Fig. 14D). It is possible that this process of forceful injection may have widened the opening that had been initiated by mechanical failure. Dissolution may also have contributed to the erosional effect of forceful injection if the fracture was open for sufficiently long. In Rattlesnake Canyon, in the wall of the host rock, within half a meter of the margins of red silt-filled fractures, the same red silt infills the moulds of several nautiloids. This suggests a multiphase process in which dissolution of the aragonite of the nautiloid shells was followed by infill of red silt into the void thus created.

If forceful injection from below was the process by which these fractures were infilled, then it is necessary to consider where the sand may have come from. The origin of the orange-brown sandstone is ambiguous; the source may have been previously deposited Yates formation sandstones, or basin sands over which the reef had prograded. The brick red dolomitic silt infill, however, is very different from the vast majority of back-reef and basin sandstones and its origin is problematic. The only other place where the author was able to find a petrographically similar material was within the proximal basin sediments of the Rader Member in McKittrick Canyon, although this is far from conclusive evidence that this was the source of the injected silts.

Solution-enlarged fractures.—

While much work has concentrated on post-Permian dissolution and speleogenesis, the possibility of syndepositional dissolution and its role in the development of fractures and cavity systems has received relatively little attention, although fractures have been reported that appear to have been solution-enlarged (Owen, 1984; Melim and Scholle, 1989; Melim, 1991). The features described here are reported for the first time. One particular field locality near Goat Cave in North Slaughter Canyon was considered by the author to be especially significant.

The cavity system in North Slaughter Canyon can be followed across and down the hillside exposure for a vertical distance of about 30 m below the contact between the Yates and Seven Rivers formations (Fig. 15A). There are also traces of sand-filled fractures for another 50 m below this. The cavity is inclined reefwards and has an irregular "stepped" profile where it cuts through bedded carbonate. The host rock is a grainstone of variable composition, containing fusulinids, algal grains, ooids, pellets, crinoid ossicles, and shell fragments. Up to 20% of the host rock is made up of medium- to fine-grained sand in a micrite matrix. Some porosity remains, and other cavities are plugged with clay minerals.

The top of the cavity occurs at the base of the Yates Formation; the lowermost Yates sandstone appears to overlie the cavity, although this sand bed thins out reefwards at this point. Unlike the vertical sand-filled fractures described above, it is possible that some of this lowermost Yates sandstone filled the cavity from above; horizontal bedding within the cavity supports this conclusion. Margins are generally abrupt; walls of the host rock are rounded with possible dissolutional scallops (Fig. 15B). The contact between the host rock and the sediment infill shows several different styles in thin section. In some parts of the cavity, sharp, truncational margins are dominant, suggesting a clean mechanical fracture, while in other parts, sandstone infills mouldic porosity along the fracture margin, suggesting solution-enlargement of the cavity. In most cases, sediment infill is in direct contact with the fractured or dissolved wall of the host rock. There is also limited evidence, however, that some parts of the cavity were empty for some time before infill by the sandstone. Some truncational margins are lined with a fringe of isopachous cement (Fig. 15C).

The petrology of the fracture fill is variable; many samples of the sandstone do not match petrographically with the lowermost Yates sandstone bed, suggesting more than one source for the sand. The sandstone is mostly very pure (95+% quartz), although carbonate grains are present in some parts. The infill also contains rounded to angular clasts of brecciated host rocks, some of which have dolomite rinds that may be the result of collapse of the cavity wall. This suggests that, unlike the vertical sand-filled fractures where injected sand filling the fracture supported the wall, parts of the cavity were initially empty.

Mechanisms for Genesis of Fractures and Cavities

The process of void formation may be subdivided into phases of initiation, development, and infill (Smart et al., 1988). Initiation controls the general morphology and distribution of fractures. Development is dependent on the time interval between initiation and infill; in a multiphase process there may be several cycles of development and infill. From a study of the different fractures and cavities presented in this paper, three possible mechanisms of initiation and development have been recognized: mechanical failure due to external forces (confining pressure) on the rock, forceful injection due to interstitial fluid (formation pressure), and dissolution. Each of these processes can act as both an initiation mechanism and as a development mechanism, although mechanical failure is more likely to be associated with void initiation while forceful injection and dissolution is more likely to be associated with void development. Forceful injection may also act as an infill mechanism. These processes may have acted in several different combinations to form a wide variety of fractures and cavities in the Capitan reef and its back-reef sediments.

Mechanical failure appears to have been a common mechanism for the generation of fractures during the depositional and diagenetic history of the Capitan Reef Complex. Since the sediment must have been sufficiently rigid for brittle deformation to occur, and not all sediments of the reef complex were cemented early, there must have been a lithological control on where mechanical failure occurred. Previous authors have suggested the cause of the fracturing to be differential compaction and settling of the massive reef core on the underlying fore-reef (Jagnow, 1979, Melim, 1991).

Forceful injection of water or liquefied sand upwards through a sedimentary sequence against gravity requires overpressured fluids. Overpressure occurs when the rate of confining pressure production due to sedimentation exceeds that of dissipation of pore fluids through the compacting sediments, a function of their hydraulic conductivity (Domenico and Schwartz, 1990). A model for the generation of overpressure within the Capitan Reef Complex (Owen, 1984) proposed that geologically rapid deposition of the reef caused excess pore fluid pressures in a bypass sand channel deposited during a lowstand. Rapid deposition and poor hydraulic conductivity of the confining layer (the reef) liquefied the sand layer. When mechanical fractures intersected this aquifer, the liquefied sand was injected upwards along the pressure gradient. The principal difficulty with this model is that overpressures developed at this depth in sandstones of moderate hydraulic conductivity are likely to have been very small. Significant overpressures are more likely to have developed within the basin where laterally pervasive calcareous mudstone units such as the Lamar Member may have effectively formed a low-permeability confining layer across a large area of the basin encouraging lateral movement of

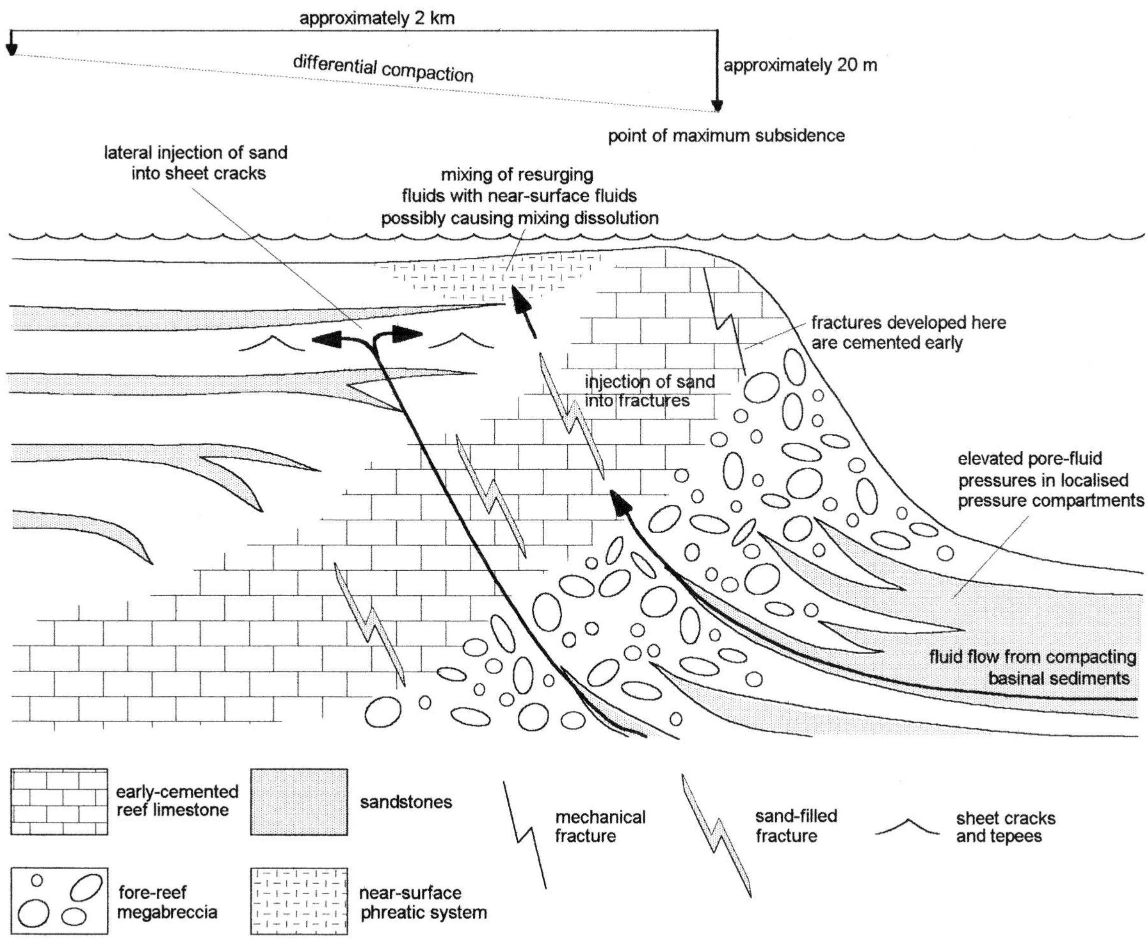

FIG. 16.—Hydrodynamic model for the syndepositional formation of fractures and cavities in the Capitan Reef Complex. Fluid flow, driven by dewatering of compacting basin sediments is driven laterally from the center of the basin to the margins. Liquefied sand is injected into fractures that develop in the lithified shelf sediments as they are tilted due to differential compaction of the basin sediments.

fluids, trapped beneath them in more permeable sands of the Bell Canyon Formation, towards the margins of the basin. If this was the case, then mechanical fractures passing through the reef and into pressure compartments in the fore-reef and basin would have released pore fluids upwards into the fractures, perhaps bringing liquefied sands from the basin.

Vadose and subaerial karstic dissolution are unlikely explanations for the cavity described from North Slaughter Canyon (Fig. 15) since they would have required prolonged exposure of the part of the shelf where dissolution took place. The cavity in North Slaughter Canyon extends through 30 m or more of stratigraphy; to achieve vadose dissolution on this scale, even to enlarge an existing fracture, a sea-level drawdown of considerable magnitude would have been required, which is unlikely given the lack of widespread karstic features across other parts of the outer shelf. Phreatic dissolution by meteoric waters in a freshwater lens or in the mixing zone of a freshwater lens would have demanded at least the occasional presence of a stable meteoric phreatic system in the back-reef. While there is evidence in the upper part of the stratigraphy for the existence of an exposed barrier island in the Tansill formation in Dark Canyon, New Mexico (Parsley and Warren, 1989; Mazzullo et al., 1989), no such evidence has been found for the lower part of the Yates Formation. Another possible explanation for the solution-enlargement of cavities that also explains the localized nature of this phenomenon is that the mixing of waters in fractures, which may have carried resurging overpressured fluids from beneath the reef, with surface waters in the lagoon, which may have been meteoric or mesosaline, may have given rise to an aggressive fluid, undersaturated with respect to calcite. Smart and Whitaker (1991) observed that where fracture flow occurs in the coastal margins of oceanic carbonate islands, then contrasts in the chemistry of recharge and phreatic lens waters may be significant.

Many of the fractures and cavities in the Capitan Reef Complex can be attributed, at least in part, to the effects of differential compaction. Solution-enlargement of fractures and forceful injection may be attributed to the effects of fluid flow resulting from the dewatering of compacting basin mudstones. Previous authors have suggested that compaction in the Capitan Reef Complex was important, not only because it brought about volume changes, but also because it resulted in the expulsion of connate fluids from the compacting sediment and was therefore one of the features that controlled fluid migration (Newell et al., 1953). It has even been suggested that water was expelled from basin limestones, upwards through the fore-reef, reef, and shelf sediments, over the reef and into the basin, contributing to density stratification of basin deposits. The large volume of connate water expelled in the differential compaction of the Devonian platforms of the Canning Basin, western Australia, is thought to have played an important role in dolomitization, mineralization, and hydrocarbon migration (Playford, 1984).

A possible hydrodynamic model for differential compaction-induced flow is suggested in Fig. 16. In this model, laterally pervasive mudstones formed a confining layer over large parts of the basin and gave rise to excess pore fluid pressures, driving fluid flow towards the margins of the basin. As the reef prograded over its own talus and fore-reef deposits, compaction of these interbedded sandstones, mudstones, and breccias took place. Mechanical failure of the reef occurred as it became gravitationally unstable and some fractures propagated through the reef, intersecting confined units of liquefied sand. Pressurized pore fluids and sand-water slurries were injected upwards through vertical fractures. Resurging fluids mixed with near-surface brines and perhaps occasionally with meteoric waters below exposed islands. Depending on the chemistry of these waters, dissolution may have occurred, enlarging mechanically initiated fractures. In this model, flow was episodic according to the build-up and release of overpressure. Flow patterns were determined by the hydraulic conductivities of the back-reef sediments with fractures acting as fluid conduits carrying much of the flow.

DISCUSSION

Many of the processes that influenced the development of the Capitan Reef Complex were intimately related to each other, although they each responded at different rates and in different ways to changes in forcing functions such as eustatic sea-level variation, tectonic subsidence, and sediment supply. Attempts to reconstruct eustasy from the geometry and cyclicity of the Capitan Reef Complex must therefore be treated with a great deal of caution. Schlager (1991) demonstrated the need to understand the "depositional bias" of each individual depositional system. Differential compaction is an important endogenic control on carbonate platform development that is part of the depositional bias of platforms such as the Capitan Reef Complex. This paper demonstrates how it may have influenced shelf margin geometry, cyclicity, and fracture development in the Capitan Reef Complex. Other authors have demonstrated that in some carbonate platforms, differential compaction is the cause of coastal pinch out and onlap (Doglioni and Goldhammer, 1988), "enhanced" unconformities, and diagenesis through fracture and karst development (Hunt et al., 1995). As awareness of the possible influence of differential compaction increases, it is possible that its effects may be recognized in many more depositional systems.

CONCLUSIONS

Evidence of differential compaction beneath the prograding carbonates of the Capitan Reef Complex has been presented in this paper from a wide-ranging study of several canyons in the Guadalupe Mountains. Measurements of stratal geometries of the outer shelf have been presented which, it is argued, could not have been produced if the basinward dips observed on the outer shelf were primary. Evidence given by other authors supports the conclusion that the fall-in beds do not represent a depositional slope (Saller, 1996; Kirkland, 1992). Observations of the lithologies of the fore-reef and basin sediments presented in this paper suggest that the most likely cause of differential subsidence at the shelf margin was compaction due to dewatering of mudstones. From this hypothesis, the following model for the development of the Capitan Reef Complex has been developed in which episodic differential compaction can be used to offer explanations for certain other features of the carbonate platform.

During Seven Rivers times, when sea level was relatively low, the dominant direction of reef accretion was progradation. At the beginning of Yates times, an apparent sea-level rise prompted a change in the direction of reef accretion and initiated a number of processes that influenced the geometry, cyclicity, sedimentology, and diagenesis of the platform. With rising sea level, aggradation of the reef took place, creating a topographic high at the shelf edge. As the reef aggraded, loading of the underlying fore-reef and basin deposits gave rise to episodic compaction-induced differential subsidence with the maximum rate of subsidence occurring at the shelf edge. This process was controlled by the variable compactability of the forereef and basin lithologies as well as by the loading of the aggrading reef. The fall-in beds of the outer shelf do not represent a depositional dip, as suggested by previous authors, but are in fact the product of differential subsidence. Differential compaction itself controlled a number of related processes on the platform. Due to the episodic nature of this process, relative sea-level changes were induced on the outer shelf, where compaction-induced subsidence occurred, that were not also experienced by the inner shelf. Shallowing-upwards cycles, which can be traced for some distances across the outer shelf, contain within them smaller-scale oscillations on the outer shelf that are entirely attributable to differential subsidence. Differential compaction also resulted in a number of vertical fractures in the reef and back-reef as lithified sediment was stressed. A regime of fluid flow may also have been generated by the dewatering of basin sediments, which was responsible for the infilling of vertical fractures with sand by forceful injection from below.

ACKNOWLEDGMENTS

This study was funded by the Natural Environment Research Council and is part of a Ph.D. thesis submitted to the University of East Anglia. The project was inspired by the late Gill Harwood who supervised and contributed to the work presented in this paper. Many thanks are also due to Alick Leslie for his contribution in the field and his comments at various stages in the development of the ideas presented here. Thanks are due to the superintendents and staff of the Carlsbad Caverns and Guadalupe Mountains National Parks for their assistance. Reviews of this paper by Dave Osleger, Leslie Melim, and Brenda Kirkland George were useful and are gratefully acknowledged. Thanks are also due to Art Saller for helpful discussions and for enabling the author to participate in the 1996 field conference.

REFERENCES

ACHAUER, C. W., 1969, Origin of Capitan Formation, Guadalupe Mountains, New Mexico and Texas: American Association of Petroleum Geologists Bulletin, v. 53, p. 2314-2323.

ADAMS, J. E., and FRENZEL, H. N., 1950, Capitan Barrier Reef, Texas and New Mexico: Journal of Geology, v. 58, p. 289-312

BABCOCK, J. A., 1977, Calcareous algae, organic boundstones, and the genesis of the Upper Capitan Limestone (Permian, Guadalupian), Guadalupe Mountains, West Texas and New Mexico, *in* Hileman, M. E., and Mazzullo, S. J., eds., Upper Guadalupian facies, Permian Reef Complex, Guadalupe Mountains, New Mexico and West Texas, Field Conference Guidebook: Midland, Society of Economic Paleontologists and Mineralogists, Permian Basin Section, Publication 77-16, v. 1, p. 3-44.

BABCOCK, J. A., and YUREWICZ, D. A., 1989, The massive facies of the Capitan Limestone, Guadalupe Mountains, Texas and New Mexico, *in* Harris, P. M., and Grover, G. A., eds., Subsurface and outcrop examination of the Capitan Shelf Margin, northern Delaware Basin: Tulsa, Society of Economic Paleontologists and Mineralogists Core Workshop 13, p. 365-371.

BEBOUT, D. G., KERANS, C., and HARRIS, P. M., 1993, Introduction, *in* Bebout, D. G., and Kerans, C., eds., Guide to the Permian Reef Geology Trail, McKittrick Canyon, Guadalupe Mountains National Park, West Texas, Guidebook 26: Austin, Texas Bureau of Economic Geology, University of Texas, p. 1-4.

BOND, G., KOMINZ, M. A., STECKLER, M. S., and GROTZINGER, J. P., 1989, Role of thermal subsidence, flexure, and eustasy in evolution of Early Paleozoic passive-margin carbonate platforms, *in* Crevello, P. D. et al., eds., Controls on Carbonate Platform and Basin Development: Tulsa, Society of Economic Paleontologists and Mineralogists, Special Publication 44, p. 39-62.

BORER, J. M., and HARRIS, P. M., 1989, Depositional facies and cycles in Yates Formation outcrops, Guadalupe Mountains, New Mexico, *in* Harris, P. M., and Grover, G. A., eds., Subsurface and outcrop examination of the Capitan Shelf Margin, northern Delaware Basin: Tulsa, Society of Economic Paleontologists and Mineralogists Core Workshop 13, p. 305-318.

BORER, J. M., and HARRIS, P. M., 1991, Lithofacies and cyclicity of the Yates Formation, Permian Basin: Implications for reservoir heterogeneity: American Association of Petroleum Geologists Bulletin, v. 75, p. 726-779.

BOSELLINI, A., 1984, Progradation geometries of carbonate platforms: Examples from the Triassic of the Dolomites, Northern Italy: Sedimentology, v. 31, p. 1-24.

BROWN, A. and LOUCKS, R. G., 1993, Toe of slope, *in* Bebout, D. G., and Kerans, C., eds., Guide to the Permian Reef Geology Trail, McKittrick Canyon, Guadalupe Mountains National Park, West Texas, Guidebook 26: Austin, Bureau of Economic Geology, University of Texas, p. 5-13.

CANDELARIA, M. P., 1982, Sedimentology and depositional environment of upper Yates Formation siliciclastics (Permian, Guadalupian), Guadalupe Mountains, southeast New Mexico: Unpublished M.S. Thesis, University of Wisconsin, Madison, 267 p.

CANDELARIA, M. P., 1985, Inferred tectonic control of Permian hemi-cyclic sandstone-carbonate sedimentation, Northwestern Shelf, Delaware Basin, New Mexico, *in* Cunningham, B. K., and Hedrick, C. L., eds., Permian Carbonate/Clastic Sedimentology, Guadalupe Mountains: Analogs for Shelf and Basin Reservoirs: Midland, Society of Economic Paleontologists and Mineralogists, Permian Basin Section, Publication 85-24, p. 67.

CANDELARIA, M. P., 1989, Shallow marine sheet sandstones, upper Yates Formation, Northwest Shelf, Delaware Basin, New Mexico, *in* Harris, P. M., and Grover, G. A., eds., Subsurface and outcrop examination of the Capitan Shelf Margin, northern Delaware Basin: Tulsa, Society of Economic Paleontologists and Mineralogists Core Workshop 13, p. 319-324.

DEVANEY, K. A., WILKINSON, B. H., and VAN DER VOO, R., 1986, Deposition and compaction of carbonate clinothems: The Silurian Pipe Creek Junior Complex of East-central Indiana: Geological Society of America Bulletin, v. 97, p. 1367-1381.

DOGLIONI, C., and GOLDHAMMER, R. K., 1988, Compaction-induced subsidence in the margin of a carbonate platform: Basin Research, v. 1, p. 237-246.

DOMENICO, P. A., and SCHWARTZ, F. W., 1990, Physical and Chemical Hydrogeology: New York, John Wiley & Sons, 640 p.

DUNHAM, R. J., 1969, Vadose pisolite in the Capitan Reef (Permian), New Mexico and Texas, *in* Friedman, G. M., ed., Depositional Environments in Carbonate Rocks: Tulsa, Society of Economic Paleontologists and Mineralogists Special Publication 14, p. 182-191.

DUNHAM, R. J., 1972, Capitan Reef, New Mexico and Texas: Facts and Questions to Aid Interpretation and Group Discussion: Midland, Society of Economic Paleontologists and Mineralogists, Permian Basin Section, Publication 72-14, 291 p.

ESTEBAN, M., and PRAY, L. C., 1977, Origin of the pisolite facies of the shelf crest, *in* Hileman, M.E., and Mazzullo, S.J., eds., Upper Guadalupian facies, Permian Reef Complex, Guadalupe Mountains, New Mexico and West Texas, Field Conference Guidebook: Midland, Society of Economic Paleontologists and Mineralogists, Permian Basin Section, Publication 77-16, v. 1, p. 451-478.

ESTEBAN, M., and PRAY, L. C., 1983, Pisoids and pisolite facies (Permian), Guadalupe Mountains, New Mexico and West Texas, *in* Peryt, T. M., ed., Coated Grains: New York, Springer-Verlag, p. 503-537.

FISCHER, A. G., and SARNTHEIN, M., 1988, Airborne silts and dune-derived sands in the Permian of the Delaware Basin: Journal of Sedimentary Petrology, v. 58, p. 637-643.

HANDFORD, C. R., KENDALL, A. C., PREZBINDOWSKI, D. R., DUNHAM, J. B., and LOGAN, B. W., 1984, Salina-margin tepees, pisoliths, and aragonite cements, Lake MacLeod, Western Australia: Their significance in interpreting ancient analogs: Geology, v. 12, p. 523-527.

HARMS, J. C., 1974, Brushy Canyon Formation, Texas: A deep-water density current deposit: Geological Society of America Bulletin, v. 85, p. 1763-1784.

HARWOOD, G. M., 1989, Large-scale channel development within the Capitan Reef Complex — evidence from Carlsbad Caverns, *in* Harris, P. M., and Grover, G. A., eds., Subsurface and outcrop examination of the Capitan Shelf Margin, northern Delaware Basin: Tulsa, Society of Economic Paleontologists and Mineralogists Core Workshop 13, p. 379-386.

HAYES, P. T., 1964, Geology of the Guadalupe Mountains, New Mexico: Washington, D.C., U.S. Geological Survey Professional Paper 446, 69 p.

HILLS, J. M., 1972, Late Paleozoic sedimentation in West Texas Permian Basin: American Association of Petroleum Geologists Bulletin, v. 56, p. 2302-2322.

HORAK, R. L., 1985, Tectonic and hydrocarbon maturation history in the Permian Basin: Oil and Gas Journal, v. 83, p. 124-129.

HULL, J. P. D., 1957, Petrogenesis of Permian Delaware Mountain Sandstone, Texas and New Mexico: American Association of Petroleum Geologists Bulletin, v. 41, p. 278-307.

HUNT, D., ALLSOP, T., and SWARBRICK, D., 1995, Compaction as a primary control on the architecture and development of depositional sequences: Conceptual framework, applications and implications, *in* Howell J. A., and Aitken, J. F., eds., High Resolution Sequence Stratigraphy: Innovations and Applications: London, Geological Society of London Special Publication 104, p 321-345.

HURLEY, N. F., 1979, Seaward primary dip of fall-in beds, lower Seven Rivers Formation (Permian), Guadalupe Mountains, New Mexico (abs.): American Association of Petroleum Geologists Bulletin, v. 63, p. 471.

HURLEY, N. F., 1989, Facies mosaic of the lower Seven Rivers Formation, McKittrick Canyon, New Mexico, *in* Harris, P. M., and Grover, G. A., eds., Subsurface and outcrop examination of the Capitan Shelf Margin, northern Delaware Basin: Tulsa, Society of Economic Paleontologists and Mineralogists Core Workshop 13, p. 325-346.

JACKA, A. D., THOMAS, C. M., BECK, R. H., WILLIAMS, K. W., and HARRISON, S. C., 1985, Guadalupian depositional cycles of the Delaware Basin and Northwest Shelf, *in* Cunningham, B. K., and Hedrick, C. L., eds., Permian Carbonate/Clastic Sedimentology, Guadalupe Mountains: Analogs for Shelf and Basin Reservoirs: Midland, Society of Economic Paleontologists and Mineralogists, Permian Basin Section, Publication 85-24, p. 81-117.

JAGNOW, D. H., 1979, Cavern Development in the Guadalupe Mountains: Cave Research Foundation: Adobe Press, 55 p.

KENDALL, C. G. St. C., 1969, An environmental re-interpretation of the Permian evaporite/carbonate shelf sediments of the Guadalupe Mountains: Geological Society of America Bulletin, v. 80, p. 2503-2526.

KERANS, C., and HARRIS, P. M., 1993, Outer shelf and shelf crest, *in* Bebout, D. G., and Kerans, C., eds., Guide to the Permian Reef Geology Trail, McKittrick Canyon, Guadalupe Mountains National Park, West Texas, Guidebook 26: Austin, Bureau of Economic Geology, University of Texas, p. 32-43.

KING, P. B., 1942, Permian of West Texas and Southeastern New Mexico: American Association of Petroleum Geologists Bulletin, v. 26, p. 535-763.

KING, P. B., 1948, Geology of the Southern Guadalupe Mountains, Texas: Washington, D.C., U.S. Geological Survey Professional Paper 215, 183 p.

KING, P. B., and NEWELL, N. D., 1956, McCombs Limestone Member of Bell Canyon Formation, Guadalupe Mountains, Texas: American Association of Petroleum Geologists Bulletin, v. 40, p. 386-387.

KIRKLAND, B. L., 1992, Distinctions between reefs and bioherms based on studies of fossil algae—*Mizzia*, Permian Capitan Reef Complex (Guadalupe Mountains, Texas and New Mexico) and *Eugonophyllum*, Pennsylvanian Holder Formation (Sacramento Mountains, New Mexico): Unpublished Ph.D. Dissertation, Louisiana State University, 156 p.

KIRKLAND, B. L., and CHAPMAN, R. L., 1990, The fossil green alga *Mizzia* (Dasycladaceae): A tool for interpretation of paleoenvironment in the Upper Permian Capitan Reef Complex, Southeastern New Mexico: Journal of Phycology, v. 26, p. 569-576.

KIRKLAND, B. L., LONGACRE, S. A., and STOUDT, E. L., 1993, Reef, *in* Bebout, D. G., and Kerans, C., eds., Guide to the Permian Reef Geology Trail, McKittrick Canyon, Guadalupe Mountains National Park, West Texas, Guidebook 26: Austin, Bureau of Economic Geology, University of Texas, p. 23-31.

LANPHERE, S., 1972, Proposed surface reference section for Yates Formation, Eddy County, New Mexico: American Association of Petroleum Geologists Bulletin, v. 56, p. 1534-1540.

LONGLEY, A. J., 1995, Differential compaction and its effects: The Upper Permian Capitan Reef Complex, New Mexico and West Texas: Unpublished Ph.D. Dissertation, University of East Anglia, 438 p.

MAZZULLO, S. J., BISCHOFF, W. D., and HEDRICK, C. L., 1989, Stacked island facies in Tansill outer-shelf platform, Dark Canyon, Guadalupe Mountains, New Mexico, *in* Harris, P. M., and Grover, G. A., eds., Subsurface and outcrop examination of the Capitan Shelf Margin, northern Delaware Basin: Tulsa, Society of Economic Paleontologists and Mineralogists Core Workshop 13, p. 287-294.

MAZZULLO, S. J., MAZZULLO, J., and HARRIS, P. M., 1985, Eolian origin of quartzose sheet sands in Permian shelf facies, Guadalupe Mountains, *in* Cunningham, B. K., and Hedrick, C.L., eds., Permian Carbonate/Clastic Sedimentology, Guadalupe Mountains: Analogs for Shelf and Basin Reservoirs: Midland, Society of Economic Paleontologists and Mineralogists, Permian Basin Section, Publication 85-24, p. 71.

MEISSNER, F. F., 1985, Cyclic sedimentation in Middle Permian strata of the Permian Basin, West Texas and New Mexico, *in* Cunningham, B. K., and Hedrick, C. L., eds., Permian Carbonate/Clastic Sedimentology, Guadalupe Mountains: Analogs for Shelf and Basin Reservoirs: Midland, Society of Economic Paleontologists and Mineralogists, Permian Basin Section, Publication 85-24, p. 118-143.

MELIM, L. A., 1991, The origin of dolomite in the Permian (Guadalupian) Capitan Formation, Delaware Basin, West Texas and New Mexico: Implications for dolomitization models: Unpublished Ph.D. Dissertation, Dedman College, Southern Methodist University, 200 p.

MELIM, L. A., and SCHOLLE, P. A., 1989, Dolomitization model for the forereef facies of the Permian Capitan Formation, Guadalupe Mountains, Texas-New Mexico, *in* Harris, P. M., and Grover, G. A., eds., Subsurface and outcrop examination of the Capitan Shelf Margin, northern Delaware Basin: Tulsa, Society of Economic Paleontologists and Mineralogists Core Workshop 13, p. 407-414.

MELIM, L. A., and SCHOLLE, P. A., 1995, The forereef facies of the Permian Capitan Formation: The role of sediment supply versus sea-level changes: Journal of Sedimentary Research, v. B65, p. 107-118

MOTTS, W. S., 1972, Geology and paleoenvironments of the Northern Segment, Capitan Shelf, New Mexico and West Texas: Geological Society of America Bulletin, v. 83, p. 701-722.

MRUK, D., and BEBOUT, D. G., 1993, Slope, *in* Bebout, D. G., and Kerans, C., eds., Guide to the Permian Reef Geology Trail, McKittrick Canyon, Guadalupe Mountains National Park, West Texas, Guidebook 26: Austin, Bureau of Economic Geology, University of Texas, p. 14-22.

NEESE, D. G., 1979, Facies mosaic of the upper Yates and lower Tansill Formations (Upper Permian), Walnut Canyon, Guadalupe Mountains, New Mexico: Unpublished M.S. Thesis, University of Wisconsin, Madison, 110 p.

NEESE, D. A., and SCHWARTZ, A. H., 1977, Facies mosaic of the upper Yates and lower Tansill Formations, Walnut and Rattlesnake Canyons, Guadalupe Mountains, New Mexico, *in* Hileman, M. E., and Mazzullo, S. J., eds., Upper Guadalupian facies, Permian Reef Complex, Guadalupe Mountains, New Mexico and West Texas, Field Conference Guidebook: Midland, Society of Economic Paleontologists and Mineralogists, Permian Basin Section, Publication 77-16, v. 1, p. 437-450.

NEWELL, N. D., RIGBY, J. K., FISCHER, A. G., WHITEMAN, A. J., HICKOX, J. E., and BRADLEY, J. S., 1953, The Permian Reef Complex of the Guadalupe Mountains Region, Texas and New Mexico: San Francisco, W. H. Freeman Co., 236 p.

OSLEGER, D. A., 1998, Sequence architecture and sea-level dynamics of Upper Permian shelfal facies, Guadalupe Mountains, Southern New Mexico: Journal of Sedimentary Research, v. 68, p. 327-346

OWEN, D. R., 1984, Hydrodynamic emplacement model for shelf-edge sandstone dikes (abs.): Geological Society of America Abstracts with Programs, v. 16, p. 185.

PARSLEY, M. J., and WARREN, J. K., 1989, Characterization of an Upper Guadalupian barrier-island complex from the middle and upper Tansill Formation (Permian), East Dark Canyon, Guadalupe Mountains, New Mexico, *in* Harris, P. M., and Grover, G. A., eds., Subsurface and outcrop examination of the Capitan Shelf Margin, northern Delaware Basin: Tulsa, Society of Economic Paleontologists and Mineralogists Core Workshop 13, p. 279-286.

PLAYFORD, P. E., 1984, Platform margin and marginal-slope relationships in Devonian reef complexes of the Canning Basin, *in* Purcell, P. G., ed., The Canning Basin Western Australia: Perth, Proceedings of the Geological Society of Australia and the Petroleum Exploration Society of Australia Symposium, p. 189-214.

PRAY, L. C., 1977, The all wet constant sea-level hypothesis of Upper Guadalupian shelf and shelf edge strata, Guadalupe Mountains, New Mexico and Texas, *in* Hileman, M. E., and Mazzullo, S. J., eds., Upper Guadalupian facies, Permian Reef Complex, Guadalupe Mountains, New Mexico and West Texas, Field Conference Guidebook: Midland, Society of Economic Paleontologists and Mineralogists, Permian Basin Section, Publication 77-16, v. 1, p. 433.

PRAY, L. C., 1985, Sea level and the Capitan Reef Complex, Southwest USA, a famous model in flux, *in* Cunningham, B. K., and Hedrick, C. L., eds., Permian Carbonate/Clastic Sedimentology, Guadalupe Mountains: Analogs for Shelf and Basin Reservoirs, Midland, Society of Economic Paleontologists and Mineralogists, Permian Basin Section, Publication 85-24, p. 68.

PRAY, L. C., and STANTON, R. J., 1992, Neptunian fossil-rich carbonate dikes in the Capitan massive (Permian), Guadalupe Mountains, New Mexico and Texas, U.S.A. (abs.): Tulsa, American Association of Petroleum Geologists Annual Convention Program, p. 106.

RANKEY, E. C., and LEHRMANN, D. J., 1996, Anatomy and origin of toplap in a mixed carbonate-clastic system, Seven Rivers Formation (Permian, Guadalupian), Guadalupe Mountains, New Mexico, USA: Sedimentology, v. 43, p. 807-826.

ROSS, C. A., and ROSS, J. R. P., 1987, Late Paleozoic sea levels and depositional sequences, *in* Ross, C. A., and Haman, D., eds., Timing and depositional history of eustatic sequences: constraints on seismic stratigraphy: Cushman Foundation for Foraminiferal Research Special Publication 24, p. 137-149.

SALLER, A. H., 1996, Differential compaction and basinward tilting of the prograding Capitan Reef Complex, Permian, West Texas and Southeast New Mexico, USA: Sedimentary Geology, v. 101, p. 21-30.

SARG, J. F., 1988, Carbonate sequence stratigraphy and controls on carbonate platform development—case study from Permian of West Texas-New Mexico (abs.): American Association of Petroleum Geologists Bulletin, v. 72, p. 1522.

SARG, J. F., 1989, Middle-Late Permian depositional sequences, Permian Basin, West Texas and New Mexico, *in* Bally, A. W., ed., Atlas of Seismic Stratigraphy: American Association of Petroleum Geologists Studies in Geology 27, v. 3, p. 140-154.

SCHLAGER, W., 1991, Depositional bias and environmental change—important factors in sequence stratigraphy: Sedimentary Geology, v. 70, p. 109-130.

SCHOLLE, P. A., and KINSMAN, D. J. J., 1974, Aragonitic and high-Mg calcite from the Persian Gulf—a modern analog for the Permian of Texas and New Mexico: Journal of Sedimentary Petrology, v. 44, p. 904-916.

SCHWARTZ, A. H., 1981, Facies mosaic of the upper Yates and lower Tansill Formations (Upper Permian), Rattlesnake Canyon, Guadalupe Mountains, New Mexico: Unpublished M.S. Thesis, University of Wisconsin, Madison, 149 p.

SHINN, E. A., and ROBBIN, D. M., 1983, Mechanical and chemical compaction in fine-grained shallow-water limestones: Journal of Sedimentary Petrology, v. 53, p. 595-618.

SILVER, B. A., and TODD, R. G., 1969, Permian cyclic strata, northern Midland and Delaware basins, West Texas and southeastern New Mexico: American Association of Petroleum Geologists Bulletin, v. 53, p. 2223-2251.

SMART, P. L., PALMER, R. J., WHITAKER, F., and WRIGHT, V. P., 1988, Neptunian dikes and fissure fills: An overview and account of some modern examples, *in* James, N. P. and Choquette, P. W., eds., Paleokarst: New York, Springer-Verlag, p. 149-163.

SMART, P. L., and WHITAKER, F. F., 1991, Karst processes,hydrology and porosity evolution, *in* Wright, V. P., Esteban, M. and Smart, P. L., eds., Palaeokarsts and Palaeokarstic Reservoirs: PRIS Contribution 152, University of Reading, p. 1-55.

SMITH, D. B., 1974, Sedimentation of upper Artesia (Guadalupian) cyclic shelf deposits of northern Guadalupe Mountains, New Mexico: American Association of Petroleum Geologists Bulletin, v. 58, p. 1699-1730.

TINKER, S. W., 1996, Reservoir-scale sequence stratigraphy: McKittrick Canyon and three-dimensional subsurface examples, West Texas and New Mexico: Unpublished Ph.D. Dissertation: The University of Colorado, Boulder, 245 p.

TYRRELL, W. W. , Jr., 1962, Petrology and stratigraphy of near-reef Tansill-Lamar strata, Guadalupe Mountains, Texas and New Mexico, *in* Permian of the Central Guadalupe Mountains, Eddy County, New Mexico, Field Trip Guidebook 62-48: West Texas Geological Society, p. 59-75.

TYRRELL, W. W. ,Jr., 1969, Criteria useful in interpreting environments of unlike but time-equivalent carbonate units, Capitan Reef Complex, West Texas and New Mexico (abs.): American Association of Petroleum Geologists Bulletin, v. 51, p. 484.

WOOD, R., DICKSON, J. A. D., and KIRKLAND-GEORGE, B., 1994, Turning the Capitan Reef upside down: A new appraisal of the ecology of the Permian Capitan Reef, Guadalupe Mountains, Texas and New Mexico: Palaios, v. 9, p. 422-427.

YUREWICZ, D. A., 1976, Sedimentology, paleoecology and diagenesis of the massive facies of the lower and middle Capitan Limestone (Permian), Guadalupe Mountains, New Mexico and West Texas: Unpublished Ph.D. Dissertation, University of Wisconsin, Madison, 278 p.

YUREWICZ, D. A., 1977, The origin of the massive facies of the lower and middle Capitan Limestone (Permian), Guadalupe Mountains, New Mexico and West Texas, *in* Hileman, M. E., and Mazzullo, S. J., eds., Upper Guadalupian facies, Permian Reef Complex, Guadalupe Mountains, New Mexico and West Texas, Field Conference Guidebook: Midland, Society of Economic Paleontologists and Mineralogists, Permian Basin Section, Publication 77-16, v. 1, p. 45-92.

PALEOENVIRONMENTS, CYCLICITY, AND DIAGENESIS IN THE OUTER SHELF TANSILL FORMATION IN THE CARLSBAD EMBAYMENT (DARK CANYON), NORTHERN GUADALUPE MOUNTAINS, NEW MEXICO

S. J. MAZZULLO

Department of Geology, Wichita State University, Wichita, Kansas 67260

ABSTRACT: Outer shelf carbonates in the Tansill Formation in Dark Canyon include the following inferred depositional environments landward of the Capitan reef: (1) subtidal: bioclastic wackestones, packstones, and graded and locally cross-stratified grainstones; and biostromes; (2) peritidal flats: fenestral and locally desiccated mudstone; and (3) shoreface: admixed subtidal and peritidal deposits. Two sequences are recognized in the section based on biotic diversity and parasequence thickness and facies-stacking patterns, and the boundary between them is at or near the base of the Ocotillo Member. Maximum flooding of the platform occurred during deposition of sequence 1 in the lower part of the middle Tansill. Environments were biostromes, mainly high energy shallow subtidal packstones and grainstones, and associated peritidal islands in the early HST in this sequence. Precipitation of abundant marine cements in these rocks may have been promoted by active seawater pumping through the sediments on this wide and shallow shelf. Microbial activity in the grainstones may have been promoted by restricted circulation around associated peritidal islands. In contrast, environments in the late HST of this sequence and in sequence 2 were more restricted in terms of shelf width and energy, and the rocks are dominantly micritic and contain little marine cement. The principal difference between depositional facies in the Carlsbad Embayment and along the Reef Escarpment appears to be the presence of patch reefs in the former area, which were deposited on an outer shelf that was wider and of more open circulation than to the south.

Inferred marine cements are dominated by prismatic calcites, with microdolomite inclusions and some recognizable radiaxial-fibrous habit, interpreted as former high-Mg calcite. The $\delta^{18}O$ and $\delta^{13}C$ compositions of the least-altered of these cements (–1.6 ‰ and 5.8 ‰ PDB, respectively) suggest precipitation from marine pore fluids. Former aragonitic cement of similar isotopic composition is a volumetrically minor, first-generation marine cement in these rocks. Dominance of inferred former high-Mg calcite cement contrasts that in coeval platform-margin patch reef facies, the Capitan reef, and in pisolites on the shelf crest in which former aragonite marine cement dominates. Mean $\delta^{18}O$ composition of the dolomite that replaced peritidal deposits (0.1 ‰ PDB) suggests that it precipitated from ambient marine fluids of elevated salinity. Mineralogic stabilization of earlier diagenetic phases likely attended precipitation of equant calcites (spar I) in the rocks, which is interpreted to have occurred in a meteoric phreatic system during the latest Guadalupian and/or early Ochoan. A second generation of equant cements (spar II) is inferred to have precipitated later, during the Ochoan or later, in a deeper meteoric-dominated groundwater system. Replacement of syndepositionally precipitated evaporites by poikilotopic calcite is the last diagenetic phase present and, based on stable oxygen-carbon isotope data, appears to have attended meteoric dedolomitization during the Tertiary.

INTRODUCTION

The Tansill Formation, exposed in the Guadalupe Mountains of west Texas and southeastern New Mexico, includes inner shelf lagoon, shelf crest (including the pisolite belt), and outer shelf deposits landward of the Capitan reef (Achauer, 1969; Tyrrell, 1969; Dunham, 1972; Pray and Esteban, 1977). Outer shelf carbonates along the Reef Escarpment (Fig. 1) are dominantly subtidal deposits (Pray, 1989; Kirkland and Moore, 1996), and interbeds of siltstone, fenestral carbonates, and pisolites largely reflect eustatic fluctuations (Neese and Schwartz, 1977; Pray, 1985; Neese, 1989). This facies tract differs in the Carlsbad Embayment to the north (Fig. 1), however, where the section includes peritidal islands, patch reefs, and biostromes (Toomey and Cys, 1977; Ward, 1988; Mazzullo et al., 1989; Noé and Mazzullo, 1992, 1994; Noé, 1996). Such deposits first appear in the lower part of the middle Tansill in Dark Canyon, and they are not known in Tansill outer shelf rocks along the Reef Escarpment (e.g., Pray, 1989). Some workers have suggested that this unique style of deposition in the Carlsbad Embayment was the result of northward shallowing along the outer shelf (e.g., Motts, 1962, 1972; Garber et al., 1989; Mazzullo et al., 1989; Pray, 1989).

The Tansill Formation in the Carlsbad Embayment thus preserves a record of environmental conditions, depositional environments, and possibly, diagenesis that may be different than in correlative strata to the south. These aspects of platform evolution in the Carlsbad Embayment, however, presently are incompletely understood. This paper extends the preliminary studies of Mazzullo et al. (1989) and focuses on describing depositional environments and diagenesis in Tansill outer shelf deposits that contain peritidal islands and biostromes in Dark Canyon, and

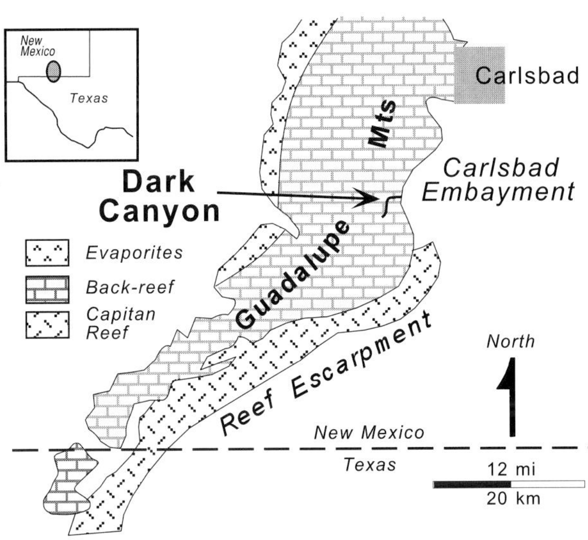

FIG. 1.—Location of study area in the Guadalupe Mountains.

Geologic Framework of the Capitan Reef, SEPM Special Publication No. 65

relates these attributes to cyclicity in the section. The overall paleobathymetry of the Carlsbad Embayment is then reconstructed in order to evaluate those factors that may have controlled deposition and diagenesis here. Study of these rocks may assist in identifying some long-term controls on sedimentologic and diagenetic aspects of platform evolution during the late Guadalupian.

LOCATION AND STRATIGRAPHY

The Tansill Formation in Dark Canyon (Fig. 2A) has been the focus of many studies, including Tyrrell (1962, 1969), Tyrrell et al. (1978), Biggers (1984), Parsley (1988), Ward (1988), Mazzullo et al. (1989), and Parsley and Warren (1989). The Tansill Formation was subdivided by Tyrrell et al. (1978) on the basis of fusulinid biostratigraphy into lower (*Yabeina texana* zone), middle (*Reichelina* zone), and upper (*Paraboultonia* zone) units, and these are the stratigraphic subdivisions recognized in this paper (Fig. 3). The outcrops described herein are exposed in the abandoned quarry, cliffs to the immediate west, and hillside outcrops along the south wall of the canyon, a short distance west of its mouth, in Eddy County, New Mexico (Fig. 2B). The oldest beds here are in the lower part of the middle Tansill, and they compose outer shelf deposits with the peritidal islands that were described by Mazzullo et al. (1989). Eleven closely spaced sections were measured in the Tansill (Fig. 3). Total thickness of the middle Tansill exposed here is about 167 ft (51 m), which includes about 33 ft (10 m) of the Ocotillo Member. The Ocotillo is about 33 ft (10 m) of poorly exposed, commonly brecciated siltstones, silty dolomites and micritic limestones. Maximum thickness of the exposed upper Tansill here is about 45 ft (13.7 m).

The Tansill dips eastward from the immediate study area, and the Ocotillo progressively loses its identity as a mixed siltstone-silty carbonate unit as it approaches the platform margin in this direction (Tyrrell et al., 1978). Ocotillo-equivalent and upper Tansill beds at the mouth of the canyon are relatively high diversity, coarse-grained bioclastic carbonate sand and patch reefs, the latter informally designated as "Toomey Reef" and "Babcock Reef" in this paper (Fig. 2A). These reefs are present in Ocotillo-equivalent strata on the north and south sides of the canyon, and probably represent a larger reef that was breached by Tertiary erosion. They were described by Babcock (1977), Toomey and Cys (1977), Noé (1996), and Weidlich and Fagerstrom (this volume). Tansill strata in the subsurface at the mouth of the canyon are Capitan reef deposits (Tyrrell et al., 1978; Parsley, 1988; Parsley and Warren, 1989). The Tansill passes westward from the studied sections into shelf crest deposits.

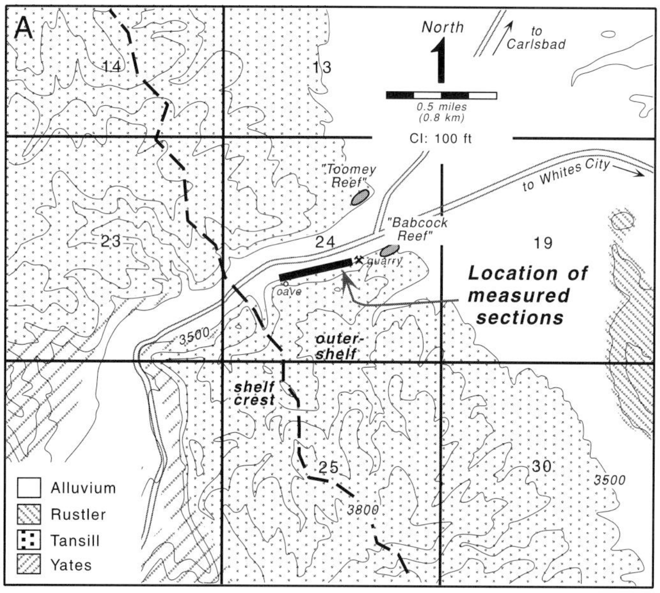

B

FIG. 2.—(A) Geologic map superimposed on topographic map (CI = 100 ft: 30.5 m) of the Dark Canyon study area, showing shelf crest-outer shelf boundary (in late middle Tansill time), patch reefs at the mouth of the canyon, and location of measured sections. (B) Photograph of the south wall of Dark Canyon showing stratigraphy and specific locations of measured sections. Sections 1-9 were measured from the canyon floor to the top of the face of the quarry; sections 10 and 12 were measured from the canyon floor to the top of the hill. Maximum thickness of section shown is about 212 ft (64.6 m); the distance between measured sections 1 and 9 is about 260 ft (79 m).

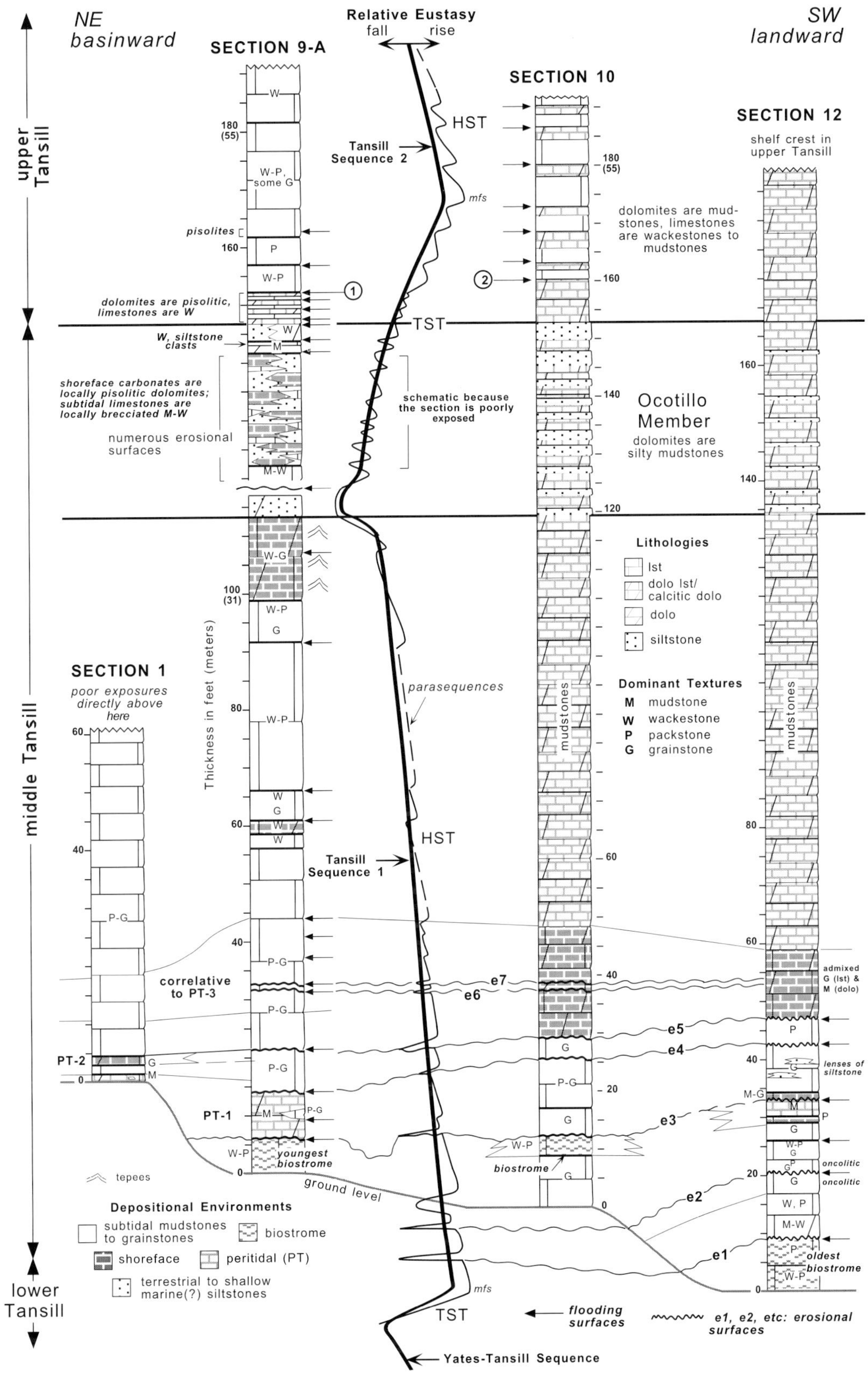

FIG. 3.—Lithostratigraphy of the Tansill in the study area; the stratigraphy of the lower part of the middle Tansill between measured sections 1 and 9 is shown in greater detail in Figs. 4A and B. Flooding surfaces bound parasequences; flooding surfaces and parasequences in most of the Ocotillo are generalized because of poor exposures. Refer to Fig. 4 for details of cyclicity in the lower 45 ft (13.7 m) at section 9-A (= measured section 9). Overlying strata (section "A") were measured beginning at the top of section 3 and drifting west, following exposures, to those beds above section 9. The section is hung on the base of the Ocotillo Member; no horizontal scale intended. The significance of flooding surfaces 1 and 2 in the basal upper Tansill is discussed in the text.

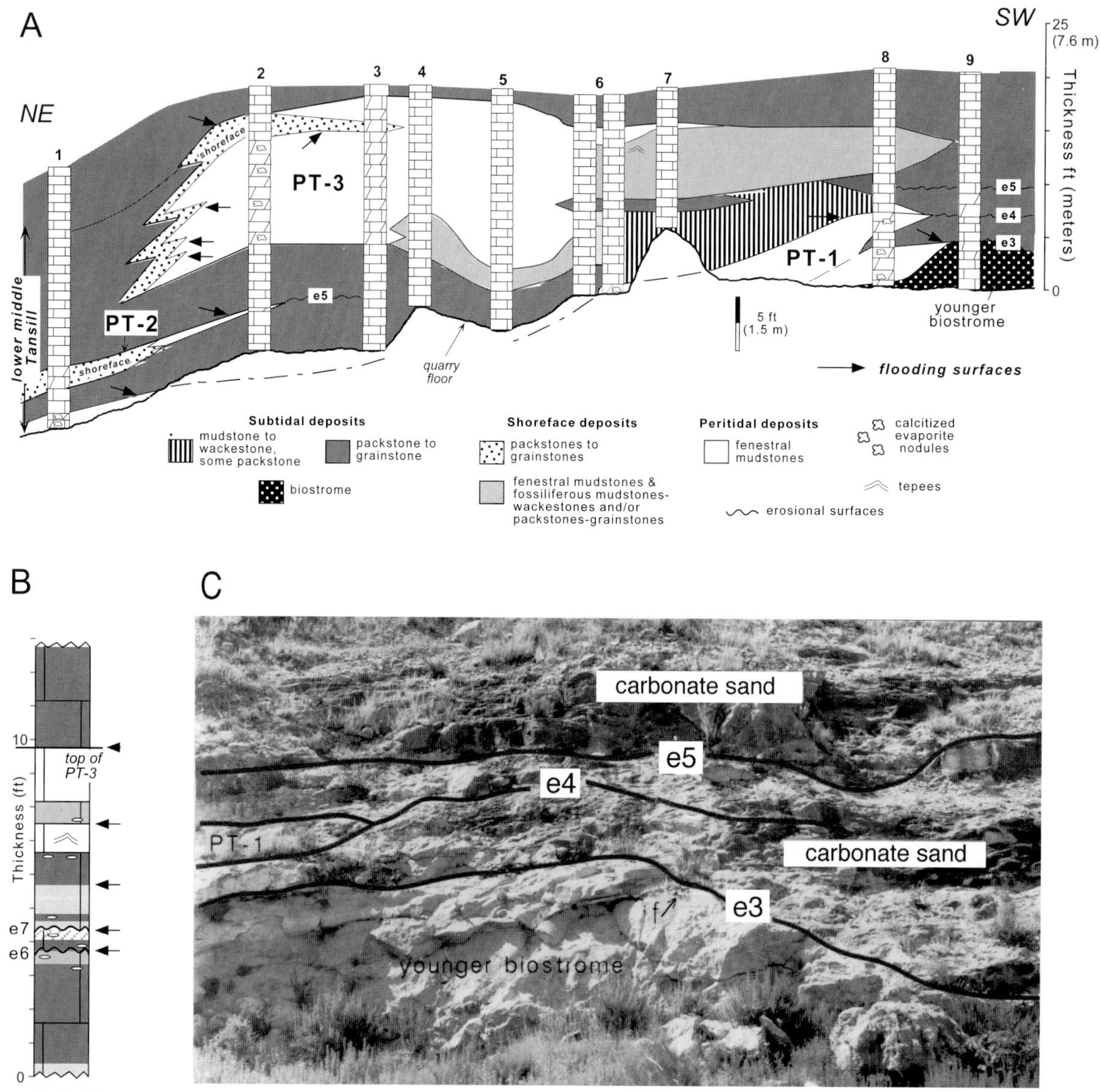

FIG. 4.—(A) Interpreted depositional environments and parasequences in the lower part of the lower middle Tansill in measured sections 1-9. This diagram is traced from a photomosaic of the outcrop; the distance between measured sections 1 and 9 is about 260 ft (79 m). The dominantly shoreface deposits shown between sections 6 and 8 are generalized because of scale limitations. Parasequences and flooding surfaces in this section, which were correlated into PT-3, are based on exposures at measured section 7 shown in (B). (C) Erosional surfaces e3-e5 and lithic relationships in the lower part of measured section 9. The upper part of the younger biostrome here is dolomitized. Note the fissure at "f".

METHODS OF STUDY

Spatial variations in texture, biotic content and diversity, sedimentary structures, and diagenetic features in the Tansill were studied in the field and laboratory. Lithologic units were correlated (Figs. 3 and 4A) by tracing beds and bedding planes along the outcrops, and samples were collected from each lithologic unit identified. Light-petrographic diagenetic studies were complemented by CL petrography and stable oxygen-carbon isotope compositional analyses of cements and dolomite. Isotope compositional data were determined at the University of Michigan, Ann Arbor, and Texas A&M University following standard analytical procedures, and are reported in uncorrected per mil notation relative to PDB. There was less than 0.3% difference in the isotopic compositions of the 30 replicate samples that were sent for analyses to each of these laboratories. Trace-element compositions of selected samples were analyzed by atomic absorption spectrophotometry at Wichita State University. Samples for isotopic and trace-element analyses were first X-rayed to ensure mineralogic purity (greater than 95% for the minerals analyzed).

DEPOSITIONAL ENVIRONMENTS AND BIOTA

Carbonate rocks of inferred peritidal, shoreface and associated deposits, subtidal, and biostromal origin are the principal

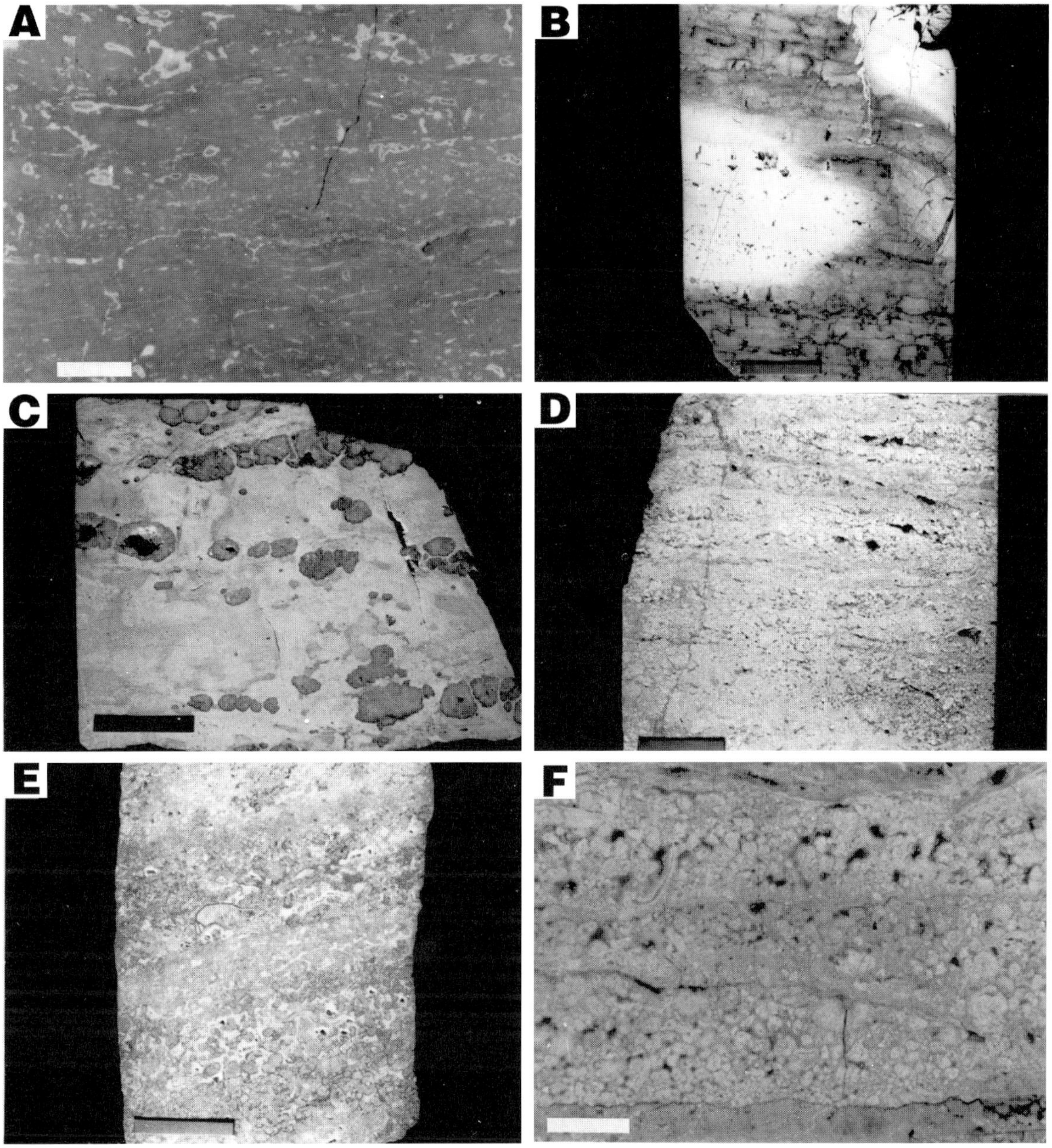

FIG. 5.—(A) Rock slab of fenestral peritidal limestone; fenestrae are partly filled with white calcite cement (inferred former high-Mg calcite) and a final pore-fill of translucent equant calcite; length of scale is 2 cm. (B) Rock slab of partly dedolomitized peritidal carbonate (gray areas are dolomite) with desiccation cracks; length of scale is 3 cm. (C) Rock slab of peritidal dolomite with inferred former evaporite nodules now partly to completely infilled with poikilotopic calcite; length of scale is 3 cm. (D) Rock slab of laminated shoreface grainstone in PT-2; length of scale is 3 cm. (E) Rock slab of typical subtidal packstone-grainstone; pores are occluded mainly by coarsely crystalline, prismatic, inferred marine cements (white). Length of scale is 3 cm. (F) Rock slab of coarse-grained, subtidal grainstone with encrusted "cerebroid" grains; length of scale is 3 cm.

depositional environments recognized in the Tansill in the study area (Figs. 3 and 4A). The lithologic and biotic attributes of these rocks, and of the siltstones in the Ocotillo Member, are described below.

Peritidal Deposits

Rocks interpreted as peritidal deposits are unfossiliferous, slightly argillaceous, commonly peloidal lime mudstone and finely crystalline dolomudstone with open and cement-filled fenestrae, desiccation cracks, inferred microbial laminites, and locally, some mudstone intraclasts (Fig. 5A,B). For reference in ensuing discussions, the successive peritidal deposits in the lower part of the middle Tansill at measured sections 1-9 are referred to as PT-1, PT-2, and PT-3 (Fig. 4A). Some of these peritidal rocks contain cauliflower-shaped nodules occluded by calcite or vugs resulting from the dissolution of this calcite. These nodules and vugs occur as nodular-mosaic masses and, locally, as enterolithically folded layers (Fig. 5C), and are interpreted as having formed syndepositionally as evaporites. The evaporites have since been dissolved from the nodules, and the calcite that now infills them is very coarsely crystalline. Some of the peritidal dolomites in the Ocotillo Member and, locally, in overlying beds, contain pisoliths or are interbedded with pisolites (Fig. 3). Silty, peritidal dolomites in the Ocotillo are interbedded with siltstones.

Shoreface and Associated Deposits

Rocks interpreted as shoreface deposits are recognized by lateral and vertical stratal relationships (that is, their location between subtidal and peritidal deposits) and diagnostic sedimentary features. The seaward margins of peritidal flats PT-2 and PT-3, for example (Fig. 4A), are composed of laminated, well-sorted carbonate sands (Fig. 5D) that are similar to shoreface deposits in modern carbonates (e.g., Hardie, 1977). Inferred shoreface deposits are also present immediately beneath and to the west of PT-3 (Fig. 4A), and between dominantly subtidal and peritidal deposits at measured sections 10 and 12 to the east (Fig. 3). These sections include thin interbeds of normally graded or locally cross-stratified carbonate sands or bioclastic wackestone-packstone and fenestral mudstone. The rocks locally are lenticular bedded and contain some tepees, cement-filled sheetcracks, and scoured surfaces. Allochems in all of these aforementioned shoreface deposits include peloids, forams, crinoids, *Mizzia*, fragments of brachiopods and pelecypods, and intraclasts. Inferred shoreface deposits immediately beneath the Ocotillo Member at measured section 9 (Fig. 3) are dominantly wackestones with lentils of grainstone; they also contain some tepees and fenestrae. Allochems include *Mizzia* and peloids. Presumed shoreface carbonates in the Ocotillo Member and in the section immediately above it at measured section 9 (Fig. 3) are interbedded fenestral dolomudstone with some pisolites, and lime mudstone to wackestone with forams; these limestones commonly are brecciated in the Ocotillo Member.

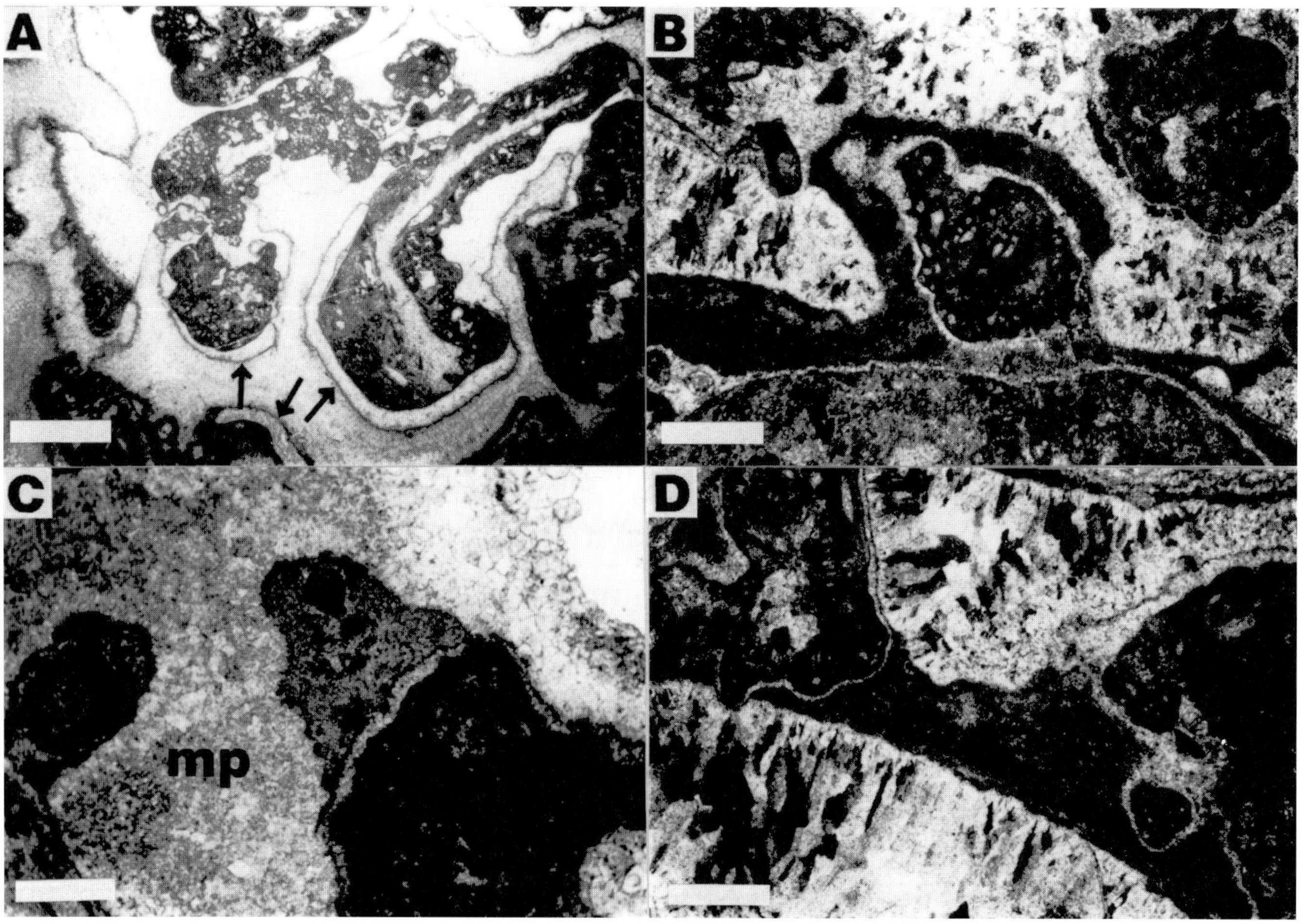

FIG. 6.—Photomicrographs of subtidal deposits in the lower part of the middle Tansill; stratigraphic top is toward the top of the page in all images. (A) Allochems (dark) encrusted by *Girvanella* and accessory forams. The light-colored encrustations (arrows) on the undersides and tops of grains are of problematic origin; plane light, scale 1.8 mm. (B) Dark microbial encrustation binds grains and coats overhanging sides of protruding grains. Thin rim of clear calcite beneath the encrustation in center of photo is presumed former marine aragonite cement; remainder of pores filled with coarse prismatic calcite. Grain at top-right is a clast derived from a microbial mat; crossed polars, scale 750 μm. (C) Encrustations that grew upward at angles from tops of grains. Micro-prismatic calcite (mp) fills most pore space; crossed polars, scale 750 μm. (D) Encrustation that bridged space between adjoining grains. Coarse prismatic cement later nucleated on this encrustation; crossed polars, scale 750 μm.

Subtidal Deposits

Inferred subtidal deposits in the lower part of the middle Tansill at measured sections 1–6 are to the immediate east of and interbedded with PT-1, PT-2, and PT-3 (Fig. 4A). The rocks are dominantly bioclastic packstone and grainstone with gravel and coarse sand-size particles (Fig. 5E). Allochems include a relatively high-diversity assemblage of *Mizzia*, forams, fusulinids, red algae, *Shamovella* (*Tubiphytes*), crinoids, fragments of brachiopods and pelecypods, and some corals, sponges, and bryozoans. Intraclasts of bioclastic wackestone to grainstone with similar component particles are present locally. The rocks occur in beds from 2–11 in (5–27 cm) thick that commonly are normally graded and cross-stratified, and wherein laminae in cross-stratified units dip at 5° or less in a basinward direction. They are interpreted to have been deposited in a high energy, shallow marine environment. Subtidal deposits to the west (Fig. 3) are slightly finer grained, locally cross-stratified, and include *Mizzia*, forams, brachiopod, and pelecypod fragments, and oncolites in some beds. These subtidal carbonates are interpreted to have been deposited in a dominantly high to moderate energy shelf environment. Lenses of yellow siltstone are present in the bioclastic grainstone unit that is about 40 ft (12 m) above the base of measured section 12. This silty unit may correlate to the siltstone immediately above the "Basal Dolomite" of the Tansill along the Reef Escarpment (Neese and Schwartz, 1977).

Inferred subtidal deposits in the overlying Tansill section (excluding the Ocotillo Member) are mostly bioclastic wackestone and packstone (Fig. 3). Biota include a relatively low-diversity assemblage of forams, *Mizzia*, fragments of brachiopods, pelecypods and gastropods, crinoids, and rare bryozoans and red algae. These rocks are interpreted as having been deposited in shallow marine environments of low to moderate energy relative to subjacent subtidal deposits. Subtidal deposits in the Ocotillo Member are mudstone and sparse wackestone with forams, and are interpreted as very shallow, low energy marine deposits.

The following features are present only in the subtidal deposits along the basinward (eastern) side of the outcrops (measured sections 1-6, Fig. 4A). Thin layers of non-isopachous, calcite microspar with inclusions of unreplaced micrite are attached to the bottoms, sides, or tops of many of the allochems

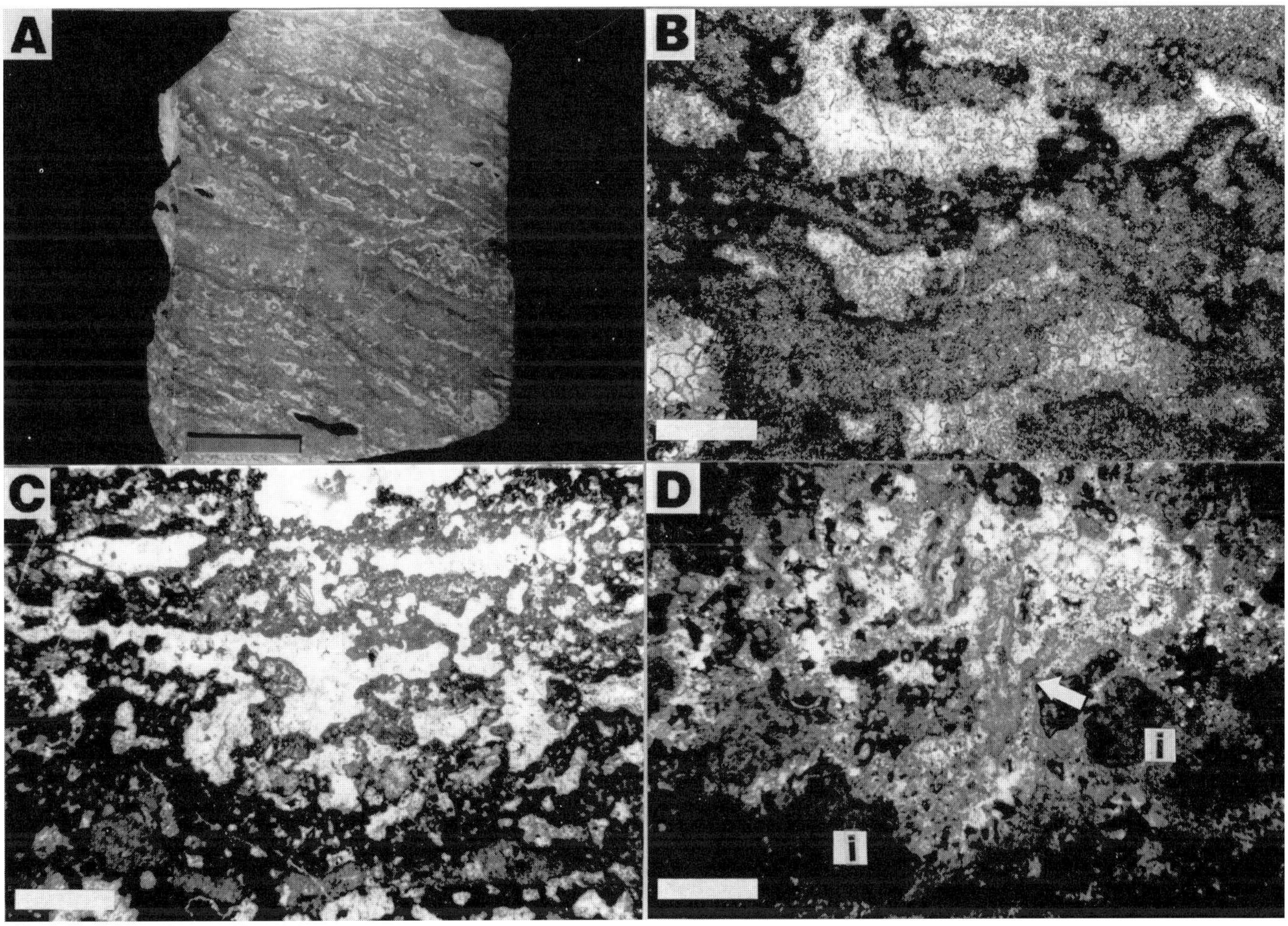

FIG. 7.—Subtidal deposits in the lower part of the middle Tansill; stratigraphic top is toward top of page in all images. (A) Rock slab of mats (dark) within subtidal grainstone; scale 3 cm. (B) Clotted fabric of mats interlaminated with inferred marine cement. *Girvanella* locally encrusts the mats; plane light, scale 750 µm. (C) Clotted appearance of mats, clots occluded by inferred marine cement; plane light, scale 1.8 mm. (D) Intraclasts of mat-derived sediment (i) and vertically disposed, branched algae (arrow) within mats; crossed polars, scale 1.8 mm.

in these rocks (Fig. 6A). This neomorphic calcite does not appear to have replaced inferred primary fibrous or prismatic marine cements. Rather, it may be of microbial origin, possibly algae, of likely cryptic, encrusting habit. Many of the allochems are heavily encrusted by *Girvanella* and accessory forams, which results in a conspicuous spongy, cerebroid appearance and accounts for the large size of the grains (Fig. 5F). Many allochems also are associated with dark-colored, micro-peloidal micrite that commonly overlies and, in turn, is overlain by, inferred marine cement. This micrite variously: (i) binds adjoining grains and encrusts the tops and overhanging sides of protruding grains (Fig. 6B); (ii) accreted at angles upwards from the tops of grains (Fig. 6C); and (iii) bridged adjoining grains prior to presumed marine cementation (Fig. 6D). These observations suggest that the dark micrite is not internal sediment. Rather, its similarity of appearance and mode of occurrence to microbial encrustations in the Capitan reef described by Wood et al. (1994, 1996) and in other Paleozoic rocks (e.g., Bosence and Bridges, 1995; Monty, 1995) suggest that it may also be of syndepositional microbial origin, likely also cryptic.

Similarly dark-colored, organic matter-rich micrite also occurs as conspicuous thin, closely spaced mats in these subtidal deposits (Fig. 7A). These mats are characterized by clotted appearance wherein the clots mostly are occluded by presumed marine cement (Fig. 7B, C), and: (i) they commonly are interlaminated with marine cements and locally encrusted by forams or *Girvanella* (Fig. 7B); and (ii) clasts derived from the mats locally are included in successive mat layers (Fig. 7D) and in associated subtidal grainstone (Fig. 6B), which suggests some degree of syndepositional lithification of the mats. Some of the mats include laminae with branched, upward-growing algae (*Girvanella?*)(Fig. 7D). Many of these features are similar to those in the cyanophyte-bacterial mats on tidal flats described by Chafetz and Buczynski (1992). Although a similar depositional setting (that is, peritidal) for the mats in the Tansill is not suggested, they likewise are believed to be of microbial origin.

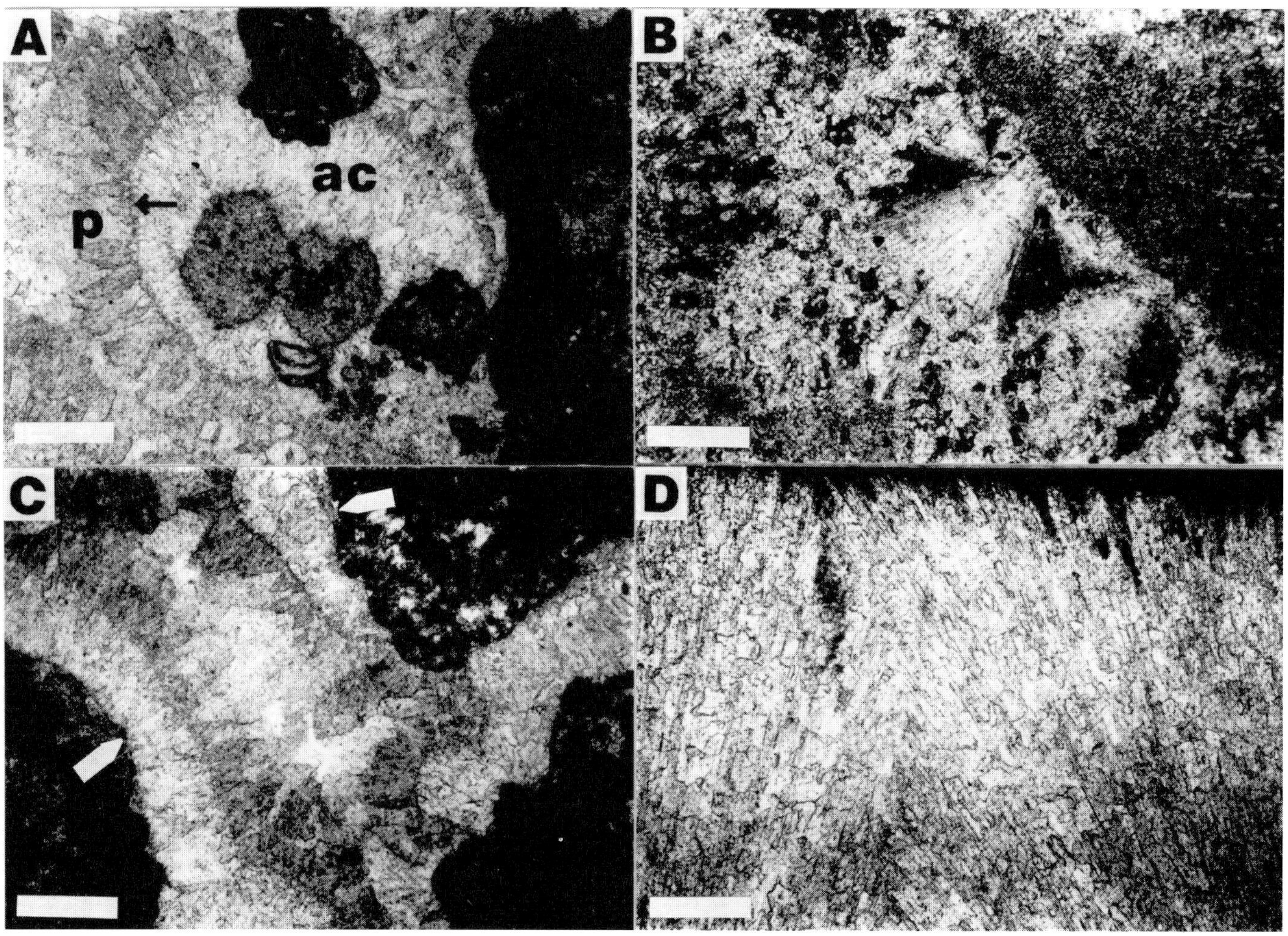

FIG. 8.—(A) Isopachous crust of former aragonite marine cement (ac). Note thin dustline (arrow) around the cement, and later coarse prismatic calcite (p); plane light, scale 750 μm. (B) Botryoids of former aragonite marine cement; crossed polars, scale 180 μm. (C) Presumed dissolution-remnants of former aragonite marine cement (arrows) in interparticle pore; plane light, scale 750 μm. (D) Former aragonite marine cement that partially occluded sheetcrack in subtidal grainstones; plane light, scale 750 μm.

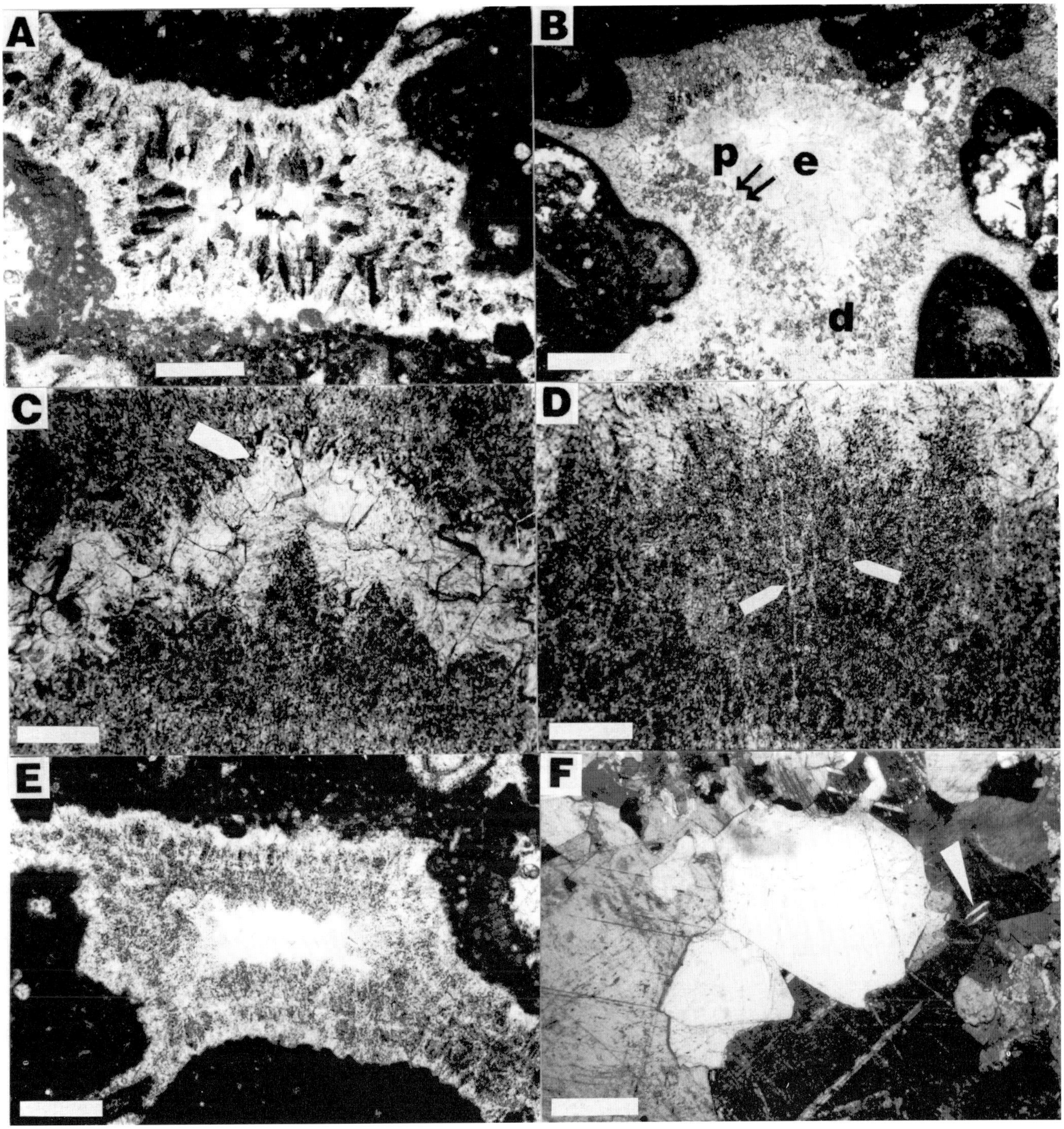

FIG. 9.—(A) Micro-prismatic cement in subtidal grainstones; crossed polars, scale 750 µm. (B) Layer of loafish, micro-prismatic cement with "necklace" of dolomite rhombs (d). Final pore-filling is clear, equant calcite (e) that has syntaxially engulfed the final generation of coarse crystalline prismatic calcite (p). Narrow veins extend from the equant calcite into the prismatic calcite (arrows); plane light, scale 750 µm. (C) Acute and obtuse terminations of coarse crystalline prismatic cements. Note dissolution-termination of crystals (arrow) at contact with later coarse, equant calcite; plane light, scale 180 µm. (D) Coarse prismatic calcite with veins of lucid calcite (arrows); plane light, scale 180 ?m. (E) Growth-banding in coarse crystalline, prismatic calcite. Final pore fill is lucid, equant calcite; plane light, scale 750 µm. (F) Poikilotopic calcite in former evaporite nodules; note anhydrite inclusions (arrow). Crossed polars, length of scale 750 µm.

Biostromes

Massive bedded limestone is present at the base of the exposed middle Tansill section around measured section 12, and this unit dips into the subsurface to the east (Fig. 3). The rocks are dominantly wackestone and packstone with a diverse biota of *Mizzia*, fusulinids, forams, crinoids, and fragments of brachiopods and pelecypods. Sponges (*Lemonia* and *Amblysiphonella*) and bryozoans in these rocks are restricted to its basinward side. Marine cements (described below) are abundant. These rocks were interpreted by Ward (1988) as a biostrome, an interpretation with which I agree. This biostrome is of Capitan-like affinity in terms of biota and abundance of marine cement, and hence, it is similar in these respects to the Toomey and Babcock reefs and also to the patch reefs described by Noé and Mazzullo (1992, 1994) and Noé (1996) farther north in the Carlsbad Embayment. Another biostrome that appears to be stratigraphically higher in the Tansill is present at measured section 9 (Figs. 3 and 4A). The rocks are wackestone and packstone with a lower diversity biota of *Mizzia*, forams, crinoids, and fragments of brachiopods and pelecypods, and no marine cements. They therefore are not of Capitan-type aspect. A small, thin biostrome with similar fossils and no marine cements is present between measured sections 9 and 10. The horizon in which these two bioherms occur appears to correlate to a section of peritidal and shoreface deposits at measured section 12 (Fig. 3).

Siltstones in the Ocotillo Member

The siltstones in the Ocotillo are poorly exposed. They are texturally homogeneous, unfossiliferous and commonly brecciated, and are interbedded with peritidal dolomudstone and shallow subtidal lime mudstones that also commonly are brecciated. The siltstones are interpreted as terrestrial deposits, possibly eolian (e.g., Kendall, 1969; Mazzullo et al., 1985; Borer and Harris, 1991), that likely were reworked in shallow marine environments during ensuing transgressions.

CEMENT PETROGRAPHY AND TRACE ELEMENT COMPOSITION

Stratigraphic Occurrence of Cements

Fibrous, prismatic, and equant calcites are very abundant, variously within subtidal grainstones and microbial mats and in shoreface grainstones in the lower part of the middle Tansill. They are also present within the oldest biostrome in this section (Ward, 1988). In contrast, these cements are uncommon in subtidal deposits in the overlying Tansill section here because the rocks are dominantly micritic, although they become more abundant in coarser grained rocks toward the platform margin (Parsley, 1988). Where grainstones are present in this section (Fig. 3), however, such cements are present in the same diagenetic order and relative abundance as described below. Fine particle size and small interparticle pores in these rocks, however, result in successive generations of cement that are thin and difficult to isolate for geochemical analyses. Prismatic and equant calcite cements are present within fenestrate in some of the peritidal lime mudstones in the lower part of the middle Tansill. The prismatic cements in these rocks decrease in abundance in a landward (to the west) direction. Such cements are much less common in peritidal deposits in overlying strata either because fenestrae are absent or very small, or more commonly, because the rocks have been pervasively dolomitized.

Hence, the following discussions focus on diagenetic and geochemical attributes of cements in only the lower part of the middle Tansill. Parsley (1988) also described similar cements in the middle and upper Tansill in Dark Canyon, but did not specify their stratigraphic distribution. The general diagenetic sequence recognized in this section is, in order: (1) precipitation of fibrous cements followed by minor dissolution; (2) precipitation of prismatic cements followed by another episode of minor dissolution; (3) minor mechanical compaction followed by precipitation of finely and coarsely crystalline, equant calcite cements; and (4) emplacement of coarsely crystalline, poikilotopic calcite in inferred former evaporite nodules. Only some of

TABLE 1.—MEAN GEOCHEMICAL COMPOSITIONS OF CEMENTS AND DOLOMITE*

Component	mole% $MgCO_3$	Sr	Fe	Mn	$\delta^{18}O$	$\delta^{13}C$
Former aragonite, Toomey reef	1.2	270	260	23	–1.3	5.5
		(n=11)			(n=6)	
Former aragonite in sheetcracks[1]	0.8	285	280	22	–2.4	4.5
		(n=8)			(n=15)	
Prismatic calcite cement	1.8	186	266	33	–2.3	4.8
		(n=11)			(n=21)	
All coarse, equant calcite cements[2]	1.7	102	247	46	–9.2	–1.2
		(n=11)			(n=21)	
Fine equant calcite cement[3]	—	—	—	—	–11.8	–3.6
					(n=3)	
Poikilotopic calcite in former evaporite nodules	0.5	94	321	22	–12.5	–14.9
		(n=10)			(n=21)	
Peritidal dolomite[4]	—	—	—	—	0.1	4.5
					(n=38)	
Dolomite replacive of allochems[5]	—	—	—	—	–0.2	5.5
					(n=12)	

*Sr, Mn, and Fe values reported in ppm, and isotopic compositions reported in per mil notation relative to PDB. Number of samples (n=X) indicated for trace-element and isotopic analyses.

[1]in subtidal deposits interbedded with peritidal deposits; [2]interparticle cement in subtidal deposits and fenestrae in peritidal deposits; [3]in biomolds in subtidal deposits; [4]whole-rock samples; [5]in subtidal deposits

the peritidal deposits in this section are pervasively dolomitized, and their interpreted site and timing of dolomitization also are discussed below.

Fibrous Cement

Thin, isopachous crusts (Figs. 6B and 8A) and, locally, thin layers of botryoidal calcite (Fig. 8B) are the first generation of cements recognized in subtidal and shoreface grainstones and in the oldest biostrome, although they are not present in all samples. These cements are not present in peritidai deposits. Botryoids also partly occlude many of the sheetcracks in shoreface deposits (Fig. 8D), wherein remaining porosity is filled with ostracod-peloid wackestone to grainstone and/or prismatic cement. The isopachous crusts are a maximum of 130 μm thick within interparticle and some primary intraparticle pores. They consist of a mosaic of replacive pseudospar with ghosts of fiber-crystals with square and/or acute-pointed tips. The botryoids are composed of aggregates of square-tipped crystals. Fibers or ghosts of fibers are 15–20 μm wide and up to 130 μm in length, and display straight extinction. The isopachous crusts and botryoids are mainly nonluminescent, although some samples display splotchy, dull-red luminescence. Similar cements are abundant in the Toomey reef (Noé and Mazzullo,

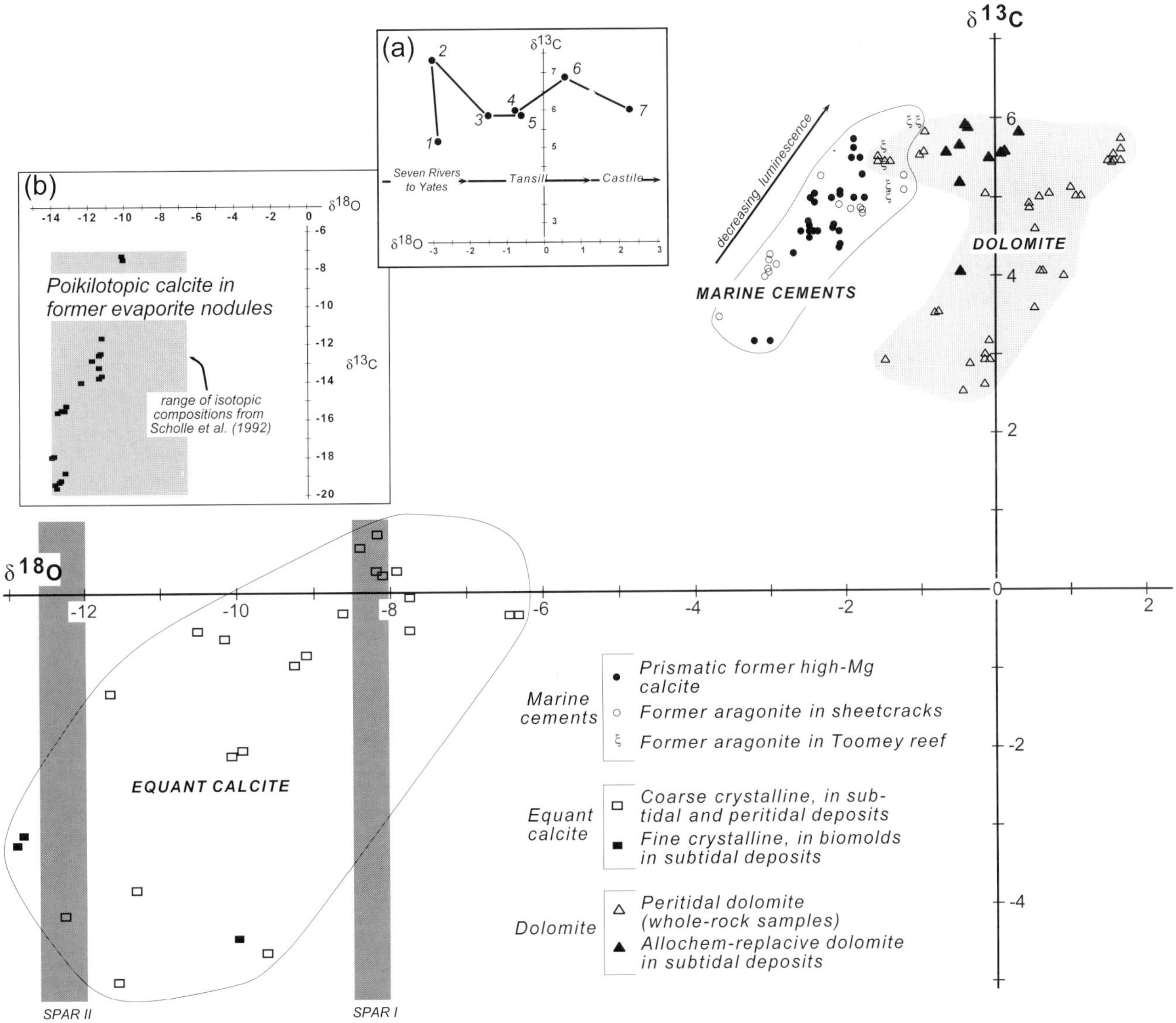

FIG. 10.—Main diagram and inset (b) show isotopic compositions of diagenetic phases in the lower part of the middle Tansill and of former aragonite marine cement in Toomey reef. Ranges of mean $\delta^{18}O$ compositions of "spar I" and "spar II" from Given and Lohmann (1986), Hill (1989, 1996), and Mruk (1989) are shown in the main diagram. Inset (a) shows trend of estimated seawater composition in late Guadalupian to Ochoan time based on data from least-altered marine cements: 1—Seven Rivers to Yates Fm. equivalent Capitan, Given and Lohmann (1985); 2—Yates Fm. equivalent Capitan, Mruk (1989); 3—lower part of the middle Tansill, and 4—Ocotillo-equivalent Toomey reef, this study; 5—middle Tansill equivalent reef (likely Toomey and/or Babcock reef), Given and Lohmann (1985); 6—middle to upper Tansill (S. Noé, pers. comm., 1997); and 7—from carbonate in the Castile Fm., Magaritz et al. (1983). Inset (b) shows range of isotopic compositions of evaporite-replacive, poikilotopic calcites from Yates and Tansill strata in Dark Canyon (this study and from Scholle et al., 1992). All isotope data are reported relative to PDB.

1994) and in the Capitan reef (Mazzullo and Cys, 1977; Mruk, 1989). There is evidence in some samples, regardless of stratigraphic location or depositional environment, of minor dissolution of these cements prior to precipitation of the prismatic cements (Fig. 8C). This dissolution event was not recognized by Parsley (1988). The cements commonly overlie and/or are overlain by encrusting organisms, and commonly, thin (30–70 μm) micritic "dustlines" (Fig. 8A) that are composed of loafish microspar crystals 9–12 μm in diameter and which include entrapped skeletal debris. The dustline commonly is thickest on the sides or undersides of grains and, where present, vague internal laminations follow the contours of subjacent grains rather than being horizontally or subhorizontally oriented. The dustlines are interpreted as microbial encrustations or microbial-induced cements rather than internal sediment; their original mineralogy is uncertain.

Based on their crystallographic similarity to cements in other rocks (e.g., Mazzullo and Cys, 1977, 1979), and their association with encrusting marine organisms, the isopachous crusts and botryoids are interpreted as syndepositional marine cement of former aragonite mineralogy. Adequate weights of samples of this cement could not be obtained from any of the subtidal grainstones for geochemical analyses. The mean Sr (285 ppm) and $MgCO_3$ concentrations (0.8 mole%) of these cements in sheetcracks, however, are close to those in Toomey reef (Table 1).

Prismatic Cements

Micro-prismatic (Figs. 6C, 9A and B) and more coarsely crystalline prismatic calcites (Figs. 6B and D, 8A, and 9C and D) are volumetrically the most abundant cements in the rocks regardless of stratigraphic location or depositional environment. Both types of prismatic cement are present with isopachous habit within some microbial mats, and in interparticle and primary intraparticle pores in some subtidal and shoreface grainstones and in the oldest biostrome. In these rocks, the coarsely crystalline cements conformably overlie the micro-prismatic cements. Only coarsely crystalline prismatic cements are present, however, in most other grainstones and microbial mats, and also in fenestrae within peritidal deposits. These cements commonly incorporate skeletal fragments, overlie encrusting marine organisms, and in turn, are encrusted by organisms, particularly in the subtidal grainstones.

The micro-prismatic cements typically are mosaics of generally non-luminescent, 15–50 μm-size crystals that are inclusion-rich. This type of cement was not recognized by Parsley (1988). In many samples the outer margins of this cement generation are somewhat finer crystalline mosaics of average 20 μm, loafish crystals with either no luminescence or splotchy, dull-red luminescence. These crystals are partially replaced by idiotopic dolomite, average crystal size of which is 45 μm, that occurs as "necklaces" around pores (Fig. 9B). Both the micro-prismatic and loafish crystals locally contain microdolomite inclusions.

The coarsely crystalline prismatic calcite crystals range in size up to 500 μm and they are very dark in thin section (opaque white in hand samples) because of abundant inclusions (Fig. 9C, D) that largely obscure recognition of their optical characteristics. Growth-banding, defined by laminae with fewer inclusions, is present locally (Fig. 9E). Crystals are generally nonluminescent, and they have both acute and obtuse terminations (Fig. 9C, D). Curved cleavage, diverging subcrystals, and convergent optic axes are present in some crystals that have fewer inclusions, and these features also are present in some of the micro-prismatic calcites. Most of the inclusions appear to be organic matter, but microdolomite rhombs and fluid inclusions also are present. This cement locally appears to have been dissolution-etched to some extent prior to precipitation of the next generation of cement, the equant calcite (Fig. 9C). This dissolution surface extends down to the micro-prismatic cement in pores where the coarsely crystalline prismatic cement locally has been removed. Evidence of this dissolution event is present in many samples regardless of stratigraphic location or depositional environment. Partial dissolution of the coarsely crystalline prismatic calcite was also described by Parsley (1988). Narrow (8–9 μm) veins of lucid calcite appear to extend from the equant calcite into the coarsely crystalline and/or the micro-prismatic calcites in many samples, where they follow intercrystalline boundaries and also transect crystals (Figs. 9B, D). Calcite in these veins displays dull-red to bright orange-yellow luminescence, and the micro-prismatic and coarser crystalline prismatic cements associated with them display splotchy, dull-red to orange luminescence. In some samples the equant calcite has syntaxially engulfed the prismatic calcite (Fig. 9B).

Both the micro-prismatic and coarsely crystalline prismatic calcites are interpreted as syndepositional marine based on their association with encrusting organisms. Mean $MgCO_3$ concentration is 1.8 mole% which is greater than in the former aragonite marine cements (Table 1). In comparison, mean Sr concentration (186 ppm) is less than in former aragonite marine cements in sheetcracks and in Toomey reef. The prismatic calcites therefore are interpreted as former high-Mg calcite on the basis of their optical properties, presence of microdolomite inclusions and similarity to other ancient prismatic calcites (e.g., Lohmann and Meyers, 1977), and relative trace-element composition. They were erroneously regarded by Mazzullo et al. (1989) and Ward (1988) to be former aragonite. Conspicuous replacement of the more loafish micro-prismatic cements by dolomite may indicate a precursor cement phase with higher mole% $MgCO_3$ than associated former high-Mg calcite cements in these rocks. Optical properties suggest that at least some of these prismatic calcites are radiaxial-fibrous calcites *sensu* Bathurst (1975), but whether all are cannot be determined because the abundant inclusions obscure optical characteristics.

Equant Calcite Cements in Subtidal to Peritidal Deposits

Coarsely crystalline, equant calcite cement occludes remaining porosity in subtidal and shoreface grainstones in the oldest biostrome, and in fenestrae in peritidal deposits (Fig. 9B, C, E). Cement precipitation appears to have post-dated, in turn, the final generation of marine high-Mg calcite cement, minor dissolution, and then, some degree of mechanical compaction.

Finely crystalline, equant calcite occludes some biomolds of former aragonitic particles in a few samples of subtidal and shoreface grainstones. The relative timing of allochem dissolution and precipitation of this calcite cannot be determined because it, and the more coarsely crystalline equant calcites, are not found to be interlayered with or overlain by fibrous or prismatic marine cements in any of the samples examined. All equant calcite crystals are lucid, with some small fluid inclusions, and have straight to slightly undulose extinction and planar intercrystalline boundaries. Crystals generally have dull-red luminescent interiors surrounded conformably by a thin band of bright orange-yellow luminescence, and then in larger crystals, a thicker band of medium red luminescence. Adequate amounts of the equant calcite in biomolds could not be obtained for trace-element analysis. Mean $MgCO_3$ and Sr concentrations in the slightly ferroan equant calcites in interparticle pores in subtidal grainstones and in fenestrae in peritidal rocks are 1.7 mole% and 102 ppm, respectively (Table 1). The origin of the equant calcites is interpreted below.

Poikilotopic Calcite in Former Evaporite Nodules

Coarsely crystalline, poikilotopic calcite infills former evaporite nodules in some of the peritidal dolomites (Figs. 5C, 9F). Crystals are lucid and with some fluid inclusions, and contain some inclusions of anhydrite or gypsum and, locally, they are partially replaced by small spherulites of lutecite. In many samples the calcite appears to have infilled interstices in brecciated zones surrounding the nodules, and also appears to be pseudomorphic after gypsum laths in the surrounding sediments. Mean $MgCO_3$ and Sr concentrations are less than in the equant calcites described above (Table 1), and the cements are only slightly ferroan (321 ppm Fe). Crystals have up to five bands of dull to medium red luminescence alternating conformably with bright orange-yellow luminescence.

ISOTOPE GEOCHEMISTRY AND TIMING OF DIAGENESIS

Marine Cements and Temporal Changes in Seawater $\delta^{18}O$

Mean isotopic compositions of the former aragonite (in sheetcracks) and high-Mg calcite marine cements in the lower part of the middle Tansill are nearly identical (Table 1). Parsley (1988) did not report isotopic compositions of inferred marine cements in the Tansill in Dark Canyon. The covariant compositional ranges of the cements in the rocks examined in the present paper overlap (Fig. 10), and together, the most enriched compositions of nonluminescent and, therefore, presumably least-altered cements (e.g., Frank and Lohmann, 1996) are interpreted to indicate an apparent average seawater isotopic composition of $\delta^{18}O \approx -1.6‰$ and $\delta^{13}C \approx 5.8‰$ at this time. Estimated seawater isotopic compositions during deposition of the overlying middle to upper Tansill based on similarly nonluminescent, former aragonite marine cements in Toomey reef agree with those of Given and Lohmann (1985) for this time (Fig. 10): seawater $\delta^{18}O$ was enriched only by about 1.0‰ relative to that in the lower part of the middle Tansill. These data points, and those of Magaritz et al. (1983), Given and Lohmann (1985), Mruk (1989) and S. Noé (pers. comm., 1997), define a trend of increasing enrichment of estimated seawater $\delta^{18}O$ from the Guadalupian into the Ochoan based on least-altered marine cements; in contrast, there is only modest change in estimated seawater $\delta^{13}C$ (Fig. 10, inset a). Progressive $\delta^{18}O$ enrichment likely reflects increasing salinity and aridity (e.g., Given and Lohmann, 1985). This trend may have been linked to regional shallowing resulting from long-term decreasing accommodation space during the Guadalupian (Ross and Ross, 1995) and/or to climatic shift coincident with the onset of the Ochoan salinity crisis (Magaritz et al., 1983).

Significance of the Equant Cements and Stabilization of Marine Cements

The marine cements with progressively more depleted isotopic compositions (Fig. 10) are those with increasing luminescence, which reflects admixtures of diagenetic low-Mg calcite (e.g., Given and Lohmann, 1985, 1986; Frank and Lohmann, 1996). Stabilization of the former aragonite marine cements to low-Mg calcite is believed to have occurred via microscale dissolution-reprecipitation and concurrent precipitation of low-Mg calcite, and resulted in characteristic replacement of original fabrics by pseudospar (Mazzullo, 1980). Stabilization of the former high-Mg marine cements likely also occurred in the same manner (e.g., Given and Lohmann, 1985, 1986). Retention of fabrics and of a near-primary marine isotopic signal indicates diagenesis in a rock-dominated system, as presumably does the lack of compositional overlap of the marine and equant calcites (Fig. 10). Frank and Lohmann (1996) noted the occurrence of fine-scale intergrowths of dolomite and low-Mg calcite within luminescent portions of partially stabilized former high-Mg calcite marine cements in Pliocene limestones from Jamaica. They similarly favored a closed diagenetic system for such replacement, wherein Mg for dolomite precipitation was supplied by microscale dissolution of the high-Mg calcite. The "necklaces" of dolomite that replaced the outer margins of the micro-prismatic marine calcites in the Tansill (Fig. 9B) may be the product of such a process, with dolomite crystal growth continuing thereafter. Rudolph (1977) invoked a similar origin for dolomite-prismatic calcite couplets in lower Tansill rocks. Parsley (1988) noted the presence of dolomite along the outer margins of coarsely crystalline prismatic calcite and, in some rocks, also as an isopachous coating around grains. He was unclear as to whether the latter dolomite pre-dated the marine cements, and was unsure whether any of the dolomites were cement or replacive.

Given and Lohmann (1986) and Mruk (1989) described zoned-luminescent, relatively $\delta^{18}O$-depleted, equant calcite cements (their spar I) that post-dated the final generation of marine cements in the Capitan reef. According to Given and Lohmann (1986), they were precipitated in a meteoric phreatic system that was emplaced during regional subaerial exposure toward the end of the Guadalupian. Mruk (1989) was less specific with regard to the timing of these spar I cements, but also believed that they post-dated marine cementation and likely were precipitated sometime during the latest Permian. Hill (1989, 1996) identified this same spar in Carlsbad Caverns. Parsley (1988) described both nonluminescent and later luminescent equant calcites that post-dated marine cements in the middle and upper Tansill, and suggested that the latter may be equivalent to the spar I of Given and Lohmann (1986) and Mruk (1989). I did not identify any nonluminescent equant calcites in the lower part of the middle Tansill. The equant calcites in fenestrae, shoreface, and subtidal deposits in the lower part of the middle Tansill also post-date marine cementation. The range in $\delta^{18}O$ compositions of the most enriched of these calcites overlaps that of the range of mean $\delta^{18}O$ compositions of spar I cements reported by Given and Lohmann (1986) and Mruk (1989) (Fig. 10). It is therefore likely that this isotopically enriched subpopulation of equant cements correlates to these spar I cements. Parsley (1988) did not report isotopic compositions of equant calcite cements in the middle and upper Tansill. In addition to the equant calcites described by these authors, equant calcites also are present in the Tansill farther north in the Carlsbad Embayment, where they display similar relationships to subjacent marine cements (Noé and Mazzullo, 1992; Noé, 1996).

The veins of luminescent calcite that extend from the equant calcite into the marine cements in the lower part of the middle Tansill (Figs. 9B, D) are similar in terms of their CL characteristics and relative timing of formation to those associated with spar I and subjacent marine cements in the samples described by Given and Lohmann (1986). These observations may suggest coincidence of the timing of mineralogic stabilization of the marine cements and precipitation of spar I equant calcites, which was also suggested by Given and Lohmann (1986) and Mruk (1989). The exact timing of precipitation of spar I equant calcite in the Tansill is unclear. It may have occurred late in the Guadalupian as suggested by, for example, Given and Lohmann (1986). However, the lack of compositional overlap of the marine and equant calcites, and the much more depleted ^{18}O compositions of these equant cements, alternatively may suggest early post-Guadalupian precipitation. Nevertheless, if the marine cements stabilized concurrently with precipitation of spar I calcites, then it would imply that stabilization occurred long after deposition. Parsley (1988) also believed that the nonluminescent and luminescent equant calcites in the middle and upper Tansill precipitated from meteoric fluids and that mineralogic stabilization of the marine cements coincided with their precipitation. He suggested, however, that these events occurred in either the deep burial environment, but did not specify the timing of such, or during exhumation of the rocks.

The range in $\delta^{18}O$ composition of the most isotopically depleted of the equant cements, including the fine crystalline cements in biomolds in subtidal deposits, overlaps that of spar II cements described by Given and Lohmann (1986), Hill (1989), and Mruk (1989)(Fig. 10). A possible interpretation is that this subpopulation of equant calcite may correlate to spar II cements, in which case it may have precipitated at slightly elevated temperatures from largely meteoric fluids in the burial environment, possibly during Ochoan time or later (e.g., Mruk, 1989). If these correlations are correct, then two distinct generations of cement, that is, spar I and II, are not recognized in these rocks as they appear to be in the samples described by Given and Lohmann (1986) and Mruk (1989), most of which were collected from along the Reef Escarpment. That is, it is assumed that spar I may be represented by the weakly luminescent cores of the equant calcite crystals, whereas spar II may be represented by their zoned-luminescent perimeters (e.g., Frank and Lohmann, 1996). Concurrent with progresive burial, cement precipitation occurred first in an oxidizing environment and then in an environment characterized by alternating oxidizing and reducing conditions. The differences between these spars in Dark Canyon and those described by Given and Lohmann (1986) and Mruk (1989) may be related to the timing and chemistry of the hydrogeologic systems in which they were precipitated. That is, subaerial exposure and establishment of a regional meteoric phreatic system may have occurred earlier in the Carlsbad Embayment than along the Reef Escarpment. If the sparry cements in these areas are correlative, then they may reflect downdip gradients or temporal changes in water chemistry. It is possible, however, that these sparry calcites are not correlative, or that their precipitation occurred in disparate groundwater systems rather than one of regional extent.

Poikilotopic Calcites

The isotopic compositions of the poikilotopic calcites within former evaporite nodules in the lower part of the middle Tansill are considerably more depleted than presumed spar II cements (Fig. 10, inset b). Their isotopic ranges do not overlap those of the spar III cements recognized by either Mruk (1989) or Hill (1989). Rather, they directly overlap those of the coarsely crystalline calcites in presumed former evaporite nodules that were described from Yates and Tansill strata in Dark Canyon by Scholle et al. (1992), with which they share similar petrographic features, for example, evaporite-mineral inclusions, coarse crystal size, and some silica replacement. These cements in the lower part of the middle Tansill therefore are similarly interpreted to have replaced evaporites as a result of meteoric flushing attending regional uplift of the Guadalupe Mountains in the Tertiary. Their relatively depleted $\delta^{13}C$ compositions may reflect admixture of light carbon from hydrocarbons that leaked out of the basin during this time (Scholle et al., 1992). Parsley (1988) also inferred meteoric calcitization of evaporite nodules in the middle and upper Tansill, but did not report isotopic compositions of these calcites.

Dolomitization

Dolomite that had replaced peritidal deposits in the lower part of the middle Tansill (Fig. 4A) typically is finely crystalline (crystals less than 15 μm), with hypidiotopic to xenotopic texture. Dolomite also is present in minor quantities in some adjoining subtidal wackestones and packstones, wherein it nonselectively replaced skeletal fragments and adjoining matrix as patches of similar fine crystalline, hypidiotopic texture. Dolomite is not observed to have nucleated on any marine cements in the rocks in Dark Canyon as suggested by Parsley

(1988). There is also no indication in the rocks that dolomite replaced any equant spar or poikilotopic calcites. Instead, these calcite cements nucleated on what are now dolomite substrate and, within former evaporite nodules, they locally enclose breccia fragments of dolomicrite. The range of $\delta^{18}O$ compositions of the dolomites defines a trend that is enriched by a maximum of about 3.3‰ relative to that estimated for seawater at this time (Fig. 10), which would suggest that the dolomite is enriched by at least 5.3‰ relative to a theoretically coprecipitated calcite. Maximum $\delta^{13}C$ composition of the dolomite is 5.9‰, whereas that of ambient seawater is estimated to have been 5.8‰. Isotopic compositions of the dolomites in peritidal rocks are not significantly different than those dolomites in subtidal rocks (Table 1 and Fig. 10). Based on these petrographic observations and isotopic compositions it appears that dolomitization essentially was syndepositional to perhaps immediately post-depositional (shallow burial), and occurred in contact with extant marine fluids of elevated salinity. Evidence of former evaporites in these rocks supports this model, wherein dolomite formed within peritidal deposits. The minor amounts of dolomite present in some subtidal deposits may have precipitated from fluids that refluxed into subtidal sediments. Increasing $\delta^{18}O$ depletion in the dolomites is interpreted to reflect later geochemical alteration by interaction with meteoric fluids, likely during the latest Guadalupian to Ochoan or, perhaps, later.

The mean $\delta^{18}O$ composition of back-reef dolomites in the uppermost Yates and Tansill in the subsurface of the northern Delaware Basin (0.15‰) reported by Garber et al. (1989) is close to that of the peritidal dolomites examined in the present study (Table 1). They likewise inferred a penecontemporaneous to perhaps shallow burial timing of dolomitization of these rocks based on the oxygen isotope composition of the dolomite. In contrast, Parsley (1988) and Parsley and Warren (1989) interpreted the dolomites in peritidal rocks in overlying middle and upper Tansill strata in Dark Canyon as having precipitated from hypersaline brines that refluxed through the sediments from extant lagoons, or later, from overlying Ochoan evaporites. They based this interpretation on the more enriched range in $\delta^{18}O$ composition of their samples (~1.5 to 5.8‰) relative to that in my samples. The trend of increasingly more enriched $\delta^{18}O$ compositions of the peritidal dolomites (Fig. 10) and those analyzed by Parsley (1988) and Parsley and Warren (1989) may reflect long-term salinity increase during the late Guadalupian. Rudolph (1977) interpreted lower Tansill dolomites to have formed in the shallow burial environment soon after deposition, but in mixed meteoric-marine fluids introduced during sea-level lowstands. He based this interpretation on the relatively depleted $\delta^{18}O$ composition of the dolomite. Although initial precipitation of these dolomites may have been essentially syndepositional, their ^{18}O-depleted compositions could indicate later geochemical alteration in meteoric fluids rather than precipitation from mixed meteoric-marine fluids.

The peritidal dolomites in the lower part of the middle Tansill have been partially dedolomitized (Fig. 5B), and at least some of the subtidal limestones also appear to have been dedolomitized (Mazzullo et al., 1989). Lucia (1961) and Parsley (1988) believed that dedolomitization occured late and was associated with meteoric influx and evaporite dissolution. I concur with this interpretation and, following Scholle et al. (1992), believe that dedolomitization is a Tertiary phenomenon.

CYCLICITY AND DEPOSITIONAL MOSAIC

Parasequences

Subtidal and peritidal parasequences are recognized in the Tansill (Figs. 3, 4A). Subtidal parasequences generally shallow upward from either bioclastic packstone and/or grainstone to less fossiliferous wackestone and mudstone, or from bioclastic wackestone and/or packstone to poorly fossiliferous, locally intraclastic mudstone. Peritidal parasequences include beds of subtidal aspect as described above that pass upward into locally intraclastic, shoreface and/or peritidal deposits. Those in the Ocotillo are overlain by siltstone. Intraclasts are common along many of the flooding surfaces that bound parasequences. Identification of all of the parasequences that may be present in the Ocotillo Member was hampered by poor exposures in this recessive weathered section. Parasequences were not identified in the thick sections of peritidal deposits at measured sections 10 and 12 (Fig. 3).

Erosional surfaces are recognized in the lower part of the middle Tansill (labeled e1–e7 in Figs. 3 and 4) and in the Ocotillo, but they are not readily obvious elsewhere in the section here (Fig. 3). However, Parsley (1988) recognized some disconformities in these rocks, but did not specify where they occurred in the section. Some of the erosional surfaces abruptly truncate subtidal parasequences in the lower part of the middle Tansill (Fig. 4). The upper part of the younger biostrome at measured section 9, for example, is dolomitized beneath erosional surface e3, which has incised as much as 4 ft (1.2 m) into the biostrome (Fig. 4C). A narrow fissure at the top of this biostrome is filled with foram grainstone derived from immediately overlying beds. This erosional surface is traced westward over the smaller biostrome and then into shoreface and peritidal deposits (Fig. 3). Erosional surface e5 locally is incised as much as 1.3 ft (0.4 m) into underlying beds (Fig. 4C). There are dissolution vugs in the older biostrome directly beneath erosional surface e1, a few of which are partly occluded by (rare) pendant calcite cement. Otherwise, cements of unequivocal meteoric (that is, vadose) origin were not identified beneath the other erosional surfaces or in the fissure-filling grainstone at the top of the younger biostrome beneath e3.

Depositional Sequences and Facies

The Ocotillo Member includes the shallowest marine carbonates in the section, and the presence of siltstones and erosional surfaces suggests siliciclastic input attending a major lowstand. Hence, I concur with Kerans et al. (1992) and Kerans and Harris (1993) in recognizing two depositional sequences in the Tansill (in ascending order, 1 and 2) in the study area, and in placing the boundary between them close to or at the base of the Ocotillo Member (Fig. 3). Both the shoreface dolomites and shallow subtidal lime mudstones in the Ocotillo commonly are brecciated *in-situ* (that is, there is little or no transport of fragments), with interstices filled with silt. Angular clasts of siltstone commonly are present in these carbonates, and some clasts of limestone are present in the siltstones. Brecciation is believed to have resulted from repeated subaerial exposure and some physical reworking of clasts along erosional parasequence boundaries.

The boundary between the Yates-Tansill sequence and overlying sequence 1 is not exposed in the outer shelf tract. This boundary and the transgressive systems tract (TST) of sequence 1 are, however, exposed in the shelf crest tract to the west, where the TST is represented by fossiliferous, cross-stratified pisolite grainstone beneath a thick section of peritidal deposits. The maximum flooding surface (mfs) of sequence 1, however, is exposed in the outer shelf tract and appears to be within the lower part of the middle Tansill (Fig. 3) based on the following evidence: (1) the oldest biostrome and associated carbonates (dominantly packstone and grainstone) include the highest diversity biotic assemblage in this section, and diversity decreases upward. The oldest biostrome is of Capitan-type affinity, and likely represents deposition on an open shelf of moderate to high energy. In contrast, the limited biota and absence of marine cements in the younger biostromes indicate slightly more restricted conditions; and (2) subtidal environments in this section had backstepped farther landward than in overlying strata in this sequence. Inferred depositional environments along the mfs and in the early highstand systems tract (HST) in this sequence define a mostly grainstone-dominated, shallow and wide, high energy shelf. The peritidal deposits and adjoining subtidal tracts in the early HST and the numerous erosional surfaces in this section suggest a major shallow water phase of deposition on the outer shelf at this time.

These peritidal-capped parasequences in the outer shelf are present only in Dark Canyon and similar deposits are not present within this section on the opposite side of the canyon (to the north) here. The peritidal flats are interpreted as islands because they are surrounded by and pass eastward, westward, and northward into subtidal deposits (Figs. 3, 4A). The shelf crest *sensu stricto* at this time is farther west. PT-1 is traced a considerable distance laterally, and also drops paleotopographically in a basinward direction (Fig. 4A). PT-2 is very thin, and may be the downdip edge of a thicker peritidal flat that does not intersect the plane of the outcrop. It appears to have been deposited basinward of and paleotopographically lower than PT-1. Although autocyclicity of these peritidal deposits was suspected initially, the stacked parasequences that are recognized in PT-1 and PT-3, the offlapping relationships between PT-1 and PT-2 (e.g., Rankey and Lehrmann, 1996), and the presence of erosional surfaces that truncate subtidal parasequences (Figs. 3, 4) instead suggest dominant eustatic forcing of deposition. It is likely that the shallow water deposits and peritidal islands in the early HST in this sequence reflect high sedimentation rates relative to accommodation increases at this time. Individual parasequences were not be readily obvious in some parts of the HST in sequence 1. Nevertheless, except for these sections and the relatively thin parasequences in the early HST, parasequence thickness and facies stacking patterns are mostly consistent with conceptual accommodation changes during sequence development (Fig. 3). Parasequences in the late HST in this sequence are dominantly more micritic carbonates deposited on a progressively narrower, shallow water outer shelf. Peritidal and associated shoreface deposits in this systems tract, and in sequence 2, compose tongues of the basinward-prograding shelf crest rather than islands.

The mfs of sequence 2 is placed in those dominantly micritic subtidal limestones in the basal upper Tansill with the highest biotic diversity (Fig. 3). Parasequence thickness and facies stacking patterns in the TST and HST are mostly consistent with conceptual trends of accommodation change during sequence development. Subtidal environments did not backstep as far as in sequence 2 during maximum transgression, and the dominantly micritic lithologies of these rocks are evidence of a lower energy and narrower shelf. Close to the edge of the platform to the east, in exposures at the mouth of the canyon, upper beds of the Ocotillo-equivalent and upper Tansill are mostly high energy bioclastic grainstones with a relatively diverse biota and patch reefs (Toomey and Babcock reefs) of Capitan-type affinity (Babcock, 1977; Toomey and Cys, 1977; Noé, 1996).

According to Kerans et al. (1992) and Kerans and Harris (1993), deposition of Tansill sequences 1 and 2 (and the older Yates-Tansill Sequence) and component parasequences was driven by glacio-eustasy. A similar interpretation has been advocated for older Guadalupian strata in the study area (e.g., Rankey and Lehrmann, 1996). A eustatic forcing mechanism also appears likely for Tansill sequences 1 and 2 in Dark Canyon. Yet, according to some workers there is little evidence for widespread continental glaciation and, hence, for pronounced glacio-eustatically forced fluctuations in sea-level, during the Guadalupian (e.g., Garber et al., 1989). According to Crowell (1995), for example, only valley glaciers may have been present locally in the Southern Hemisphere because Earth already had entered into a greenhouse state by this time. Analysis of cycle-driving mechanisms in this section is beyond the scope of this paper.

POSSIBLE RELATIONSHIPS BETWEEN CYCLICITY, DIAGENESIS, AND DEPOSITION

Marine Cementation and Dolomitization

The notion that marine cements are not very abundant in outer shelf deposits was championed by Schmidt (1977). Yet, former aragonite and, especially, former high-Mg calcite marine cements are very abundant in some of the outer shelf deposits in the Tansill in Dark Canyon. Depositional texture and environment appear to have imparted a first-order control on the abundance of marine cements and, in turn, the specific types of deposits in which these cements are present may have been controlled by the comparative history of relative eustasy of Tansill sequences 1 and 2. That is, the cements are very abundant in moderate- to high-energy subtidal grainstones, the oldest biostrome, which is of Capitan-type open shelf character, and in fenestrae in the peritidal islands that were surrounded by subtidal deposits in the mfs and early HST of sequence 1, which inundated the platform farther landward than sequence 2.

Extensive marine cementation at this time may have been facilitated by the existence of a wide, shallow, high energy shelf where large volumes of seawater were pumped through the porous, coarse-grained carbonate sands (see also Parsley, 1988), and apparently, also the more micritic (older) biostrome and peritidal islands. In contrast, sediments in equivalent systems tracts in sequence 2 and in the late HST of sequence 1 are dominantly micritic and less porous, and marine cements are not abundant. They were deposited in environments of low to moderate energy on narrower shelves that apparently did not promote the flow of requisite volumes of seawater through the sediments. Similar relationships of cement abundance are evident in the Tansill along the Reef Escarpment.

Dolomitization of peritidal deposits, and possibly, also of the upper part of the younger biostrome in the early HST in sequence 1 (Fig. 4C), appears to have essentially been syndepositional. Dolomitization may have occurred either during lowstands or, perhaps more likely, during ensuing flooding.

Dominance of Former High-Mg Calcite

The dominance of former high-Mg calcite marine cement in the outer shelf deposits contrasts with the aragonite cements that dominate in near-platform margin patch reefs (Noé and Mazzullo, 1992, 1994; Noé, 1996) and in the Capitan reef (e.g., Mazzullo and Cys, 1978; Kirkland et al., 1993). High-Mg calcite cements likewise dominate in outer shelf deposits in middle and upper Tansill strata (this paper and Noé, 1996) and, based on other workers (e.g., Parsley, 1988; Ward, 1988; Parsley and Warren, 1989; Noé and Mazzullo, 1994) and my as-yet published studies, in many outer shelf grainstones in the Yates and Seven Rivers Formations as well. Hence, there is no difference in the mineralogies of marine cements in Tansill outer shelf deposits between the Carlsbad Embayment and along the Reef Escarpment. The dominance of high-Mg calcite marine cements in the Tansill outer shelf deposits examined in the present paper may reflect increasing salinity in a landward direction. Based on $\delta^{18}O$ compositions of least-altered marine cements from platform margin reef deposits, including those in Dark Canyon (presumably from Toomey and Babcock reefs), Given and Lohmann (1985) suggested that salinity may have been 70–75‰ along the margin of the Delaware Basin during the late Guadalupian. The mean $\delta^{18}O$ composition of former aragonite marine cements from the Toomey reef analyzed in the present study are close to the compositions reported by Given and Lohmann (1985). Isotope data from the former aragonite and high-Mg calcite marine cements in the slightly older outer shelf deposits described in the present study suggest that the $\delta^{18}O$ composition of seawater was not as enriched, and hence, salinity was somewhat lower than later in Tansill time (Fig. 10, inset a). The relatively high salinity postulated by Given and Lohmann (1985) during the late middle Tansill (Ocotillo-equivalent) corresponds to the prominent lowstand between sequences 1 and 2, when most of the platform was exposed. Sequential precipitation of aragonite and high-Mg calcite cements is the typical style of marine cementation in both platform margin and outer shelf deposits, and may reflect salinity increase in interstitial fluids as cementation progressively reduced permeability (e.g., Given and Lohmann, 1985).

It is possible that precipitation of high-Mg calcite cements in Tansill outer shelf deposits may have been promoted by higher temperature of ambient seawater as salinity on the platform increased, although other effects such as fluid Mg:Ca ratio and saturation states relative to the different carbonate polymorphs also may affect the resulting mineralogy of the marine precipitate (e.g., Morse and Mackenzie, 1990). Sequential precipitation of micro-prismatic followed by more coarsely crystalline prismatic, former high-Mg calcite cements is typical in these rocks, and may reflect decreasing rates of cement precipitation and/or decreasing supersaturation levels of the interstitial fluids. The abundant organic matter inclusions in the high-Mg calcite cements, particularly in the coarsely crystalline prismatic calcites, may suggest some degree of microbial mediation of pore fluid chemistry that facilitated precipitation of these cements. A more perplexing question, perhaps, and one that also remains to be addressed, is why cement mineralogies once again became dominated by aragonite in the pisolite facies of the shelf crest farther updip (e.g., Garber et al., 1989).

Rarity of Syndepositional Meteoric Cements

Pendant calcite cements are present but rare in the oldest biostrome in Tansill sequence 1, but they are not associated with marine cements in these micritic rocks. Parsley (1988) noted some gravitational cements in subtidal deposits in the middle and upper Tansill, although he was unsure of their timing of precipitation and did not specify their abundance and stratigraphic distribution. Except for these occurrences, there is no evidence of unequivocal syndepositional meteoric (e.g., vadose) cements, interlayering of marine and meteoric cements, or of attending stabilization of marine cements prior to precipitation of the equant calcite cements in the rocks, despite inferred cyclicity and evidence of exposure in the section. Demonstrably meteoric cements have not even been identified within the micritic carbonates or in interstices between carbonate clasts in the Ocotillo Member. The only evidence of periodic subaerial exposure in these rocks may be the dissolution surfaces that truncate the first and second generations of former aragonite and high-Mg calcite marine cements in the carbonate in the Tansill, and the erosional surfaces that were readily identified in the lower part of the middle Tansill. Parsley (1988) suggested that the dissolution surfaces he noted between the prismatic marine cement and later nonluminescent equant calcite may have formed either in the deep burial environment or attending uplift.

The fact is that cements of unequivocal syndepositional meteoric origin are extremely rare in outer shelf and reef deposits in the Capitan complex (Babcock, 1977; Garber et al., 1989; Mruk, 1989; Pray, 1989; Noé, 1996).

Although there are several possible reasons for this rarity, perhaps the simplest explanations may be that the durations of periods of subaerial exposure were insufficient to expose sediments to meteoric fluids, and/or there may have been only limited supplies of meteoric water in the arid environment that existed during the late Guadalupian. Even though the peritidal islands were at least partly subaerially exposed, any meteoric lenses on them may not have been thick enough, especially in an arid climate, to have affected diagenesis in subjacent grainstones (e.g., Budd and Vacher, 1991). Studies have in fact suggested that both the magnitudes and durations of relatively high frequency sea-level fluctuations declined during the Guadalupian (Garber et al., 1989; Borer and Harris, 1991; Kerans et al., 1992; Ross and Ross, 1995). Likewise, the magnitudes of sea-level drops might not have been sufficient to result in prolonged periods of subaerial exposure on the outer shelf relative to, for example, on the shelf crest (e.g., Garber et al., 1989). Even in shelf crest deposits, however, meteoric cements are not common (Garber et al., 1989). Although erosional surfaces are present in outer shelf deposits in the Tansill (Figs. 3 and 4), they are not traced into deposits close to the platform edge and in the Capitan reef in the subsurface at the mouth of the canyon (Parsley, 1988). This distribution suggests that only proximal areas on the outer shelf may have been periodically subaerially exposed during deposition. Similar relationships were described in Tansill strata elsewhere in the Guadalupe Mountains (e.g., Neese and Schwartz, 1977; Noé, 1996).

Peritidal Islands and Microbial Deposits

The peritidal islands in sequence 1 (Figs. 3 and 4A) are interbedded with and overlie coarse-grained, subtidal grainstones with micritic deposits of inferred microbial origin. Such microbial deposits are not present elsewhere in the section. Microbial activity in shallow shelf environments typically is promoted by high organic matter content in sediments (Chafetz and Buczynski, 1992), and in order to form the mats that are present the sediments must have been immobile for some time. It is possible that the islands baffled currents such that organic matter was deposited, and its ensuing degradation by microbes effected precipitation of micritic cements within mats and encrustations on grains. Similarly coarse-grained skeletal sands with locally abundant organic matter compose shallow subtidal deposits around low-lying peritidal flats in the large lagoon within Ambergris Caye in Belize (Mazzullo et al., 1987), and thin microbial mats are present as surficial deposits along the shorelines of these islands and in small depressions on the subtidal substratum. Although circulation is restricted here, the lagoon is not hypersaline. This occurrence may be a modern analog of the microbial-rich subtidal deposits in the Tansill.

PLATFORM EVOLUTION IN THE CARLSBAD EMBAYMENT

The general facies mosaic of the Tansill in Dark Canyon is a west-to-east transition along the outcrop from dominantly peritidal deposits of the shelf crest to dominantly subtidal outer shelf deposits (Fig. 3) which, to the immediate lee of the Capitan reef, are higher energy subtidal packstone and grainstone and patch reefs (Babcock, 1977; Toomey and Cys, 1977; Tyrrell et al., 1978; Parsley, 1988; Parsley and Warren, 1989; Noé and Mazzullo, 1994; Noé, 1996). There is no evidence that the Toomey and Babcock reefs or the buried Capitan reef were subaerially exposed during deposition (e.g., Babcock, 1977; Toomey and Cys, 1977; Parsley, 1988; Parsley and Warren, 1989). Erosional surfaces in shelf crest and proximal outer shelf deposits do not appear to extend to the edge of the platform or into the Capitan reef (Parsley, 1988). This mosaic suggests that environments on the outer shelf in sequences 1 and 2 here generally deepened in a basinward direction, which is consistent with inferred paleobathymetry of the Tansill outer shelf along the Reef Escarpment as well (e.g., Neese and Schwartz, 1977; Rudolph, 1977; Pray, 1985, 1989; Neese, 1989). A similar paleobathymetry was inferred by Noé (1996) for the middle and upper Tansill farther north in the Carlsbad Embayment (Sheep Draw Canyon). Peritidal islands are present in the lower part of the middle Tansill only in Dark Canyon and patch reefs, many of which are of Capitan-type aspect, and in the middle and upper Tansill farther north (Noé and Mazzullo, 1992, 1994; Noé, 1996). Such deposits are conspicuous by their absence in the outer shelf Tansill along the Reef Escarpment (Pray, 1989), and some workers have attributed such a contrast in depositional environments to northward shallowing into the Carlsbad Embayment (e.g., Motts, 1962, 1972; Garber et al., 1989; Mazzullo et al., 1989; Pray, 1989).

To test this idea, depositional slopes and the width of the Tansill outer shelf in the Carlsbad Embayment versus those parameters along the Reef Escarpment were compared based on published studies and the measured sections described in this paper (Table 2). Data from along the Reef Escarpment were restricted to its northern portion because upper beds of the Tansill are progressively eroded to the south. The slope of the entire outer shelf during deposition of Tansill sequence 1 in Dark Canyon is difficult to estimate because the rocks dip east-

TABLE 2.—ESTIMATES OF WIDTH AND SLOPE OF THE OUTER SHELF IN THE CARLSBAD EMBAYMENT AND ALONG THE NORTHERN REEF ESCARPMENT*

Location	Tansill Unit	Maximum Width (mi)	Slope
Northern Reef Escarpment	upper	—	—
	middle	—	—
	lower	0.3–0.6	5–10°
Dark Canyon	upper		
		0.5	0.3–3°
	middle		
		1.0	3–4°
	lower		
Sheep Draw Canyon	upper	~0.3	4–5°
	middle		
		~1.0	max 5°
	lower		

*(sources cited in text).

ward into the subsurface (Fig. 2B). Based on the downstepping of PT-1 and of the dip of erosional surface e5 (Fig. 4A), however, it may have been as much as 3–4° in the lower part of the middle Tansill, at least along the proximal outer shelf close to the shelf crest. This estimate is in agreement with that of Parsley (1988) for this section. According to data in Noé (1996), maximum slope of the Middle Tansill farther north was about 5°. Slope was less in that area where peritidal islands are present in Dark Canyon and progressively decreased to 0.3–0.5° (Parsley, 1988) at least along the proximal outer shelf as the shelf crest prograded basinward in the late HST of sequence 1 (Fig. 3). Slope may have increased toward the platform margin accompanying fall-in of outer shelf strata. According to data in Rudolph (1977), maximum slope of the outer shelf in the lower Tansill along the northern Reef Escarpment is estimated to have been about 6–7°. Pray (1985) estimated slopes of 5–10° along the distal outer shelf in the lower Tansill, along the contact with the Capitan reef, at the mouth of Walnut Canyon. Maximum slope here likely reflects fall-in of distal outer shelf strata, whereas the lower range of values may be in more proximal strata. According to Parsley (1988), the slope of the outer shelf in sequence 2 in Dark Canyon was about 0.3–0.5°. Based on correlation of flooding surfaces 1 and 2 between measured sections 9-A and 10 (Fig. 3), however, slope at least in the proximal outer shelf in the TST and early HST of sequence 2 was about 2-3°, and likely increased toward the platform margin. Data in Noé (1996) suggest that maximum slope at this time farther north in the Carlsbad Embayment was 4–5°. Data are not available for correlative rocks in the northern Reef Escarpment.

Based on these estimates (Table 2), slope of the outer shelf in the Carlsbad Embayment does not appear to have differed significantly from that along the Reef Escarpment, at least during deposition of Tansill sequence 1. The peritidal islands in sequence 1 in Dark Canyon are not recognized elsewhere in outcrops in the Carlsbad Embayment, and therefore, they appear to indicate only local shallowing. Other than these islands, subtidal carbonates of similar biotic and lithologic aspect dominate in both areas. None of the patch reefs examined by Noé and Mazzullo (1992, 1994) and Noé (1996) in the Carlsbad Embayment were truncated by unconformities. Hence, it does not appear that Tansill outer shelf facies in the Carlsbad Embayment were deposited in water that was significantly shallower than along the northern Reef Escarpment. In both areas water depth decreased toward the shelf crest and increased toward the platform margin.

The most important difference in environments between the Carlsbad Embayment and along the Reef Escarpment appears to be the presence of patch reefs and biostromes in the former area, and their occurrence may correlate to shelf width. Maximum width of the outer shelf during maximum flooding of the platform in sequence 1 for example, was about 1 mile (1.6 km) in Dark Canyon, which agrees with Parsley's (1988) estimate, and it may have approached this width to the north (Noé, 1996). Patch reefs and bioherms are common in sequence 1 and become more abundant to the north (Noé and Mazzullo, 1992, 1994; Noé, 1996; this paper). In contrast, data in Rudolph (1977) and Pray (1985) suggest that maximum width of the outer shelf in sequence 1 was only about 0.3–0.6 mi (0.5-1 km) along the northern Reef Escarpment, where such buildups have not been identified behind the Capitan reef. The occurrence of patch reefs and biostromes in the Carlsbad Embayment may therefore reflect a wider, more open shelf with greater circulation than along the Reef Escarpment (Noé and Mazzullo, 1994). If a Tansill-age Capitan reef did not exist in the northern Carlsbad Embayment (e.g., Motts, 1962, 1972; Noé and Mazzullo, 1994), then its absence may have enhanced platform circulation and promoted patch reef growth in this area, where sequence 1 includes more buildups than are present in this sequence in the southern Carlsbad Embayment (Noé, 1996). In contrast, the Capitan reef is present in the subsurface at the mouth of Dark Canyon (Tyrrell et al., 1978) where, although present on this wide shelf, patch reefs and biostromes are less common.

Shelf widths decreased to 0.5 mi (0.8 km) and less in the late HST of sequence 1 and in sequence 2 in Dark Canyon and farther north (Figs. 2B and 3G; also Parsley, 1988); data are not available to estimate width in the northern Reef Escarpment. Patch reefs are less common in sequence 2 (Noé, 1996), accompanying the decrease in shelf width at this time.

CONCLUSIONS

Two sequences are recognized in outer shelf deposits in the Tansill Formation in Dark Canyon on the basis of inferred depositional environments, biotic diversity, and thickness and facies-stacking patterns of component parasequences. The boundary between these sequences is at or near the base of the Ocotillo Member, which is interbedded, very shallow marine and shoreface to peritidal carbonates and terrestrial (eolian?) to marine-reworked siltstones. Sequence 1 in the middle Tansill represents maximum flooding of the platform. Subtidal bioclastic grainstones, biostromes, shoreface deposits, and micritic peritidal islands compose inferred depositional environments in the mfs and early HST of this sequence. Precipitation of the abundant marine cements in these rocks may have been promoted by active pumping of seawater through the sediments on this wide and shallow, high energy shelf. Evidence of microbial activity is present only in the subtidal deposits adjoining the peritidal islands. Surficial mats, cryptic encrustations, and micritic cements on and around grains are present in many of these grainstones. Microbial activity may have attended restricted circulation around peritidal islands, and also may have promoted marine cementation to some extent. Depositional environments in the late HST of this sequence, and in sequence 2 in the middle to upper Tansill, are more micritic subtidal carbonates deposited on narrower and lower energy shelves, and are associated basinward-prograding tongues of shelf crest peritidal and shoreface deposits. Marine cements are uncommon in these rocks. Dominantly subtidal carbonates are present in Tansill outer shelf deposits in the Carlsbad Embayment and along the Reef Escarpment; peritidal islands, which are local features, and patch reefs are present only in the latter area. There does not appear to have been a significant difference in water depth between these areas. It is the presence of patch reefs that makes the Carlsbad Embayment unique in terms of depositional environments. They apparently were deposited on a wider, open shelf with more circulation than the narrower shelf regimes along the Reef Escarpment.

Syndepositional marine cements are dominated by former high-Mg calcites that succeeded an earlier generation of minor fibrous and botryoidal, former aragonite cement. Mean isotopic compositions of these former aragonite and high-Mg calcite cements overlap (–2.3 - –2.4‰ $\delta^{18}O$, 4.5 – 4.8‰ $\delta^{13}C$), and their least-altered compositions indicate precipitation from seawater that had a mean isotopic composition of –1.6‰ $\delta^{18}O$ and 5.8‰ $\delta^{13}C$. This estimate of seawater isotopic composition, and those of others, define a trend of increasing salinity of seawater from the Guadalupian into the Ochoan. Dominance of high-Mg calcite marine cements may reflect increasing salinity and/or temperature in a landward direction. Maximum value of $\delta^{18}O$ in fine crystalline dolomite in peritidal and associated subtidal deposits in sequence 1 is enriched by about 3.3‰ relative to estimated seawater and even more enriched relative to a theoretically coprecipitated calcite. The mean $\delta^{13}C$ of the dolomite is close to that of seawater (5.9‰ versus 5.8‰, respectively). These compositions suggest that dolomitization occurred essentially syndepositionally, either in extant or shallow buried deposits, in marine fluids of elevated salinity.

There is no evidence of syndepositional meteoric cementation or stabilization of unstable cement mineralogies despite high frequency cyclicity and evidence of exposure. Absence of such cements may reflect short periods of subaerial exposure and limited supplies of meteoric fluids. Mineralogic stabilization of earlier diagenetic phases likely attended precipitation of coarsely crystalline, equant calcites in a meteoric phreatic system during the latest Guadalupian and/or early Ochoan. A second generation of coarsely crystalline, equant calcite cements may have been precipitated in a deeper, meteoric-dominated groundwater system. The ranges in isotopic compositions of these equant calcite cements overlap those of spar I and II cements recognized by Given and Lohmann (1986), Hill (1989, 1996), and Mruk (1989) and with which they may correlate. Poikiloptopic calcite in former evaporite nodules is the last diagenetic phase recognized in the rocks. Their petrographic similarity to, and isotopic compositional overlap with, late-stage calcites recognized by Scholle et al. (1992) suggest they are calcitized evaporites, the timing of which is related to meteoric influx attending regional uplift of the Guadalupe Mountains during the Tertiary.

ACKNOWLEDGMENTS

I thank K.C. Lohmann (University of Michigan, Ann Arbor) and Ethan Grossman (Texas A&M University) for analyzing isotopic compositions, and William D. Bischoff (Wichita State University) for analyzing trace elements. Carroll Hedrick (College Station, Texas) and Chellie Teal (Wichita State University) assisted in measuring the sections in Dark Canyon, and Roland Fech (Carlsbad, New Mexico) allowed access to outcrops on his land. This research was supported by faculty enhancement grants to the author from the Office of Research Administration, Wichita State University. Bill Fitchen, Brenda Kirkland, and Gene Rankey reviewed the original manuscript and offered many suggestions for its improvement.

REFERENCES

Achauer, C. W., 1969, Origin of Capitan Formation, Guadalupe Mountains, New Mexico and Texas: American Association of Petroleum Geologists Bulletin, v. 53, p. 2314-2323.

Babcock, J. A., 1977, Calcareous algae, organic boundstones, and the genesis of the Upper Capitan Limestone (Permian, Guadalupian), Guadalupe Mountains, west Texas and New Mexico, *in* Hileman, M. E., and Mazzullo, S. J., eds., Upper Guadalupian Facies, Permian Reef Complex, Guadalupe Mountains, New Mexico and West Texas: Midland, Society of Economic Paleontologists and Mineralogists, Permian Basin Section, Publication 77-16, p. 3-44.

Bathurst, R. G. C., 1975, Carbonate Sediments and Their Diagenesis: Amsterdam, Elsevier, 658 p.

Biggers, B., 1984, Reef to back-reef facies and diagenesis of the Permian Tansill-Capitan Formations in Dark Canyon, Guadalupe Mountains, New Mexico: Unpublished M.S. Thesis, Texas Tech University, Lubbock, 113 p.

Borer, J. M., and Harris, P. M., 1991, Lithofacies and cyclicity of the Yates Formation, Permian Basin: implications for reservoir heterogeneity: American Association of Petroleum Geologists Bulletin, v. 75, p. 726-779.

Bosence, D. W. J., and Bridges, P. H., 1995, A review of the origin and evolution of carbonate mud-mounds, *in* Monty, C. L. V., Bosence, D. W. J., Bridges, P. H., and Pratt, B. R., eds., Carbonate Mud-Mounds, Their Origin and Evolution: Oxford, International Association of Sedimentologists, Special Publication 23, p. 3-9.

Budd, D. A., and Vacher, H. L., 1991, Predicting the thickness of fresh-water lenses in carbonate paleo-islands: Journal of Sedimentary Petrology, v. 61, p. 43-53.

Chafetz, H. S., and Buczynski, C., 1992, Bacterially induced lithification of microbial mats: Palaios, v. 7, p. 277-293.

Crowell, J. C., 1995, The ending of the Late Paleozoic ice age during the Permian Period, *in* Scholle, P. A., Peryt, T. M. and Ulmer-Scholle, D. S., eds., The Permian of Northern Pangea, v. 1: Berlin, Springer-Verlag, p. 62-74.

Dunham, R. J., 1972, Capitan Reef, New Mexico and Texas: Facts and Questions to Aid Interpretation and Group Discussion: Midland, Society of Economic Paleontologists and Mineralogists, Permian Basin Section, Publication 72-14, 296 p.

Frank, T. D., and Lohmann, K. C., 1996, Diagenesis of fibrous magnesian calcite marine cement: Implications for the interpretation of $\delta^{18}O$ and $\delta^{13}C$ values of ancient equivalents: Geochimica et Cosmochimica Acta, v. 60, p. 2427-2436.

Garber, R. A., Grover, G. A., and Harris, P. M., 1989, Geology of the Capitan shelf margin - subsurface data from the northern Delaware Basin, *in* Harris, P.M. and Grover, G.A., eds., Subsurface and Outcrop Examination of the Capitan Shelf Margin, Northern Delaware Basin: Tulsa, Society of Economic Paleontologists and Mineralogists Core Workshop 13, p. 3-269.

Given, R. K., and Lohmann, K. C., 1985, Derivation of the original isotopic composition of Permian marine cements: Journal of Sedimentary Petrology, v. 55, p. 430-439.

Given, R. K., and Lohmann, K.C., 1986, Isotopic evidence for the early meteoric diagenesis of the reef facies, Permian reef complex of west Texas and New Mexico: Journal of Sedimentary Petrology, v. 56, p. 183-193.

Hardie, L. A., 1977, Sedimentation on the modern carbonate tidal flats of northwest Andros Island, Bahamas: Baltimore, Johns Hopkins University Press, Studies in Geology 22, 202 p.

Hill, C. A., 1989, The evolution of the Delaware Basin—preliminary results, *in* Harris, P. M., and Grover, G. A., eds., Subsurface and Outcrop Examination of the Capitan Shelf Margin, Northern Delaware Basin: Tulsa, Society of Economic Paleontologists and Mineralogists Core Workshop 13, p. 467-473.

Hill, C. A., 1996, Geology of the Delaware Basin, Guadalupe, Apache, and Glass Mountains, New Mexico and West Texas: Midland, Society of Economic Paleontologists and Mineralogists, Permian Basin Section, Publication 96-39, 480 p.

KENDALL, C. G. ST. C., 1969, An environmental reinterpretation of the Permian evaporite/carbonate shelf sediments of the Guadalupe Mountains: Geological Society of America Bulletin, v. 80, p. 2503-2526.

KERANS, C., FITCHEN, W. M., GARDNER, M. H., SONNENFELD, M. D., TINKER, S. W., and WARDLAW, B. R., 1992, Styles of sequence development within uppermost Leonardian through Guadalupian strata of the Guadalupe Mountains, Texas and New Mexico, *in* Mruk, D. H., and Curran, B. C., eds., Permian Basin Exploration and Production Strategies: Application of Sequence Stratigraphic and Reservoir Characterization Concepts: Midland, West Texas Geological Society Publication 92-91, p. 1-6.

KERANS, C., and HARRIS, P. M., 1993, Outer shelf and shelf crest, *in* Bebout, D. G., and Kerans, C., eds., Guide to the Permian Reef Geology Trail, McKittrick Canyon, Guadalupe Mountains National Park, West Texas, Guidebook 26: Austin, Bureau of Economic Geology, University of Texas, p. 32-43.

KIRKLAND, B. L., LONGACRE, S. A., and STOUDT, E. L., 1993, Reef, *in* Bebout, D. G., and Kerans, C., eds., Guide to the Permian Reef Geology Trail, McKittrick Canyon, Guadalupe Mountains National Park, West Texas, Guidebook 26: Austin, Bureau of Economic Geology, University of Texas, p. 23-31.

KIRKLAND, B. L., and MOORE, C. H., 1996, Microfacies analysis of the Tansill outer shelf, Permian Capitan reef complex, *in* Martin, R. L., ed., Permian Basin Oil and Gas Fields: Keys to Success that Unlock Future Reserves: Midland, West Texas Geological Society, Publication 96-101, p. 99-106.

LOHMANN, K. C., and MEYERS, W. J., 1977, Microdolomite inclusions in cloudy prismatic calcites: A proposed criterion for former high magnesium calcites: Journal of Sedimentary Petrology, v. 47, p. 1078-1088.

LUCIA, F. J., 1961, Dedolomitization in the Tansill (Permian) Formation: Geological Society of America Bulletin, v. 72, p. 1107-1110.

MAGARITZ, M., ANDERSON, R. Y., HOLSER, W. T., SALTZMAN, E. S., and GARBER, J., 1983, Isotope shifts in the Late Permian of the Delaware Basin, Texas, precisely timed by varved sediments: Earth and Planetary Science Letters, v. 66, p. 111-124.

MAZZULLO, S. J., 1980, Calcite pseudospar replacive of marine acicular aragonite, and implications for aragonite cement diagenesis: Journal of Sedimentary Petrology, v. 50, p. 409-422.

MAZZULLO, S. J., BISCHOFF, W. D., and HEDRICK, C. L., 1989, Stacked island facies in the outer-shelf platform, Dark Canyon, Guadalupe Mountains, New Mexico, *in* Harris, P. M., and Grover, G. A., eds., Subsurface and Outcrop Examination of the Capitan Shelf Margin, Northern Delaware Basin: Tulsa, Society of Economic Paleontologists and Mineralogists Core Workshop 13, p. 287-293.

MAZZULLO, S. J., and CYS, J. M., 1977, Submarine cements in Permian boundstones and reef-associated rocks, Guadalupe Mountains, West Texas and southeastern New Mexico, *in* Hileman, M. E., and Mazzullo, S. J., eds., Upper Guadalupian Facies, Permian Reef Complex, Guadalupe Mountains, New Mexico and West Texas: Midland, Society of Economic Paleontologists and Mineralogists, Permian Basin Section, Publication 77-16, p. 151-200.

MAZZULLO, S. J., and CYS, J. M., 1978, *Archaeolithporella*-boundstones and marine aragonite cements, Permian Capitan reef, New Mexico and Texas, USA: Neue Jahrbuch fur Geologie und Palaontologie, Monatshafte 1978, p. 600-611.

MAZZULLO, S. J., and CYS, J. M., 1979, Marine aragonite sea-floor growths and cements in Permian phylloid algal mounds, Sacramento Mountains, New Mexico: Journal of Sedimentary Petrology, v. 49, p. 917-936.

MAZZULLO, S. J., MAZZULLO, J., and HARRIS, P. M., 1985, Eolian origin of quartzose sheet sands in Permian shelf facies, Guadalupe Mountains, *in* Cunningham, B. K., and Hedrick, C. L., eds., Permian Carbonate/Clastic Sedimentology, Guadalupe Mountains—Analogs for Shelf and Basin Reservoirs (abs.): Midland, Society of Economic Paleontologists and Mineralogists, Permian Basin Section, Publication 85-24, p. 71.

MAZZULLO, S. J., REID, A. M., AND GREGG, J. M., 1987, Dolomitization of high-Mg calcite supratidal deposits, Ambergris Cay, Belize: Geological Society of America Bulletin, v. 98, p. 224-231.

MONTY, C. L. V., 1995, The rise and nature of carbonate mud-mounds: an introductory actualistic approach, *in* Monty, C. L. V., Bosence, D. W. J., Bridges, P. H., and Pratt, B. R., eds., Carbonate Mud-Mounds, Their Origin and Evolution: Oxford, International Association of Sedimentologists, Special Publication 23, p. 11-48.

MORSE, J. W., and MACKENZIE, F. T., 1990, Geochemistry of Sedimentary Carbonates: Elsevier, Amsterdam, Developments in Sedimentology 48, 707 p.

MOTTS, W. S., 1962, Geology of the West Carlsbad Quadrangle, New Mexico: Washington, D.C., U.S. Geological Survey, Geologic Quadrangle Map GQ-167. Motts, W. S., 1972, Geology and paleoenvironments of the northern segment, Capitan shelf, New Mexico and west Texas: Geological Society of America Bulletin, v. 83, p. 701-722.

MRUK, D. H., 1989, Diagenesis of the Capitan Limestone, upper Permian, McKittrick Canyon, west Texas, *in* Harris, P. M., and Grover, G. A., eds., Subsurface and Outcrop Examination of the Capitan Shelf Margin, Northern Delaware Basin: Tulsa, Society of Economic Paleontologists and Mineralogists, Core Workshop 13, p. 387-406.

NEESE, D. A., 1989, Peritidal facies of the Guadalupian shelf crest, Walnut Canyon, New Mexico, *in* Harris, P. M., and Grover, G. A., eds., Subsurface and Outcrop Examination of the Capitan Shelf Margin, Northern Delaware Basin: Tulsa, Society of Economic Paleontologists and Mineralogists, Core Workshop 13, p. 295-301.

NEESE, D. A., and SCHWARTZ, A. H., 1977, Facies mosaic of the upper Yates and lower Tansill Formations, Walnut and Rattlesnake Canyons, Guadalupe Mountains, New Mexico, *in* Hileman, M. E., and Mazzullo, S. J., eds., Upper Guadalupian Facies, Permian Reef Complex, Guadalupe Mountains, New Mexico and West Texas: Midland, Society of Economic Paleontologists and Mineralogists, Permian Basin Section, Publication 77-16, p. 437-450.

NOÉ, S.U., 1996, Late-stage reef evolution of the Permian reef complex: Shelf margin and outer-shelf development of the Tansill Formation (Late Permian), northern Guadalupe Mountains, New Mexico, USA, *in* Reitner, J., Neuweiler, F., and Gunkel, F., eds., Global and Regional Controls on Biogenic Sedimentation 1. Reef Evolution Research Papers: Gottinger Arb. Geologie Palaontologie, Sb2, p. 317-324.

NOÉ, S. U., and MAZZULLO, S. J., 1992, Upper Tansill patch reef facies, Sheep Draw Canyon, Guadalupe Mountains, Eddy County, New Mexico: West Texas Geological Society Bulletin, v. 32, p.5-11.

NOÉ, S. U., and MAZZULLO, S. J., 1994, Patch reef dominated outer-shelf facies along a non-rimmed platform, middle to upper Tansill Formation, northern Guadalupe Mountains, New Mexico: West Texas Geological Society Bulletin, v. 33, p. 5-11.

PARSLEY, M. J., 1988, Deposition and diagenesis of a Late Guadalupian barrier-island complex from the middle and upper Tansill Formation (Permian), east Dark Canyon, Guadalupe Mountains, New Mexico: Unpublished M.A. Thesis, University of Texas, Austin, 247 p.

PARSLEY, M. J., and WARREN, J. K., 1989, Characterization of an upper Guadalupian barrier-island complex from the middle and upper Tansill Formation (Permian), East Dark Canyon, Guadalupe Mountains, New Mexico, *in* Harris, P. M., and Grover, G. A., eds., Subsurface and Outcrop Examination of the Capitan Shelf Margin, Northern Delaware Basin: Tulsa, Society of Economic Paleontologists and Mineralogists, Core Workshop 13, p. 279-285.

PRAY, L. C., 1985, The Capitan-massive and its proximal carbonate, and minor siliciclastics of the Capitan back-reef, Walnut Canyon area, Carlsbad Caverns National Park, New Mexico, *in* Cunningham, B. K., and Hedrick, C. L., eds., Permian Carbonate/Clastic Sedimentology, Guadalupe Mountains: Analogs for Shelf and Basin Reservoirs: Midland, Society of Economic Paleontologists and Mineralogists, Permian Basin Section, Publication 85-24, p. 25-41.

PRAY, L.C., 1989, Lateral variability of the Capitan reef complex, west Texas and New Mexico, *in* Harris, P. M., and Grover, G. A., eds., Subsurface and Outcrop Examination of the Capitan Shelf Margin, Northern Delaware Basin: Tulsa, Society of Economic Paleontologists and Mineralogists, Core Workshop 13, p. 273-278.

PRAY, L. C., and ESTEBAN, M., 1977, Road Logs and Locality Guides: Midland, Society of Economic Paleontologists and Mineralogists, Permian Basin Section, Publication 77-16, v. 2, 194 p.

RANKEY, E. C., and LEHRMANN, D. J., 1996, Anatomy and origin of toplap in a mixed carbonate-clastic system, Seven Rivers Formation (Permian, Guadalupian), Guadalupe Mountains, New Mexico, USA: Sedimentology, v. 43, p. 807-826.

ROSS, C. A., and ROSS, J. R. P., 1995, Permian sequence stratigraphy, *in* Scholle, P. A., Peryt, T. M., and Ulmer-Scholle, D. S., eds., The Permian of Northern Pangea, Volume 1: Berlin, Springer-Verlag, p. 98-123.

RUDOLPH, K. W., 1977, Diagenesis of back-reef carbonates: An example from the Capitan complex: Unpublished M.S. Thesis, University of Texas, Austin, 159 p.

SCHMIDT, V., 1977, Inorganic and organic reef growth and subsequent diagenesis in the Permian Capitan reef complex, Guadalupe Mountains, Texas, New Mexico, *in* Hileman, M. E., and Mazzullo, S. J., eds., Upper Guadalupian Facies, Permian Reef Complex, Guadalupe Mountains, New Mexico and West Texas: Midland, Society of Economic Paleontologists and Mineralogists, Permian Basin Section, Publication 77-16, p. 93-131.

SCHOLLE, P. A., ULMER, D. S., and MELIM, L. A., 1992, Late-stage calcites in the Permian Capitan Formation and its equivalents, Delaware Basin margin, west Texas and New Mexico: evidence for replacement of precursor evaporites: Sedimentology, v. 39, p. 207-234.

TOOMEY, D. F., and CYS, J. M., 1977, Rock/biotic relationships of the Permian Tansill-Capitan facies exposed on the north side of the entrance to Dark Canyon, Guadalupe Mountains southeastern New Mexico, *in* Hileman, M. E., and Mazzullo, S. J., eds., Upper Guadalupian Facies, Permian Reef Complex, Guadalupe Mountains, New Mexico and West Texas: Tulsa, Society of Economic Paleontologists and Mineralogists, Permian Basin Section, Publication 77-16, p. 133-150.

TYRRELL, W. W., 1962, Petrology and stratigraphy of near-reef Tansill-Lamar strata, Guadalupe Mountains, Texas and New Mexico, *in* Permian of the Central Guadalupe Mountains: West Texas, Roswell and Hobbs Geological Societies Publication 62-48, p. 59-75.

TYRRELL, W. W., 1969, Criteria useful in interpreting environments of unlike but time-equivalent carbonate units (Tansill-Capitan-Lamar), Capitan reef complex, west Texas and New Mexico, *in* Friedman, G. M., ed., Depositional Environments in Carbonate Rocks: Tulsa, Society of Economic Paleontologists and Mineralogists, Special Publication 14, p. 80-97.

TYRRELL, W. W., LOKKE, D. H., SANDERSON, G. A., and VERVILLE, G. J., 1978, Late Guadalupian correlations, Permian reef complex, west Texas and New Mexico, *in* Austin, G. S., ed., Geology and Mineral Deposits of Ochoan Rocks in Delaware Basin and Adjacent Areas: New Mexico Bureau of Mines and Mineral Resources, Circular 159, p. 85.

WARD, R. J., 1988, Isolated bioherms in the Tansill Formation (Guadalupian), Dark Canyon, Guadalupe Mountains, southeastern New Mexico: Unpublished M.S. Thesis, Texas Christian University, Fort Worth, 54 p.

WOOD, R., DICKSON, J. A. D., and KIRKLAND-GEORGE, B., 1994, Turning the Capitan reef upside down: A new appraisal of the ecology of the Permian Capitan reef Guadalupe Mountains, Texas and New Mexico: Palaios, v. 9, p. 422-427.

WOOD, R., DICKSON, J. A. D., and KIRKLAND, B. L., 1996, New observations on the ecology of the Permian Capitan reef, Texas and New Mexico: Palaeontology, v. 39, p. 733-762.

PALEOECOLOGY OF THE CAPITAN REEF

RACHEL WOOD

Department of Earth Sciences, University of Cambridge, Downing Street, Cambridge CB2 3EQ, United Kingdom

ABSTRACT: Although the rate of accumulation, topography, and stratigraphic setting of the Permian Capitan reef are similar to that of modern coral reefs, both the ecology of the reef communities and the relative sources of carbonate in their final geological expression are profoundly different. Like modern coral reefs, the Capitan reef was constructed by a patchy mosaic of many different communities, but most were dominated by relatively short-lived aggregations of metazoans (particularly calcified sponges and bryozoans), and some were strongly differentiated into distinct open surface and cryptic subcommunities. Although our understanding of the distribution of these communities is poor, much of the deeper water Middle Capitan reef framework appears to have been constructed initially by a scaffolding of large frondose bryozoans. This framework created cryptic niches that were colonized by a diverse community dominated by solitary (cylindrical) sphinctozoan sponges and bryozoans that grew downwards from the roofs of the cavities. Bathymetrically shallow areas of the Middle and Upper Capitan reef were characterized by large, platy calcified sponges. In the Upper Capitan, the platy inozoan sponge *Gigantospongia discoforma* reached up to 2 m in diameter, and individuals projected from the reef slope to form the ceilings of substantial open cavities. The undersurfaces of these large sponges were colonized by an extensive cryptic community, including downward-growing branching sphinctozoans up to 0.5 m long.

In the absence of both the physical and biological destructive forces prevalent on modern reefs, much of the relatively fragile Capitan reef-building community remained in growth position. It is unlikely, however, that such a community could have withstood highly turbulent and agitated waters. Additional strength was imparted to the framebuilding community by encrustations of *Archaeolithoporella* and *Tubiphytes*, followed by the precipitation of abundant micrite of probable microbial origin. This cavernous framework was subsequently partially filled with internal sediment and *Archeolithoporella*, and botryoidal aragonite and other submarine cements. These cements and sediments account for up to 70% of the reef rock. So unlike modern coral reef carbonates, most of the Capitan Reef was derived from probably inorganic sources, and moreover none of the organic carbonate—with the exception of phylloid algae and possibly *Archaeolithoporella*—is interpreted to have formed as a result of light-enhanced (photosynthetic) calcification. Such observations suggest that models that simulate modern coral reef growth, which is dependent upon the distribution of light, may not be applicable generally to Paleozoic reefs.

INTRODUCTION

Reef communities are commonly assumed to have achieved a modern ecological cast by the Early Cambrian, when skeletal metazoans (multicellular animals), open frameworks, and cryptic organisms (those living within hidden or otherwise protected areas) became established (e.g., Fagerstrom, 1987). Yet many of the organisms responsible for the unique ecology of modern coral reef communities were absent from Cambrian, and indeed later Paleozoic, reef ecosystems. These include metazoans that have entered into symbioses with algae, and herbivores, predators, and boring organisms that are capable of significant skeletal destruction. Some have argued, therefore, that the absence of these organisms in Paleozoic reef communities produced ecologies that were radically different from modern coral reefs, and that these factors are responsible for the profound differences in their geological expression (e.g., Wood, 1995; Wood et al., 1996).

Much has been written on the paleoecology of the Permian Capitan reef, as expressed in the Capitan Limestone (e.g., Babcock, 1974, 1977; Babcock et al., 1977; Babcock and Yurewicz, 1989; Yurewicz, 1976, 1977; Fagerstrom, 1987; Kirkland et al., 1993). Established ecological reconstructions have emphasized the role of various upward-growing and baffling organisms (sponges, bryozoans, and the problematicum *Tubiphytes*) and massive putative algae (*Collenella*, *Parachaetetes*, and *Solenopora*) in the construction of the reef, with the binding and encrusting contribution of *Archaeolithoporella* and extensive early marine cementation. Many have used such reconstructions to infer the depth at which the Capitan reef grew (e.g., Crandall, 1929; Lloyd, 1929; Johnson, 1942; Newell et al., 1953; Newell, 1955; Klement, 1966; Cys, 1971; Cronoble, 1974; Mazzullo and Cys, 1978; Mazzullo, 1979; Kirkland et al., 1993).

Like modern coral reefs, topography and accommodation space were probably important controls on development of the Capitan reef (see Kerans and Tinker, this volume). Likewise, the depth at which the Capitan Limestone formed —which includes both the reef itself and the detrital foreslope—was also critical to late Guadalupian physiography, but remains a matter of controversy, as exemplified by the number of competing bathymetric profile models proposed for the shelf margin (e.g., compare King, 1948; Newell et al., 1953; Dunham, 1972; Kirkland-George, 1992). The Capitan Limestone is one of the steepest known ancient prograding carbonate shelf margins with existing erosional slope angles of 20-40°, but there is no consensus as to how much this apparent depositional relief has been exaggerated by post-depositional compaction. For example, Bebout and Kerans (1993) support an original steep reef front on the basis of interbed geometry, but Saller (1996) recorded a consistent basinward tilt of the Capitan Limestone of 11°, suggesting the original depositional slope of the Capitan reef to have been only approximately 10-30°. Further data in support of a post-depositional compaction origin for Capitan topography is offered by Hunt and Fitchen (1999, pers. comm.).

Although the reef clearly extended over a considerable range of depths, whether its uppermost part grew to sea level in turbulent surface waters or was restricted entirely to more tranquil, deeper waters is of fundamental ecological and geological importance. Some consider the Capitan reef to represent a wave-resistant, shallow framework similar to that of a modern barrier reef (Newell et al., 1953; Kirkland George, 1992); others have

Geologic Framework of the Capitan Reef, SEPM Special Publication No. 65

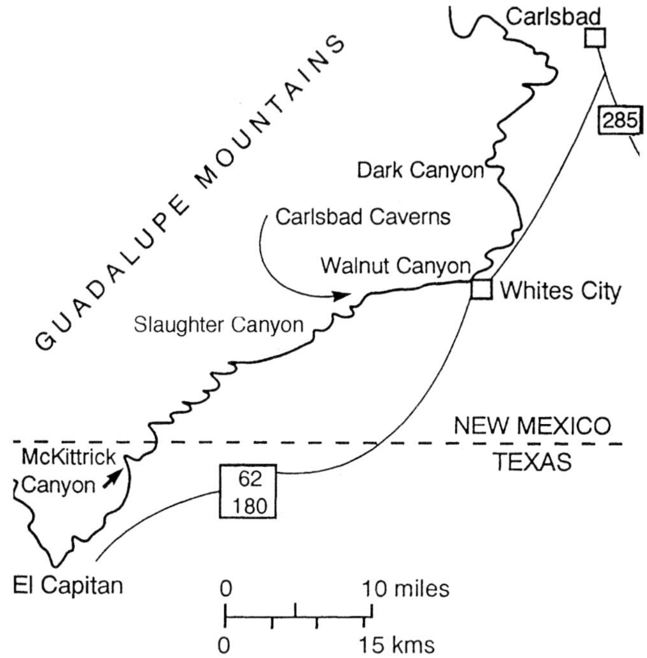

FIG. 1.—The Permian Capitan reef front, Guadalupe Mountains with localities studied (modified from King, 1948).

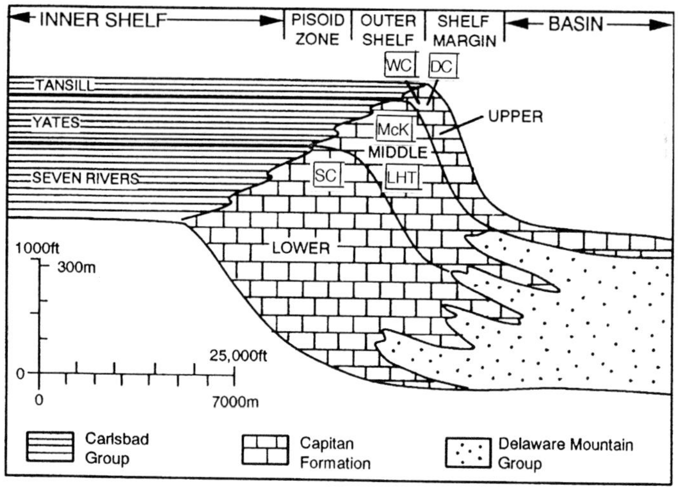

FIG. 2.—Schematic cross section showing shelf-to-basin correlation of the Capitan Formation and equivalents (after Babcock and Yurewicz, 1989) and subdivision of the shelf (after Esteban and Pray, 1983). The approximate position of the localities studied is given: McK: McKittrick Canyon; DC: Dark Canyon; SC: Slaughter Canyon; WC: Walnut Canyon: LTH: Left Hand Tunnel, Carlsbad Caverns. (modified from Wood et al., 1996).

stressed the constructional role of extensive algal encrustation and early cementation in the formation of a rigid framework in deeper water (e.g., Babcock, 1974, 1977; Yurewicz, 1976, 1977; Schmidt, 1977; Mazzullo and Cys, 1978). Others still have found little evidence for preservation of the community in growth position, and so consider the reef to have formed as a linear bank of baffling metazoans growing in quiet and deep waters (e.g., Dunham, 1972).

This paper considers: 1) the ecology of the reef-building communities of the Capitan reef; 2) how this ecology influenced the eventual geological fabric of the reef rock, and 3) to what extent modern coral reefs are appropriate analogues for understanding the development of Paleozoic reef margins.

LOCALITIES STUDIED

The paleoecology of the Capitan reef is poorly understood relative to the size of the reef. This is due to two reasons. Firstly, the boundstone fabrics are far larger in scale than can be described from hand samples alone, necessitating the study of extensive outcrop surfaces. These must be parallel to reef growth to reveal the growth orientation of the *in situ* reef community. However, extensive surface alteration of the Capitan Limestone has resulted in only a limited number of outcrops with clearly observable depositional fabrics. As a consequence, much of the massive reef may appear to consist of detrital carbonate with little discernible *in-situ* biota.

Detailed field observations of the paleoecology of boundstones were made on naturally weathered surfaces on the Permian Reef Geology Trail, McKittrick Canyon, which exposes approximately 140 m of the Middle Capitan Limestone (middle and upper Yates-equivalent). Acid-etched surfaces of approximately 43 m of Upper Capitan Limestone (lower Tansill-equivalent) were studied at the mouth of Walnut Canyon, Whites City, and at Dark Canyon. Additional observations of reef fabrics were made in the Left Hand Tunnel of Carlsbad Caverns, thought to be lower Middle Capitan in age (B. L. Kirkland, pers. comm., 1997). A limited number of well-preserved vertical reef fabrics were studied in the Lower Capitan Limestone outcrops in Slaughter Canyon (Figs. 1 and 2).

PALEOECOLOGY

At least six reef-building communities have been described from the Middle and Upper Capitan Limestone (Babcock, 1974, 1977; Yurewicz, 1976, 1977; Babcock and Yurewicz, 1989; Kirkland et al., 1993; Wood et al., 1994, 1996; Senowbari-Daryan and Rigby, 1996): 1) phylloid algae; 2) *Tubiphytes*-sponge; 3) *Tubiphytes-Acanthocladia*; 4) frondose bryozoan; 5) platy sponge; and 6) richthofenid brachiopod aggregations. Determining the relative abundance and importance of these communities is difficult, but the frondose bryozoan and platy sponge communities are described below to illustrate some of the variable ecologies found within the Capitan reef.

Frondose Bryozoan Community

The frondose bryozoan community has been identified from 1) the Middle Capitan Permian Reef Geology Trail in McKittrick Canyon, where it forms the bulk of discernible reef fabric, 2) the Upper Capitan Limestone at the mouth of Walnut Canyon by Whites City, where it outcrops at the base of the exposure some 30 m below the contact between the reef and backreef, 3) in allochthonous blocks in slope strata from the Left Hand Tunnel of Carlsbad Cavern, thought to be of lower Middle Capitan age, and 4) in a limited number of exposures in the Lower Capitan Limestone, Slaughter Canyon.

The vast majority of calcified sponges (sphinctozoans) at these locations grew downwards into cavities (crypts) (Fig. 3). Sphinctozoan sponges grow by the incremental addition of hemispherical chambers that indicate the direction of growth to form cylindrical, conical, massive, or branching growth forms. Most sponges are attached to the upper surface of crypts by often small holdfasts, while a few project from the walls or floors. Where holdfasts are discernible, the sponges are frequently attached to robust frondose bryozoans (*Goniocladia* sp., *Fenestella* sp., and *Polypora* sp.). The ceilings and walls of crypts are often supported by these highly convoluted fan- (frondose) or plate-shaped metazoans, forming large crypts up to 0.5 m in width and 0.4 m in height. Bryozoan fronds appear to have grown in varied orientations and are up to 0.25 m in any one dimension, but because these framebuilders are very thin and widely spaced, samples taken from the Capitan reef may not include any obvious examples.

Frondose bryozoans and *Guadalupia zitteliana* thus formed a community of erect, or horizontally projecting fans, whose undersurfaces were colonized by an abundant, attached cryptic community (the cryptos) (Fig. 4). This cryptos was dominated by diverse solitary sphinctozoan sponges, especially *Lemonia cylindrica*, *Amblysiphonella* spp. and *Cystauletes* spp. Common, small (up to 7 mm diameter) nodular bryozoans (*Girtypora* sp. and *Fistulipora* spp.). Rare, nestling rugose corals and crinoids have also been identified within crypts (Wood et al., 1996).

Much of the reef community shows evidence of multiple biological encrustation and microboring. Chains of pendant sphinctozoan individuals are common, and both the open surface community and cryptos are often covered by encrustations up to 4 cm thick of intergrown *Archaeolithoporella*, *Tubiphytes*, and foraminifera with rare weakly calcified encrusting sponges, and soft-bodied encrusters.

The metazoan reef community accounts, however, for only a minor proportion of the total boundstone, occupying approximately 7–17% of the volume, of which 6–15% of the metazoans are cryptic sponges. Approximately 25–50% of the boundstone is represented by pale- to mid-gray laminated micrite [Fig. 3(b)](based on field tracings from vertical outcrop; Wood et al., 1996). This micrite includes detrital internal sediment (Babcock, 1974, 1977; Babcock et al., 1977; Kirkland et al., 1993), and an encrustation around both the frondose bryozoans and pendant cryptos, which has a layered, thrombolitic to peloidal texture in thin-section. These encrustations are suggested to be of microbial origin. On the basis of an enveloping form around many skeletal elements of the reef fabric, this encrustation is proposed to have developed after the death of the primary framework community. This putative microbial micrite grew incrementally and became lithified during deposition, lending considerable strength to the relatively fragile initial reef framework, modifying the original reef framework to form a network of connected open cavities and tunnels.

Remaining cavity space (up to approximately 70% of the volume) was filled by bioclastic sediment, cement botryoids of aragonite, *Archaeolithoporella*, crusts of prismatic calcite (including radiaxial) cements, and various late stage spars (Mruk, 1989). Botryoidal aragonite forms gray and brown fans 1–3 cm in radius and nucleated on crypt ceilings, walls, and floors as well as from

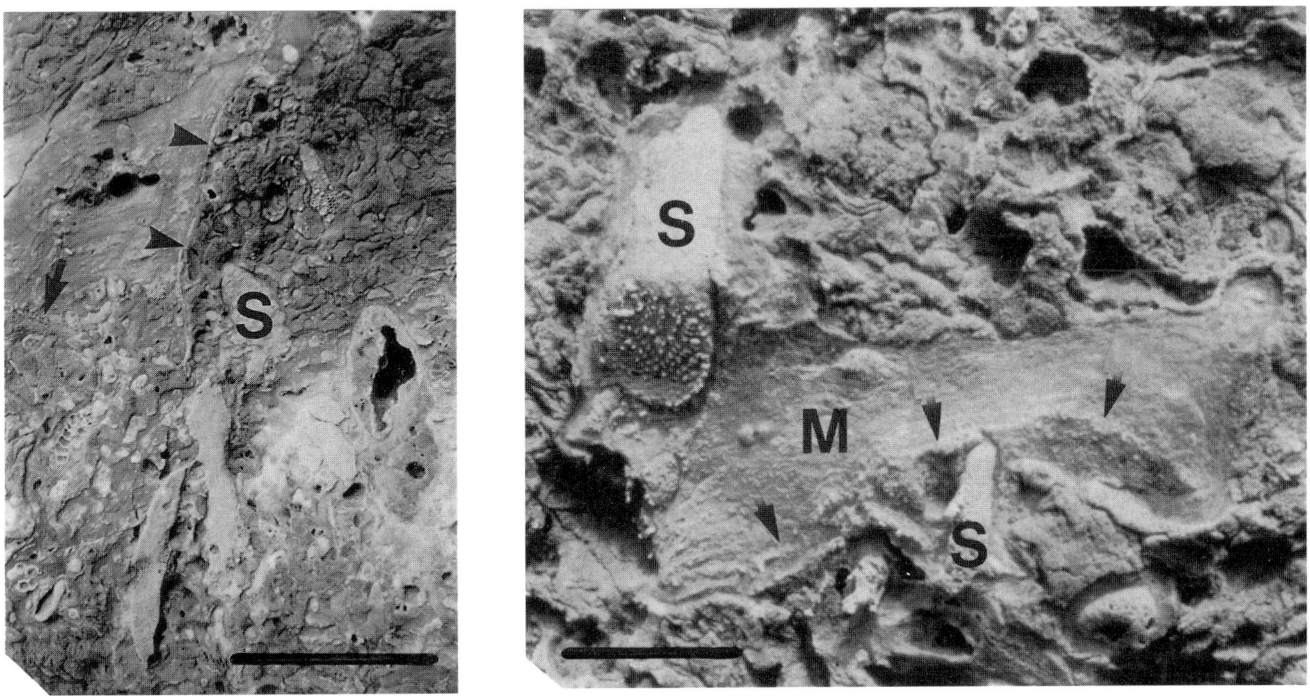

FIG. 3.—Weathered surfaces parallel to reef growth from the Middle Capitan Limestone, McKittrick Canyon, showing latticework of bryozoans (arrowed) forming cavity boundaries which support a cryptos of sphinctozoan sponges (S). Remaining space is infilled by intergrown *Archaeolithoporella* and cement botryoids. (a) Scale = 5 cm. (b) Note post-mortem encrustation of micrite (M) around a bryozoan frond (arrowed). Scale = 2 cm.

the projecting cryptos (Fig. 3). Several generations of isopachous calcite cements (0.5–4 mm thick) also encrust the surfaces of pendant cryptos and botryoidal aragonite, indicating that they postdate botryoid growth (see Rahnis and Kirkland, this volume). Remaining cavity space was filled by dolomites, late-stage sparry calcites, evaporite minerals, and kaolinite (Scholle et al., 1992).

These observations allow reconstruction of the sequence of development of the frondose bryozoan community (Fig. 4):

1. Large, convoluted fan-shaped bryozoan colonies formed the initial reef framework, along with small thickets of branching acanthocladid bryozoans and *Tubiphytes*.
2. The undersurfaces of these bryozoans were colonized by an abundant cryptic biota dominated by pendant sphinctozoan sponges and small bryozoans.
3. The cryptos was encrusted by foraminifera, *Tubiphytes* and *Archaeolithoporella*.
4. The entire community was encrusted post-mortem by microbes that caused precipitation of micrite, further lithifying and partially filling the reef framework.
5. Remaining cavity space was initially filled by sediment, intergrowths of aragonitic botryoids and *Archaeolithoporella*, followed by large volumes of botryoidal aragonite. Some cavities remained entirely open, or were filled by late diagenetic cements, including coarse calcite and anhydrite.

Platy Sponge Community

Platy sponge-dominated communities, found in the uppermost part of the Capitan reef, are interpreted to be within the bathymetrically shallow parts of the reef. They have been identified from the upper 10 m of the Middle Capitan reef on the Permian Reef Geology Trail, McKittrick Canyon, and in fallen blocks within the reef foreslope (thought to be equivalent to the Upper Capitan (Tansill); Bebout and Kerans, 1993). Here, platy sponges are represented by the sphinctozoan *Guadalupia zitteliana*, which develops a convex morphology and may reach 0.25 m in diameter; they are associated with uncommon, small massive growths (up to 0.15 m high) of *Collenella*.

A second platy sponge community is well developed in the lower Upper Capitan Limestone (lower Tansill equivalent) some 5–10 m below the topmost exposure of the Capitan Limestone at Walnut Canyon, Whites City. Here, the reef framework was dominated by large individuals of the sphinctozoan *Guadalupia explanata* that reached a maximum of 0.5 m in diameter, and the inozoan (non-chambered calcified sponge), *Gigantospongia discoforma*, which although never more than 20 mm thick, grew up to 2 m in diameter. This latter species has also been found at Chinaberry and Hackberry Draws (see Rigby and Senowbari-Daryan, 1996). All these platy sponges display laminar, convex, or convoluted morphologies, sometimes with marked downturned edges forming tunnel-like structures. Upper surfaces may bear traces of astrorhizal systems, which represent the remains

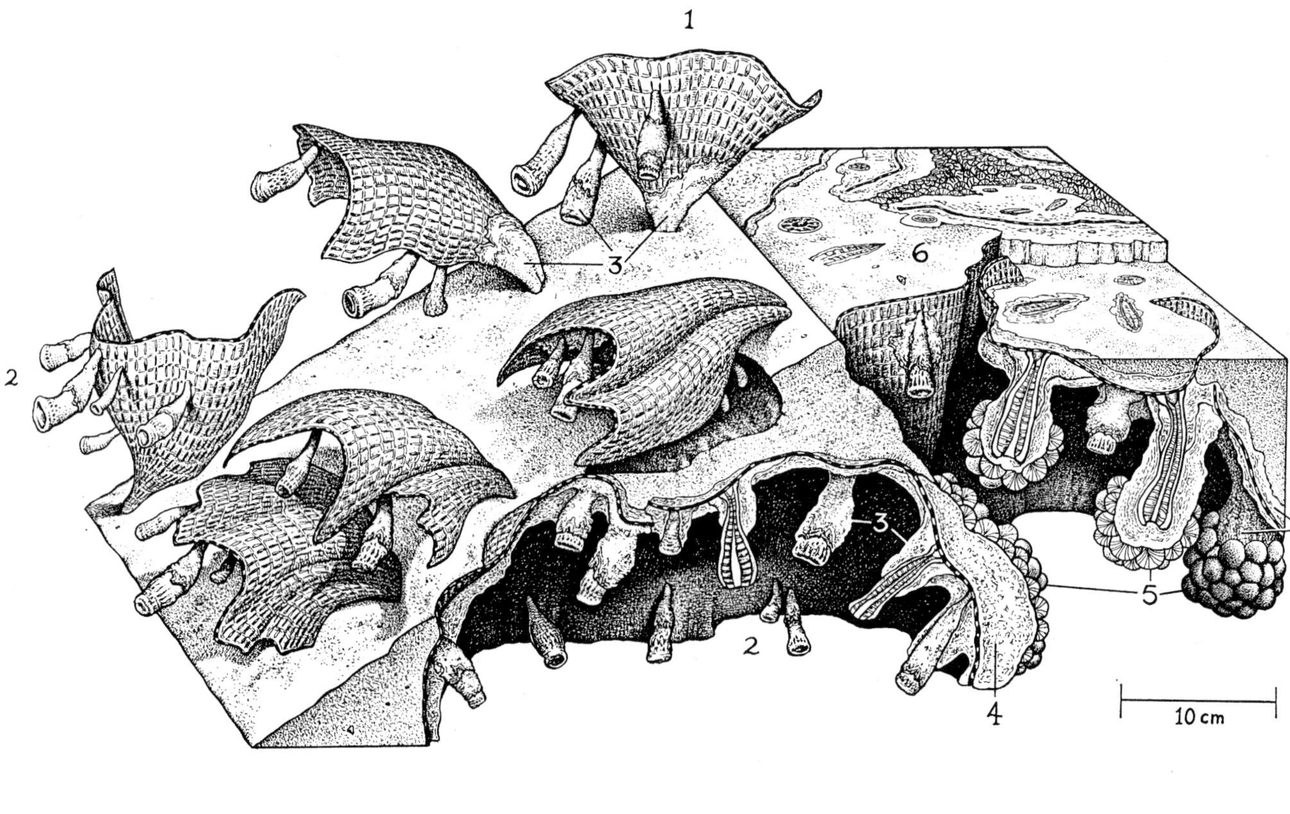

FIG. 4.—Schematic community reconstruction of the frondose bryozoan community: 1: frondose bryozoans (*Polypora* sp., *Goniopora* sp. and *Fenestella* sp.); 2: cryptic solitary sphinctozoans; 3: *Archaeolithoporella* encrustation; 4: microbial micrite with entrapped bioclastic sediment; 5: aragonitic cement botryoids; 6: carbonate sediment. (From R. A. Wood, 1999).

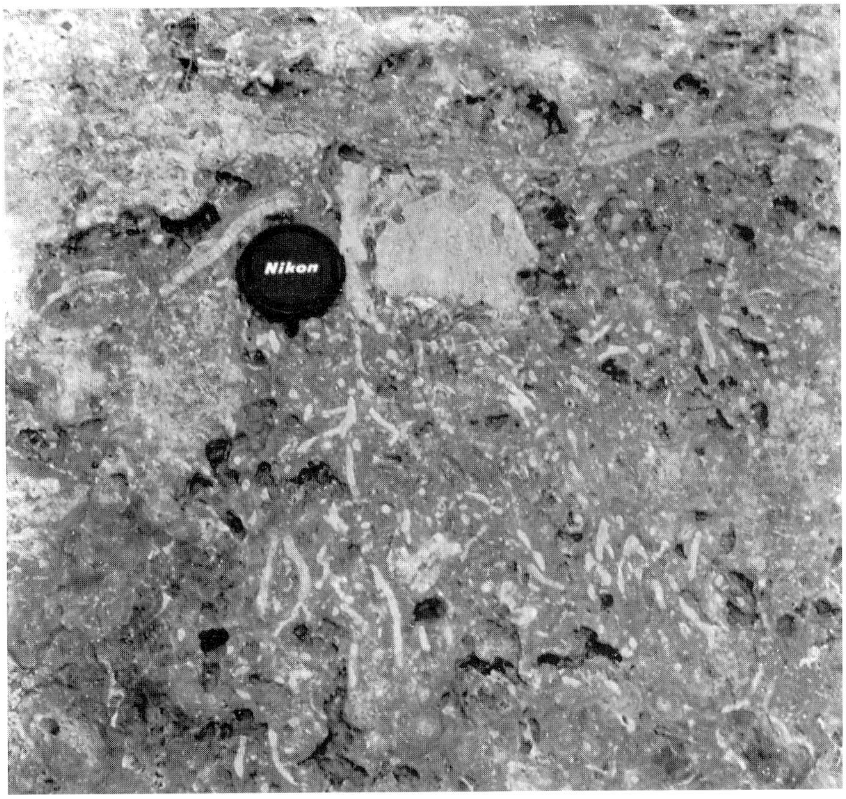

FIG. 5.—Extensive cavity ceiling formed by the platy sponge *Gigantospongia discoforma* supporting an extensive cryptos, including a branching individual of *Lemonea* sp. (left) and the compound *Guadalupia explanata* (right). Lens cap = 52 mm.

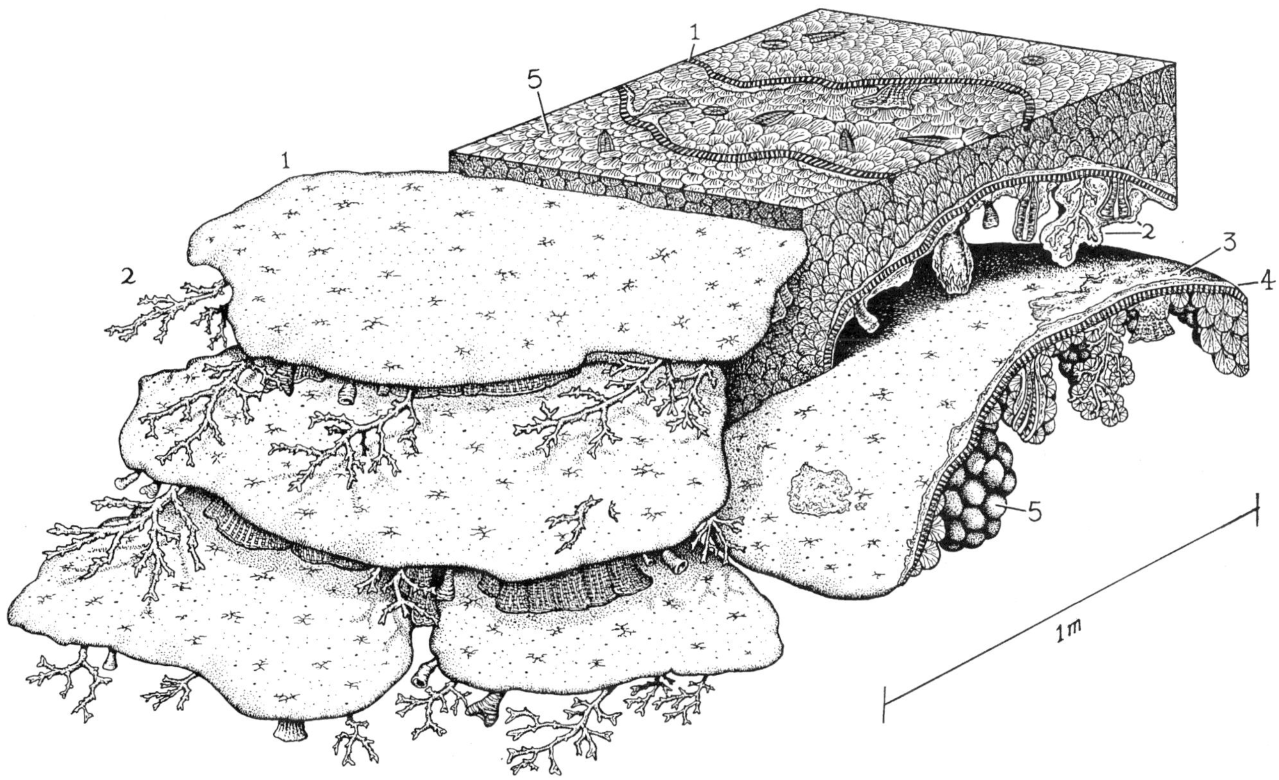

FIG. 6.—Schematic reconstruction of the platy sponge community: 1: platy sponges (*Gigantospongia discoforma* and *Guadalupia zitteliana*); 2: cryptic sphinctozoans (solitary, branching and compound); 3: *Archaeolithoporella* encrustation; 4: microbial micrite; 5: aragonitic cement botryoids. (From R. A. Wood, 1999).

of the exhalant aquiferous filtration system (Senowbari-Daryan, 1990; Rigby and Senowbari-Daryan, 1996). These platy sponges were presumably attached by a site of limited size; no evidence of stacking or mutual attachment has been found.

The platy sponges appear to have grown either perpendicular to the reef front out into ambient currents, or perhaps formed a series of domes or tunnels towards the top of the reef. These sponges formed the ceilings of huge open crypts up to 2 m in height and width (Figs. 5 and 6). In McKittrick Canyon, the smaller, platy sponges were colonized by a cryptic community of small, solitary sphinctozoan sponges and small, nodular bryozoans. Rare macroborings are found within the cryptos. At Walnut Canyon, the undersurfaces of the *Gigantospongia discoforma* were colonized by a diverse cryptic community dominated by large sphinctozoan sponges (Wood et al., 1996), including massive sphinctozoans (for example, *Lemonea conica*), which grew up to 0.25 m in diameter, and branching sphinctozoans (for example, *Lemonea* sp.) up to 0.5-m long (Fig. 5). Branching acanthocladid bryozoan colonies up to 0.4 m in length are also present. Small, nodular bryozoans are found attached either directly to the undersurfaces of the platy sponges or nestled amongst the larger members of the cryptos with small rugose corals and crinoids.

The upper surfaces of the platy sphinctozoans and cryptic community were thickly encrusted by generations of often intergrown foraminifera, *Tubiphytes* and *Archaeolithoporella*. At Walnut Canyon, the community was patchily enveloped on both upper, open, and lower cryptic surfaces by a poorly preserved greenish-grey micrite with a peloidal fabric resembling the calcified cyanobacterium *Renalcis*. This micrite is estimated to occupy less than 10% of the total boundstone by volume.

The considerable remaining pore space was infilled by small botryoids of aragonite (7–20 mm radius) and intergrown *Archaeolithoporella*, followed by very large botryoids of aragonite (30–60 mm radius; now translucent brown on outcrop). As in the frondose bryozoan-sponge community, the botryoids nucleated on the crypt ceilings, walls, and floors as well as from the projecting cryptos. Some minor geopetal infilling of coarse, bioclastic sediment is present.

From these observations we can infer the sequence of development of the platy sponge community (Fig. 6):

1. The growth of large, platy sponges that projected horizontally from the reef slope or formed tunnels toward the reef crest.
2. The undersurfaces of the largest platy sponges were colonized by an abundant cryptos dominated by large, pendant sphinctozoan sponges, including solitary, branching, and massive forms.
3. The upper surfaces of the sponges and cryptic community were thickly encrusted by foraminifera, *Tubiphytes* and *Archaeolithoporella*.
4. The entire reef community was encrusted by patchy microbial micrite.
5. Cavity space was filled mainly by the synsedimentary intergrowth of aragonitic botryoids and *Archaeolithoporella*, and later simply by very coarse botryoidal aragonite.

DISCUSSION

Most fenestellid bryozoans lived in moderately energetic to relatively quiet waters, either near or below normal wavebase or in sheltered localities (McKinney and Gault, 1980). Those with lyre-like growth morphologies could withstand more vigorous unidirectional currents. Indeed, Elias and Condra (1957) considered that an intergrown lattice of fenestellids would be able to withstand turbulence when stiffened by a partial filling or coating of submarine cement or micrite. The frondose bryozoans that construct the framework of the Middle Capitan show considerable secondary thickening and were capable of lateral fusion and connection. Although some frondose bryozoan debris is noted in the Capitan reef, most of the bryozoan fronds appear to be intact. Such a latticework would be able to withstand moderate turbulence, especially when further strengthened by biological encrustation and early cements.

There is no morphological or isotopic evidence to support the presence of symbiotic algae within either Permian sphinctozoan sponges or frondose bryozoans. It is therefore assumed that the majority of the sessile Capitan reef metazoans were primary consuming suspension-feeders (Wood et al., 1996). This infers the presence of suspended food matter in the Delaware Basin, presumably mainly in the form of plankton, supported by reasonably abundant nutrients. The calcified benthic community was volumetrically dominated by these consumers, most of which were cryptic: indeed, most biodiversity of the skeletal attached benthos of the Capitan reef was housed within the cryptic community. Possible benthic primary producers were represented by *Archaeolithoporella*, although the photosynthetic nature of this organism has been questioned (see Grotzinger and Knoll, 1995). As such, only a small proportion of overall primary productivity appears to have been present in the preservable benthic photosynthetic community, although the presence of large stands of micro- or macroalgae is possible. Therefore, unlike modern coral reefs, there is no evidence to suggest that benthic primary producers were closely linked to primary consumers, although some gastropod species may have been herbivores. Few potential predators of the sessile reef biota have been identified, and bioerosion—with the exception of microboring activity—appears to have been relatively insignificant. By contrast, the intense bioerosion prevalent on many modern coral reefs can reduce reef frameworks to piles of rubble and sand (Hubbard et al., 1990).

Although the Capitan Limestone shows a mean rate of accumulation that compares with that of modern carbonate platforms (see Garber et al., 1989; Bosscher and Schlager, 1993), this masks the fact that organic and inorganic carbonate make substantially different contributions to the final reef fabrics (Fig. 7). This is a consequence of profound differences in ecology and setting. Original internal cavities were initially preserved by biological encrustation (calcified encrusters and microbial micrite) and possibly pervasive microscopic cementation, and, later, infilling by cement. In contrast to modern reefs, where the overwhelming bulk of carbonate is organically derived and photosynthetically boosted (Barnes and Chalker, 1990), a substantial proportion of Capitan reef carbonate (up to 70% as exposed) was thus contributed by cement processes. Unlike the inferred setting of the Capitan reef, modern coral reefs usually thrive in low nutrient seas where net planktonic primary productivity is low—often only one-tenth of the net benthic primary productivity (Grigg et al., 1984). Benthic productivity is derived from various algae and photosymbiotic scleractinian corals. This benthically dominated ecosystem is regulated by intense predation, which also controls production higher in the food web; some estimates suggest that between 50-85% of a species' annual production is consumed by predation or herbivory (Grigg et al., 1984).

The apparently fragile and cavernous Permian reef framework shows no evidence of physical or biological destruction. This suggests that the reef withstood the ambient hydrodynamic regime, whether it was one of moderate wave action in the most shallow, landward parts of the reef or in the relatively quiet waters at depth. However, such an observation does not reveal whether the reef grew to sea level; the shores of the relatively small and enclosed Delaware Basin may not have presented the highly energetic conditions of substantial wave impact under which modern oceanic coral reefs grow.

Many authors have suggested that oceans, including the partially isolated Delaware Basin, were stratified with anoxic bottom waters during the Upper Permian (e.g., Margaritz and Turner, 1982; Erwin, 1994). This is supported by the unusual enrichment of ^{13}C in Late Permian carbonates, including those from the Capitan reef itself (e.g., Garber et al., 1989; Margaritz et al., 1992). The presence of an alkalinity pump (which introduced alkaline deep water into the shallows) has also been inferred to have controlled the amount of carbonate ions in the upper water masses (Margaritz and Turner, 1982). All these phenomena would have favored the precipitation of carbonate cements at the margins of the basins (Grotzinger and Knoll, 1995), but perhaps suggest that the processes of cementation that contributed to the formation of the Capitan Limestone were not analogous to those operating in most modern coral reef settings.

Moreover, prior to the evolutionary rise of excavatory herbivory and predation (which was part of the wider Mesozoic Marine Revolution; Vermeij, 1977), relatively fragile reefs could be preserved if intense early cementation aided early lithi-

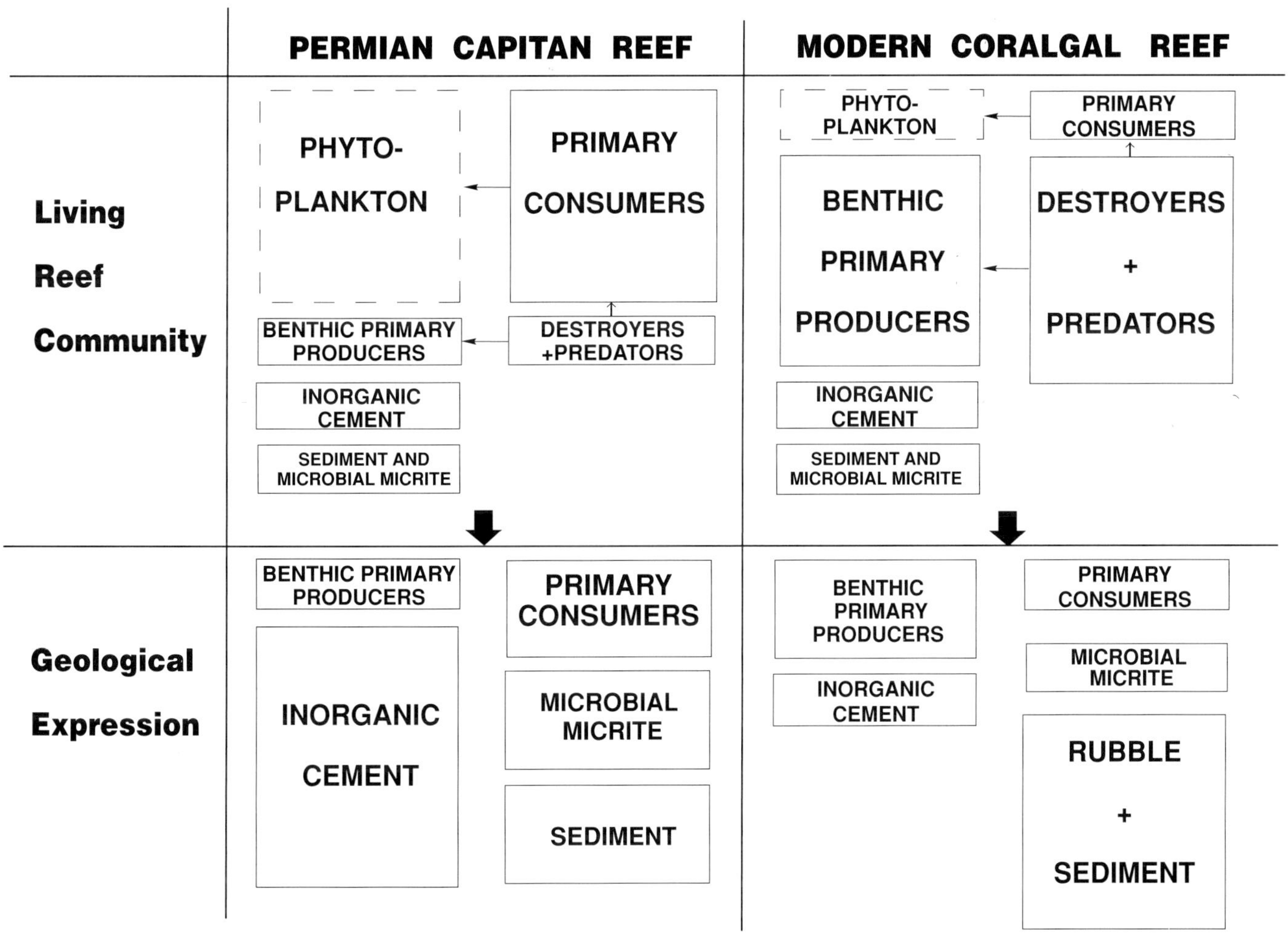

FIG. 7.—Comparison of the relative contribution of organic and inorganic sources of carbonate in the Permian Capitan and modern coral reefs, in both the living reef communities and their final geological expression. The Permian reef framework, whilst showing no evidence of a high degree of wave resistance, remains intact due to the absence of significant bioerosion. Original crypts are thus preserved which become infilled with substantial volumes of inorganic cement. In contrast, the abundant predators and destroyers on modern coral reefs reduce the framework to mainly rubble and sediment. (From Wood et al., 1996).

fication. Studies of ancient reef ecology may have overemphasized the importance of massive metazoan framebuilders. While necessary for reef growth in areas exposed to high levels of physical (waves and storms) and biological destruction (bioerosion), massive framebuilders are not required for reef formation. The Capitan reef appears to have developed in the absence of substantial destructive forces, and its sustained growth and preservation was due to post-mortem microbial encrustation and extensive, rapid early cementation. Without early lithification, the reef may have developed as a bioclastic bank. Microbial encrustation may have been crucial for the preservation of the relatively fragile frondose bryozoan-sponge community. Although no methodology is currently available to determine the rate of ancient cement growth, if cements within the Capitan grew at the rapid rates recorded from modern aragonitic botryoids (e.g., Grammar et al., 1993) then most primary pore space within the Capitan reef could have been totally occluded in a few tens of thousands of years. Such reasoning suggests that Paleozoic reef frameworks should be considered in a new light. The presence of either an intact reef framework or similar rates of carbonate accumulation should not be criteria used to infer that an ancient community grew under the same conditions as those required by modern coral reefs.

CONCLUSIONS

The ecology of the Permian Capitan reef was profoundly different from that of modern coral reefs, and we cannot infer that the community was under the same environmental conditions that govern modern reef ecosystems and associated carbonate platform accumulation. There is a clear need for further paleoecological work on the communities that formed the Capitan Limestone, especially on the degree to which the boundstone fabrics currently described are typical, and how they may vary in relation to stratigraphy, inferred bathymetry, and geographical location.

ACKNOWLEDGMENTS

I thank Charlie Kerans, Oliver Weidlich, and Art Saller for their constructive reading of the manuscript, and the Royal Society for its continued financial support. John Sibbick produced the artistic reconstructions, and Hilary Alberti drafted the figures. Earth Sciences Publication No. 5236.

REFERENCES

Babcock, J. A., 1974, The role of algae in the formation of the Capitan Limestone (Permian, Guadalupian) Guadalupe Mountains, West Texas-New Mexico: Unpublished Ph.D. Dissertation, University of Wisconsin, Madison, 241 p.

Babcock, J. A., 1977, Calcareous algae, organic boundstones, and the genesis of the upper Capitan Limestone (Permian, Guadalupian), Guadalupe Mountains, West Texas and New Mexico, *in* Hileman, M. E., and Mazzullo, S. J., eds., Upper Guadalupian facies, Permian Reef Complex, Guadalupe Mountains, New Mexico and West Texas, Field Conference Guidebook 77-16: Midland, Society of Economic Paleontologists and Mineralogists, Permian Basin Section, p. 3-44.

Babcock, J. A., Pray, L., and Yurewicz, D., 1977, The shelf edge Capitan-masssive and outer shelf facies, *in* Pray, L .C., and Esteban, M. , eds., Upper Guadalupian Facies, Permian Reef Complex, Guadalupe Mountains, New Mexico and Texas, Volume 2, Midland, Society of Economic Paleontologists and Mineralogists, Permian Basin Section, Publication 77-16: p. 11-40.

Babcock, J. A., and Yurewicz, D. A., 1989, The massive facies of the Capitan Limestone, Guadalupe Mountains, Texas and New Mexico, *in* Harris, P. M., and Grover, G. A. , eds., Subsurface and outcrop examination of the Capitan Shelf Margin, northern Delaware Basin: Tulsa, Society of Economic Paleontologists and Mineralogists Core Workshop 13, p. 365-371.

Barnes, D. J., and Chalker, B. E., 1990, Calcification and photosynthesis in reef-building corals and algae, *in* Dubinsky, Z., ed., Coral Reefs: Ecosystems of the World, v. 25, p. 109-131.

Bebout, D. G., and Kerans, C., eds., 1993, Guide to the Permian Reef Geology Trail, McKittrick Canyon, Guadalupe Mountains National Park, West Texas, Guidebook 26: Austin, Bureau of Economic Gcology, University of Texas, p. 1-48.

Bosscher, H., and Schlager, W., 1993, Accumulation rates of carbonate platforms: Journal of Geology, v. 10, p. 345-355.

Crandall, K. H., 1929, Permian stratigraphy and origin of facies in the Capitan reef, New Mexico and Texas: Discussion: American Association of Petroleum Geologists Bulletin, v. 11, p. 95-108.

Cronoble, J. M., 1974, Biotic constituents and origin of facies in Capitan reef, New Mexico and Texas: Discusssion: Mountain Geologist, v. 11, p. 95-108.

Cys, J. M., 1971, Origin of Capitan Formation, Guadalupe Mountains, New Mexico and Texas: American Association Petroleum Geologists Bulletin, v. 55, p. 310-315.

Dunham, R. J., 1972, Capitan Reef, New Mexico and Texas: Facts and Questions to Aid Interpretation and Group Discussion, Midland, Society of Economic Paleontologists and Mineralogists, Permian Basin Section, Publication 72-14: 291 p.

Elias, M. K., and Condra, G. E., 1957, *Fenestella* from the Permian of west Texas: Memoir of the Geological Society of America, v. 70, p. 1-158.

Erwin, D. H., 1994, The Permo-Triassic extinction: Nature, v. 367, p. 231-236.

Esteban, M., and Pray, L. C., 1983, Pisoids and pisolite facies (Permian), Guadalupe Mountains, New Mexico and West Texas, *in* Peryt, T., ed., Coated Grains: Berlin, Springer-Verlag, p. 505-537.

Fagerstrom, J. A., 1987, The Evolution of Reef Communities: New York, John Wiley and Sons, 600 p.

Garber, R. A., Grover, G. A., and Harris, P. M., 1989, Geology of the Capitan Shelf Margin — subsurface data from the northern Delaware Basin, *in* Harris, P. M., and Grover, G. A. , eds., Subsurface and outcrop examination of the Capitan Shelf Margin, Northern Delaware Basin: Tulsa, Society of Economic Paleontologists and Mineralogists Core Workshop 13, p. 3-269.

Grammar, G. M., Ginsburg, R. N., Swart, P. W., McNeill, D. F., Jull, A. J. T., and Prezbinowski, D. R., 1993, Rapid growth rates of syndepositional marine aragonite cements in steep marginal slope deposits, Bahamas and Belize: Journal of Sedimentary Petrology, v. 63, p. 983-989.

Grigg, R. W., Polovina, J. L., and Atkinson, M. J., 1984, Model of a coral reef ecosystem: 111. Resource Limitation, Community Regulation, Fisheries Yield and Resource Management: Coral Reefs, v. 3, p. 23-27.

Grotzinger, J. H., and Knoll, A. H., 1995, Anomalous carbonate precipitates: Is the Precambrian the key to the Permian?: Palaios , v. 10, p. 578-596.

Hubbard, D. K, Miller, A. I., and Scaturo, D., 1990, Production and cycling of calcium carbonate in a shelf-edge reef system (St. Croix, U.S. Virgin Islands): Applications to the nature of reef systems in the fossil record: Journal of Sedimentology, v. 60, p. 335-360.

Johnson, J. H., 1942, Permian lime-secreting algae from the Guadalupe Mountains, New Mexico: Bulletin of the Geological Society of America, v. 53, p. 195-226.

King, P. B., 1948, Geology of the Southern Guadalupe Mountains, Texas: Washington, D.C., U.S. Geological Survey Professional Paper 215, 183 p.

Kirkland, B. L., Longacre, S. A., and Stoudt, E. L., 1993, Reef, *in* Bebout, D. G., and Kerans, C., eds., Guide to the Permian Reef Geology Trail, McKittrick Canyon, Guadalupe Mountains National Park, West Texas, Guidebook 26: Austin, Bureau of Economic Geology, University of Texas, p. 23-31.

Kirkland George, B., 1992, Distinctions between reefs and bioherms based on studies of fossil algae-*Mizzia*, Permian Capitan reef complex (Guadalupe Mountains, Texas and New Mexico) and *Eugonophyllum*, Pennsylvanian Holder Formation (Sacremento Mountains, New Mexico): Unpublished Ph.D. Dissertation, Louisiana State University, Baton Rouge, 156 p.

KLEMENT, K. W., 1966, Studies on the ecological distribution of lime-secreting and sediment-trapping algae in reefs and associated environments: Neues Jahrbuch für Geologie und Paläontologie, Abhandlungen, v. 125, p. 363-381.

LLOYD, E. R., 1929, Capitan Limestone and associated formations of New Mexico and Texas: American Association of Petroleum Geologists Bulletin, v. 13, p. 645-658.

MAZZULLO, S. J., and CYS, J. M., 1978, *Archaeolithoporella*-boundstones and marine aragonite cements, Permian Capitan reef, New Mexico and Texas, USA: Neues Jahrbuch für Geologie und Palaontologie, Monatshafte, v. 1978, p. 600-611.

MAZZULLO, S. J., 1979, Marine aragonite sea-floor growths and cements in Permian phylloid algal mounds, Sacramento Mountains, New Mexico: Journal of Sedimentary Petrology, v. 49, p. 917-936.

MCKINNEY, F. K., and GAULT, H. W., 1980, Paleoenvironment of late Mississippian fenestrate bryozoans, eastern United States: Lethaia, v. 13, p. 127-146.

MARGARITZ, M., and TURNER, P., 1982, Carbon cycle changes of the Zechstein Sea: Isotopic transition zone in the Marl Slate: Nature, v. 297, p. 1982.

MARGARITZ, M., KRISHNAMURPHY, R. V., and HOLSER, W. T., 1992, Parallel trends in organic and inorganic carbon isotopes across the Permian-Triassic boundary: American Journal of Science, v. 292, p. 727-739.

MRUK, D. H., 1989, Diagenesis of the Capitan Limestone, Upper Permian, McKittrick Canyon west Texas, *in* Harris, P. M. , and Grover, G. A. , eds., Subsurface and Outcrop Examination of the Capitan Shelf Margin, Northern Delaware Basin: Tulsa, Society of Economic Paleontologists and Mineralogists Core Workshop 13, p. 387-406.

NEWELL, N. D., 1955, Depositional fabric in Pemian reef limestones: Journal of Geology, v. 63, p. 301-317.

NEWELL, N. D., RIGBY, J. K., FISCHER, A. G., WHITEMAN, A. J., HICKCOX, J. E., and BRADEY, J. S., 1953, The Permian Reef Complex of the Guadalupe Mountains Region, Texas and New Mexico: San Francisco, W. H. Freeman and Company, 236 p.

RIGBY, J. K., and SENOWBARI-DARYAN, B., 1996, Discovery and description of the largest known Permian sponge, Capitan Limestone, Guadalupe Mountains, New Mexico: Journal of Paleontology, v. 70, p. 347-355.

SALLER, A. H., 1996, Differential compaction and tilting of a prograding carbonate shelf margin, Capitan reef complex, Permian, west Texas and southeast New Mexico: Sedimentary Geology, v. 101, p. 21-30.

SCHMIDT, V., 1977, Inorganic and organic growth and subsequent diagenesis in the Permian Capitan reef complex, Guadalupe Mountains, New Mexico, *in* Hileman, M. E., and Mazzullo, S. J., eds., Upper Guadalupian Facies, Permian Reef Complex, Guadalupe Mountains, New Mexico and West Texas, Field Conference Guidebook 77-16: Midland, Society of Economic Paleontologists and Mineralogists, Permian Basin Section, p. 93-132.

SCHOLLE, D. A. L., ULMER, D. S., and MELIM, L. A., 1992, Late stage calcite in the Permian Capitan Formation and its equivalent, Delaware Basin Margin, West Texas and New Mexico: Evidence for replacement of precursor evaporites: Sedimentology, v. 39, p. 207-234.

SENOWBARI-DARYAN, B., 1990, Die systematische Stellung der thalamiden Schwamme und ihre Bedeutung in der Erdgeschichte: Münchner Geowissenschaftliche Abhandlungen, Reihe A, Geologie und Palaontologie, v. 21, p. 1-325.

SENOWBARI-DARYAN, B., and RIGBY, J. K., 1996, Brachiopod mounds not sponge reefs, Permian Capitan-Tansill Formations, Guadalupe Mountains, New Mexico: Journal of Paleontology, v. 70, p. 697-701.

VERMEIJ, G. J., 1977, The Mesozoic Marine Revolution: Evidence from snails, predators and grazers: Paleobiology, v. 3, p. 245-58.

WOOD, R. A., 1995, The changing biology of reef-building: Palaios, v. 10, p. 517-529.

WOOD, R., 1999, Reef Evolution: Oxford, Oxford University Press, 410 p.

WOOD, R. A., DICKSON, J. A. D., and KIRKLAND-GEORGE, B., 1994, Turning the Capitan reef upside down: A new appraisal of the ecology of the Permian Capitan reef, Guadalupe Mountains, Texas and New Mexico: Palaios, v. 9, p. 422-427.

WOOD, R. A., DICKSON, J. A. D., AND KIRKLAND-GEORGE, B., 1996, New observations on the ecology of the Permian Capitan reef, Texas and New Mexico: Palaeontology, v. 39, p. 733-762.

YUREWICZ, D. A., 1976, Sedimentology, paleoecology, and diagenesis of the massive facies of the Lower and Middle Capitan Limestone (Permian), Guadalupe Mountains, New Mexico and Texas: Unpublished Ph.D. Dissertation, University of Wisconsin, Madison, 248 p.

YUREWICZ, D. A., 1977, The origin of the massive facies of the Lower and Middle Capitan Limestone (Permian), Guadalupe Mountains, New Mexico and west Texas, *in* Hileman, M. E., and Mazzullo, S. J., eds., Upper Guadalupian Facies, Permian Reef Complex, Guadalupe Mountains, New Mexico and West Texas, Field Conference Guidebook 77-16: Midland, Society of Economic Paleontologists and Mineralogists, Permian Basin Section, p. 45-92.

INFLUENCE OF SEA-LEVEL CHANGES ON DEVELOPMENT, COMMUNITY STRUCTURE, AND QUANTITATIVE COMPOSITION OF THE UPPER CAPITAN MASSIVE (PERMIAN), GUADALUPE MOUNTAINS, TEXAS AND NEW MEXICO

OLIVER WEIDLICH

Institut für Angewandte Geowissenschaften II, Technische Universität Berlin, Sekr. EB 10, Ernst-Reuter-Platz 1, 10587 Berlin, Germany

AND

J. A. FAGERSTROM

Department of Geology, Oregon State University, Corvallis, Oregon 97331-5506, U.S.A.;

URA CNRS 1453, Centre de Recherches et Observatoire de l'Environment-CRIOBE-EPHE, Moorea, French Polynesia

Abstract: Facies relations determined from outcrop studies and photomosaics have been used in conjunction with reef maps, polished slabs, and microfacies data to better understand the depositional facies, quantitative composition, and control mechanisms of the upper Capitan massive, which differs from the outer shelf in its biotic composition and unbedded nature. A three-stage model of the seaward shelf is established comprising (1) a sponge reef/algal cement/phylloid algal stage, (2) a *Tubiphytes* stage, and (3) prograding cyclic outer shelf beds with isolated reefbuilders. Stage 1 is characterized by progressive shallowing as evidenced by a shift from aggradation to progradation, by changes in boundstone composition, and finally by disintegration of the biogenic framework. Stage 2, dominated by *Tubiphytes obscurus*, bryozoans, microbes, and small reefbuilders, exhibits a lateral zonation triggered by the seaward dipping outer shelf. Stage 3 is composed of cyclic outer shelf grainstones; scattered sponges, and *Tubiphytes* microbial-level bottom communities are restricted to few horizons.

The entire depositional sequence is controlled by a third-order sea-level fall as evidenced by exposure horizons. This sequence is in turn superimposed by three high-frequency cycles. Water was only deep enough for typical Permian reef types during stage 1, which produced the thickest depositional unit. Relative water depth decreased significantly during stage 2, leading to a small, tabular reef constructed by an impoverished fauna. During stage 3, water was too shallow for reef communities. In contrast to modern reefs, shallowing beyond a distinct level in subtidal depth limited reef growth. In addition, salinity fluctuations indicated by cyclic cementation caused the demise of the *Tubiphytes* microbial-level bottom communities.

Our quantitative data from the reef maps suggest a reinterpretation of existing models. The most important constructional element is the microframework (76.7% average coverage), a consortium of low-growing reefbuilders and marine-phreatic cements. In contrast to previous interpretations, neither macro-sponges nor syndepositional cement predominated, making the upper Capitan massive a poor analog for both modern well-skeletonized metazoan and Precambrian cement reefs.

INTRODUCTION

The Capitan reef complex that rims the Delaware basin is an uninterrupted platform-basin transect and therefore an important cornerstone for the interpretation of Permian platform architecture and reef evolution (Fig. 1). Analysis of the various reef facies has resulted in contradictory interpretations despite the presence of excellent natural exposures that highlight depositional geometries and composition of the framework. Since the pioneering work of Newell et al. (1953) and others (summarized in Hileman and Mazzullo, 1977), the Capitan reef has attracted considerable attention, especially with the completion of the Permian Reef Trail in McKittrick Canyon (Bebout and Kerans, 1993). Various groups have concentrated on important aspects of reef research, including sponge taxonomy, paleoecology, sedimentology, and sequence stratigraphy (Kirkland et al., 1993; Wood et al., 1994, 1996; Grotzinger and Knoll, 1995; Noé, 1996; Rigby and Senowbari-Daryan, 1996; Rigby et al., 1998; Weidlich and Fagerstrom, 1998; Fagerstrom and Weidlich, 1998 a, b).

This contribution focuses on the upper Capitan massive (age equivalent to the Tansill Formation on the shelf below the reef), the interval of most prominent reef growth. Based on digitized maps covering 17 m^2 of reef surface near Whites City and resulting quantitative data, the roles of extrinsic and intrinsic control mechanisms as well as the composition and evolution of the reef are discussed. The following topics are crucial: (1) characterization of the upper Capitan massive with quantitative, field-derived data and (2) recognition of lateral or vertical zonations, and (3) consideration of relative sea-level changes as control mechanisms. Sea-level changes are well documented for platform and basin settings, but have not been previously documented for the upper Capitan massive.

GEOLOGICAL SETTING AND PREVIOUS WORK

Outcrops in the Guadalupe and Delaware Mountains have been recognized since Girty (1908) as excellent exposures of the Capitan reef complex and are among the most important localities of marine Permian deposits. The Capitan reef complex rims the Delaware basin and provides an uninterrupted transition from the platform to the reef and the basin with important clues for Permian shelf architecture (Figs. 1A, B, and 2A). The discovery of considerable quantities of oil and gas in the subsurface Permian section substantially increased the interest of geologists from geological surveys, oil industry, and universities in outcrops of Guadalupian sedimentary rocks (Ward et al., 1986).

The Capitan reef complex is generally characterized by persistent growth with significant changes in architecture (Fig. 1B). Carbonate production commenced along a ramp during the lower Guadalupian and evolved into a carbonate platform rimmed by a massive reef tract (King, 1948). Shelf margin boundstones and reef communities exhibit a distinct large-scale

Geologic Framework of the Capitan Reef, SEPM Special Publication No. 65

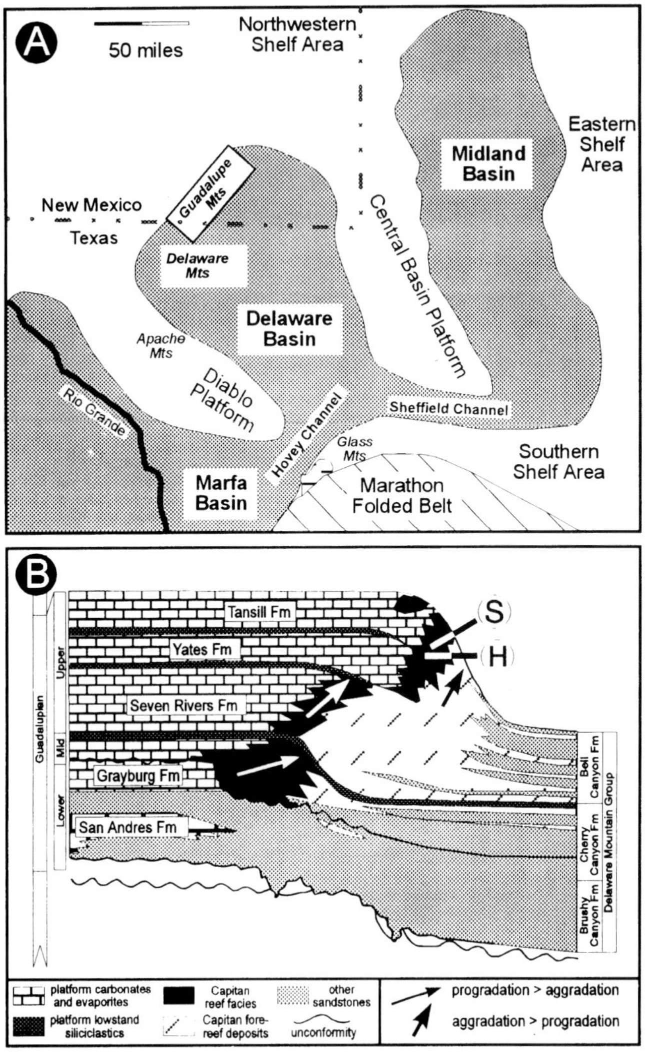

FIG. 1.—Guadalupian platform-to-basin stratigraphy and paleogeography of the Capitan reef complex (modified after King, 1948).

evolutionary history influenced by changes in paleobathymetry and platform architecture from sediment-rich, algal-poor associations in the lower/middle Capitan massive to cement-rich, algal-rich associations in the upper Capitan massive (Babcock, 1974, 1977; Yurewicz, 1976, 1977). The youngest stage, studied by Toomey and Cys (1977), Mazullo et al. (1977), Weidlich et al. (1995), Noé (1996), and Senowbari-Daryan and Rigby (1996b) is characterized by a disintegration of the massive reef tract into small patch reefs.

Several shelf margin carbonate system geometries were proposed and subsequently summarized by Cys et al. (1977) and Bebout et al., (1993). They include an uninterupted slope from shelf to basin (King, 1948), a shallow water shelf-margin barrier reef (Newell et al., 1953), a marginal mound with a shallow landward pisolite shelf crest (Lang, 1937; Dunham, 1972), and a shelf margin barrier reef with a landward pisolite shelf (Kirkland-George, 1992). From the proposed paleobathymetric shelf profiles, the marginal mound hypothesis gathered wide acceptance for the upper Capitan massive. The most important aspect of this model is the fall-in bed geometry, that is, the seaward dip of outer shelf beds from the pisolite crest to the reef rim. Recently, however, Saller (1996) described inner to outer shelf dips between 8–10° and attributed them to differential compaction and basinward tilting rather than syndepositional dip of the shelf surface.

Water depth also decreased during the evolution of the Capitan reef complex from a depth of 45 m at the crest during lower Capitan deposition (Seven Rivers equivalent) to 10 m during Upper Capitan (Tansill age equivalent), Babcock and Yurewicz (1989).

According to the fusulinacean zonation of Ross and Ross (1987, 1994) and estimates of Garber et al. (1989), duration of the Capitan shelf margin deposition was only about 2.5 m.y., comprising a 1.5-m.y. phase of progradation during the deposition of the Seven Rivers Formation and a 1.0 m.y. phase for aggradation and steepening of the margin during the depositon of the Yates and Tansill formations. Different orders of orbitally forced, eustatic sea-level fluctuations affected the shelf profile. The cyclic sea-level chart for the Guadalupian of Ross and Ross (1987) points to two third-order sea-level cycles for the Capitanian interval with the Seven Rivers phase below and the Yates-Tansill phase above. This agrees with the two-phase model established by Garber et al. (1989) and modifies the assumption of equal rates of progradation for the Capitan Limestone proposed by Tyrrell (1969) and subsequent researchers. Evidence for higher frequency sea-level changes is obvious in different parts of the Capitan reef complex and the adjacent Delaware Basin (e.g., Borer and Harris 1989, 1991a, b; Garber et al., 1989).

Esteban and Pray (1977) were among the first to provide detailed field descriptions and an informal terminology for outcrops of upper Yates and lowermost Tansill formations, which formed the basis for further research. It was later observed that Capitan back-reef strata of the upper Yates and lowermost Tansill formations are characterized by cycles composed of a lower siliciclastic unit and an upper carbonate unit. The cycles are bounded by sharp and erosive surfaces on the shelf that disappear towards the outer-shelf and the reef. Breccias are occasionally associated with cycle boundaries (Borer and Harris, 1991 a, b; Mutti and Simo, 1993). Borer and Harris (1989) discussed the cyclicity in the Yates formation and interpreted them as Milankovitch cycles. Neese (1989) described the peritidal facies of the Guadalupian shelf crest at Walnut Canyon and recognized significant facies changes in a direction perpendicular to the depositional strike. According to descriptions of Kerans and Harris (1993), the outer shelf facies is composed of vertically stacked small-scale, upward-coarsening cycles. Silt beds on the shelf that typically mark high-frequency cycles were used for a further subdivison of the Yates Formation (Newell et al., 1953; Candelaria, 1989; Borer and Harris, 1991 a,b).

The combination of brecciation, erosional relief, and karstification in the back-reef deposits suggests that the position of the reef to outer shelf areas underwent repeated subaerial exposure. Low-amplitude, high-frequency oscillations of relative sea-level are apparent throughout the Permian shelf-top strata except for the reef zone. As a consequence, most researchers

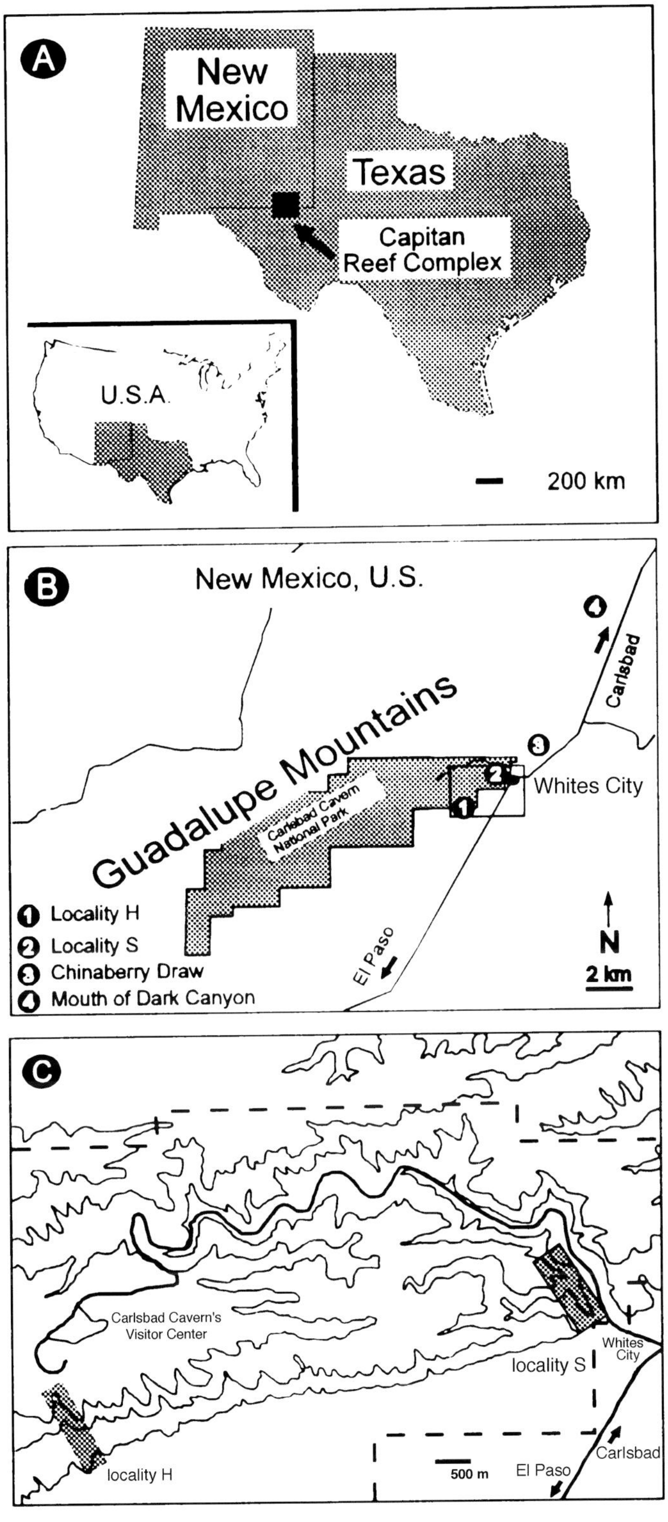

FIG. 2.—Location of the Capitan reef complex. (A) Location of the Capitan reef complex in southeastern New Mexico and western Texas. (B) Distribution of localities; 1 = locality H, Hackberry Draw; 2 = locality S, mouths of Bat Cave Draw and Walnut Canyon; 3 = Chinaberry Draw; 4 = mouth of Dark Canyon; coordinates were presented by Rigby et al. (1998). (C) Topographic map of the study area with locality S (mouth of Walnut Canyon, mouth of Bat Cave Draw) and locality H (mouth of Hackberry Draw).

now question a subaerial exposure of the reef during lowstands (see Mazzullo, 1995 for discussion), but give no evidence from detailed field work (e.g., Babcock, 1977; Yurewicz, 1977; Kirkland et al., 1993).

STUDY AREAS AND DATA BASE

Study Areas

Although the northeast-southwest-trending Guadalupe Mountains escarpment is 50 km long and includes such famous landmarks as Guadalupe Peak and El Capitan in the south, most outcrop research has concentrated on canyons and draws transecting the southeast-facing escarpment (Fig. 2B). This situation resulted in numerous intensely studied localities, beginning with King (1948) and including McKittrick, Slaughter, Rattlesnake, Walnut, and Dark Canyons. Recently, additional reef data have come from Sheep Draw (e.g., Noé, 1996) and Hackberry Draw (Rigby et al., 1998). Our research concentrated on mouth of Bat Cave Draw and Hackberry Draw (Fig. 2C).

Locality S: Cliff between mouth of Bat Cave Draw and mouth of Walnut Canyon at the entrance to Carlsbad Caverns National Park near Whites City, New Mexico.—

This show-and-tell locality has been a cornerstone for many excursions to the Capitan Reef, especially since the 1970s (e.g., Dunham, 1972, Stop 2-1; Babcock et al., 1977, Stop 1; Toomey and Babcock, 1983, Stop 10; Reid et al., 1988, Stop 4-1) and has been discussed as important locality of Permian reefs and their communities (e.g., James, 1983, Figs. 153, 154, and 157; Fagerstrom, 1987, Plates 1b, 29b, 33 44b, 45, 46; Senowbari-Daryan, 1990, Plate 55).

Locality S forms the spur between the mouths of Bat Cave Draw and Walnut Canyon near Whites City, New Mexico, Figs 3B and 3C. It is easily accessible and its partly acid-cleaned outcrop surfaces on the southwest-facing bluff allow detailed study of the framework and reefbuilders.

Locality H: Southwest-facing slope in Hackberry Draw.—

This locality was recently investigated by Rigby et al. (1998) and J. A. Fagerstrom and O. Weidlich (1998). The site is situated directly south of and below the Carlsbad Caverns Visitor Center and is accessible by a rough road from Whites City.

The complete vertical transect at locality H provided important data for the interpretation of the evolution of the upper Capitan massive (Fig. 3A).

Data Base

The field-oriented quantitative methodology of modern reef research was adapted to the upper Capitan massive following Weidlich et al. (1993).

Data evaluation.—

In decreasing order of area, our study comprises five data sets: (1) Facies boundaries and depositional geometries were

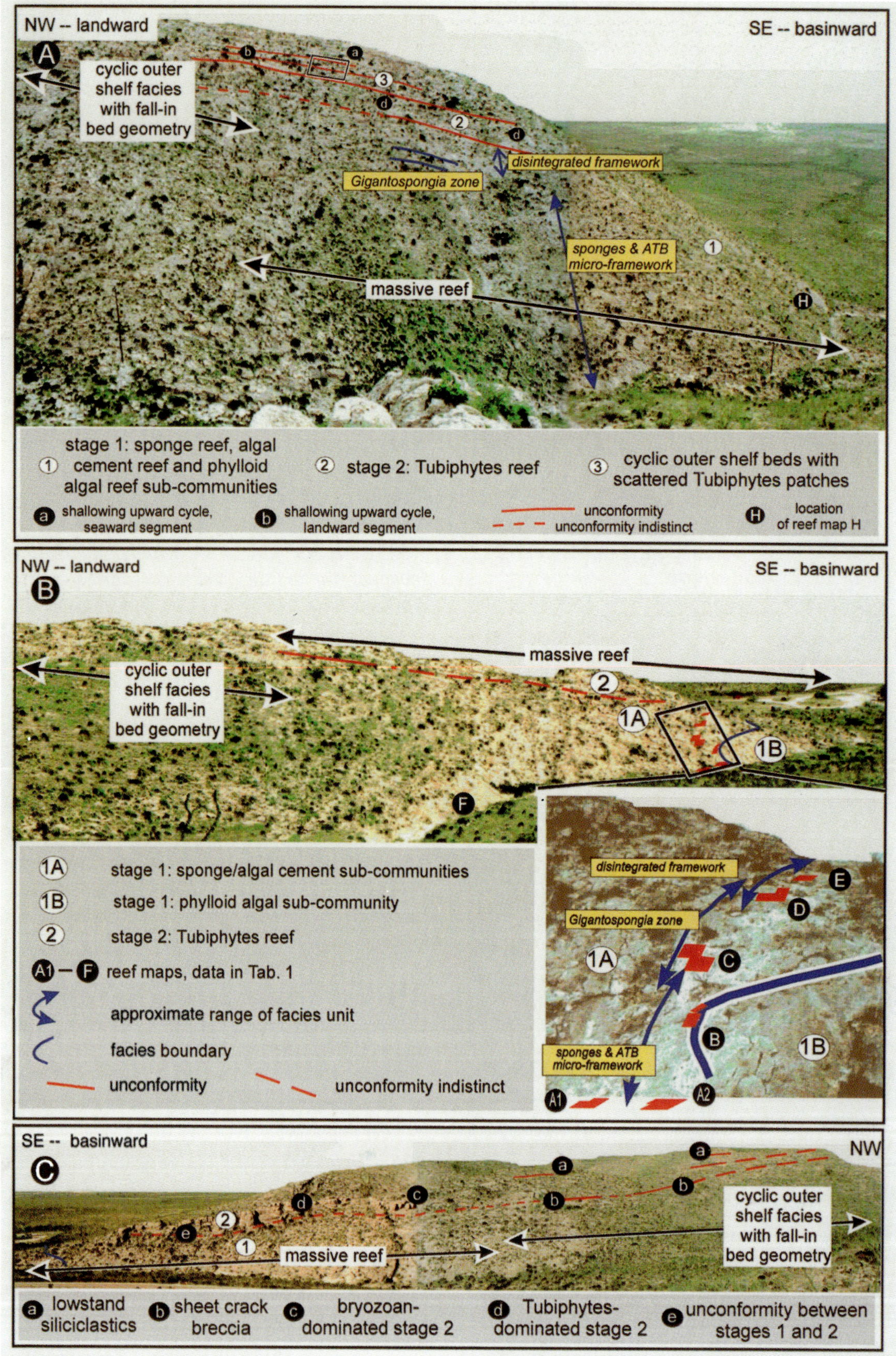

FIG. 3.—Photomosaics of the study areas. (A) Northeastern wall of locality H (Hackberry Draw) showing complete succession of the upper Capitan massive from stage 1 to 3. (B) Northeastern wall of locality S (mouth of Bat Cave Draw) showing depositional geometries, facies distribution, and location of the reef maps. Length of transect is about 1000 m. Note progradation of stage 1A over 1B. (C) Southwestern wall of locality S (mouth of Walnut Canyon) showing fall-in bed geometry and unconformities. Mixed siliciclastic/carbonate cycles exposed on the shelf. Length of transect is about 800 m.

mapped in photomosaics; these data were derived predominantly from locality H and only subordinately from S. (2) Based on these field studies, a selection of acid-cleaned surfaces was chosen to map the reef fabric on acetate sheets in natural scale. The total mapped area is 15 m^2 and consists of individual 40 cm x 40 cm quadrats. Locality S comprises seven digitized maps covering 14.3 m^2 (six of them are arranged in a vertical transect, Figs. 3B, 4) and locality H provided one reef map with 0.7 m^2. Criteria and measurement parameters for the quantitative data set will be discussed in the next chapter. (3) Detailed field photographs complement the data set. Reef maps were photographed in total and every quadrat was documented with one to four pictures depending on the complexity of the framework. (4) A total of 105 polished quadrat slabs (averaging 18 cm x 18 cm) were collected from locations 1 to 4 (see Fig. 2C) using a stratified random procedure to characterize the community structure and taphonomy of the sponges. Sampling technique, locations, and interpretation of the data are described by J. A. Fagerstrom and Weidlich (1998 a,b). (5) About 250 thin-sections (7.5 x 10 cm; 6 x 9 cm; 5 x 5 cm) were used for microfacies analysis to supplement data from the field and polished quadrat samples.

Data processing and management.—

Color prints of the reef maps were used to (1) monitor the evolution of the upper Capitan massive and to (2) provide the quantitive data set for the reconstruction of the areal coverage of major reef-building components of the upper Capitan massive through Tansill times. The reef maps were digitized with a color video camera and for each 40 cm x 40 cm quadrat, the quantitative analysis was performed using the image anylsis system Lucia M (Nikon). The areal coverage data in Tables 1 and 2 were grouped into (i) macro-reefbuilders at genus or species levels (eight criteria), (ii) the type of micro-framework (two criteria), (iii) the type of internal sediment (three criteria), and (iv) pore-filling marine-phreatic cements (Weidlich and Fagerstrom, 1998).

To better understand the proposed categories, the terminology is briefly explained. "Macro-reefbuilders" are organsims that were identified and mapped in the field (diameter > 1 cm). The term "micro-framework" describes the complex consortium of low-growing organisms (e.g., *Tubiphytes* [= *Shamovella*, Riding, 1993]), *Archaeolithoporella*, small sponges, and bryozoans) and syndepositional marine-phreatic cement. Most components of the micro-framework can only be recognized in polished slabs or thin-sections. The "internal sediment" and "syndepositional cement" consist of finely occluded growth framework pores filled after the accretion of macro-reefbuilders and micro-framework. All categories are discussed in detail by Weidlich and Fagerstrom (1998).

FACIES DESCRIPTION AND INTERPRETATION

A three-stage model for the seaward shelf of the upper Capitan massive is established at locations S and H based on analysis of photo-mosaics, reef maps, quantitative data, and thin-sections (Fig. 3).

Stage 1: Sponge Reef, Algal Cement Reef, and Phylloid Algal Reef Subcommunities

Depositional geometries and size.—

The exact length of the upper Capitan massive can be traced best at locality S along the road from Whites City to the visitor center of the Carlsbad Caverns National Park. The distance from the boundstones near Whites City to the first occurrence of transitional bedded, cyclic outer shelf facies at the entrance to the National Park is about 700 m. Locality H provides data for the reef thickness.

Depositional geometries in the massive reef were reconstructed along subcommunity boundaries. Previously, Babcock et al. (1977, Fig. 1-3 A) differentiated at locality S several facies types including "phylloid algal boundstones," and "*Tubiphytes* sponge boundstone/packstone-*Archaeolithoporella*/nodular boundstones." The boundary between both subcommunities shifts seaward, indicating a change in the depositional geometries from aggradation to progradation at least for the seaward part of the platform rim (Fig. 3B). The dominance of sponge/algal cement subcommunities and the absence of phylloid algal boundstones impeded an interpretation of the depositional geometries in the massive reef at locality H.

Reefbuilders and guild structure.—

The taxonomic composition, abundance, and guild structure of the reefbuilding organisms of the upper Capitan massive were recently investigated by Rigby et al. (1998) and J. A. Fagerstrom and O. Weidlich (1998). Thirty-four species of reefbuilders were identified, including 17 calcisponges, nine bryozoans, one richthofenid brachiopod, the phylloid alga *Eugonophyllum* sp., four problematica, and crinoids (J. A. Fagerstrom and O. Weidlich, 1998, Table 1, Fig. 4). Based on the polished slabs of quadrat samples, sponge abundance was determined; the predominant taxa are *Minispongia constricta, Sollasia ostiolata, Parauvanella minima, Lemonea micra*, and *Lemonea cylindrica*. Only members of the reef-building guilds *sensu* Fagerstrom (1987, 1991) were considered for this calculation. Most species belong to either the constructor or the baffler guilds. Members of the constructor guild comprise sponges (for example, *Lemonea conica, L. cylindrica*, and *Minispongia constricta*), the richthofenid brachiopod *Sestropoma*(?), and the problematicum *Permosoma* sp. Important members of the baffler guild are bryozoans, most specimens of *Tubiphytes obscurus*, the phylloid alga *Eugonophyllum* sp., and sponges (e.g., *Sollasia ostiolata, Girtyocoelia* spp., and *Discosiphonella mammilosa*). The binder guild is represented chiefly by *Archaeolithoporella hidensis*. Rare "reefbuilders" like solitary Rugosa, *Lercaritubus problematicus* (Senowbari-Daryan and Rigby, 1996a) as well as problematic algae and microbes (*Parachaetetes, Hedstroemia*, and *Collenella*) were assigned to the dweller guild and therefore did not contribute significantly to reef-building.

Composition and zonation.—

The zonation at locality S is based on excellently cleaned outcrop surfaces (Figs. 3B and 4). The same general pattern was also recognized at locality H, especially zones 1 (sponges and ATB

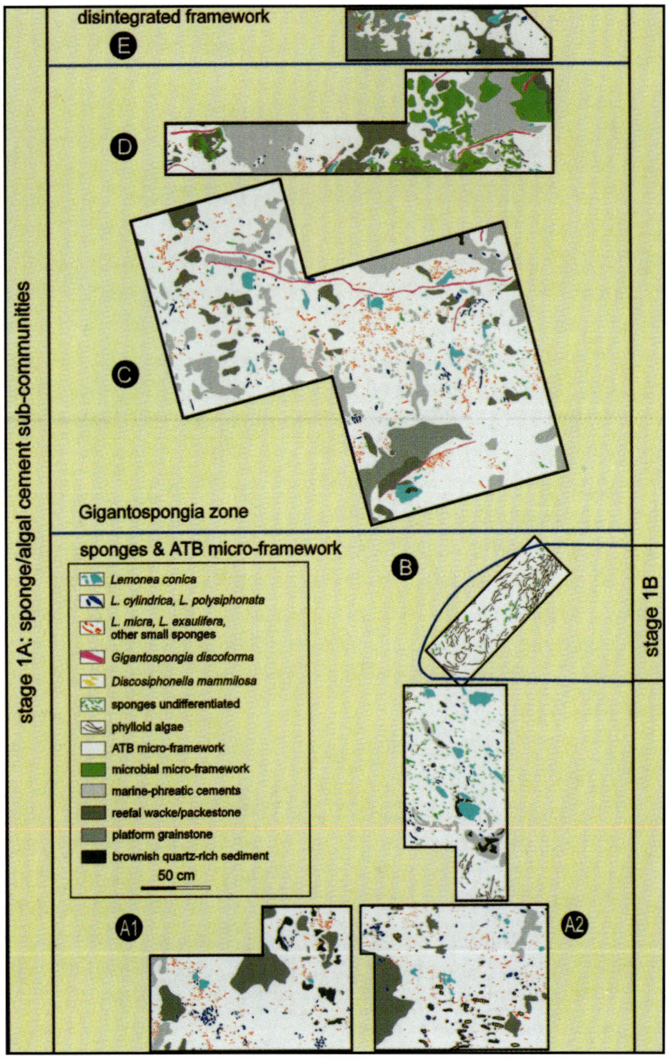

FIG. 4.—Distribution of reef components of stage 1 based on reef maps from locality S, see Fig. 3B for overview. Resulting quantitative data in Tables 1 and 2.

micro-framework) and 3 (disintegrated framework). The intervening *Gigantospongia* zone forms only a thin unit. Stage 1 is constructed by cosmopolitan reef types defined by Flügel (1994), including sponge reefs, algal cement reefs, and phylloid algal reefs. Considerable variation in the composition and abundance of the constructional elements (macro-reefbuilders, micro-framework, internal sediments, and syndepositional cement) is displayed within these reef types:

(1) Zone with sponges and ATB (Archaeolithoporella-Tubiphytes-botryoid) micro-framework. — Zone 1 reef maps (H, F, A1, A2, and B) are dominated by ATB micro-framework. Important macro-reefbuilders are *Lemonea polysiphonata, L. cylindrica, L. conica*, as well as small-branched species of *Lemonea* (reef maps H, A1, and B); subordinately, *Discosiphonella mammilosa* (map A2) and *Guadalupia explanata* (map H) occur. The sponges are either in growth position as evidenced by erect branch clusters (maps F, A1) and horizontal sheets (map H) or toppled (maps A2, B). Transport of fragments by platform-to-basin currents was not observed. Reef map B shows interfingering of phylloid algal and sponge-algal/cement subcommunities with a sharp, nonerosional contact. Diversity and areal coverage of the phylloid algal boundstones are comparatively low. Both the *Eugonophyllum* sp. and large sponges show an alignment of their elongate skeletons. Phylloid thalli are oriented perpendicular to platform-to-basin currents and sponges parallel to them. The phenomenon may be explained either by current oriented growth of reefbuilders or post-mortem alignment of the toppled sponges.

Growth framework pores up to 50 cm in diameter were filled with reef wackestone/packstone, smaller pores contain brownish quartz-rich sediment or cements; platform grainstones are absent (Weidlich and Fagerstrom, 1998).

(2) Gigantospongia zone. — Zone 2 (reef maps C and D) is dominated by the sheet-like inozoid sponge *Gigantospongia discoforma* spanning up to 2.5 m of outcrop surface (Rigby and Senowbari-Daryan, 1996). Reef pores up to 2 m in diameter contain reef wackestones/packstones, platform grainstones, and brownish quartz-rich sediment. In contrast to zone 1, the occlusion of growth framework pores reveals a complex pattern of sedimentation and cementation. Abundance and size of the framework pores suggests a very irregular reef surface during this stage of growth. This zone contains abundant, partly large crypts recognized by Wood et al. (1994, 1996) and Rigby and Senowbari-Daryan (1996). Low-growing organisms, chiefly *Tubiphytes* and *Archaeolithoporella*, and marine-phreatic cements dominate the micro-framework; microbialites are not important. The elongate axis of large growth framework pores, lateral extent of *Gigantospongia explanata*, and basinward growth of pendant and erect sponges provide evidence for strong progradation of this zone (Weidlich and Fagerstrom, 1998, Fig. 8).

(3) Zone of framework disintegration. — The youngest part of the sponge/algal-cement reef succession of reef stage 1 is characterized by an important change in reef architecture. In contrast to the previous zones, the microbial micro-framework is abundant. This change is accompanied with an increase in bioclastic sediment relative to the other zones. The internal sediments are predominantly grainstones exported from the platform. These changes are indicative of disintegration of the framework.

FIG. 5.—Photomicrographs of stage 2, *Tubiphytes* reef; (A)-(D) micro-framework composition of the landward segment, (E)-(F) micro-framework composition of the seaward segment. (A) Bryozoan encrusted by *Archaeolithoporella hidensis*; growth framework pores partly filled with microbial carbonate; scale = 2.5 mm; thin section IX 2. (B) Complex micro-framework with several recrystallized sponge (S) and bryozoans (B); note microbial carbonate (M); scale = 2.5 mm; thin section IX 1. (C) Pseudopunctate brachiopod attacked by microborers of unknown affinity (arrows); note microbial carbonate (M); scale = 2.5 mm; thin section H 5.2. (D) Growth framework pore rimmed by radiaxial fibrous calcite (1) and filled with calcite spar (2); note microbial carbonate; scale = 2.5 mm; thin section H 5.2. (E) Recrystallized small botryoids (B) are encrusted by *Archaeolithoporella hidensis* and *Tubiyphytes obscurus*; scale = 2.5 mm; thin section X 1. (F) *Tubiphytes obscurus* with irregular base (arrows) grown upon marine-phreatic cements; scale = 2.5; thin section X 1. (G) ATB micro-framework with recrystallized botryoids (1), *Tubiphytes obscurus* (2), and *Archaeolithoporella hidensis* (3); note microbial carbonate; scale = 2.5; thin-section X 1. (H), (I) grainstone with peloidal, filamentous cynobacteria filling open pores in the micro-framework; scale = 1.0 mm; thin section IX 2.

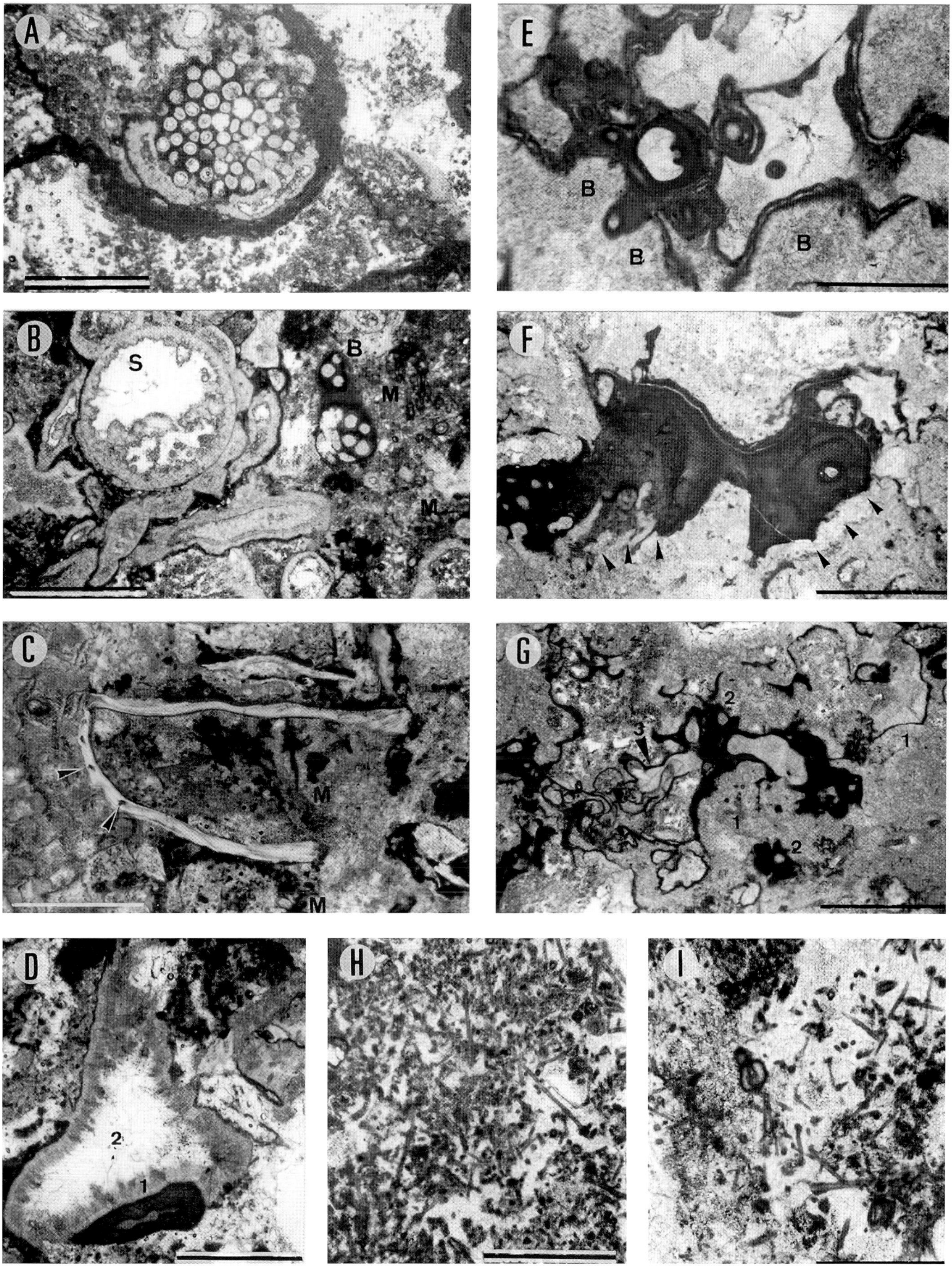
A
E
B
B
B
B
S
B
M
M
F
C
M
M
G
2
3
1
1
2
D
2
1
H
I

Interpretation.—

We suggest the following interpretation for reef stage 1 and modification of the hypothesis that the upper Capitan massive was "a period of minor progradation but of pronounced aggradation" (e.g., Garber et al., 1989). The boundary between the phylloid algal and the sponge/algal cement subcommunities at locality S gives evidence of a change from aggradation to progradation in the younger part of the upper Capitan massive. The lack of erosion at the boundary indicates the contemporaneous occurrence of both subcommunities. Progradation of the reef front is even better documented by the reef maps. Zone 1 (sponges and ATB micro-framework) with numerous sponges in erect growth position is typical for an aggrading subcommunity. By contrast, the elongate *Gigantospongia discoforma* of zone 2, basinward growth of attached sponges, and numerous, basinward-extending reef cavities partly filled with grainstones shed from the platform suggest progradation.

The succession from ATB micro-framework with sponges (zone 1) to the *Gigantospongia* zone (zone 2) and to the disintegrated reef (zone 3) indicates shallowing and is accompanied with the change from aggrading to prograding reef communities. Decreasing depth is also reflected by changes in the composition of internal sediments in growth framework pores. While reef wackestones dominate in zone 1, "shallow" platform grainstones are increasingly abundant in zones 2 and 3. Keeping in mind that (1) the Capitan reef flourished in deeper water than modern reefs (marginal mound hypothesis of Lang, 1937) and that (2) Permian reefbuilders probably lacked algal symbionts, we suggest that shallowing was the chief factor causing a reduction in the carbonate productivity of the reef community as indicated in zone 3 (disintegrated framework).

The reef maps also modify the hypothesis of Wood et al. (1994, 1996). Large reef cavities were not ubiquitous throughout the upper Capitan massive and were here restricted to a relatively small interval, that is, the *Gigantospongia* zone.

Stage 2: Tubiphytes Reef and Lateral Transition

Reef stage 2 is exposed at localities S and H. Locality S is important for the interpretation of the lateral variation and locality H for determining the thickness of the stage.

Depositional geometries and size based on outcrop data.—

The measured length of stage 2 at locality S is 700 m. At locality H the thickness is only 20 m; at locality S the unit was removed by post-uplift erosion and only lateral equivalents are exposed. This unit forms a tabular body of massive, unstratified boundstones that lack facies boundaries needed to interpret the depositional geometries.

Reefbuilders and guild structure.—

Stage 2 is characterized by dominance of low-growing organsims, including *Tubiphytes obscurus* (Fig. 5D-G), bryozoans (*Acanthocladia* sp., *Goniocladia* sp, and encrusting taxa, Fig. 5A, B), small sponges (Fig. 5B), filamentous cyanobacteria similar to *Orthonella* (Fig. 5H, I), and microbialites (Fig. 5A, B). The dominance of *Tubiphytes obscurus* was already recognized by Babcock (1977, Fig. I-3 A), who described this unit as *Tubiphytes* boundstone. In addition, pseudopunctate brachiopods were attacked by bioeroders (Fig. 5C). *Archaeolithoporella hidensis* is present but not abundant (Fig. 5E). Macro-reefbuilders are absent at locality S except for some specimens in isolated patches. At locality H, solitary Rugosa in growth position were found with toppled sponges (*Lemonea* sp.).

Reefbuilder diversity is moderate as compared to the previous reef stage due to the absence of macro-sponges. The guild structure is characterized by an impoverished constructor guild (only small and well-skeletonized sponge taxa like *Minispongia constricta*) and dominance of the baffler guild. Many specimens of *Tubiphytes obscurus* and bryozoans are assigned to the baffler guild or — as a consequence of morphologic plasticity — to the binder guild.

The taphonomic history is difficult to reconstruct. Field and thin-section data suggest that a few reefbuilders are oriented upside-down, but do not indicate the presence of "ubiquitous" cryptic habitats. The absence of reefbuilder attachments to the substrate and absence of sedimentological criteria indicative of crypts suggest that many specimens are toppled rather than in growth position.

Composition and zonation.—

Due to the absence of acid-cleaned windows in stage 2, neither reef maps nor quantitative data were evaluated. Such data are not crucial in this case, because of the rarity or absence of large reefbuilding components; that is, macro-reefbuilders or growth framework pores. However, microfacies analysis of thin sections indicates the presence of abundant ATB micro-framework and microbial micro-framework. Macro-reefbuilders, internal sediments, and syndepositional cements are rare. An enrichment in peloidal sediment or microbial micrite is characteristic of stage 2. Complex intergrowth of *Tubiphytes*, bryozoans, and marine-phreatic cements are present. In contrast to stage 1, thin-section observations suggest that synsedimentary diagenesis is dominated by isopachous fibrous and radiaxial fibrous cement, while botryoids are rare.

FIG. 6.—Photomicrographs of lowstand sediments; (A)-(C) karst breccia, (E)-(G) lowstand siliciclastics, (D), (H), (I) karst phenomena. (A) calcite spar of karst breccia; note crystal corrosion in the upper left of photomicrograph (arrows); scale = 1.0 mm; thin section III 2.2. (B) Lithoclastic rudstone from karst breccia with pisolite clast (1) and dasycladacean fragment (2), scale = 2.5 mm; thin section III 2.2. (C) Corroded calcite spar of karst breccia. Fitting of some "clasts" suggests that no transport occurred (arrows); scale = 1.0 mm; thin section III 2.2. (D) Incipient karst dissolution of outer shelf dasycladacean grainstone beginning from interparticle porosity (arrows); scale = 2.5 mm; thin section III 3.1. (E) Photomicrograph showing typical textures of the brownish silt- and sandstones. Lamination resulted from changes in grain-size and packing density; scale = 2.5 mm; thin section I 2.2 A. (F) Marine depositional environment, probably developed during the next transgression, is indicated by "oncoids" and bioclasts; scale = 2.5 mm; thin section I 2.2 A. (G) Hypersaline environment of the lowstand siliciclastics is suggested by intense cementation; scale = 2.5 mm; thin section II 2 A. (H) Outer shelf dasycladacean grainstone with distinct karst cavity cemented by blocky calcite; scale = 2.5 mm; thin section H 2.2. (I) Outer shelf wackestone affected by karst dissolution; scale = 2.5 mm; thin section III 1.

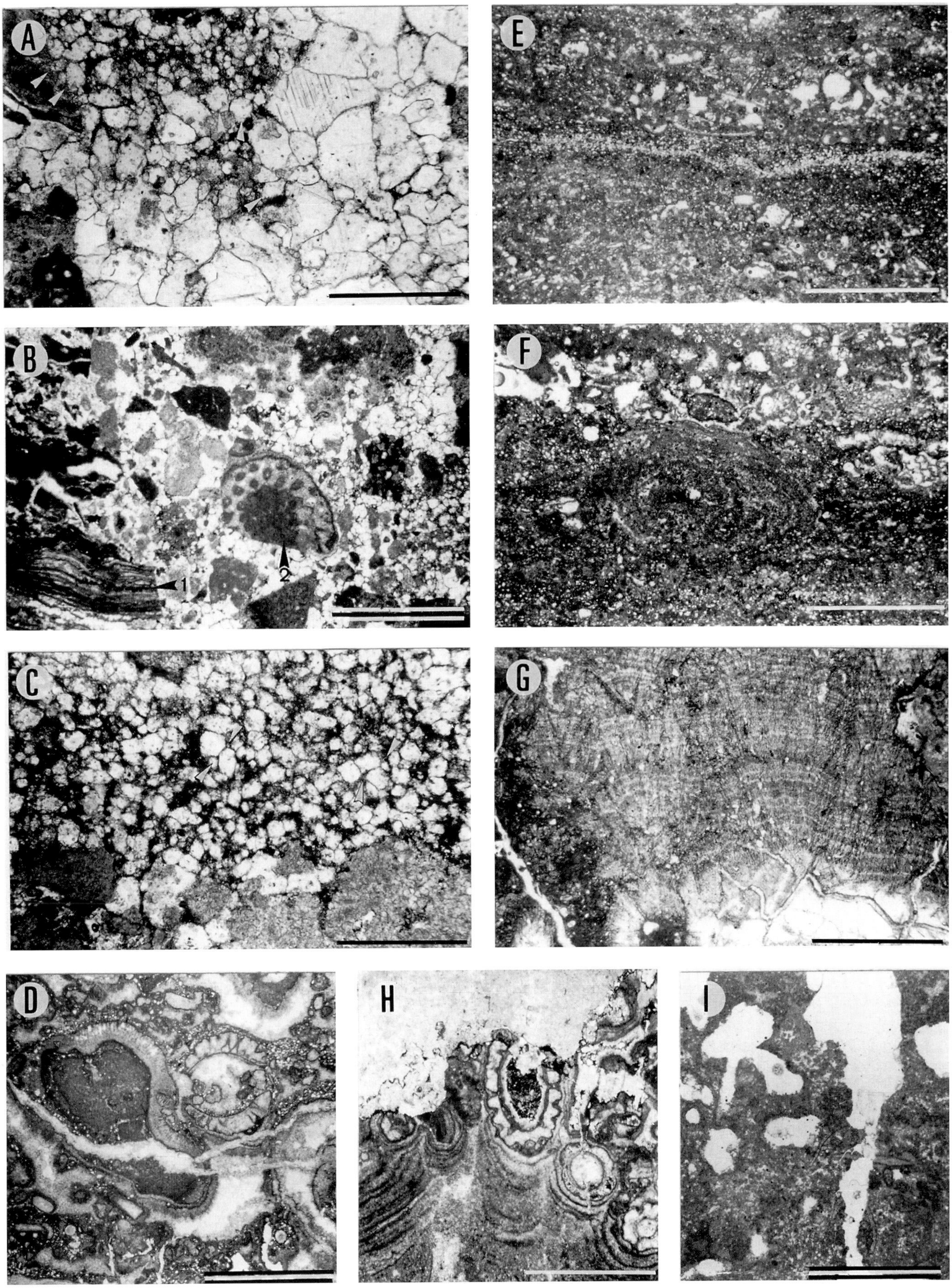
A
E
B
1
2
F
C
G
D
H
I

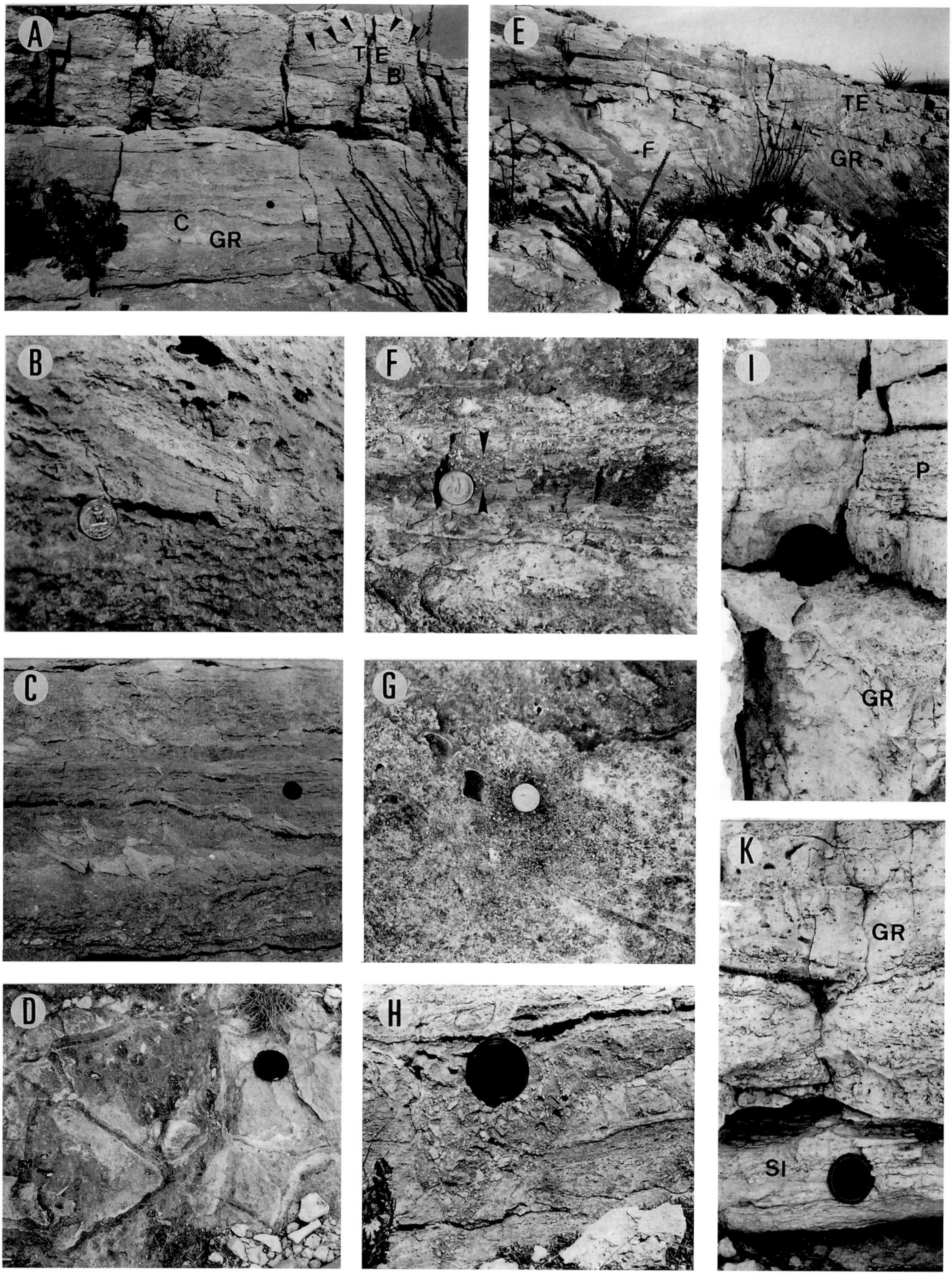
A
TE
B
C
GR
E
TE
F
GR
B
F
I
P
GR
C
G
K
GR
D
H
SI

No vertical zonation was observed, but a lateral zonation — following the fall-in bed geometry — is present at both localities. In the seaward segment, the community is dominated by *Tubiphytes obscurus* passing landward to bryozoan subcommunities dominated by *Acanthocladia*.

Interpretation.—

Moderate thickness and great lateral extension suggest that reef stage 2 grew during a short time interval. Progradation of the reef front is likely, but was not confirmed. The dominance of *Tubiphytes* and bryozoans in conjunction with the absence of macro-reefbuilders is unusual and resembles neither reef stage 1 nor other Permian reefs of the Tethys. Therefore, it is not part of the reef classification of Flügel (1994). This enigmatic reef type hardly resulted from autoecological processes alone. Our data suggest a low-amplitude sea-level rise compared to that of reef stage 1. The transgression was short and relative water depth too little to establish a typical Permian reef community. The lateral zonation might be either a function of water depth or salinity. Diversity of bryozoans, however, suggests normal salinity favoring water depth as a control mechanism.

The Unconformity Separating Stages 1 and 2

Description.—

Exposure horizons as evidenced by lowstand siliciclastics, sheet-crack breccias, and meteoric diagenesis are well-known from the shelf crest (e.g., Esteban and Pray, 1977) and the outer shelf (e.g., Kerans and Harris, 1993; Noé, 1996), but could not be traced downslope from the outer shelf into the reef.

The landward segment of the outer shelf at locality S is composed of thick bedded, partly cyclic carbonates comparable to the facies reported by Kerans and Harris (1993) from McKittrick Canyon. Small-scale, coarsening-upward cycles develop from bioclastic grainstones to partly laminated, fenestral packstones. Dolomitic silt- to sandstones occur sporadically at the cycle bases. The siliciclastic input was generally low to moderate at localities S and H and the silt- to sandstone beds pinch out seaward. The last seaward counterparts of the siliciclastics occur approximately 100 m landward of the upper Capitan massive (Fig. 3C, a) as fenestral, laminated wackestones with quartz silt.

Even more striking is a polymict breccia exposed below the siliciclastic unit (Fig. 3C, b), which cuts erosionally into the underlying outer-shelf grainstones. It contains clasts up to 10 cm in diameter derived from the outer shelf and the shelf crest; reef-derived clasts are absent. Microfacies types of the clasts consist of pisolites, mudstones with birds eyes, laminated mudstones, peloidal wackestones, and dasycladacean wackestones (Fig. 6B). One clast yields fecal pellets probably belonging to *Favreina* (Senowbari-Daryan, pers. comm., 1998). Fabric selective dolomitization of the clasts prior to erosion also is observed. The breccia fabric is grain-supported and cemented exclusively by blocky spar without marine-phreatic cements (Fig. 6A). Sand-sized quartz grains indicative of sediment transport during lowstand were partly incorporated into the spar. Calcite spar was subsequently replaced inward from the crystal boundaries by dolomite and other minerals, probably during a second stage of karstification. This phenomenon is the first step of calcite crystal corrosion, which occurs in several stages from unaltered cement to sand with coated crystals and finally fills fissures and growth framework pores as youngest generation (Fig. 6A, C; brownish quartz-rich sediment, see Weidlich and Fagerstrom, 1998, Plate 4, Figs. 8-10).

Interpretation.—

Dolomitic silt- to sandstones and thick sheet-crack breccia record significant shallowing or even subaerial exposure. The siliciclastic sediments described from the outer shelf cannot be traced into the reef because they pinch out further landward. Equivalents of the lowstand siliciclastics are preserved in reef pores and fissures in the upper Capitan massive. The material was described by Weidlich and Fagerstrom (1998) as "brownish quartz-rich sediment." The composition is heterogenous with variable amounts of detrital quartz grains and crystal silt from calcite spar. This siliciclastic sediment, the youngest generation of growth framework pore-fills, is restricted to reef stage 1. We interpret it as sediment transported from the clastic inner-shelf towards the basin during sea-level lowstand. It occurs in the upper Capitan massive either in small pockets or as channel-fills (light gray dolomitic siltstone/sandstone of Garber et al., 1989, Fig. 50 G, J) and forms thick downslope wedges in the Delaware basin.

The breccia at the landward segment of locality S clearly indicates lowstand of sea-level due to the erosive character and abundant meteoric spar. The absence of reef material suggests a unidirectional transport from the shelf crest toward the platform rim. The polymict composition and large sizes of clasts indicate transport over short distances and thus longer subaerial exposure as compared to the sheet-crack breccias. Downslope projection of the unconformity correlates with the boundary separating reef stages 1 and 2 at localitiy S. Our data suggest that the breccia represents a third- or fourth-order lowstand that never

FIG. 7.—Field photographs of cyclic outer shelf facies. (A)-(C) Landward shallowing-upward cycle; (E)-(G) seaward shallowing upward cycle; (I)-(K) small scale mixed siliciclastic-carbonate cycle. (A) Shallowing upward cycle passing from stratified grainstones (GR) into incipient tepee structure (TE). Tepee marked by arrows. (B) and (C) = position of close-up views figured below. Exact position of field photograph is location (b) of Fig. 5. Cap lense = 5.5 cm. (B) Close-up view of tepee structure; see coin (quarter) for scale. (C) Close-up view of cycle base; thin-bedded outer shelf grainstone with rip-up clasts probably indicating minor sea-level fluctuations. (D) surface view of a sheet crack breccia related to sea-level lowstand; cap lens = 5.5 cm. (E) Shallowing-upward cycle passing from stratified grainstones (GR) into tepee structure (TE). (F) = position of close-up views mentioned below. Exact position of field photograph is location (a) of Fig. 5. (F) Close-up view stratified outer shelf grainstone with *Tubiphytes* microbial level-bottom community (arrows) of Fig. 11; see coin (quarter) for scale. (G) Close-up view of underlying dasycladacean grainstone; see coin (quarter) for scale. (H) Small-scale incised valley related to sea-level lowstand. (I) Upper part of a mixed siliciclastic carbonate cycle with fenestral, partly laminated packstones (P) on-top of bioclastic grainstones (GR). Location corresponds to (a) of Fig. 3.; cap lens = 5.5 cm. (K) Lower part of a mixed siliciclastic carbonate cycle with bioclastic grainstones (GR) on top of siliciclastic silt-to sandstone (SI). Location corresponds to (a) of Fig. 3.; cap lens = 5.5 cm.

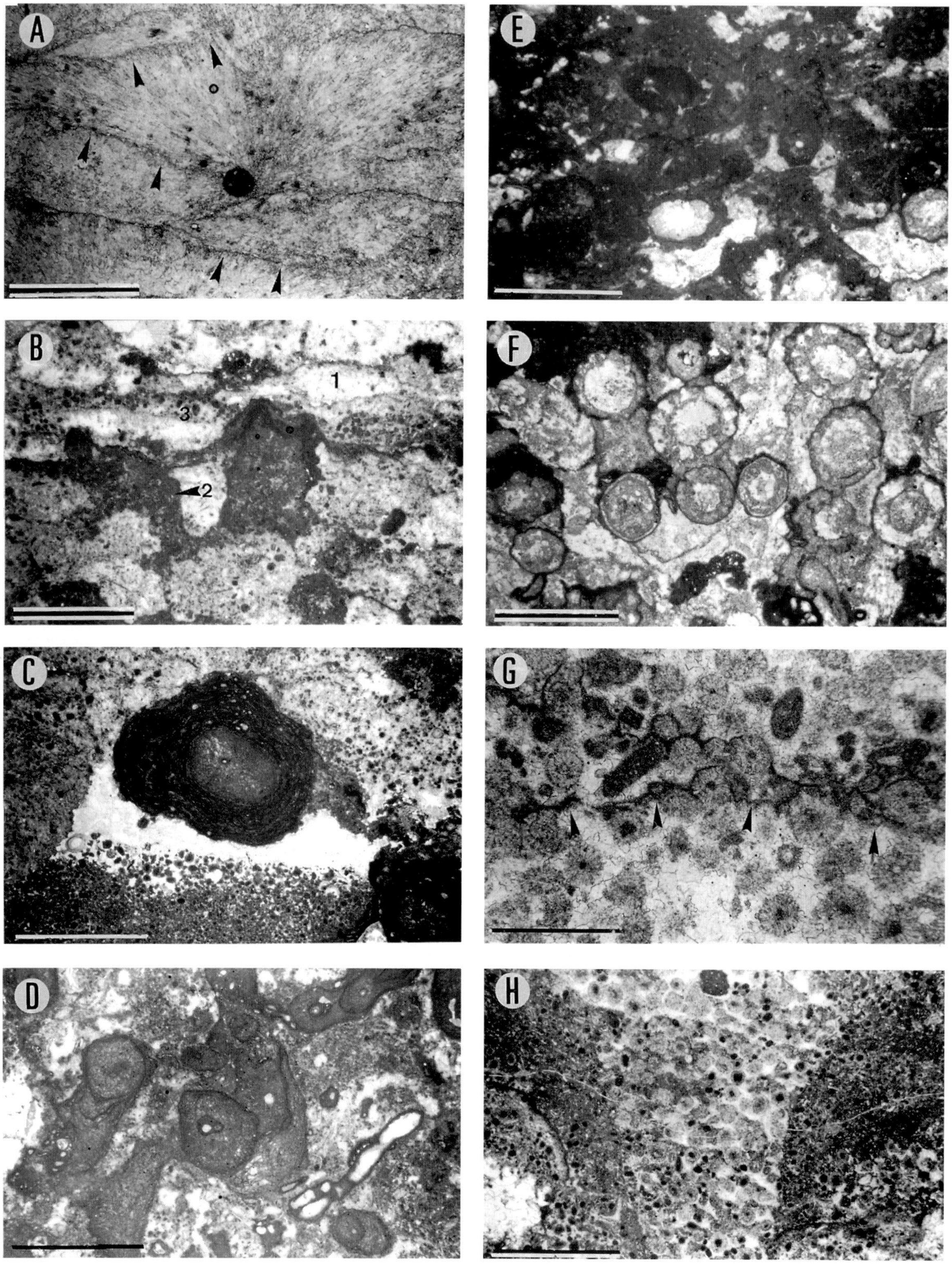

FIG. 8.—Photomicrographs of cyclic outer-shelf facies. (A)-(F) Sequence of a salinity-triggered cycle with *Tubiphytes* microbial level-bottom community (from top to base); (G)-(H) oolitic grainstones of stage 3. (A) Termination of cycle with botryoids; micrite envelopes of botryoids indicate discontinous growth of cements (arrows). Note fecal pellet in the center of botryoid; scale = 2.5 mm; thin-section H 3.2. (B) Top of *Tubiphytes* microbial subcommunity with recrystallized cements (1), microbial carbonate (2), and scattered peloids (3); scale = 2.5 mm; thin-section H 3.2. (C) *Tubiphytes obscurus* with thick encrustation of *Archaeolithoporella hidensis* forms the roof of a micro-cavity filled with peloids; scale = 2.5 mm; thin section H 3.2. (D) Microphotograph showing high packing density of *Tubiphytes obscurus;* scale = 2.5 mm; thin section H 3.1. (E) Base of level-bottom community with microbes and first occurrence of *Tubiphytes;* scale = 2.5 mm; thin section H 3.1. (F) Base of cycle composed of typical outer-shelf dasycladacean grainstone *(Macroporella sp.)* and scatterd microbial carbonate; scale = 2.5 mm; thin section H 3.2. (G) Oolitic grainstone of outer shelf facies with scattered peloids. Thin microbial layer suggest periods of nondeposition (arrows); scale = 1.0 mm; thin section H 2.1. (H) Microphotograph showing small columnar stromatolites embedded in oolitic grainstone; scale = 1.0 mm; thin section H 2.1.

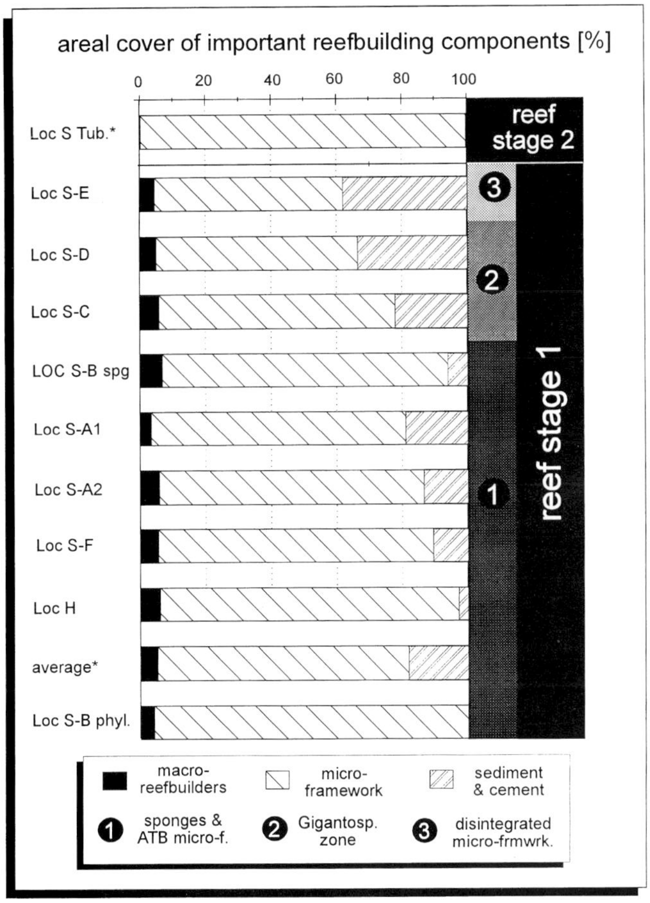

FIG. 9.—Quantitative data of important reefbuilding components in reef stages 1 and 2.

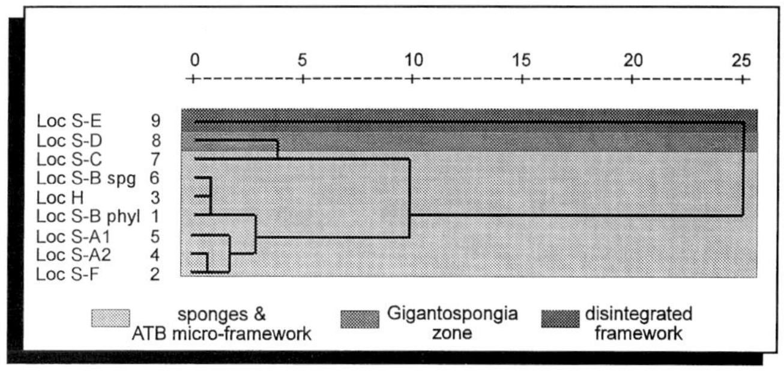

FIG. 10.—Dendrogram of the quantitative data (hierarchical cluster analysis; criteria: area percentage of sponges, algae, micro-framework, reef wackestone/packstone, platform grainstone, brownish sediment).

reached the reef. Landward, the surface is developed as a karst horizon where seaward marine conditions prevailed. In the reef, these bathymetric changes produced impoverished reef communities and also perhaps a breakdown in productivity of the reef community.

Our data correspond to observations of Kerans and Harris (1993), who studied an older unconformity (*Polydiexodina* zone?) at McKittrick Canyon. Following their studies, the exposure surface also never reached the reef. The absence of continuous siltstone beds and unconformities traceable in the reef suggest that most unconformities on the platform were related to third- or fourth-order sequence boundaries, which did not affect the platform margin. In the Late Permian greenhouse world, such exposure horizons would not be expected to occur in deeper water reefs.

Stage 3: Cyclic Outer Shelf Fall-in Beds with Scattered Reef-builders

Description.—

Evidence of sea-level changes as the control mechanism on outer shelf and shelf crest sedimentation is abundant in the Capitan reef complex (e.g., Borer and Harris, 1991 a,b; Garber et al., 1989; Parsley and Warren, 1989; Mazzullo et al., 1989; Noé, 1996; Kerans and Harris, 1993). However, except for Noé (1996), there are no published data indicative of the control of sea-level changes on reef-building communities and boundstone fabrics.

Progradational outer-shelf rocks above reef stage 2 are only exposed at locality H (Fig. 3A), while they have been removed by post-uplift erosion at locality S. Here, only the landward equivalents of reef stage 2 are preserved. The fall-in bed geometry is evident at both localities (Fig. 3).

The upper part of locality H (Fig. 3A a, b), represents a shallowing-upward sequence passing from outer shelf grainstones sporadically interrupted by sheet-crack breccias (Fig. 7D), into shelf-crest pisolites (Fig. 7H) and tepee structures. The grainstones exhibit a distinct lateral zonation parallel to the depositional dip of the outer shelf. The landward segment of the outer shelf grainstones is of low biotic diversity, containing only microbes that stabilized skeletal grains (Fig 7A-C). The seaward segment contains abundant *Tubiphytes* microbial communities showing evidence of small-scale cycles (Fig. 7E-G). Bioclastic grainstones, commonly dominated by broken thalli of alligned *Macroporella* sp., form the base (Fig. 8F). The skeletal grains were stabilized by microbial mats and isolated *Tubiphytes* (Fig. 8E). Later, an impoverished community of *Tubiphytes obscurus, Archaeolithoporella hidensis*, sessile foraminifers, and microbes became established (Fig. 8C, D). This level-bottom community passes into botryoid-rich pelletal grainstone with layers of microbial carbonate (Fig. 8B). The sequence is terminated by large botryoids (Fig. 9A). Comparable patches dominated by either *Tubiphytes* or *Archaeolithoporella* are also present at the mouth of Dark Canyon (Weidlich et al., 1995). Smaller areas of oolites and small stromatolites are also present (Fig. 8G, H). Various degrees of karst dissolution also occur (Fig. 6D, H, I).

At locality S, the landward equivalent of stage 2 exhibits small-scale coarsening-upward cycles from bioclastic grainstones to fenestral, partly laminated packstones (Fig. 7I-K). Dolomitic silt- to sandstones occur sporadically at the base and vary considerably in composition including monotonous packstones with detrital quartz and laminated, fenestral wackestones to packstones with layers of detrital quartz, and botryoids (Fig. 6E-G). The composition of the clastic-rich layers includes detrital quartz, mica, and K-feldspar. Sediment lamination, fenestral porosity in association with

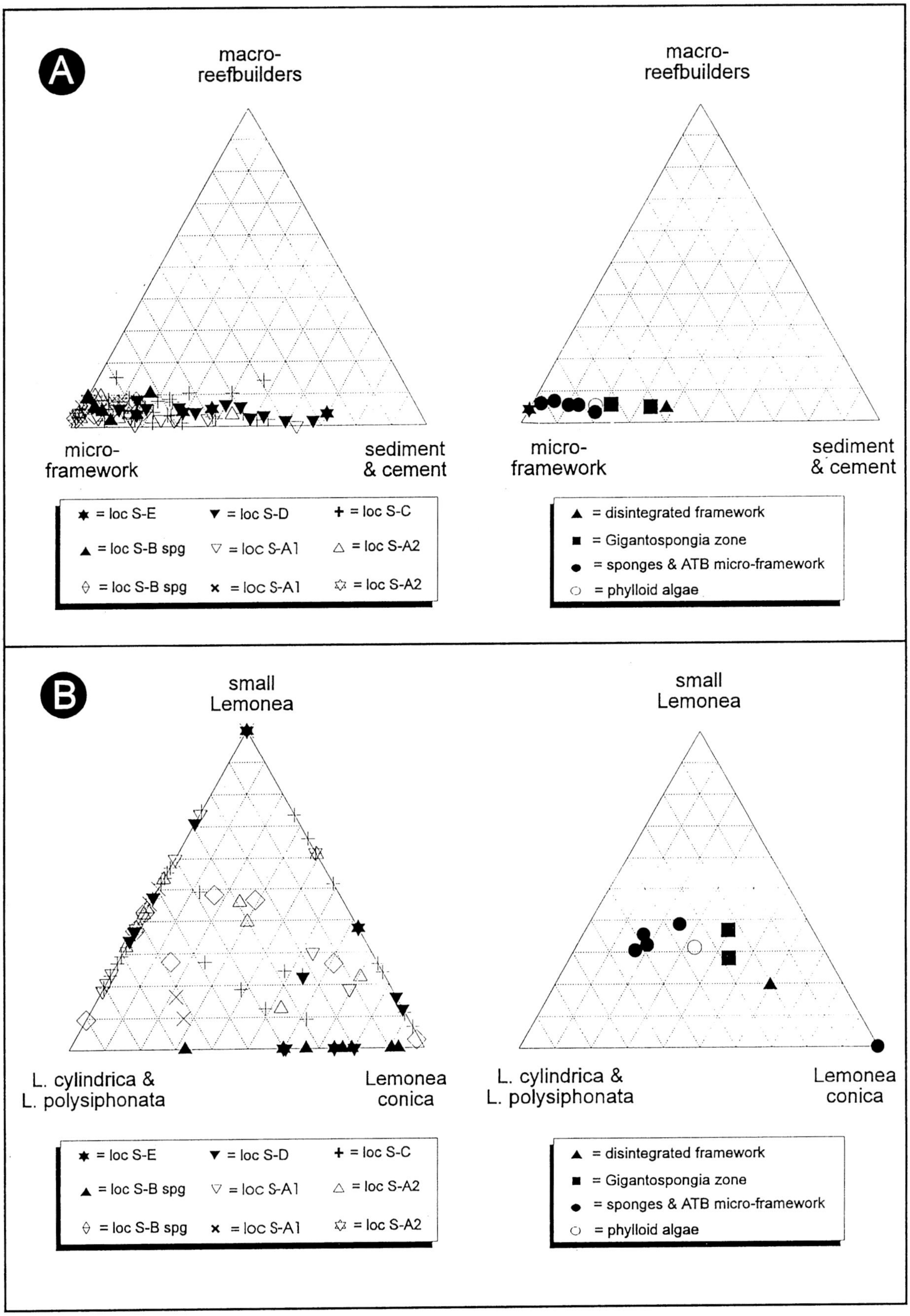

FIG. 11.—Ternary diagrams summarizing (A) important reefbuilding components (micro-framework, sediment/cement, macro-reefbuilders) and (B) sponge taxa (*Lemonea cylindrica*/*L. polysiphonata, Lemonea conica*, small *Lemonea* species) throughout stage 1. Studies with a minimum area of 0.5 m^2 are requested for reliable data.

fossils (shell debris and smaller foraminifers), and oncoids suggest overprinting of the sediments during the next transgression. Potential reef-builders are rare and restricted to one horizon below the unconformity. Recrystallization and dolomitization obliterated the original aragonite of the sponge skeletons. Thus, they could only be recognized in the field and not in thin sections. *Tubiphytes obscurus, Archaeolithoporella hidensis*, bryozoans, and microbes are absent.

Interpretation.—

According to the definitions of Fagerstrom (1987) and Flügel and Flügel-Kahler (1992), the presence of isolated reef-builders does not fullfill the criteria of a reef. The potential reef-builders described above lived in level-bottom communities. The occurrence of isolated and reworked sponges at locality S suggests that they may have been transported by exceptional events like storms, or that larvae could have settled there for a short period of time. The *Tubiphytes* patches are clearly autochthonous. At this time, they were the last representatives of the former reef community. The proximity of exposure horizons indicates the direct control of sea-level whereas extensive cementation by botryoids above the succession points to salinity fluctuations as a possible control mechanism.

Hunt et al. (1995) reported that postdepositional tilting and differential compaction of the prograding shelf margin system affected the stratal geometries. According to Saller (1996), geopetals of reef facies at McKittrick Canyon have angles up to 10°, which favors postdepositional tilting instead of depositional relief. Saller (1996) consequently concluded that "existing stratal geometries should not be used to reconstruct original bathymetries in the Capitan system." Geopetals were too rare at localities S and H to use for statistical analysis. However, facies changes are another useful tool for the ongoing discussion of the depositional dip of the outer shelf. They are well known from the landward to the seaward portion at several localities including McKittrick Canyon (e.g., Kerans and Harris, 1993), Dark Canyon (e.g., Toomey and Cys, 1977; Mazzullo et al., 1989), and Sheep Draw (Noé, 1996). Our study provides additional data in support of increasing water depth across the platform transect toward the reef (marginal mound hypothesis) and includes facies changes, pinchout of lowstand siliciclastics, and karst horizons. Even more interesting are changes in the composition and structure of the platform level-bottom communities from bryozoan in the landward segment to *Tubiphytes* in the seaward segment.

QUANTITATIVE DATA SET

Interpretation of the Quantitative Composition

Babcock (1977) long ago emphasized that "further work is also needed to document the quantitative contribution to Capitan fabrics of carbonate components such as inorganic cements, organic cements, organic frameworks, and clastic carbonate debris." The only study that has provided quantitative data is Schmidt (1977, discussed below).

Our quantitative data from reef maps concentrate on stage 1, the most important reef phase in the entire upper Capitan massive; stage 2 contains too few macro-reefbuilders for a reliable quantitative field-data evaluation but was analyzed from thin sections. Stage 3 is interpreted as a level-bottom community and therefore excluded from the following analysis.

Eight reef maps of stage 1 (Fig. 4) cover 15 m^2 with 15 recorded criteria including important taxa of macro-reef-builders (one alga, eight sponge taxa), micro-framework (two types), internal sediments (three types), and syndepositional cements. The data plotted in Tables 1 and 2 clearly indicate that the micro-framework, with a cover of 57.5% to 91.4% (mean = 76.7%), formed the most important constructional element of the sponge/algal cement subcommunities (stage 1A). Macro-reefbuilders with 3.4–6.9% (mean = 5.4%), internal sediments with 0–38% (mean = 12.1%), and cements with 0–22.2% (mean = 5.1%) were less crucial for reef accretion. The coverage by micro-framework (95.7%) and macro-reefbuilders (4.3%) in the phylloid algal subcommunity (stage 1B) is similar to the other subcommunities except for the absence of sponges, growth framework pores with internal sediments, and cements. In contrast to reef stage 1, coverage (based on visual estimates) in reef stage 2 suggests that the micro-framework formed nearly 100% of the reef due to the rarity of macro-reefbuilders and growth framework pores. As mentioned above, field observations provided no detailed data for this zone.

On the basis of the data from stages 1 and 2, we suggest that reef construction was largely maintained by low-growing reefbuilders, including small sponges, *Tubiphytes*, bryozoans, *Archaeolithoporella*, and syndepositional marine-phreatic cements ("micro-framework"). Our quantitative approach modifies the existing models for the origin of the upper Capitan massive. On the basis of slabs and thin sections, Schmidt (1977) concluded that "organic binding and growth of attached framework affected only a minor percentage of the rock volume in any part of the reef profile. Inorganic processes quantitatively as well as qualitatively dominated sediment binding and growth of attached framework along the entire reef profile" (*Archaeolithoporella hidensis* is interpreted as inorganic in this case). Grotzinger and Knoll (1995) used the data from Schmidt to conclude that "the Capitan limestone, particularly its youngest part, contains vast quantities of marine cement that locally form 50% to 80%." From interpreting our data, we certainly agree that syndepositional marine-phreatic cements are important for the upper Capitan massive. However, the low-growing organisms largely participate in the formation of the micro-framework, which is the most important constructional element of the upper Capitan massive. As evidenced by the reef maps, the dimension and distribution of the main constructional elements (macro-reef-builders, micro-framework, and syndepositional growth framework pores with sediment or cement) exceed thin sections and polished slabs in size. Quantitative data from thin sections may be consequently biased, which explains the difference between Schmidt's and our results.

Our alternative interpretation makes the upper Capitan massive unique, even compared to Permian counterparts (Weidlich et al., 1993; Weidlich, 1996), and both Precambrian and modern

TABLE 1.—QUANTITATIVE DATA OF THE REEF MAPS BASED ON 40 CM X 40 CM UNITS

		Meas.	major reef components				important taxa		sphinctozoans & inozoans								growth framework pores			micro-framework	
Window	Qua. Nr.	area [cm2]	Macro-reefb.	Micro-framew.	sedim.	cement	sponges	phylloid algae	small Lemon.	Lemonea pol. & cyl.	Lemon. conica	Amb. spp.	Disco. mamm.	Giga. disc.	Guad. expl.	other sponge	reef wacke/ packstone	platform grainstone	brownish sediment	ATB micr-fr.	micro. micr-fr.
S-E	a1	1600	3,75	26,51	69,75	0	3,74	0	1,14	0	0	0	0	0	0	2,61	0	69,75	0	26,51	0
	a2	1600	4,69	66,2	29,12	0	4,68	0	0,73	0	1,23	0	0	0	0	2,72	0	29,12	0	66,2	0
	a3	1600	5,55	57,76	36,69	0	5,55	0	0	1,11	3,23	0	0	0	0	1,21	36,69	0	0	57,76	0
	a4	1600	3,93	79,61	16,46	0	3,93	0	0	0,89	1,34	0	0	0	0	1,70	16,46	0	0	79,61	0
S-D	1a1	1600	8,5	77,0	14,48	0,0	8,51	0	1,75	0,71	0	0	0	5,53	0	0,52	14,48	0	0	65,1	11,91
	1a2	1600	1,8	31,4	3,37	63,4	1,82	0	0,67	1,15	0	0	0	0	0	0	3,37	0	0	28,04	3,35
	1a3	1600	3,3	44,5	0	52,3	3,27	0	1,3	1,43	0	0	0	0,54	0	0	0	0	0	44,47	0
	1a4	1600	6,2	49,5	44,27	0	6,24	0	0,45	0	3,2	0	0	1,88	0	0,71	38,03	0	0	49,49	0
	1a5	1600	2,7	48,5	48,78	0	2,68	0	0	0,77	1,18	0	0	0	0	0,73	46,1	0	0	29,13	19,41
	1a6	1600	6,1	66,6	12,89	14,4	6,1	0	0	0,77	3,09	0	0	1,62	0	0,62	6,79	0	0	22,62	44
	1a7	1600	6,9	53,2	11,24	28,6	6,9	0	0,57	0,97	0	0	0	4,52	0	0,84	4,34	0	0	32,66	20,57
	1a8	913	5,9	75,8	7,66	10,6	5,917	0	0,65	0	3,34	0	0	1,13	0	0,8	1,74	0	0	75,83	0
	2a1	1600	5,7	83,7	10,61	0	5,71	0	0,64	0,66	1,54	0	0	2,28	0	0,59	4,9	0	0	35,86	47,82
	2a2	1600	1,8	39,2	1,78	57,3	1,78	0	0,42	0,81	0	0	0	0	0	0,55	0	0	0	11,52	27,66
	2a3	1300	4,5	62,9	14,85	17,8	4,46	0	0,9	0	0	0	0	2,96	0	0,6	14,58	0	0	7,11	55,74
S-C	a1	1600	3	81,5	15,5	0	3	0	1,9	0	0	0	0	0,8	0	0,3	15,5	0	0	81,5	0
	a2	1600	5	82,5	12,5	0	5	0	1,2	0,9	0	0	0	1,7	0	1,2	12,5	0	0	82,5	0
	a3	1600	1,3	76,4	0,8	21,5	1,3	0	0,7	0,6	0	0	0	0	0	0	0,8	0	0	76,4	0
	a4	1600	4,2	78,6	10,7	6,5	4,2	0	1,7	1,9	0	0	0	0	0	0,6	10,7	0	0	49,8	0
	a5	1600	8,1	71,6	6,4	13,9	8,1	0	0,6	0,6	1,4	0	0	4,6	0	0,9	6,4	0	0	71,6	0
	a6	1600	8,6	68	0	23,4	8,6	0	1	2,1	0	0	0	5,5	0	0	0	0	0	68,0	0
	a7	1600	2,5	80,8	3,2	13,5	2,5	0	1,3	1,2	0	0	0	0	0	0	3,2	0	0	80,8	0
	a8	1600	3,3	70,3	24	2,4	3,3	0	1,5	1,2	0	0	0	0	0	0,6	24	0	0	70,3	0
	a9	1600	7,9	82,4	0	9,7	7,9	0	1,8	0	5	0	0	1,1	0	0	0	0	0	82,4	0
	a10	1600	8	82,7	9,3	0	8	0	0,9	1,6	0	0	0	5,5	0	0	9,3	0	0	82,7	0
	a11	1600	10,5	61,6	0	27,9	10,5	0	0,8	0	6,3	0	0	3,4	0	0	0	0	0	61,6	0
	a12	1600	10,3	49,8	6	33,6	10,3	0	1,5	0	4,8	0	0	4	0	0	6,3	0	0	49,8	0
	a13	1600	4	49,8	2	44,2	4	0	0,5	1,2	0	0	0	2,3	0	0	0	0	2	49,8	0
	a14	1600	2,42	58,31	0,97	38,3	2,42	0	0,33	0,88	0	0	0	0	0	1,21	0,97	0	0	58,31	0
	a15	1600	6,19	76,71	6,6	10,5	6,19	0	0,69	2,1	2,64	0	0	0	0	0,76	6,6	0	0	76,71	0
	a16	1600	1,88	47,02	2,5	48,6	1,88	0	1,38	0	0,5	0	0	0	0	0	2,5	0	0	47,02	0
	a17	1600	4,95	82,75	7,5	4,8	4,95	0	4,54	0	0	0	0	0	0	0,41	7,5	0	0	82,75	0
	a18	1600	6,92	85,91	0,47	6,7	6,92	0	4,93	0	0	0	0	0	0	1,99	0,47	0	0	85,91	0
	a19	1600	9,12	79,08	10,6	1,2	9,12	0	1,39	2,44	1,24	0	0	2,3	0	1,75	10,6	0	0	79,08	0
	a20	1600	7,8	67,6	10	14,6	7,8	0	0,81	1,84	1,68	0	0	2,3	0	1,17	10	0	0	67,6	0
	a21	1600	4,58	74,54	0	20,88	4,58	0	0,91	2,98	0	0	0	0	0	0,69	0	0	0	74,54	0
	a22	1600	3,63	71,12	0	25,25	3,63	0	1,53	1,2	0,38	0	0	0	0	0,52	0	0	0	71,12	0
	a23	1600	15,47	79,47	0	5,06	15,47	0	3,54	4	7	0	0	0	0	0,93	0	0	0	79,47	0
	a24	1600	5,84	88,5	3,85	1,81	5,84	0	1,93	0	1	0	0	0	0	2,91	3,85	0	0	88,5	0
	a25	1600	3,12	58,88	38	0	3,12	0	1,96	0,66	0	0	0	0	0	0,5	7,3	30,7	0	58,88	0
	a26	1600	2,04	43,84	52,51	1,61	2,04	0	0,62	0	0,39	0	0	0,39	0	0,64	0,81	51,7	0	43,84	0
	a27	1600	3,7	86,39	9,91	0	3,7	0	0,19	0,59	1,26	0	0	1,26	0	0,4	2,24	7,67	0	86,39	0
	a28	1600	2,5	97,5	0	0	2,5	0	0,19	0,78	0	0	0	0	0	1,53	0	0	0	97,5	0
	a29	1600	14,25	38,95	46,8	0	14,25	0	0,84	0	11,6	0	0	1,81	0	0	26,1	20,7	0	38,95	0
	a30	1600	8,43	85,29	6,28	0	8,43	0	2,94	0	2,77	0	0	2,1	0	0,62	5,1	1,18	0	85,29	0
	a31	1600	2,07	89,56	0	8,37	2,07	0	1,42	0	0	0	0	0	0	0,65	0	0	0	89,56	0
	a32	1600	1,22	68,18	0	30,6	1,22	0	0,41	0	0	0	0	0	0	0,81	0	0	0	68,18	0
S-B spg.	a4	1600	5,95	90,09	1,10	2,86	5,4	0,55	0	0,69	2,65	0,97	0	0	0	1,09	0	0	1,10	90,09	0
	a5	1600	9,59	90,41	0	0	9,59	0	0	0,65	8,06	0,29	0	0	0	0,59	0	0	0	90,41	0
	a6	1600	6,44	90,08	1,42	2,06	6,13	0,31	0	1,26	2,48	1,31	0	0	0	1,09	0	0	1,42	90,08	0
	a7	1600	8,93	91,07	0	0	8,93	0	0	1,63	5,33	0,39	0	0	0	1,58	0	0	0	91,07	0
	a8	1600	4,96	88,59	0	6,45	4,96	0	0	2,36	1,14	0,51	0	0	0	0,94	0	0	0	88,59	0
	a9	1600	10,54	72,46	5,74	11,26	10,54	0	0	0,9	8,71	0,3	0	0	0	0,62	0	0	5,74	72,46	0
	a10	1600	2,10	87,84	2,41	7,65	0,68	1,41	0	0	0	0	0	0	0	0,68	0	0	2,41	87,84	0
S-A1	1a1	1600	2,5	60,4	23,7	13,4	2,5	0	1,5	1	0	0	0	0	0	0	23,7	0	0	60,4	0
	1a2	1600	8,9	90,7	0,4	0	8,9	0	1,6	6,9	0	0	0	0	0	0,4	0,4	0	0	90,7	0
	1a3	1600	3,3	96,7	0	0	3,3	0	1,1	1,8	0	0	0	0	0	0,4	0	0	0	96,7	0
	1a4	1600	4,3	94,5	1,2	0	4,3	0	0,8	3	0	0	0	0	0	0,5	0	0	1,2	94,5	0
	1a5	930	3	90,8	6,2	0	3	0	1	1,4	0	0	0	0	0	0,6	0	0	6,2	90,8	0
	2a1	1600	2	70	12,7	15,3	2	0	0,8	0	0,5	0	0	0	0	0,7	12,7	0	0	70	0
	2a2	1600	8,3	87	4,7	0	8,3	0	1,4	0,9	5,2	0	0	0	0	0,8	4,7	0	0	87	0
	2a3	1600	0	37	63	0	0	0	0	0	0	0	0	0	0	0	63	0	0	37	0
	2a4	1600	3	81,9	15,1	0	3	0	0,9	0,5	1,6	0	0	0	0	0	9,1	0	6	81,9	0
	2a5	930	0,8	81,6	17,6	0	0,8	0	0,8	0	0	0	0	0	0	0	7,3	0	10,3	81,6	0
	3a1	1600	2,5	69,34	28,16	0	2,5	0	0,6	1,9	0	0	0	0	0	0	19,4	0	8,76	69,34	0
	3a2	1600	2,3	73,6	24,1	0	2,3	0	1,7	0,6	0	0	0	0	0	0	16	0	8,1	73,6	0
S-A2	1a1	1600	6,6	80	13,4	0	6,6	0	2,15	0	0	0	4,45	0	0	0	12,81	0	0,59	80	0
	1a2	1600	9,6	86,83	3,57	0	9,6	0	1,18	1,03	0	0	7,4	0	0	0	0,86	0	2,71	86,83	0
	1a3	1600	3,9	81,86	14,24	0	3,9	0	0,86	1,4	0	0	1,64	0	0	0	12,37	0	1,88	81,86	0
	2a1	1600	6,72	86,76	5,12	1,4	6,72	0	2,59	3,54	0	0	0,59	0	0	0	4,50	0	0,62	86,76	0
	2a2	1600	2,92	95,85	1,23	0	2,92	0	1,4	0	0,9	0	0,62	0	0	0	0	0	1,23	95,85	0
	2a3	1600	9,73	88,3	1,96	0	9,73	0	1,85	0,58	5,83	0	1,46	0	0	0	1,46	0	0,51	88,3	0
	2a4	1600	3,63	53,19	43,18	0	3,63	0	1,19	1,51	0	0	0,93	0	0	0	43,18	0	0	53,19	0
	3a1	1600	5,56	91,84	0	2,6	5,56	0	0,66	1,74	2,68	0	0,48	0	0	0	0	0	0	91,84	0
	3a2	1600	4,37	95,63	0	0	4,37	0	1,74	1,32	1,32	0	0	0	0	0	0	0	0	95,63	0
	3a3	1600	2,81	85,16	12,03	0	2,81	0	1,57	1,01	0	0	0	0	0	0	12,03	0	0	85,16	0
	3a4	1600	6,92	85,62	7,46	0	6,92	0	1,58	3,32	0	0	2,02	0	0	0	7,46	0	0	85,62	0
S-F	a1	1600	7,95	77,69	14,36	0	7,95	0	1,55	0,72	3,54	0	0,93	0	0	1,21	14,36	0	0	77,69	0
	a2	1600	6,95	83,44	9,61	0	6,95	0	0,55	5,2	0	0	0,63	0	0	0,57	9,61	0	0	83,44	0
	a3	1600	5	77,23	17,77	0	5	0	1,57	0,83	0,98	0	1,05	0	0	0,57	17,77	0	0	77,23	0
	a4	1600	4,21	91,35	4,44	0	4,21	0	1,48	2,49	0	0	0	0	0	0,24	4,44	0	0	91,35	0
	a5	1600	5,60	77,47	16,93	0	5,6	0	2,35	1,71	0,82	0	0	0	0	0,73	16,93	0	0	77,47	0
	a6	1600	4,81	86,92	8,27	0	4,81	0	1,96	2,6	0	0	0	0	0	0,24	8,27	0	0	86,92	0
	a7	1600	2,87	97,13	0	0	2,87	0	2,15	0,72	0	0	0	0	0	0	0	0	0	97,13	0
	a8	1600	7,37	80,66	11,97	0	7,37	0	1,76	3,7	0,95	0	0	0	0	0,96	11,97	0	0	80,66	0
H	a1	1600	6,47	92,5	0	1,03	6,47	0	1,51	0	0	0	0	0	4,54	0,43	0	0	0	92,5	0
	a2	1600	7,87	91,13	0	1,01	7,87	0	0,52	0,62	0	0	0	0	5,81	0,91	0	0	0	91,13	0
	a3	1384	7,14	82,59	0	10,27	7,14	0	0,69	2,55	0,9	0	0	0	3	0	0	0	0	82,59	0
	a4	1600	2,88	97,12	0	0	2,88	0	0,65	0,64	0	0	0,51	0	0	1,09	0	0	0	97,12	0
	a5	1600	6,32	93,68	0	0	6,32	0	0,43	2,72	1,18	0	0,47	0	1,09	0,43	0	0	0	93,68	0
S-B phyl.	a1	1600	4,63	95,37	0	0	0,67	3,96	0	0	0	0	0	0	0	0,67	0	0	0	95,37	0
	a2	1600	4,13	95,87	0	0	2,02	2,11	0	0	0	0,53	0	0	0	1,5	0	0	0	95,87	0
	a3	1600	4,23	95,77	0	0	1,41	2,81	0	0	0	0,66	0	0	0	0,75	0	0	0	95,77	0

TABLE 1.—SUMMARY DATA OF THE REEF MAPS

Window	Meas. area [cm2]	major reef components: Macro-reefb.	Micro-framew.	sedm.	cement	important taxa: sponges	phylloid algae	sphinctozoans & inozoans: small Lemon.	Lemonea pol. & cyl.	Lemon. conica	Amb. spp.	Disco. mamm.	Giga. disc.	Guad. expl.	other sponge	growth framework pores: reef wacke/ packstone	platform grainstone	brownish sedment	micro-framework: ATB micr-fr.	micro. micr-fr.
S-E	6400	4,5	57,5	38,0	0	4,48	0	0,47	0,5	1,45	0	0	0	0	2,06	13,29	24,71	0	57,52	0
S-D	16613	4,9	61,8	11,1	22,2	4,9	0	0,7	0,7	1,1	0	0	1,9	0	0,5	11,1	0	0	60,7	18,1
S-C	51200	5,7	72,3	9,0	13,0	5,7	0	1,4	0,9	1,5	0	0	1,2	0	0,7	5,4	3,5	0,1	72,4	0
S-B spg.	11200	6,9	87,2	1,5	4,3	6,6	0,32	0	0	1,07	4,05	0,54	0	0	0,94	0	0	1,52	87,93	0
S-A2	17860	3,4	77,8	16,4	2,4	3,4	0	1	1,5	0,6	0	0	0	0	0,3	13	0	3,4	77,8	0
S-A1	17600	5,7	81,2	11,1	2,0	5,68	0	1,5	1,4	0,98	0	1,8	0	0	0	11,1	0	0	83,4	0
S-F	12800	5,6	84,0	10,4	0	5,6	0	1,67	2,25	0,79	0	0,33	0	0	0,56	10,42	0	0	83,98	0
H	7784	6,1	91,4	0	2,5	6,14	0	0,76	1,31	0,42	0	0,2	0	2,89	0,57	0	0	0	91,4	0
average*	141457	5,4	76,7	12,1	5,8	5,31	0,04	0,94	1,07	0,99	0,51	0,36	0,39	0,36	0,70	8,04	3,50	0,63	76,89	2,25
S-B phyl.	4800	4,3	95,7	0	0	1,37	2,96	0	0	0	0,4	0	0	0	0,97	0	0	0	95,67	0

reefs are poor analogs. The dominance of micro-framework is one important reason fueling the longlasting reef versus nonreef discussion. Many researchers focused on macro-reefbuilders and growth framework pores with syndepositional cement and neglected the large volume of the low-growing organisms (for example, micro-framework), which acted as a major constructor. The upper Capitan massive is clearly an organic reef and not an inorganic pile of cement.

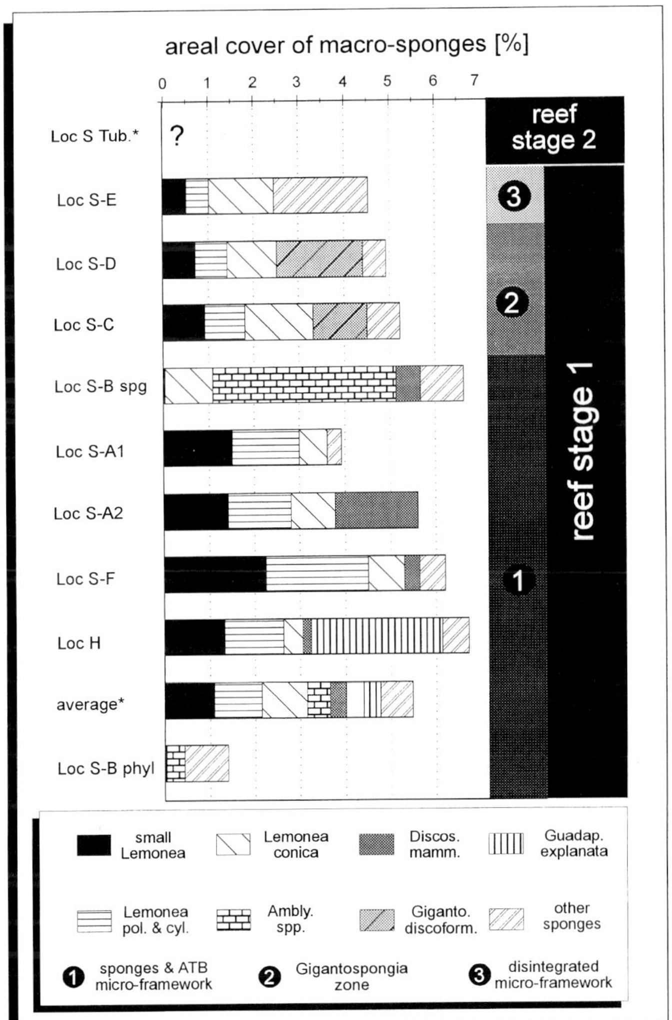

FIG. 12.—Quantitative distribution of macro-reefbuilders throughout reef stage 1.

Quantitative Data of Reef Components Versus Vertical Evolution of the Seaward Rim

The most significant changes in the quantitative composition of the upper Capitan massive involve the principle constructional elements, that is, macro-reefbuilders, micro-framework, and growth framework pores with syndepositional cement and/or sediments (Fig. 9). The zone with sponges and ATB micro-framework (zone 1) exhibits minor variations in the percentage of micro-framework, while zones 2 to 3 are characterized by a significant decrease (72.3–57.5%). These changes point to a decrease in the percentage of micro-framework and an increase in internal sediment reflecting the initial disintegration of the framework. In contrast, stage 2 is exclusively dominated by micro-framework. A cluster analysis based on the distribution of sponges, algae, micro-framework, reef wackestone/packstone, platform grainstone, and brownish quartz-rich sediment, confirms the proposed zonation of the upper Capitan massive (Fig. 10). However, the recognition of facies changes requires a minimum size for the reef maps (Fig. 11A). A ternary diagram of the data shows progressive disintegration of the framework zones 1–3 but is only evident from maps exceeding 0.5 m^2, whereas maps of "standard cells" with coverage of 0.16 m^2 (40 cm x 40 cm) produced no obvious trend.

The taxonomic diversity of the macro-sponges reflects less perfectly the environmental changes visible from maps and quantitative data of the principal components.

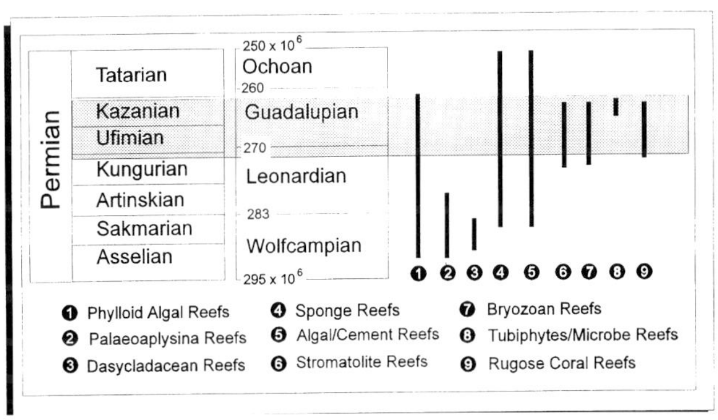

FIG. 13.—Distribution pattern of Permian reef types (modified from Flügel, 1994). Cosmopolitical Middle/Late Permian reef types including phylloid algal reefs, sponge reefs, and algal/cement reefs requested a minimum water depth. Above that depth, *Tubiphytes* reefs flourished.

Sponge Composition Versus Vertical Zonation

On the basis of sponge presence/absence data from 105 polished quadrat slabs (see topic 4 of chapter "data assessment"), the sponge community appears uniform throughout the upper Capitan massive. Quantitative field-derived data describing sponge areal cover and species diversity of reef maps at localities S and H do not reflect environmental changes as well as the quantitative reef fabric data do (Fig. 11). The only exception is *Gigantospongia discoforma*, which characterizes zone 2 of reef stage 1. A ternary diagram of important *Lemonea* species shows a trend from *Lemonea cylindrica* and *L. polysiphonata* toward *L. conica* from zone 1 to 3 during stage 1 (Fig. 11B). As in the ternary diagram of reef fabrics data, a minimum study area is required and again the data from single 40 cm x 40 cm quadrats were not reliable. We conclude that supposed azooxanthellate filter- or suspension-feeding sponges (Senowbari-Daryan, 1990) were limited recorders of environmental change.

Quantitative Macro-reefbuilder Data and Their Interpretation

Macro-reefbuilders are dominated by sponges including sphinctozoans or inozoids and locally by the phylloid alga *Eugonophyllum* sp. occurring at the seaward segment of locality S and Chinaberry Draw. Subordinately, crinoids (J. A. Fagerstrom and O. Weidlich, 1998) and richthofenid brachipods flourished also at mouth of Dark Canyon (Senowbari-Daryan and Rigby, 1996b) and solitary Rugosa during stage 2 at Hackberry Draw.

Differences in diversity and quantitative macro-sponge composition reflect the endemism of the Capitan sponges. Except for a few taxa, that is, *Amblysiphonella* spp., *Discosiphonella mammilosa*, and *Parauvanella minima*, most species of macro-sponges *Lemonea, Guadalupia*, and *Gigantospongia* are exclusively restricted to the Capitan reef complex. By contrast, only *Eugonophyllum* sp., *Tubiphytes obscurus, Archaeolithoporella hidensis*, bryozoans, and rare *Lercaritubus problematicus* are cosmopolitan. The abundance of macro-reef-builders during stage 1 remained with 3.4–6.9% relatively constant throughout the upper Capitan massive in both the sponge and phylloid algal subcommunities (Figs. 9 and 12). The strikingly different sizes of the study areas as well as variations in coverage by micro-framework and cement/sediment suggest that the quantitative data of the macro-reefbuilders of the upper Capitan massive are unbiased and not artifacts of data assessment. Interpretation of the data from stages 2 and 3 indicates a minor decrease in sponge coverage (from 5.7%–4.5%) and may be monitoring changes in the evolution of the reef front.

The phylloid algal coverage (4.3%) in zone 1 is similar to the cover in the sponge subcommunities (3.4–6.9%; mean 5.4%). However, an impoverished sponge fauna (*Amblysiphonella* spp. plus undeterminable taxa) associated with the phylloids makes up only 1.4% of the total coverage. This suggests that the phylloid algae grew, rapidly surpassing the slower growing sponges, as described by West (1988) and Samankassou (1997).

The zone of sponges and ATB micro-framework (zone 1) represented by reef maps H, S-F, S-A1, S-A2, and S-B spg is characterized by the highest species diversity but with the greatest variation in sponge coverage. Such variation is typical for diverse reef communities, which were triggered by autecological processes or minor variations in abiotic control mechanisms like topography, water energy, and substrate. In contrast, zones 2 and 3, represented by reef maps S-C to S-E, are characterized by a slight but continuous decrease in the abundance of *Lemonea conica, Lemonea polysiphonata*, and small *Lemonea* species perhaps due to decreasing water depth during the upper Capitan massive deposition.

The percentage of macro-reefbuilders is low compared to most ancient and modern reefs. Macro-reefbuilders were an important constructional element of the initial reef framework (J. A. Fagerstrom and O. Weidlich, 1998) but were not volumetrically significant in the overall reef budget.

CONCLUSIONS

Although extrinsic control mechanisms are important factors influencing sedimentation on the platform and in the adjacent Delaware basin, little research has been directed to assessment of their influence on the upper Capitan massive. Combined data from depositional geometries, detailed reef maps, quantitative data, and thin sections provide a framework for interpreting the effects of sea-level fluctuation on reef growth and sedimentation patterns.

The seaward segment of the upper Capitan massive is composed of three stages. Stage 1, up to 100-m thick, is composed of sponge/algal cement subcommunities (stage 1A) and phylloid algal subcommunities (stage 1B). Counterparts of stage 1A are known from Tunisia (Newell et al., 1976; M'Rabet et al., 1996), China, and Oman (Weidlich et al., 1993; Weidlich, 1996), while representatives of stage 1B in the southwestern United States (Toomey and Babcock, 1983) are commonly older. Development of stage 1 began with aggradational geometries during the zone of sponges and ATB micro-framework. The overlying progradational *Gigantospongia* zone is unique for sponge reefs; it represents the growth optimum of the entire Capitan reef. Rapid accretion during this period is suggested by numerous growth framework pores filled with different generations of sediment, and large quantities of marine-phreatic cements in growth framework pores. The youngest zone of stage 1 (zone of framework disintegration, zone 3) is characterized by an increase in platform grainstones at the expense of organic boundstones and macro-reefbuilders.

Stage 2 differs from stage 1 and other Permian counterparts in its tabular geometry, reduced thickness (about 20 m), and its impoverished fauna. Although vertical zonation is absent, a lateral shift from *Tubiphytes* microbial assemblages on the seaward segment to mixed bryozoan and *Tubiphytes* communities on the landward portion is apparent. Macro-sponges are rare. This reef type is exclusively dominated by low-growing organisms. Although bryozoans are moderately abundant, this reef type does not correspond to the "bryozoans reefs" of Flügel (1994), which

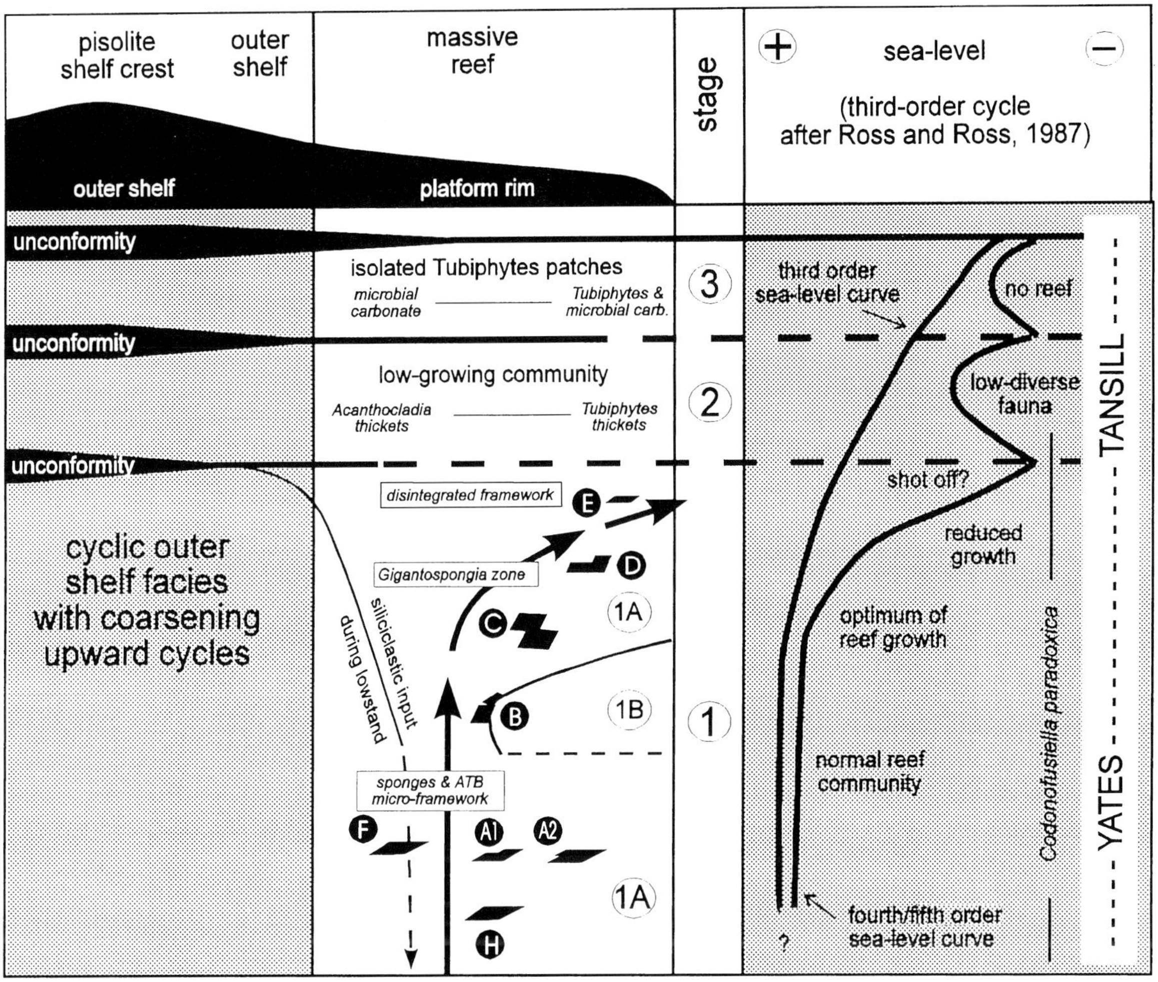

FIG. 14.—Conceptual model explaining reef growth under consideration of sea-level fluctuations.

form bioclastic rudstones rather than organic boundstones.

The youngest "potential reefbuilders" in stage 3 are *Tubiphytes* microbe communities or scattered, toppled sponges, which flourished in cyclic outer shelf grainstone intervals intercalated between layers of botryoids. Their low relief of less than 0.5 m and their considerable lateral extent suggest that these associations are level-bottom communities rather than reefs.

Ross and Ross (1987, 1994) recognized two third-order depositional sequences for the Capitanian. In the younger sequence, the upper Yates and the Tansill form the regressive phase. This agrees with our observations that reef thickness and prograding depositional geometries resulted from falling sea level during the second third-order cycle. Stages 1 to 3 modify this general pattern and point to superimposed high-frequency sea-level fluctuations that may correspond to 400 ky eccentricity cycles described by Borer and Harris (1991) for Yates platform deposits. A detailed correlation of the observed stages with published data is difficult due to the lack of datum levels and good biostratigraphic control in the upper Capitan massive.

We suggest that sea-level fluctuations directly controlled reef growth. A minimum depth — only attained during stage 1 — was crucial for sponge reefs, algal cement reefs, and phylloid algal reefs and probably controlled their global distribution in connection with climatic parameters (Fig. 14). Shallowing caused a disintegration of the reef framework, but, in contrast to modern analogs, the communities never reached sea level. Stage 2, constructed by endemic *Tubiphytes* microbe communities, represents reef growth above the bathymetric limit for sponges and algae, that is, near sea level (Fig. 14).

Control mechanisms other than sea-level fluctuations are also indicated. They include small-scale changes in stage 3 from *Tubiphytes* microbe, level-bottom communities to cement layers on the sea floor that were probably caused by salinity fluctuation. During middle to late Guadalupian time, the salinity of deep oceanic water in the Delaware basin increased due to restricted circulation caused by the topographically high "Hovey Channel" (e.g., Newell et al., 1953). Occasional upwelling of saline waters on the outer shelf may have produced large quantities of cement (Grotzinger and Knoll, 1995, Fig. 8) and also limited the vertical distribution of the level bottom communities.

Comparison of photomosaics, reef maps, and quantitative data indicate that modifications of reef growth due to extrinsic control mechanisms are recorded best by major reef constructional elements. Sponges alone were important for the initial framework of stage 1 but were limited as recorders of sea-level change.

ACKNOWLEGMENTS

Financial support for the study was provided by Grant EAR 9321726 from the National Science Foundation to J. A. Fagerstrom. O. Weidlich was kindly supported during the Priority Program "Global and Regional Controls on Biogenic Sedimentation" by the German Research Foundation with the project We 1804/2-1,2. Special thanks are addressed to Erik Flügel (Erlangen, Germany), the reef-coordinator of the Priority Program. We are grateful to the staff of Guadalupe Mountains National Park (Texas) and Carlsbad Caverns National Park (New Mexico) for logistic support and sampling permits. The joint project benefitted from the enthusiastic support of Michaela Bernecker (Erlangen), Keith Rigby (Provo), Baba Senowbari-Daryan (Erlangen), and Sibylle Noé (Bremen). We appreciate technical support and assistance of Constanze von Engelhardt, Silke Becker, Kahlo Thraenhardt, Bernd Kleeberg, Hanne Glowa, Markus Stoewer, André Gardei, and Heinz Holl (from the Technical University Berlin). Johannes Schroeder (Berlin, Germany), Hans Machel (Edmonton, Canada), Art Saller, Brenda Kirkland George, Emily Stoudt, and Ronald A. Johns helped with critical discussion and improvement of the manuscript. J. A. Fagerstrom is especially grateful to the Geological Sciences Department of the University of Canterbury, Christchurch, New Zealand, for providing office space and the use of their facilities during the austral summers of 1995, 1996, and 1997.

REFERENCES

Babcock, J. A., 1974, The role of algae in the formation of the Capitan Limestone (Permian, Guadalupian), Guadalupe Mountains, West Texas and New Mexico: Unpublished Ph.D. Dissertation, University of Wisconsin, Madison, 241 p.

Babcock, J. A., 1977, Calcareous algae, organic boundstones, and the genesis of the Upper Capitan Limestone (Permian, Guadalupian), Guadalupe Mountains, west Texas and New Mexico, *in* Hileman, M. E., and Mazzullo, S. J., eds., Upper Guadalupian Facies, Permian Reef Complex, Guadalupe Mountains, New Mexico and West Texas: Midland, Society of Economic Paleontologists and Mineralogists, Permian Basin Section, Publication 77-16, v. 1, p. 3-44.

Babcock, J. A., Pray, L. C., and Yurewicz, D. A. 1977, Locality Guide I, Stop I, Upper Capitan massive, Mouth of Walnut Canyon, *in* Pray, L. C., and Esteban, M., eds., Upper Guadalupian Facies, Permian Reef Complex, Guadalupe Mountains, New Mexico and west Texas, 1977 Field Conference Guidebook, v. 2: Midland, Society of Economic Paleontologists and Mineralogists, Permian Basin Section, Publication, 77-16., p. 17-40.

Babcock, J. A., and Yurewicz, D. A., 1989, The massive facies of the Capitan Limestone, Guadalupe Mountains, Texas and New Mexico, *in* Harris, P. M., and Grover, G. A., eds., Subsurface and outcrop examination of the Capitan Shelf Margin, northern Delaware Basin: Tulsa, Society of Economic Paleontologists and Mineralogists Core Workshop 13, p. 365-371.

Bebout, D. G., and Kerans, C., eds., 1993, Guide to the Permian reef geology Trail, McKittrick Canyon, Guadalupe Mountains National Park, West Texas, Guidebook 26: Austin, Bureau of Economic Geology, University of Texas, 48 p.

Bebout, D. G., Kerans, C., and Harris, P. M. 1993, Introduction, *in* Bebout, D. G., and Kerans, C., eds., Guide to the Permian reef geology Trail, McKittrick Canyon, Guadalupe Mountains National Park, West Texas, Guidebook 26: Austin, Bureau of Economic Geology, University of Texas, p. 1-4.

Borer, J. M., and Harris, P. M., 1989, Depositional facies and cycles in Yates Formation outcrops, Guadalupe Mountains, New Mexico, *in* Harris, P. M., and Grover, G. A., eds., Subsurface and outcrop examination of the Capitan shelf margin, northern Delaware Basin: Tulsa, Society of Economic Paleontologists and Mineralogists Core Workshop 13, p. 305-317.

Borer, J. M.,and Harris, P. M.,1991a, Depositional facies and model for mixed siliciclastics and carbonates of the Yates Formation, Permian Basin: Tulsa, Society of Economic Paleontologists and Mineralogists Core Workshop 15, p. 1-133.

Borer, J. M., and Harris, P. M., 1991b, Lithofacies and cyclicity of the Yates Formation: American Association of Petroleum Geologists Bulletin, v. 75/4, p. 726-779.

Candelaria, M. P., 1989, Shallow marine sheet sandstones, upper Yates Formation, northwest shelf, Delaware Basin, New Mexico, *in* Harris, P. M., and Grover, G. A., eds., Subsurface and outcrop examination of the Capitan shelf margin, northern Delaware Basin: Tulsa, Society of Economic Paleontologists and Mineralogists Core Workshop 13, p. 319-324.

Cys, J. M., Toomey, D. F., Brezina, J. L., Greenwood, E., Groves, D. B., Klement, K. W., Kullman, J. D., McMillan, T. L., Schmidt, V., Sneed, E. D., and Wagner, L. H., 1977, Capitan Reef — Evolution of a concept, *in* Hileman, M. E., and Mazzullo, S. J., eds., Upper Guadalupian Facies, Permian Reef Complex, Guadalupe Mountains, New Mexico and West Texas, v. 1: Midland, Society of Economic Paleontologists and Mineralogists, Permian Basin Section, Publication 77-16, p. 201-322.

Dunham, R. J., 1972, Capitan Reef, New Mexico and Texas: Facts and questions to aid interpretation and group discussion: Midland, Society of Economic Paleontologists and Mineralogists, Permian Basin Section, Publication 72-14, 291 p.

Esteban, M., and Pray, L. C., 1977, Carbonate shoal facies of the shelf crest, *in* Upper Guadalupian facies, Permian Reef Complex, Guadalupe Mountains, New Mexico and west Texas, 1977 Conference Guidebook, v.2, Road Logs and Locality Guides: Midland, Society of Economic Paleontologists and Mineralogists, Permian Basin Section, Publication 77-16, p. G79-G118.

Fagerstrom, J. A., 1987, The evolution of reef communities: New York, John Wiley and Sons, 600 p.

Fagerstrom, J. A., 1991, Reefbuilding guilds and a checklist for determining guild membership: Coral Reefs, v. 10, p. 4752.

Flügel, E., 1994, Pangean shelf carbonates: Controls and paleoclimatic significance of Permian and Triassic reefs: Boulder, Geological Society of America Special Paper 288, p. 247-266.

Flügel, E., and Kahler, E., 1992, Phanerozoic reef evolution: Basic questions and data base: Facies, v. 26, p. 167-278.

Garber, R. A., Grover, G. A., and Harris, P. M., eds., 1989, Geology of the Capitan shelf margin-subsurface data from the northern Delaware basin, *in* Harris, P. M., and Grover, G. A., eds., Subsurface and outcrop examination of the Capitan shelf margin, northern Delaware Basin: Tulsa, Society of Economic Paleontologists and Mineralogists Core Workshop 13, p. 3-269.

Girty, G. H., 1908, The Guadalupian fauna: Washington, D. C., U.S. Geological Survey Professional Paper 58, 651 p.

Grotzinger, J. P., and Knoll, A. H., 1995, Anomolous carbonate precipitates: Is the Precambrian the key to the Permian?: Palaios, v. 10, p. 578-596.

Hileman, M. E., and Mazzullo, S. J., eds., 1977, Upper Guadalupian facies, Permian reef complex, Guadalupe Mountains, New Mexico and west Texas, *in* 1977 Field Conference Book Guidebook, v.1: Midland, Society of Economic Paleontologists and Mineralogists, Permian Basin Section, Publication 77-16, 508 p.

Hunt, D., Allsop, T., and Swarbrick, R., 1995, Compaction as a primary control on the architecture and development of depositional sequences: Conceptual framework, applications and implications, *in* Howell, J., and Aitken, J. F, eds., High resolution sequence startigraphy: Innovations and applications: London, Geological Society of London Special Publication 104, p. 321-345.

James, N. P., 1983, Reef environments, *in* Scholle, P. A., Bebout, D. G., and Moore, C. H., eds., Carbonate Depositional Environments: Tulsa, American Association of Petroleum Geologists Memoir 33, p. 345-462.

KERANS, C., and HARRIS, P. M., 1993, Outer shelf and shelf crest, *in* Bebout, D. G., and Kerans, C., eds., Guide to the Permian reef geology trail, McKittrick Canyon, Guadalupe Mountains National Park, West Texas, Guidebook 26: Austin, Bureau of Economic Geology, University of Texas, p. 32-43.

KING, P. B., 1948, Geology of the southern Guadalupe Mountains, Texas: Washington, D.C., U.S. Geological Survey Professional Paper 215, 183 p.

KIRKLAND-GEORGE, B., 1992, Distinctions between reefs and bioherms based on studies of fossil algae-*Mizzia*, Permian Capitan reef complex (Guadalupe Mountains, Texas and New Mexico) and *Eugonophyllum*, Pennsylvanian Holder Formation (Sacramento Mountains, New Mexico): Unpublished Ph.D. Dissertation, Louisiana State University, 156 p.

KIRKLAND, B. L., LONGACRE, S. A., and STOUDT, E. L. 1993, Reef, *in* Bebout, D. C., and Kerans, C., eds., Guide to the Permian Reef geology trail, McKittrick Canyon, Guadalupe Mountains National Park, West Texas Guidebook 26: Austin, Bureau of Economic Geology, University of Texas, p. 23-31.

LANG, W. B., 1937, The Permian formations of the Permian Basin: American Association of Petroleum Geologists Bulletin, v. 21, p. 833-898.

MAZZULLO, S. J., 1995, Permian stratigraphy and facies, Permian Basin (Texas-New Mexico) and adjoining areas in the midcontinent United States, *in* Scholle, P. A., Peryt, T. M., and Ulmer-Scholle, D. J., eds., The Permian Northern Pangea, v. 2: Berlin, Springer, p. 41-60.

MAZZULLO, S. J, BISCHOFF, W. D., and HEDRICK, C. L., 1989, Stacked island facies in Tansill outer-shelf platform, Dark Canyon, Guadalupe Mountains, New Mexico, *in* Harris, P. M., and Grover, G. A., eds., Subsurface and outcrop examination of the Capitan Shelf Margin, northern Delaware Basin: Tulsa, Society of Economic Paleontologists and Mineralogists Core Workshop 13, p. 287-293.

MAZZULLO, S. J., and CYS, J. M., 1977, Submarine cements in Permian boundstones and reef-associated rocks, Guadalupe Mountains, west Texas and southeastern New Mexico, *in* Hileman, M. E., and Mazzullo, S. J., eds., Upper Guadalupian Facies, Permian Reef Complex, Guadalupe Mountains, New Mexico and West Texas, v. 1: Midland, Society of Economic Paleontologists and Mineralogists, Permian Basin Section, Publication 77-16, p. 151-200.

M'RABET, A., RAZGALLAH, S., DRIDI, M., and CHAOUACHI, M. C., 1996, Upper Permian reefs and associated siliciclastics at Jebel Tebaga, Southern Tunisia, Field Trips Guidebook: 17th Regional African European Meeting of Sedimentology, p. 3-37.

MUTTI, M., and SIMO, J. A., 1993, Stratigraphic patterns and cycle-related diagenesis of Upper Yates Formation, Permian, Guadalupe Mountains, *in* Loucks, R. G., and Sarg, J. F., Carbonate Sequence Stratigraphy: Tulsa, American Association of Petroleum Geologists Memoir 57, p. 515-523.

NEESE, D. G., 1989, Peritidal facies of the Guadalupe shelf crest, Walnut Canyon, New Mexico, *in* Harris, P. M., and Grover, G. A., eds., Subsurface and outcrop examination of the Capitan shelf margin, northern Delaware Basin: Tulsa, Society of Economic Paleontologists and Mineralogists Core Workshop 13, p. 295-303.

NEWELL, N. D., RIGBY, J. K., FISCHER, A. G., WHITEMAN, A. J., HICKCOX, J. E., and BRADEY, J. S., 1953, The Permian Reef Complex of the Guadalupe Mountains Region, Texas and New Mexico: San Francisco, W. H. Freeman and Company, 236 p.

NEWELL, N. W., RIGBY, K., DRIGGS, A., BOYD, D. W., and STEHLI, F. G., 1976, Permian reef complex, Tunisia: Brigham Young University Geology Studies 23/1, p. 75-112.

NOÉ, S., 1996, Late-stage evolution of the Permian reef complex: Shelf margin and outer-shelf development of the Tansill Formation (Late Permian), northern Guadalupe Mountains, New Mexico, USA, *in* Reitner, J., Neuweiler, F., and Gunkel, F. eds., Global and regional controls on biogenic sedimentation, I. Reef evolution, research reports: Goettinger Arbeiten zur Geologie und Palaeontologie, Sb2., p. 317-324.

PARSLEY, M. J., and WARREN, J. K., 1989, Characterization of an upper Guadalupian barrier-island complex from the middle and upper Tansill Formation (Permian), east Dark Canyon, Guadalupe Mountains, New Mexico, *in* Harris, P. M., and Grover, G. A., eds., Subsurface and outcrop examination of the Capitan shelf margin, northern Delaware Basin: Tulsa, Society of Economic Paleontologists and Mineralogists Core Workshop 13, p. 279-285.

PRAY, L. C., and SCHWARTZ, A., 1977, Locality guide, Stop II, Mouth of Walnut Canyon, outer shelf facies, Yates and Tansill formations, *in* Upper Guadalupian facies, Permian Reef Complex, Guadalupe Mountains, New Mexico and west Texas, 1977 Conference Guidebook, v. 2, Road Logs and Locality Guides: Midland, Society of Economic Paleontologists and Mineralogists, Permian Basin Section, Publication 77-16, p. G41-G46.

REID, S. T., BASS, R. O., and WELCH, P., 1988, Guadalupe Mountains revisited, Texas and New Mexico, West Texas Geological Society, 1988 Field Seminar: Midland, West Texas Geological Society Publications 88-84, 123 p.

RIDING, R., 1993, *Shamovella obscura*: The correct name for *Tubiphytes obscurus* (Fossil): Taxon, 42, p.7173.

RIGBY, J. K., and SENOWBARI-DARYAN, B., 1996, *Gigantospongia,* new genus, the largest known Permian sponge, Capitan Limestone, Guadalupe Mountains, New Mexico: Journal of Paleontology, v. 70, p. 347-355.

RIGBY, J. K., SENOWBARI-DARYAN, B., and LIU, H., 1998, Sponges of the Permian Upper Capitan Limestone, Guadalupe Mountains, New Mexico and Texas: Brigham Young University, Geology Studies, v. 42, p. 19-118.

ROSS, C. A., and ROSS, R. P., 1987, Late Paleozoic sea levels and depositional sequences: Cushman Foundation for Foraminiferal Research Special Publication 24, p.137-149.

ROSS, C. A., and ROSS, J. R. P., 1994, Permian sequence stratigraphy and fossil zonation, *in* Embry, A. F., Beauchamp, B, and Glass, D. J. , eds., Pangea: Global Environments and Resources: Calgary, Canadian Society of Petroleum Geologists Memoir 17, p. 219-231.

SALLER, A. H., 1996, Differential compaction and basinward tilting of the prograding Capitan reef complex, Permian, west Texas and southeast New Mexico, USA: Sedimentary Geology, v. 101, p. 21-30.

SAMANKASSOU, E., 1997, Paloekologie der dasycladalen Alge *Anthracoporella* Pia 1920. Daten aus dem Jungpaläozoikum der Karnischen Alpen, Österreich: Terra Nostra 97/6, Schriften der Alfred Wegener-Stiftung, 67. Jahrestagung der Paläontologischen Gesellschaft, Vortrags- und Posterkurzfassungen, p.103-104.

SCHMIDT, V., 1977, Inorganic and organic reef growth and subsequent diagenesis in the Permian Capitan Reef Complex, Guadalupe Mountains, Texas, New Mexico, *in* Hileman, M. E., and Mazzullo, S. J., eds., Upper Guadalupian Facies, Permian Reef Complex, Guadalupe Mountains, New Mexico and West Texas, v. 1: Midland, Society of Economic Paleontologists and Mineralogists, Permian Basin Section, Publication 77-16, p. 93-131.

SENOWBARI-DARYAN, B., 1990, Die systematische Stellung der thalamiden Schwaemme und ihre Bedeutung in der Erdgeschichte: Muenchner Geowissenschaftliche Abhandlungen, Reihe A, Geologie, v. 21, 326 p.

SENOWBARI-DARYAN, B., and RIGBY, J. K., 1996a, First report of *Lercaritubus* in North America, from the Permian Capitan Limestone, Guadalupe Mountains, New Mexico: Journal of Paleontology, v. 70, p. 22-26.

SENOWBARI-DARYAN, B., and RIGBY, J. K., 1996b, Brachiopod mounds not sponge reefs, Permian Capitan-Tansill Formations, Guadalupe Mountains, New Mexico: Journal of Paleontology, v. 70, p. 697-701.

TOOMEY, D. F., and BABCOCK, J. A., 1983, Precambrian and Paleozoic algal carbonates, west Texas–southern New Mexico: Colorado School of Mines, Professional Contributions 11, 345 p.

TOOMEY, F. T., and CYS, J. M., 1977, Rock/biotic relationships of the Permian Tansill-Capitan facies exposed on the north side of the entrance to Dark Canyon, Guadalupe Mountains, southeastern New Mexico, *in* Hileman, M. E., and Mazzullo, S. J., eds., Upper Guadalupian Facies, Permian Reef Complex, Guadalupe Mountains, New Mexico and West Texas, v. 1: Midland, Society of Economic Paleontologists and Mineralogists, Permian Basin Section, Publication 77-16, p. 133-150.

TYRRELL, W.W., 1969, Criteria useful in interpreting environments of unlike but time equivalent carbonate units (Tansill, Capitan, Lamar), Capitan reef complex, West Texas and New Mexico, *in* Friedman, G., ed., Depositional Environments in Carbonate Rocks: Tulsa, Society of Economic Paleontologists and Mineralogists Special Publication 14, p. 80-97.

WARD, R. F., KENDALL, C. St. C., and HARRIS, P. M., 1986, Upper Permian (Guadalupian) facies and their association with hydrocarbons — Permian basin, West Texas and New Mexico: American Association of Petroleum Geologists Bulletin, v. 70/3, p. 239-262.

WEIDLICH, O., 1996, Comparative analysis of Late Permian reefal limestones from the Capitan Reef (New Mexico, USA) and the Oman Mountains, *in* Reitner, J., Neuweiler, F., and Gunkel, F., eds., Global and regional controls on biogenic sedimentation, I. Reef evolution, research reports: Goettinger

Arbeiten zur Geologie und Palaeontologie, Sb2, p. 329-332.

WEIDLICH, O., BERNECKER, M., and FLÜGEL, E., 1993, Combined quantitative analysis and microfacies studies of ancient reefs: An integrated approach to Upper Permian and Upper Triassic reef carbonates (Sultanate of Oman): Facies, v. 28, p. 115-144.

WEIDLICH, O., and FAGERSTROM, J. A., 1998, Evolution of the Upper Capitan Massive (Late Permian), Guadalupe Mountains, New Mexico: Brigham Young University Geology Studies, v. 42, p. 167-188.

WEIDLICH, O., FAGERSTROM, J. A., RIGBY, K., and SENOWBARI-DARYAN, B., 1995, Capitan reef litho and biofacies: Implications for community structure, colonization strategy and carbonate nettoproduction (abs.): VII International Symposium on Fossil Cnidaria and Porifera, Abstracts, p.100-101.

WEST, R., 1988, Temporal changes in Carboniferous reef mound communities: Palaios, v. 3, p. 152-169.

WOOD, R., DICKSON, J. A. D., and KIRKLAND, B. L., 1994, Turning the Capitan reef upside down: A new reappraisal of the ecology of the Permian Capitan reef, Guadalupe Mountains, Texas and New Mexico: Palaios, v. 9, p. 422-427.

WOOD, R., DICKSON, J. A. D., and KIRKLAND, B. L., 1996, New observations on the ecology of the Permian Capitan Reef, Texas and New Mexico: Paleontology, v. 39, p. 733-762.

YUREWICZ, D. A., 1976, Sedimentology, paleoecology, and diagenesis of the massive facies of the Lower and Middle Capitan Limestone (Permian), Guadalupe Mountains, New Mexico and Texas: Unpublished Ph D. Dissertation, University of Wisconsin, Madison, 278 p.

YUREWICZ, D. A. ,1977, Origin of the massive facies of the Lower and Middle Capitan Limestone (Permian), Guadalupe Mountains, New Mexico and west Texas, *in* Hileman, M. E., and Mazzullo, S. J. (eds.), Upper Guadalupian Facies, Permian Reef Complex, Guadalupe Mountains, New Mexico and West Texas, v. 1: Midland, Society of Economic Paleontologists and Mineralogists, Permian Basin Section, Publication 77-16.

THE DYNAMIC CAPITAN REEF: AN IMAGE OF AN ANCIENT REEF AND SUGGESTIONS FOR FUTURE RESEARCH

B. L. KIRKLAND

Department of Geological Sciences, University of Texas at Austin, Austin, Texas 78712

S. A. LONGACRE

Texaco E&P Technology Department, 3901 Briarpark, Houston, Texas 77042

AND

E. L. STOUDT

Texaco E&P, 500 North Lorraine, Midland, Texas 79701

ABSTRACT: The origin of the Massive Member of the Capitan Formation in the Guadalupe Mountains of West Texas and New Mexico is controversial. It has been interpreted in many different ways, including as a barrier reef, a deep water skeletal wackestone mound, a cement boundstone mound, and a linear complex of buildups. The reality is that it is not simply one or the other, but a complex with characteristics of all of the above — a reefal complex that changes both in time and space.

The Massive Member of the Capitan Formation contains the components of a framework reef: frame-building organisms — bryozoans, platy sponges, and *Tubiphytes*; binding organisms — *Archaeolithoporella* and microbialite; infilling internal sediment; and marine cement. Much of the Capitan was probably deposited below wave base as a diverse framework reef. The profile of the Capitan reef varied throughout its deposition. At the mid-point of Capitan development, the slope of the reef was very steep (80° or more) and the living reef extended to a depth of approximately 140 m. The organisms that lived on the reef varied with depth. The shallowest portions of the reef were inhabited by *Collenella* and a diverse assemblage of other organisms. In slightly deeper water, large platy sponges formed meter-scale over-hangs and cavities, which were inhabited by a cryptic community dominated by sponges. In the deepest parts of the reef, frondose bryozoa created decimeter-scale framework and internal cavities inhabited by cryptic pendant sponges. These internal cavities were filled by a succession of encrustations and cements, beginning with thick layers (1–3 cm) of microbialite followed by profuse botryoidal aragonite (up to tens of centimeters thick), thin (1–4 mm) layers of radiaxial fibrous calcite, and much later, large crystals of meteoric spar.

The Capitan reef influenced deposition in surrounding environments. Throughout its growth the Capitan was a "sediment factory" that generated prolific amounts of sediment that accumulated in the fore-reef. Study of the upper Capitan at Walnut Canyon suggests that the living Capitan reef probably created a depositional high that modified sedimentation in the lee of the reef on the outer shelf. Study of the middle Capitan and equivalents at McKittrick Canyon suggest that at other times circulation on the outer shelf was unrestricted by the reef escarpment and its living surfaces.

Many questions about the Capitan remain unresolved, for example, what was the response of the reef to sea-level change? How did the reef profile change through time? Is there consistent evidence of exposure in the reef? Does diagenesis vary throughout the reef? How diverse is the biota in the lower and middle Capitan, and does biotic diversity change through time?

INTRODUCTION

The Capitan Formation and associated units (Fig. 1) constitute one of the best preserved ancient reef complexes in the world. The reef complex is particularly well exposed in the Guadalupe Mountains of Texas and New Mexico, where it has suffered little structural deformation and outcrops are easily accessible (Fig. 2). The objectives of this paper are to 1) review studies of the Capitan Formation prior to 1990, 2) synthesize recent observations, and weld them together to create an image of the Capitan reef through time along the shelf margin of the Delaware basin, and 3) summarize important remaining questions regarding the Capitan.

In the 1860s the Shumard brothers first recognized that the strata within the Guadalupe Mountains were of Permian age (Adams, 1988). Early studies documented the rich Capitan biota (Girty, 1908) and supported the interpretation of the Capitan as a reef (Lloyd, 1929; Crandall, 1929; Blanchard and Davis, 1929; Adams and Fresnel, 1950). King's (1948) inclusive study of the Guadalupe Mountains became the foundation for further detailed stratigraphic, paleontologic, and depositional studies (Newell et al., 1953; Hayes, 1964; and Kerans and Fitchen, 1995). In particular the work of Newell et al. (1953) remains one of the strongest studies of the Capitan reef to date. Many aspects of their work, for example descriptions of the biota, remain unsurpassed. The Capitan Formation is traditionally divided into lower, middle, and upper units correlative to the Seven Rivers, Yates, and Tansill formations, respectively (Babcock and Yurewicz, 1989). The Capitan is also divided into the Massive Member, the reef, and the Breccia Member, the fore-reef (Fig. 1) (Hayes, 1964).

In the 1960s the nature of the Massive Member of the Capitan Formation became the subject of an ongoing debate. A marginal mound hypothesis for deposition of the Capitan was proposed (Baars, 1964; Achauer, 1969; Dunham, 1972; Pray, 1986), suggesting that sponges and bryozoans acted as sediment bafflers and that the shelf-margin build-up formed below wave base. Dunham (1970) applied Achauer's (1969) organic bank model to the Capitan in his discussion of the terminology for reefs versus bioherms, and he expanded on the idea in his guidebook on the Guadalupe Mountains (Dunham, 1972).

This hypothesis was explored by a series of studies. A study of algae and organisms of uncertain affinities in the upper Capitan led to the conclusion that the algae act as both framebuilders and, along with cement, as binders, and the interpretation that the Capitan formed a submerged organic reef (Babcock, 1974, 1977). Hypotheses were also put forth suggesting that a significant portion of the reef framework consist-

Geologic Framework of the Capitan Reef, SEPM Special Publication No. 65

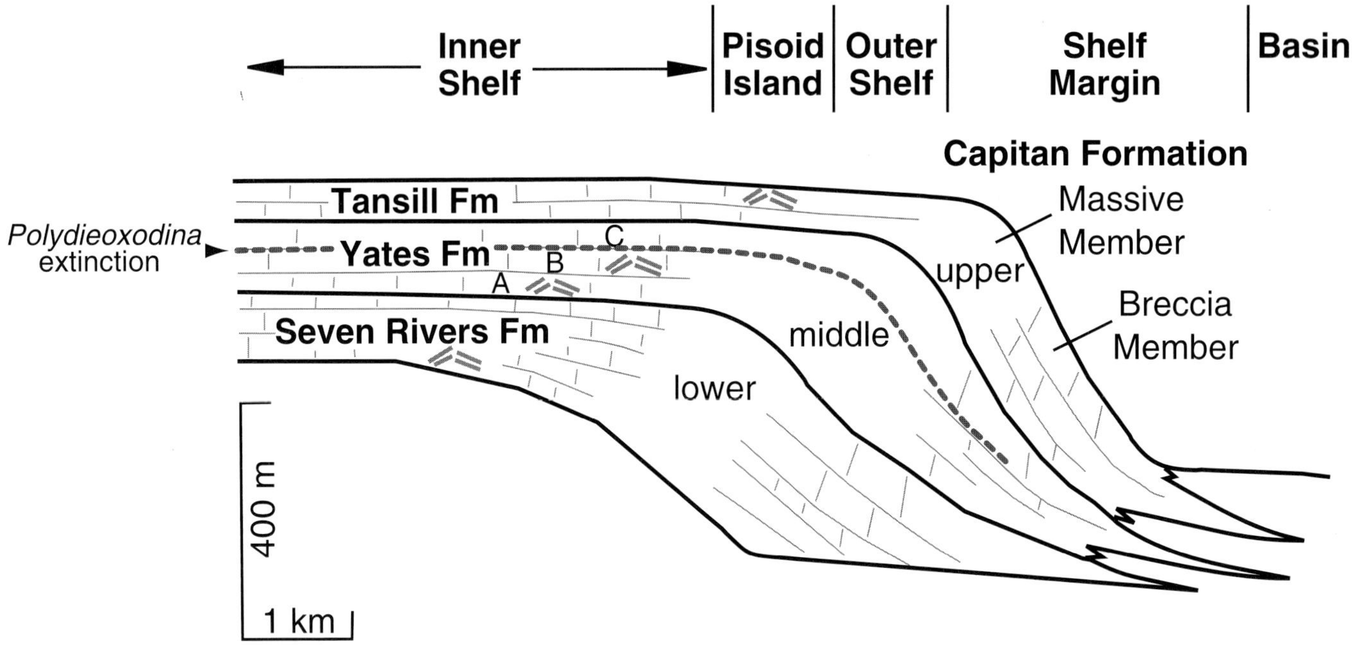

Fig. 1.—Summary diagram of stratigraphic nomenclature and terminology used in reference to the Capitan Reef and associated units.

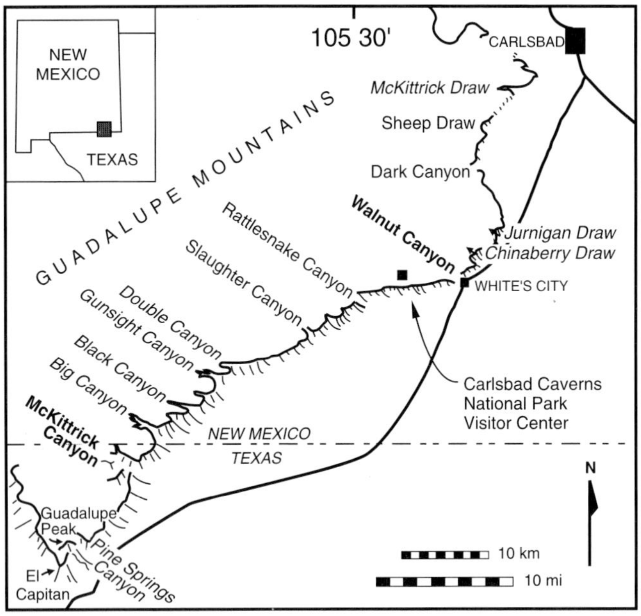

FIG. 2.—Location map showing localities that have received intensive study (bold), localities that have received more limited study (plain), and localities that remain virtually unexplored (italics).

ed of massive submarine cement boundstone (Mazzullo, 1977; Mazzullo and Cys, 1977; Schmidt, 1977). Botryoidal aragonite was recognized as dominating many parts of the Massive Member (Yurewicz, 1977) and was interpreted as forming framework, playing a role in binding, and as having the potential to form a wave-resistant reef (Mazzullo and Cys, 1977; Cys, et al., 1977, Schmidt, 1977). Studies of aragonite from the Capitan Formation and associated units were important for recognition of the significance of marine cementation in ancient buildups (Loucks and Folk, 1976; Mazzullo, 1980).

Another study of the lower and middle Capitan (Yurewicz, 1976, 1977) revealed variations with respect to marine cement type (botryoidal aragonite versus radiaxial calcite) and abundance vertically through the section. In addition, Yurewicz (1977, 1976) proposed that the reef was never emergent, based on the present-day dip of beds. Hurley (1978, 1989) studied the shelf sediments that are equivalent to the lower Capitan — those of the Seven Rivers Formation. Based on study of eight geopetal structures and the present dip of bedding planes, he suggested that the shelf beds nearest the Capitan had a primary seaward dip of 8° ± 2°. Hurley's description of these "fall in beds" provided strong evidence in support of the marginal mound model. This long-term research effort was summarized by Pray (1986), who at that time described the Capitan as "an outstanding example of massive limestone formed at a submerged shelf edge," consisting largely of sponge wackestones and massive submarine-cement boundstones.

Although the marginal mound model is still widely accepted, the idea that the Capitan was deposited as a barrier reef was never completely abandoned, and has recently been revived. As early as 1955, Newell noted the similarity between the Capitan and modern barrier reefs and recognized within the Capitan the presence of frame-builders, binders, and internal sediment. Studies of the Capitan Formation that focused on algae and drew comparisons to modern reefs also led to the conclusion that the Capitan was a barrier reef (Johnson, 1942; Klement, 1966). Field relationships, recognition of frame-building organisms, and the presence of low-energy, restricted deposits on the shelf leeward of the reef were listed as evidence in support of a barrier-reef model for deposition of the Capitan Formation (Cys, 1971). Based on the biotic constituents of the reef and the nature of associated facies, Cronoble (1974) also concluded that the Capitan represented a complex ecologic reef system.

Many studies recognized the presence of frame-building and binding organisms in the Capitan (Newell, 1955, Babcock, 1974, 1977; Yurewicz, 1976, 1977; Cys, 1971; Kirkland et al. 1993; Wood et al., 1994, 1996). The recognition of cavities with geopetal infillings of contemporaneous sediment provided further evidence consistent with the existence of framework (Babcock, 1977; Yurewicz, 1977; Harwood, 1989). Some studies suggested that the reef was capable of wave resistance, but did not grow to sea level (Yurewicz, 1976, 1977); others suggested that it did grow to sea level (Cys, 1971; Kirkland George, 1992). The most recent studies suggest that most, but not all, of the reef was too delicate to be resistant to wave action (Wood et al., 1996, Kirkland et al., 1998).

Studies of the Massive Member of the Capitan alone cannot answer the question of whether the Capitan reef grew to sea level. Evidence for modification of the environments of the outer shelf by an emergent reef is subtle and variable. In the upper Capitan, field relationships and the presence of low-energy, restricted facies on the outer shelf leeward of the reef support the hypothesis that the Capitan was a barrier reef (Cys, 1971; Cronoble, 1974; Kirkland and Moore, 1996). Other studies, largely those focusing on the lower and middle Capitan, suggest that the outer shelf was open to current energy, and, thus, the reef did not act as a shelf-margin barrier (Hurley, 1978; Kerans and Harris, 1993; Tinker, 1996). Studies focusing on the uppermost upper Capitan suggest deposition as a series of patch reefs (Noé, 1996; Rigby and Senowbari-Daryan, 1996a). Taken collectively these studies indicate that water depth and shelf margin position of the reef varied significantly through time, and that many questions remain unanswered.

METHODOLOGY

The Permian Reef Geology Trail, constructed in the early 1980s in McKittrick Canyon in Guadalupe Mountains National Park (Mruk and Bebout, 1993), has been the focus of much of our research. Observations have also been made in Dark, Walnut, Rattlesnake, and Slaughter canyons (Fig. 2). More than 200 samples were collected and examined in conjunction with this work. All samples were taken with permission of the National Park Service in conjunction with previous studies (Bebout and Kerans, 1993; Kirkland-George, 1992; Rahnis, 1995; Wood et al., 1994, 1996, Kirkland et al., 1998). Observations were made on outcrop surfaces, polished slabs, and thin sections.

OBSERVATIONS ON THE CAPITAN FORMATION

Biotic and Binding Components of the Capitan Massive

Many of the organisms found in the Capitan Formation had the capacity to act as frame-builders. Sphinctozoan sponges are volumetrically the dominant organisms in the reef, and are, by far, the largest and most abundant organisms in the Capitan Massive. Sponges with a plate-like morphology are found throughout the reef. Small sponges (a typical specimen is 2 cm thick by 8 cm wide) form the roofs of decimeter-scale cavities in the deepest part of the reef (near the forereef transition). Large (up to 2 m wide), plate-like sponges, for example

FIG. 3A.— Sponges on an upper Capitan outcrop surface at the mouth of Walnut Canyon. *Gigantospongia* (G) (2 cm thick) with cryptic sponges attached to its underside.

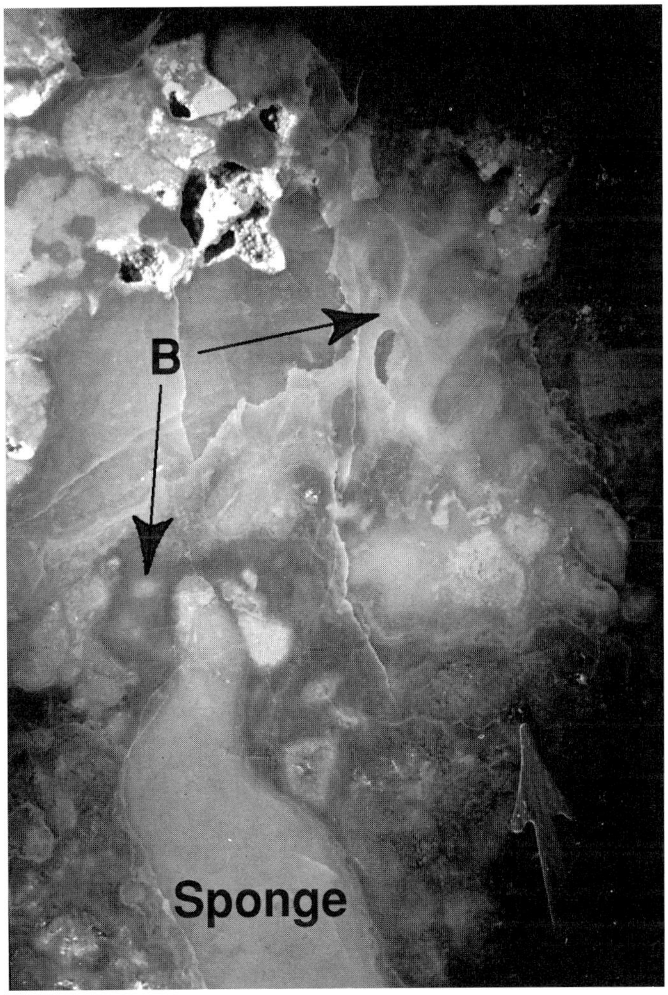

FIG. 3B.— Fenestrate bryozoan (B) and attached sponge in a polished slab from the middle Capitan at McKittrick Canyon. Arrow is 1.6-cm long.

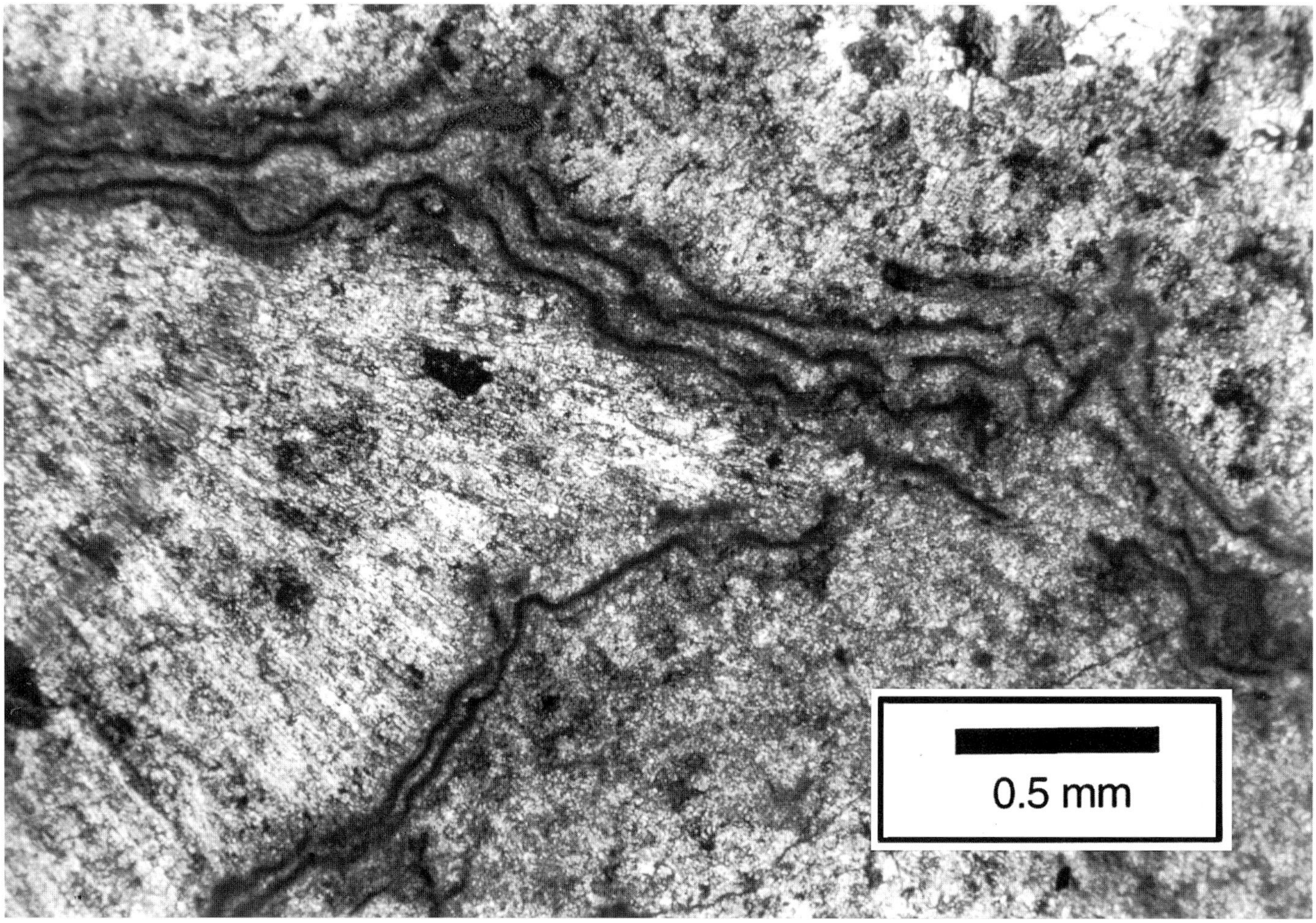

Fig. 4.—*Archaeolithoporella* encrusting previously aragonitic botryoids in thin-section photomicrograph.

Gigantospongia (Fig. 3A) (Senowbari-Daryan and Rigby, 1996), occur about 30 m below the reef-to-outer shelf transition in the upper Capitan at Walnut Canyon and in the middle Capitan at McKittrick Canyon. These large, platy sponges form the ceilings of meter-scale cavities. In many localities, tubular and segmented sponges are so densely packed that they functioned as framework, and some support cavity infillings, which accumulated above them.

Fenestrate bryozoans, while volumetrically a minor constituent of the reef (usually <5%), form the roofs and floors of most of the decimeter-scale cavities (Wood et al., 1994, 1996) (Fig. 3B). Other organisms attach to the underside of bryozoan fronds and accumulations of fine-grained sediment occur above the bryozoan fronds. Framework cavities formed by bryozoans are most common in the deepest part of the reef (near the reef-to-forereef transition).

Other organisms had the capacity to function as framebuilders. In samples of the upper Capitan at Walnut Canyon, *Tubiphytes* and *Leracaritubus problematicus* (Senowbari-Daryan and Rigby, 1996b), both organisms of uncertain affinities, form the substrate for encrustation by other organisms. Rugose corals are also present, but, perhaps because they are rare, they have not been clearly recognized in a frame-building capacity.

The tubular sphinctozoan sponges, which numerically and volumetrically dominate the reef, are consistently found in a position that suggests that they grew downward, that is, with their holdfasts and narrow, juvenile stages positioned above their larger segments and arms or branches pointed downwards towards the base of the outcrop. In what were the deepest parts of the reef (near the reef-to-forereef transition), sponges are commonly attached to and hang from bryozoan fronds. In other localities the sponges attach to one another, sometimes growing in chains of up to four organisms, one below the other. Nodular bryozoans and rare rugose corals are also found inhabiting the roofs of cavities, but are volumetrically minor compared to the sphinctozoans. Along the Permian Reef Geology Trail in McKittrick Canyon, near the reef-to-outer-shelf transition, sphinctozoan sponges are also oriented downward. These sponges are surrounded by coarse skeletal debris, but how or where they are attached to a substrate is not visible in outcrop. In Sheep Draw Canyon, in localities near the reef-to-outer-shelf transition, sponges are found growing downward from roofs of cavities.

Micritic encrustations of various origins are volumetrically significant within the Capitan. *Archaeolithoporella* (an organism most commonly thought to be algal) is typically the first encrustation around framework organisms or cryptic biota (Fig. 4). In thin section, *Archaeolithoporella* is composed of thin, subparallel laminae of homogeneous micrite separated by layers of fine microspar. *Archaeolithoporella* may or may not be interlaminated with layers of peloidal micrite, but it is consistently encrusted by layers of peloidal, probably microbial, micrite. Layers of peloidal micrite drape the upper surfaces of organisms and also encrust the undersides of framework and cryptic biota. Individual accumulations of peloids may be steep-sided (up to 90° from horizontal). In thin section, micritic encrustations are clotted, contain rare cyanophyte fossils, and, when viewed under high-resolution SEM, contain abundant nannobacterial textures (Kirkland et al., 1998; Folk, 1993). Clotted layers of micrite are commonly bounded either by denser layers of clotted micrite or by *Archaeolithoporella* (Fig. 4).

Sediment-producing biota for the Capitan reef includes brachiopods, crinoids, articulated algae such as *Eugonophyllum* and *Mizzia* (Walnut Canyon) (Babcock, 1974; 1977), as well as occasional trilobites and cephalopods. Internal sediment is found throughout the Capitan, commonly as geopetal infilling within framework voids or between organisms; however, the nature and distribution of the sediment varies. In the decimeter-scale cavities in the deeper parts of the reef, internal sediment is dominated by fine-grained peloids, but other allochems include larger peloids, dasycladacean algal fragments, micrite-encrusted grains, echinoderm fragments, and foraminifera (including fusulinids). Allochems found in neptunian dikes include brachiopods, foraminifera, fragments of *Tubiphytes*, as well as both dasycladacean and probable eudoteacean (phylloid) algae.

Near the reef to outer shelf transition, cement-filled cavities are absent and grain-rich sediment dominates what were the shallowest parts of the reef. This sediment is poorly sorted, angular to subangular, and contains a diverse array of allochems, including probable sponge fragments, bryozoan fragments, brachiopods, mollusks, dasycladacean algae, diverse foraminifera, fragments of *Tubiphytes*, and peloids. Some allochems have micritic, probably microbial, encrustations that contain cyanophyte fossils. Evidence of bioerosion is rare throughout much of the Capitan (Fig. 5), but extensive borings and micritization are reported from the uppermost Capitan Formation (Noé and Mazzullo, 1992).

Marine cement is volumetrically the most significant component through much of the Massive Member of the Capitan Formation. Both the percentage of cement and the volume of each different type of marine cement (relative to one another) vary between the lower, middle, and upper Capitan Formation (Yurewicz, 1976, Babcock and Yurewicz, 1989) and between the Massive (reef) and Breccia (forereef) Members (Mruk, 1989). The

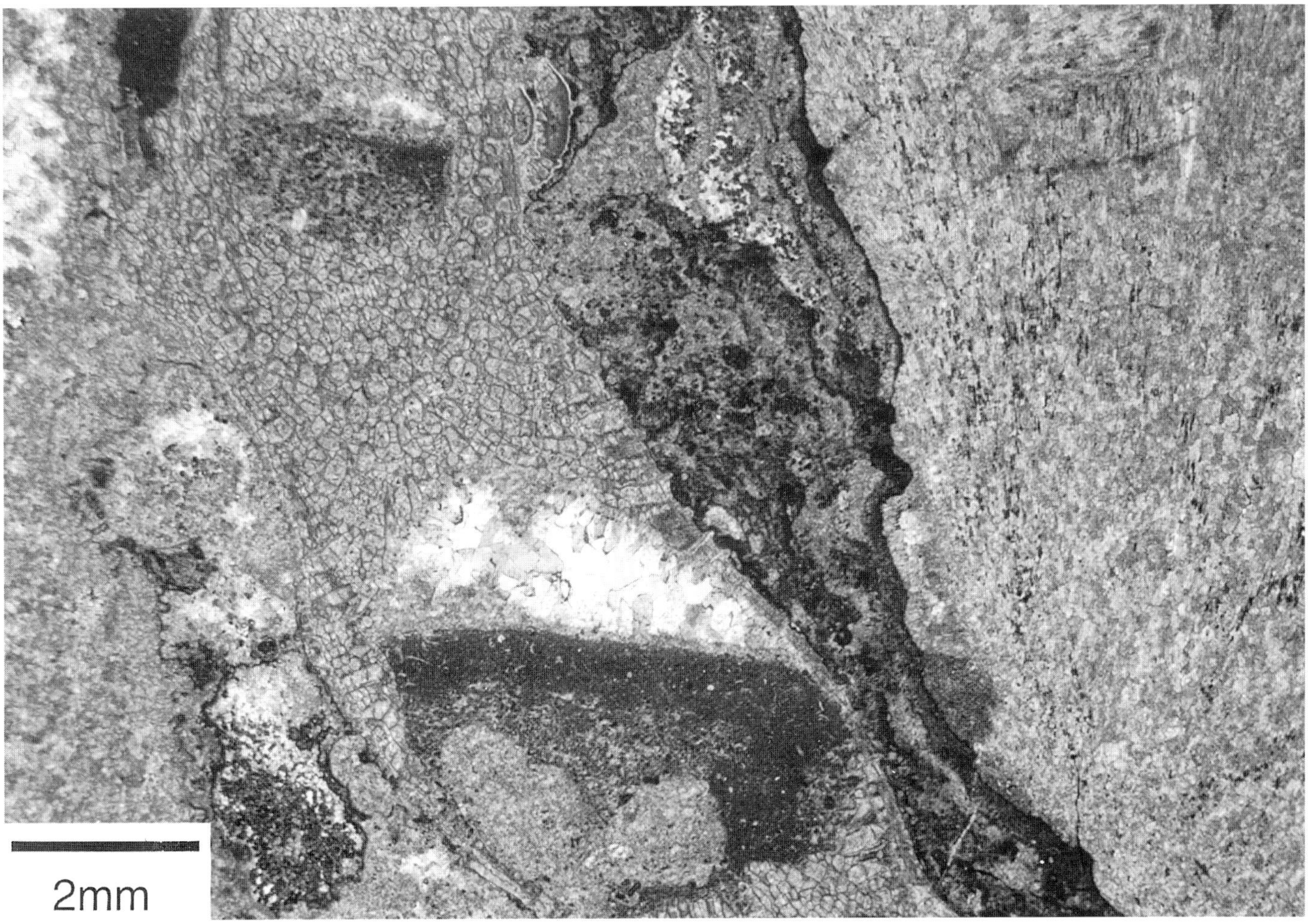

Fig. 5.—Photomicrograph of sample from the middle Capitan along the Permian Reef Geology Trail showing a bryozoan, which has been bored and refilled by internal sediment. A thin layer of microbial micrite encrusts the outside of the bryozoan (center of photomicrograph).

initial and dominant marine cement through much of the Capitan was botryoidal aragonite (Mruk, 1989, Kirkland et al., 1993) (Fig. 6). Initial layers of aragonite are interlaminated with *Archaeolithoporella* and layers of clotted, probably microbial, micrite. Subsequent layers generally consist of larger botryoids (1–4 cm in diameter). Botryoids may comprise over 90% of the reef massive and are more abundant in the upper Capitan (Yurewicz, 1976). Botryoids have not been found as a primary precipitate in the forereef, or Breccia Member (Mruk, 1989), although isopachous rims of aragonite are present in those facies.

Radial calcite is rare in the Massive Member. In contrast, radiaxial fibrous calcite and fasicular optic calcite are consistently the last marine pore-filling cements in the Massive Member and form isopachous layers up to 4-mm thick (Kendall, 1983; Mruk, 1985; Tucker and Wright, 1990; Kirkland et al., 1993; Rahnis, 1995). In the Breccia Member (forereef), radiaxial fibrous calcite is the first abundant marine cement (Mruk, 1989). Radiaxial prismatic calcite commonly grades into meteoric spar, although in some samples, a layer of dolomite separates the calcites (See Rahnis and Kirkland, this volume).

In the subsurface, anhydrite is the final pore-filling phase (Garber et al., 1989; Scholle et al., 1992). In outcrop, large volumes of meteoric spar are present, probably due to replacement of evaporites during the Tertiary. These spars are divided into three groups based on appearance in cathodoluminescence and stable isotopic signature (Mruk, 1989; Hill, 1996). The volume of meteoric spar varies laterally through the Guadalupe Mountains. Small amounts of pyrite are associated with this late stage spar. Late fluorite is also present in a few samples.

Bathymetric Profile and Facies Trends Within the Reef

Careful reconstruction of time lines and documentation of biotic differences form a basis for reconstruction of the bathy-

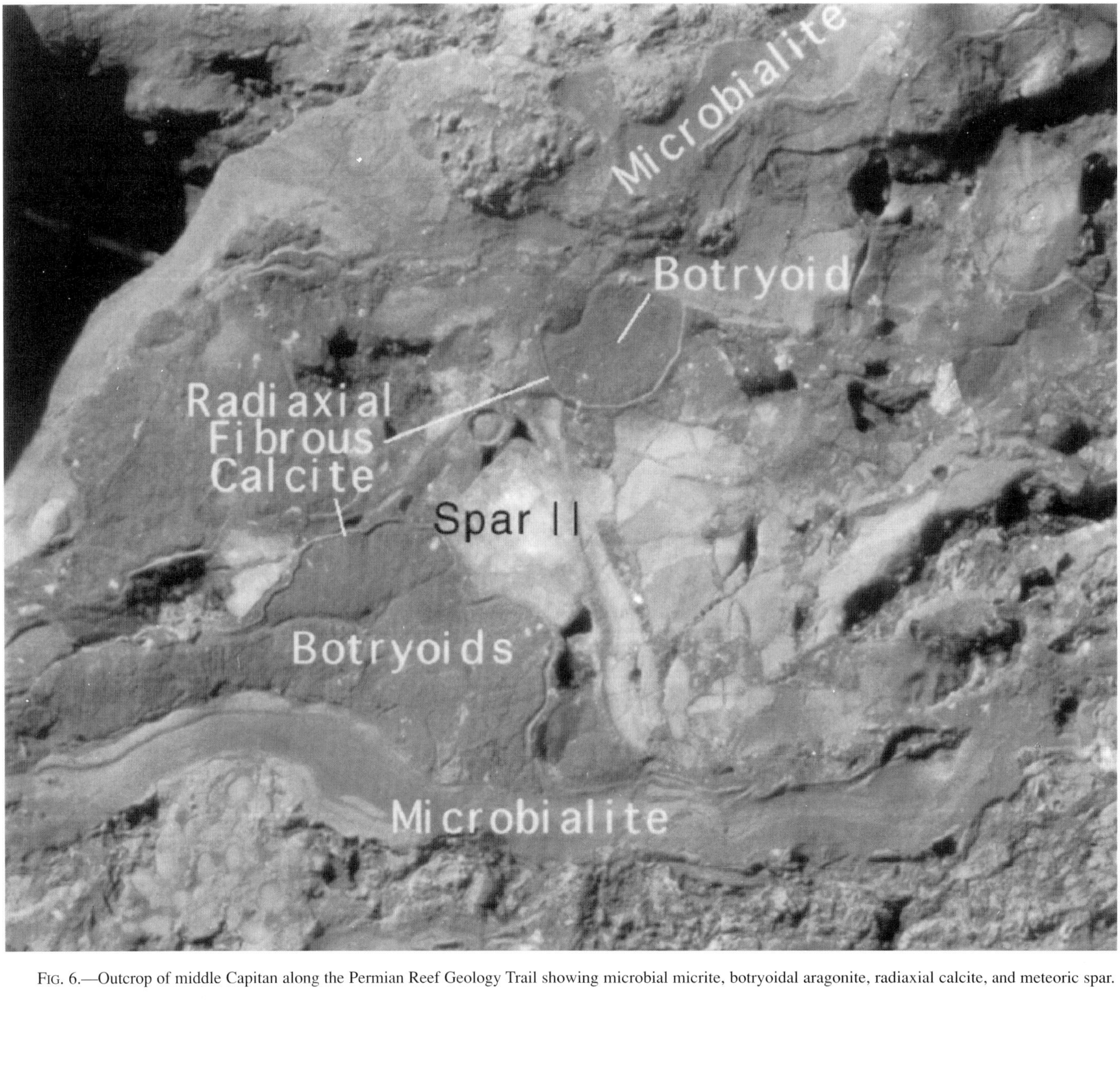

FIG. 6.—Outcrop of middle Capitan along the Permian Reef Geology Trail showing microbial micrite, botryoidal aragonite, radiaxial calcite, and meteoric spar.

metric profile of the marine reef. Based on measurements made on a balanced cross-section constructed by Tinker (1996), the angle of the basin-facing profile of the Capitan reef becomes steeper over time, varying from 40° in the lower Capitan to 80° in the middle Capitan (Hurley, 1978, 1989; Kirkland et al., 1993). The extinction of the fusulinid *Polydieoxodina* marks the boundary between Yates B and the younger Yates C (Newell, 1955). Switchbacks along the Permian Geology Reef Trail allowed us to trace this timeline through the middle Capitan reef, where it marks a steep (80°) plane through the reef (Fig. 7) (Kirkland et al., 1993). A study of weathering patterns, distribution of fossils, and lithologies, along with comparison to the Magnesian Limestone of County Durham, England, also led to the conclusion that the front of the upper Capitan also dipped steeply (60–90°) (Smith, 1973). Outcrop photos of the north walls of Slaughter and McKittrick Canyons show that the Capitan Massive weathers to a steep face on the basinward margin (Tinker, 1996; Kerans and Tinker, 1997). The orientation of Neptunian dikes is also essentially vertical and dominantly parallel to the front of the reef (Stanton and Pray, 1993; Kirkland et al., 1993).

Recognition of the timelines and the steep profile allows us to distinguish subtle variations in the distribution of biologic and lithologic facies. In the middle Capitan at McKittrick Canyon and in the upper Capitan at Dark Canyon, near the reef to outer-shelf transition, articulated, aligned crinoids are preserved. In the shallowest part of the reef (0–10 m below the reef to outer-shelf transition), *Collenella* is common (Fig. 8, 9). *Collenella* is also found near the reef-to-outer-shelf transition in the Middle Capitan at McKittrick Canyon and the Upper Capitan at Walnut and Dark canyons (Babcock, 1974, 1977; Toomey and Babcock, 1983), but it has not been reported from what would have been deeper parts of the reef. *Collenella* is an organism of uncertain affinities, but an algal affinity or an algal-foraminiferal symbiosis, and therefore a dependence on light, have been suggested. The upper reef is sediment-rich (Dunham, 1972) at all of the upper reef-outer shelf localities studied. In the upper Capitan at the mouth of Walnut Canyon, Babcock (1974) documented four zones, listed here in order of what would have been shallowest to deepest: 1) *Tubiphytes*, Sponge Boundstone/ Packstone; 2) *Tubiphytes*, Sponge Boundstone/ Packstone-*Archaeolithoporella*/Nodular Boundstone; 3) Phylloid algal boundstone; and 4) *Archaeolithoporella*/Nodular Boundstone. In a study of fabric development, samples from Babcock's uppermost zone contain a millimeter-scale, laminated fabric that is tightly bound by encrusting organisms and microbial micrite (Kirkland et al., 1998).

In the upper Capitan at Walnut Canyon and the middle Capitan along the Permian Geology Reef Trail in McKittrick Canyon a zone of large platy sponges (*Gigantospongia*) occurs about 30 m below the reef-to-outer-shelf transition. Surveys along the Permian Reef Geology Trail suggest that between 30 m below the reef-to-outer-shelf transition and 140 m below the

FIG. 7.—Switchbacks along the Permian Reef Geology Trail, north of McKittrick Canyon. On the basinward side (left) of the line, the fusulinid *Polydieoxodina* is absent. On the shelfward side (right) of the line, it is abundant.

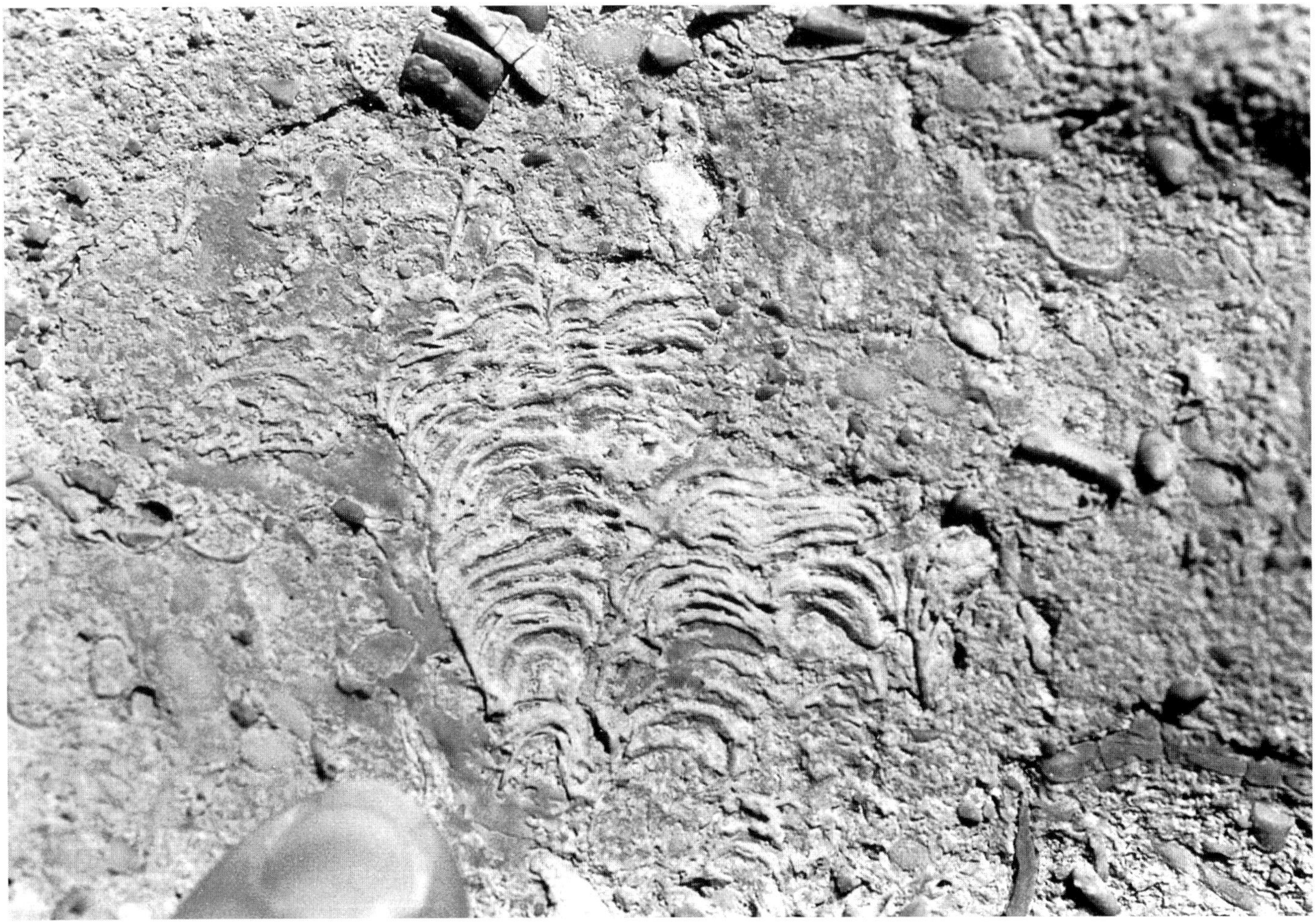

FIG. 8.—*Collenella* in a weathered surface along the Permian Reef Geology Trail in McKittrick Canyon.

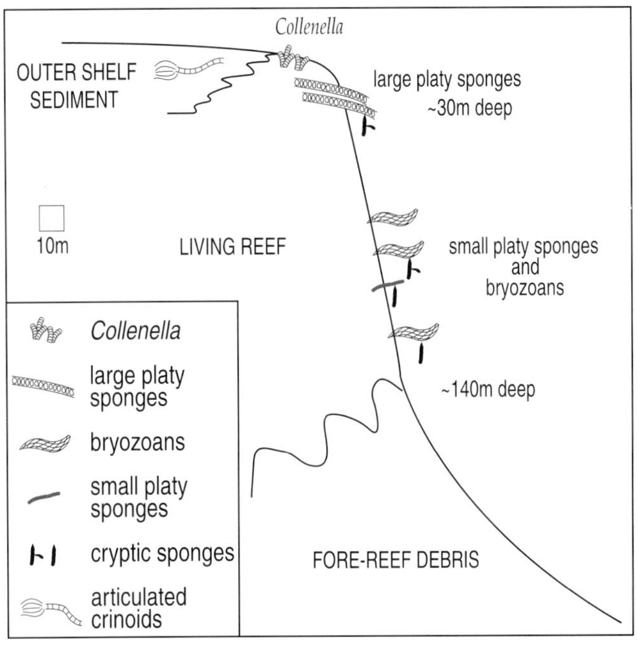

FIG. 9.—Biotic distribution along a reconstructed profile of the middle Capitan Formation, McKittrick Canyon.

reef-to outer-shelf transition (at the reef-to-forereef transition), zonation within the reef is either imperceptibly subtle or absent. Fenestrate bryozoans form much of the initial framework for the lower part of the reef. Fine-grained, largely microbial micrite is abundant, species diversity is low (Johns and Kirkland, 1997), organisms are small (sponges are generally 2–5 cm long), framework voids are small, and bryozoans and platy sponges form cavity floors and ceilings. Bryozoans, particularly fenestrate bryozoans with attached sponges, may be more abundant in lower parts of the reef.

The Forereef

Because the Breccia Member has recently been well documented (Mruk and Bebout, 1993; Melim and Scholle, 1995), our discussion will be brief. The forereef is volumetrically much larger than the reef (Fig. 10). The average thickness of the Breccia Member is about 381 m (1,250 ft) compared to 122 m (400 ft) for the Massive Member (Hayes, 1964). Digital quantification of the relative area of reef versus forereef as diagramatically represented on a cross-section of McKittrick Canyon (Tinker, 1996) and a photomosaic of the north wall of Slaughter Canyon, show that the middle Capitan Breccia

Member probably contains two to three times the amount of sediment as the middle Capitan Massive Member. Relative to the middle Capitan, the lower Capitan forereef contained less debris and the upper Capitan forereef contained more debris. In contrast to the Massive Member, the Breccia Member is bedded and virtually entirely allochthonous. Sediment in the forereef is diverse, ranging from blocks of biolithite (a few centimeters to several meters in diameter) to individual allochems typical of those found in the outer shelf (fusulinids, crinoid fragments, dasycladacean algae, bryozoan fragments, brachiopods, and molluscs). Some layers of the forereef are dominated by micrite or siliciclastics. The sediments of the forereef provide a chronicle of variation in production of sediment in the shelf margin reef and outer shelf (Melim and Scholle, 1995). For example, in one locality along the Permian Reef Geology Trail, hexactinellid sponges are preserved apparently *in situ* and are surrounded by sediment rich in silt grains, suggesting that hexactinellids might have inhabited the forereef at a time of relative sea-level low-stand.

DISCUSSION—A LIST OF KNOWNS AND UNKNOWNS

A Reef

The Massive Member of the Capitan was clearly deposited as an organic reef. It fulfills criteria defined through study of both modern and ancient reefs: the presence of organisms that functioned as frame-builders and binders as well as internal sediment and cement that fills framework cavities (Ginsburg and Lowenstam, 1958). Bryozoans and sponges, however unlikely candidates as they may seem to be, filled the roles of frame-builders. The cryptic fauna also filled the role of frame-builder by virtue of sheer abundance. While the cryptic fauna is well preserved, the lack of a clearly defined surface biota is perplexing (Wood et al., 1996). We suggest that the platy sponges and frond-like bryozoans probably inhabited the surface of the steep, basinward-facing reef wall. The lack of clearly defined biota inhabiting upward-facing surfaces may be a function of preservation. Many of the organisms that are more common in the forereef, for example brachiopods, foraminifera, bryozoans, and molluscs, could have been part of the open surface biota of the reef. Even with further study, this question may remain difficult to resolve.

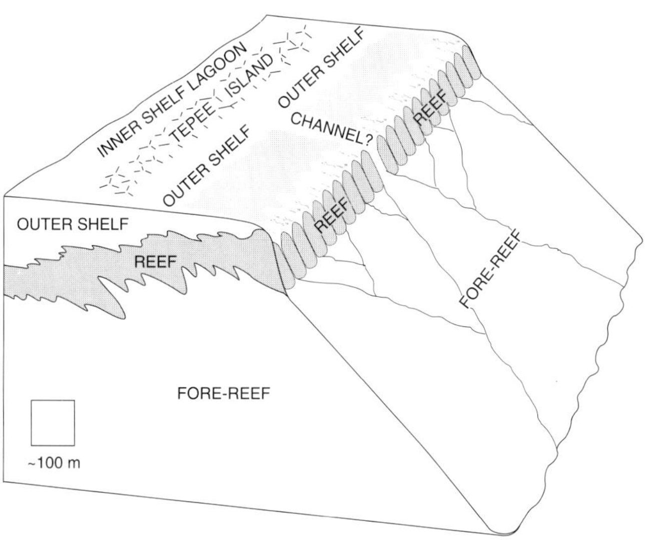

Fig. 10.—Schematic diagram of the Capitan reef at middle Permian time (near top of Yates B).

The role of binder was filled by *Archaeolithoporella* and microbial micrite. An imaginary diver looking into the mouth of dimly lit cavities on the reef wall would probably have seen *Archaeolithoporella* encrusting, and thereby binding, any substrate to which it could attach, which included both the frame-building and cryptic biota. Sometimes *Archaeolithoporella* was interlaminated with millimeter-scale layers of aragonite, which surely augmented its capacity to bind the reef. In the darkest part of the cavities, where other organisms could not survive, microbial mucilage trapped and probably precipitated micrite (Reitner, 1993). Unlikely as it may seem, this gelatinous mucilage functioned as a binder, and these centimeters-thick layers of micrite were perhaps the most important agent in lithification/stabilization of the reef. In the deepest parts of the reef, the decimeter-scale cavities that these frame-builders and binders created were filled with fine-grained, penecontemporaneous internal sediment. In shallower parts of the reef, layers of microbial micrite are thin, and geopetal infillings, where they are present, fill small (less than centimeter-scale) cavities. Coarse sediment surrounded and probably choked organisms.

If our diver were to blast away part of the reef wall, she would have found that marine cements lined cavities and filled pores. Aragonite botryoids (1–3 cm in diameter) grew up from sediment-filled floors. The large volumes of aragonite that precipitated suggest that a large volume of marine fluid moved through the reef. Over time, aragonite precipitation progressively filled porosity and probably slowed flow rates. After the reef had lost significant porosity to microbial micrite, internal sediment, and abundant botryoidal aragonite, precipitation of radiaxial fibrous and fasicular optic calcite filled most of the remaining void space. The cause of this fundamental shift in saturation state remains perplexing despite detailed study of the calcites (Rahnis and Kirkland, this volume). A greater understanding of details of the hydrologic system within the platform is essential for geochemical studies seeking information about the composition of waters in the Delaware Basin and the global Permian ocean.

Reconstructing a Bathymetric Profile and Biotic Zonation

The basis for reconstructing the bathymetric profile of the Capitan Reef is a timeline based on the last occurrence of *Polydieoxodina* (Fig. 7). This profile is the basis for reconstruction of biotic zonation within the reef. Unfortunately the timeline and the zonation are based on negative evidence—our inability to find the fossil in rocks where we hypothesize it should be present. This time line is so distinct, however, that finding and drawing it has become a student exercise. The observations that this time line parallels neptunian dikes and is echoed throughout the section in weathering profiles further supports the interpretation. This time line is extraordinarily significant, because it suggests that the Capitan, like modern reefs, had a steep front at depth. This interpretation provides an essential framework for organizing observations about the reef biota.

The imaginary diver visiting the reef at middle Capitan time would have dropped straight down (80–90° slope) the front of the reef wall. At a minimum, 90% of the living reef would have existed below fair-weather wave base. Our diver might have been jostled slightly by waves while viewing *Collenella* in the shallowest part of the reef, but by the time she descended to the depth where huge platy sponges like *Gigantospongia* thrived she would have been below wave base. These large sponges provided attachment points and sheltered cavities, or crypts, populated by large cylindrical sponges (15 cm long). As the diver descended further she would have seen the steep reef wall inhabited by bryozoans and platy sponges. She would have been able to look in on decimeter-scale cavities filled with centimeter-scale cryptic sponges. In the dimly lit parts of these cavities she would have seen encrusting sheets of *Archaeolithoporella*, probably bright red, and in the deepest, darkest parts of the cavities she would have found layers of microbial mucilage.

At about 140 m depth our diver (who by virtue of being imaginary would not have been constrained by diving tables) would have reached the point where the steep wall of the reef gave way to the huge (two to three times bigger than the reef) talus pile composed of reef blocks and sediment (Fig. 10). Our diver might have encountered sediment that originated on the outer shelf and certainly would have seen blocks, some of them immense, of biolithite and skeletal fragments that fell from the reef front. The reef was continually oversteepened by growth of organisms. Neptunian dikes within the reef suggest synsedimentary fracturing. This fracturing may have been related to progradation of the reef and forereef over poorly lithified basinal sands (Newell et al., 1953). Fracturing, either related to oversteepening or differential subsidence, or a combination of both, probably contributed to spalling of large blocks off of the reef front. In any case, the reef was a prolific carbonate factory, and most of the sediment it produced was deposited in the forereef.

While descending, our diver might have noticed the environmental changes that leave an elusive signature in the rock record at best: a thermocline, a halocline, or perhaps subtle biotic changes we have yet to recognize. At some point, the concentration of oxygen in the sea water dropped (Babcock, 1974; Melim and Scholle, 1995) — perhaps gradually or perhaps abruptly in association with a distinct halocline. Certainly by the time that the upper Capitan was being deposited, but probably through much of the Guadalupian, the waters of the deepest part of the Delaware Basin were virtually anoxic, as suggested by the lack of burrowing in the basinal siliciclastics and carbonates (Babcock, 1974).

Topics for Future Research

The debate over whether the Capitan reef existed as a "shallow" or "deep" now concerns only the topmost 10% of the Massive Member in the lower, middle, and upper Capitan. The question of "shallow barrier" versus "deep mound" is significant, because it affects our understanding of controls over platform development, the effects of subaerial exposure, and porosity development in reefs throughout the rock record. If the shallowest 10% of the Capitan reef were at times within the zone of wave action it should contain distinct lithologic or biotic zones that change over a short distance; these zones should vary through time with respect to relative sea level. To further complicate the issue, if the uppermost part of the reef was within the reach of storm wave base, it would have had a lower preservation-potential than the rest of the reef. The shallowest 10% of the Capitan reef probably holds the answers to many questions. This promises to be the most complex part of the reef and remains, for the present, the least understood part.

The answer to the question of whether the Capitan reef grew to sea-level lies, in part, in analysis of whether the reef modified environments in its lee and thereby controlled deposition of the sediments of the outer shelf. At some times and in some localities, the reef did little to break waves and thereby modify sedimentary processes in the back reef/outer shelf. For example, along the Permian Reef Geology Trail in McKittrick Canyon, outer shelf sediments of the middle Capitan show evidence of increased environmental energy shelfward from the Capitan shelf margin (Hurley, 1989; Kerans and Harris, 1993). A modern analog for this type of shelf margin/outershelf would be the western margin of the Bahamian platform. At other localities, in other parts of the section, the shallow top of the reef probably took the brunt of wave energy and significantly modified environments in the outer shelf. For example, in the shallowest part of the upper Capitan at Walnut Canyon, the reef fabric is composed of millimeter-scale laminae that are similar in appearance to samples from modern Caribbean boilers or Pacific reef crest laminae; these facies were deposited in high-energy environments. The sediments of the adjacent outer shelf show evidence of decreased environmental energy away from the shelf margin (Kirkland and Moore, 1996). A modern analog for this type of margin would be the Belize reef or, with a more substantial drop in relative sea-level, the McLeod Basin, Australia (Logan, 1987). The conflicting data surrounding this issue underscore the need for further research.

Large portions of the Capitan Formation, particularly the lower Capitan, have not been studied in adequate detail and outcrops in most canyons remain virtually unexplored (Fig. 2). From the studies that have been done, however, the reef seems to have evolved from build-up that thrived below wave base during deposition of the middle Capitan, to a build-up that may have lived within the zone of wave action during deposition of at least part of the upper Capitan, to a series of patch reefs by the time the uppermost Capitan was deposited. Before we can develop a more accurate model, we must examine more of the outcrop, establish the pattern of variation in the uppermost reef with respect to cyclical changes in sea level, and perform a detailed fabric and facies study within a well-defined sequence stratigraphic framework. A detailed stratigraphic framework for the Capitan Massive and its shelfal equivalents has been developed in North McKittrick and Slaughter canyons (Tinker, 1996; Osleger, 1998). Within their frameworks, the Capitan Massive and associated units (lower Capitan/Seven Rivers and middle Capitan/Yates) display several aggradational and progradational phases, but based on stratal geometries, the reef never seems to have grown into water depths less than 30 m. A model for development of those stratal geometries, using data only from the Middle Capitan, suggests that they might result from compaction of poorly lithified basinal sands and mud (Saller, 1996). To further complicate the issue, a single preliminary study suggests that a chemical signature of exposure might be present in the middle Capitan (Tinker, 1996). No previous study of diagenesis in the reef has been detailed enough to thoroughly document the chemical signature of early meteoric diagenesis, if it did occur.

Several researchers have suggested that the gross morphology of the outcrop north of Rattlesnake Canyon (Fig. 2) resembles spur and groove structures or buttress and channel structures of modern reefs (P. Davies, pers. comm., 1986; M. Queen, pers. comm., 1987; Kirkland George, 1992). In contrast, the outcrop south of Rattlesnake Canyon has a different expression, which is reminiscent of a talus slope below an escarpment. Preliminary observations from the area just below the Visitors Center at Carlsbad Caverns National Park suggest that this is a complex problem worthy of further study. As the record of the steep-walled reef front is suggested in canyon walls, it is not unreasonable to suspect that spur and groove morphology, if it was indeed present, influenced erosion patterns.

The Guadalupe Mountains provide us access to a large, tropical Permian ecosystem where paleoecologic interactions are preserved *in situ*. These interactions, however, have only been studied in great detail in two localities. Most of the lower and middle Capitan remain largely unexplored. The Capitan of the Guadalupe Mountains is remarkable in that large sections of the reef are preserved intact and have undergone little structural deformation. Patterns of, and variations in, reef growth through time are preserved. The basic sequence stratigraphic framework has now been established (Tinker, 1996; Osleger, 1998), but the details of reef growth in relation to sea-level change are not well understood. This reef existed during what was probably a time of change, just before a major biotic extinction. The extinction at the end of the Capitanian rivals the Cretaceous/Tertiary extinction in scale and was a prelude to the larger extinction at the end of the Permian (Stanley and Yang, 1994). Efforts to understand the events leading up to this extinction might be facilitated by a greater understanding of how this ecosystem changed through time. While the upper Capitan biota is currently being documented and updated (Senowbari-Daryan and Rigby, 1996a; Rigby and Senowbari-Daryan, 1998), the last major study of lower and middle Capitan biota was published in 1908. Clearly it is worth revisiting the Capitan, and much work remains to be done.

ACKNOWLEDGMENTS

Acknowledgment is made to the donors of the Petroleum Research Fund, administered by the American Chemical Society, for partial support of this research. B.K. gratefully acknowledges the support of the DuPont Foundation and the Stokes Teaching Fellowship of the Geology Foundation, University of Texas. We also wish to thank Guadalupe Mountains National Park and Carlsbad Caverns National Park for granting permission to collect samples used for this study. D. Kirkland, A. Lodato, M. Rahnis, D. Stoudt, and G. Vadillos served as field assistants. M. Harris, D. Kirkland, J. Lewis, L. Lynch, and A. Saller read the manuscript and made valuable suggestions. J. Horowitz, M. Rahnis, and T. Pierce assisted with preparation of the figures.

REFERENCES

Achauer, C. W., 1969, Origin of Capitan Formation, Guadalupe Mountains, New Mexico and Texas: The American Association of Petroleum Geologists Bulletin: v. 53, p. 2314-2323.

Adams J. E., and Fresnel, H. N., 1950, Capitan Barrier Reef, Texas and New Mexico, Journal of Geology, v. 58, p. 289-312

Adams, J. E., 1988, Historic Guadalupe Pass: Midland, West Texas Geological Society Publication, 88-84, p. 191-196.

Baars, D. L., 1964, Modern carbonate sediments as a guide to old limestones: World Oil, v. 158, p. 95-100.

Babcock, J. A., 1974, The Role of Algae in the Formation of the Capitan Limestone (Permian, Guadalupian) Guadalupe Mountains, West Texas-New Mexico: Unpublished Ph.D. Dissertation, University of Wisconsin, Madison, 241 p.

Babcock, J. A., 1977, Calcareous algae, organic boundstones, and the genesis of the upper Capitan Limestone (Permian, Guadalupian), Guadalupe Mountains, West Texas and New Mexico, *in* Hileman, M. E., and Mazzullo, S. J., eds., Upper Guadalupian Facies, Permian Reef Complex, Guadalupe Mountains, New Mexico and West Texas, Field Conference Guidebook: Midland, Society of Economic Paleontologists and Mineralogists, Permian Basin Section, Publication 77-16, p. 3-44.

Babcock, J. A., and Yurewicz, D. A., 1989, The massive facies of the Capitan Limestone Guadalupe Mountains, Texas and New Mexico: Tulsa, Society of Economic Paleontologists and Mineralogists Core Workshop 13, p. 365-371.

Babcock, L. C., 1974, Statistical Approaches to the Conodont Paleoecology of the Lamar Limestone, Permian Reef Complex, West Texas: Unpublished Ph.D. Dissertation, University of Wisconsin, Madison, 175 p.

Bebout , D., and Kerans, C., 1993, The Permian Reef Geology Trail, McKittrick Canyon, Guadalupe Mountains National Park, West Texas, Guidebook 26: Austin, Bureau of Economic Geology, University of Texas, 48 p.

Blanchard, W. G., and Davis, M. J., 1929, Permian stratigraphy and structure of parts of southeastern New Mexico and southwest Texas: American Association of Petroleum Geologists Bulletin, v. 13, p. 957-995.

Crandall, K. H., 1929, Permian stratigraphy of New Mexico and Texas: American Association of Petroleum Geologists Bulletin, v. 13, p. 927-944.

Cronoble, J. M., 1974, Biotic constituents and origin of facies in Capitan reef, New Mexico and Texas: Discussion: Mountain Geologist, v. 11, p. 95-108.

Cys, J. M., 1971, Origin of Capitan Formation, Guadalupe Mountains, New Mexico and Texas: Discussion: American Association of Petroleum Geologists Bulletin, v. 55, p. 310-315.

Cys, J. M., Toomey, D. F., Brezina, J. L., Greenwood, E., Groves, D. B., Klement, K. W., Kullmann, J. D., McMillan, T. L., Schmidt, V., Sneed, E. D., and Wagner, L. H., 1977, Capitan reef—evolution of a concept, *in* Hileman, M. E., and Mazzullo, S. J., Upper Guadalupian Facies, Permian Reef Complex, Guadalupe Mountains, New Mexico and West Texas, Permian Basin Section SEPM, Publ. 77-16, 201 -322.

Dunham, R. J., 1970, Stratigraphic reefs versus ecologic reefs: American Association of Petroleum Geologists Bulletin, v. 54, p. 1931-1932.

DUNHAM, R. J., 1972, Facts and Questions to Aid Interpretation and Group Discussion: Midland, Society of Economic Paleontologists and Mineralogists, Permian Basin Section, 352 p.

FOLK, R. L., 1993, SEM imaging of bacteria and nannobacteria in carbonate sediments and rocks: Journal of Sedimentary Petrology., v. 63, p. 990-999.

GARBER, R. A., GROVER, G. A., and HARRIS, P. M., 1989, Geology of the Capitan shelf margin—subsurface data from the northern Delaware Basin: Tulsa, Society of Economic Paleontologists and Mineralogists Core Workshop 13, 481 p.

GINSBURG, R. N., and LOWENSTAMM, H. A., 1958, The influence of marine bottom communities on the depositional environment of sediments: Journal of Geology, v. 66, p. 310-318.

GIRTY, G. H., 1908, The Guadalupian Fauna: Washington, D.C., U. S. Geological Survey Professional Paper 58, 651 p.

HARWOOD, G. M., 1989, Large-scale channel development within the Capitan reef complex—evidence from Carlsbad Caverns, *in* Harris, P. M., and Grover, G. A., eds., Subsurface and Outcrop Examination of the Capitan Shelf Margin, Northern Delaware Basin: SEPM Core Workshop no. 13, p. 379-386.

HAYES, P.T., 1964, Geology of the Guadalupe Mountains, New Mexico: United States Geological Survey Professional Paper 446, 69 p.

HILL, C. A., 1996, Geology of the Delaware Basin, Guadalupe, Apache, and Glass Mountains, New Mexico and West Texas: Permian Basin Section, SEPM, Publication No., 96-39, 480 p.

HURLEY, N.F., 1978, Facies mosaic of the lower Seven Rivers Formation (Permian), North McKittrick Canyon, Guadalupe Mountains, New Mexico [Master's Thesis]: University of Wisconsin, Madison, 198 p.

HURLEY, N.F., 1989, Facies mosaic of the lower Seven Rivers Formation, McKittrick Canyon, New Mexico, *in* Harris, P.M. and Grover, G.A. eds., Subsurface and Outcrop Examination of the Capitan Shelf Margin, Northern Delaware Basin: Tulsa, Society of Economic Paleontologists and Mineralogists Core Workshop 13, p. 325-346.

JOHNS, R. A., and KIRKLAND, B. L., 1998, Patterns of sponge diversity in the middle Capitan Reef of the Guadalupe Mountains, Texas, *in* DeMis, W. D. and Nelis, M. K., eds., The search continues into the 21st century: West Texas Geological Society, Publ. 98-105, p. 77-87.

JOHNSON, J. H., 1942, Permian lime-secreting algae from the Guadalupe Mountains, New Mexico: Bulletin of the Geological Society of America, v. 53, p. 195-226.

KENDALL, A. C., 1983; Radiaxial fibrous calcite: A reappraisal, *in* Schneidermann, N., and Harris, P. M., eds., Carbonate Cements: Tulsa, Society of Economic Paleontologists and Mineralogists Special Publication 36, p. 59-95.

KERANS, C., and FITCHEN, W. M., 1995, Sequence hierarchy and facies architecture of a carbonate-ramp system: San Andres Formation of Algerita Escarpment and western Guadalupe Mountains, West Texas and New Mexico: Austin, Bureau of Economic Geology, University of Texas, Report Inv. 235, 86 p.

KERANS, C., and HARRIS, P. M., 1993, Outer shelf and shelf crest, *in* Bebout, D.G., and Kerans, C., 1993, The Permian Reef Geology Trail, McKittrick Canyon, Guadalupe Mountains National Park, West Texas, Guidebook 26: Austin, Bureau of Economic Geology, University of Texas, p. 32-43.

KERANS, C., and TINKER, S.W., 1997, Sequence Stratigraphy and Characterization of Carbonate Reservoirs: Tulsa, Society of Economic Paleontologists and Mineralogists Short Course Notes 40, 130 p.

KING, P. B., 1948, Geology of the Southern Guadalupe Mountains Texas: Washington, D.C., U. S. Geological Survey Professional Paper 215, 183 p.

KIRKLAND GEORGE, B. L., 1992, Distinctions between reefs and bioherms based on studies of fossil algae: *Mizzia*, Permian Capitan Reef complex (Guadalupe Mountains, Texas and New Mexico) and *Eugonophyllum*, Pennsylvanian Holder Formation (Sacramento Mountains, New Mexico): Unpublished Ph.D. Dissertation, Louisiana State University, Baton Rouge, 156 p.

KIRKLAND, B. L., and MOORE, C. H., 1996, Microfacies analysis of the Tansill outer shelf, Permian Capitan Reef Complex: Midland, West Texas Geological Society Publication 96-101, p. 99-106.

KIRKLAND, B.L., DICKSON, J.A.D., WOOD, R.A., and LAND, L.S., 1998, Microbialite and Microstratigraphy: The origin of encrustations in the Capitan Formation, Guadalupe Mountains, Texas and New Mexico: Journal of Sedimentary Petrography, v. 68, p. 956-969.

KIRKLAND, B. L., LONGACRE, S. A., and STOUDT, E. L., 1993, Reef, *in* Bebout, D. G., and Kerans, C., eds., The Permian Reef Geology Trail, McKittrick Canyon, Guadalupe Mountains National Park, West Texas, Guidebook 26: Austin, Bureau of Economic Geology, University of Texas, p. 23-30.

KLEMENT, K.W., 1966, Studies on the ecological distribution of lime-secreting and sediment-trapping algae in reefs and associated environments: Neues Jahrbuch für Geologische Paläontologische Abhandlungen, v. 125, p. 363-381.

LLOYD, E. R., 1929, Capitan Limestone and associated formations of New Mexico and Texas: American Association of Petroleum Geologists Bulletin, v. 13, p. 645-658.

LOGAN, B. W., 1987, The MacLeod Evaporite Basin Western Australia: Tulsa, American Association of Petroleum Geologists Memoir 44, 140 p.

LOUCKS, R. G., and FOLK, R. L., 1976, Fan-like rays of former aragonite in Permian Capitan reef pisolite: Journal of Sedimentary Petrology, v. 46, p. 481-485.

MAZZULLO, S. J., 1977, Synsedimentary diagenesis of reefs, *in* Hileman, M. E., and Mazzullo, S. J., eds., Upper Guadalupian Facies, Permian Reef Complex, Guadalupe Mountains, New Mexico and West Texas, 1977 Field Conference Guidebook: Midland, Society of Economic Paleontologists and Mineralogists Publication 77-16, p. 323-354.

MAZZULLO, S. J., 1980, Calcite pseudospar replacive of acicular aragonite, and implications for aragonite cement diagenesis: Journal of Sedimentary Petrology, v. 50, p. 409-422.

MAZZULLO, S. J., and CYS, J. M., 1977, Submarine cements in Permian boundstones and reef-associated rocks, *in* Hileman, M. E., and Mazzullo, S. J., eds., Upper Guadalupian Facies, Permian Reef Complex, Guadalupe Mountains, New Mexico and West Texas, Field Conference Guidebook: Midland, Society of Economic Paleontologists and Mineralogists, Permian Basin Section, Publication 77-16, p. 151-200.

MELIM, L. A., and SCHOLLE, P. A., 1995, The forereef facies of the Permian Capitan Formation: The role of sediment supply versus sea-level changes, Journal of Sedimentary Research, v. B65, p. 107-118.

MRUK, D., and BEBOUT, D. G., 1993, Slope, *in* Bebout, D. G., and Kerans, C., eds., 1993, The Permian Reef Geology Trail, McKittrick Canyon, Guadalupe Mountains National Park, West Texas, Guidebook 26: Austin, Bureau of Economic Geology, University of Texas, p. 14-22.

MRUK, D. H., 1985, Cementation and dolomitization of the Capitan Limestone (Permian) McKittrick Canyon, West Texas: Unpublished Master's Thesis, University of Colorado, Boulder, 155 p.

MRUK, D. H., 1989, Diagenesis of the Capitan Limestone, Upper Permian, McKittrick Canyon, West Texas, *in* Harris, P. M., and Grover, G. A., eds., Subsurface and Outcrop Examination of the Capitan Shelf Margin, Northern Delaware Basin: Tulsa, Society of Economic Paleontologists and Mineralogists Core Workshop 13, p. 387-406.

NEWELL, N. D., 1955, Depositional fabric in Permian reef limestones, Journal of Geology, v. 63, p. 301-309.

NEWELL, N. D., RIGBY, J. K., FISCHER, A. G., WHITEMAN, A. J., HICKOX, J. E. and BRADLEY, J. S., 1953, The Permian Reef Complex of the Guadalupe Mountains Region, Texas and New Mexico: A Study in Paleoecology: San Francisco, Freeman Publishing Company, 236 p.

NOÉ, S. U., 1996, Late-stage evolution of the Permian reef complex: Shelf margin and outer-shelf development of the Tansill Formation (Late Permian), northern Guadalupe Mountains, New Mexico, USA, *in* Reitner, J., Neuwieler, R., and Gunkel, F. eds., Global and Regional Controls on Biogenic Sedimentation, I, Reef Evolution, Research Reports. Göttinger Arbeiten zur Geologie und Paläontologie, Sonderband 2, p. 317-324.

NOÉ, S. U., and MAZZULLO, S. J., 1992, Upper Tansill patch reef facies, Sheep Draw Canyon, Guadalupe Mountains, Eddy County, New Mexico, West Texas Geological Society: West Texas Geological Society Bulletin, v. 32, p. 5-11.

OSLEGER, D. A., 1998, Sequence architecture and sea-level dynamics of Late Permian shelfal facies Guadalupe Mountains, southern New Mexico: Journal of Sedimentary Research, v. 68, p. 327-346.

PRAY, L. C., 1986, Capitan Reef Complex (Permian), Guadalupe Mountains, southwestern United States: Reinterpretation of deposition and diagenesis of a classic sedimentologic model: American Association of Petroleum Geologists Bulletin, v. 70, p. 1762.

RAHNIS, M. A., 1995, Radiaxial Calcite in the Capitan Formation, Texas and New Mexico: Unpublished Master's Thesis, University of Texas, Austin, 131 p.

REITNER, J., 1993, Modern cryptic microbialite/metazoan facies from Lizard Island (Great Barrier Reef, Australia) formation and concepts: Facies, v. 29, p. 3-40.

RIGBY, J. K., SENOWBARI-DARYAN, B., and LIU, H., 1998, Sponges of the Permian Upper Capitan Limestone, Guadalupe Mountains, New Mexico and Texas: Salt Lake City, Brigham Young University Geology Studies, p.19-89.

RIGBY, K., and SENOWBARI-DARYAN, B., 1996, *Gigantospongia*, n. gen., the largest known Permian sponge, Capitan Limestone, Guadalupe Mountains, New Mexico: Journal of Paleontology, v. 70, p. 347-355.

SALLER, A.H., 1996, Differential compaction and basinward tilting of the prograding Capitan reef complex, Permian, West Texas and southeast New Mexico: Sedimentary Geology, v. 101, p. 21-30.

SCHMIDT, V., 1977, Inorganic and organic reef growth and subsequent diagenesis in the Permian Capitan reef complex, Guadalupe Mountains, Texas and New Mexico, *in* Hileman, M. E., and Mazzullo, S. J., eds., Upper Guadalupian Facies, Permian Reef Complex, Guadalupe Mountains, New Mexico and West Texas, Field Conference Guidebook: Midland, Society of Economic Paleontologists and Mineralogists, Permian Basin Section, Publication 77-16, p. 93-132.

SCHOLLE, P. A., ULMER, D. S., and MELIM, L. A., 1992, Late-stage calcites in the Permian Capitan Formation and its equivalents, Delaware Basin margin, west Texas and New Mexico: Evidence for replacement of precursor evaporites: Sedimentology, v. 39, p. 207-234.

SENOWBARI-DARYAN, B., and RIGBY, K., 1996a, Brachiopod mounds not sponge reefs, Permian Capitan-Tansill Formations, Guadalupe Mountains, New Mexico: Journal of Paleontology, v. 70, p. 697-701.

SENOWBARI-DARYAN, B., and RIGBY, K., 1996b, First report of *Leracaritubus* in North America from the Permian Capitan Limestone, Guadalupe Mountains, New Mexico: Journal of Paleontology, v. 70, p. 22-26.

SMITH, D. B., 1973, Geometry and correlation along Permian Capitan Escarpment, New Mexico and Texas: Discussion: American Association of Petroleum Geologists, Bulletin, v. 57, p. 940-943.

STANLEY, S. M. and YANG, X., 1994, A double mass extinction at the end of the Paleozoic Era: Science, v. 266, p. 1340-1344.

STANTON, R. J., and PRAY, L. C., 1993, Significance of neptunian fossil-rich dikes in the Capitan-Massive (Permian) Guadalupe Mountains, New Mexico and Texas: American Association of Petroleum Geologists Bulletin, p. 142.

TINKER, S.W., 1996, Reservoir-scale sequence stratigraphy: McKittrick Canyon and three-dimensional subsurface examples, West Texas and New Mexico: Unpublished Ph.D. Dissertation, University of Colorado, Boulder, 245 p.

TOOMEY, D. F., and BABCOCK, J. A., 1983, Precambrian and Paleozoic Algal Carbonates, West Texas and Southern New Mexico: Third International Symposium on Fossil Algae, Colorado School of Mines Professional Contribution, 345 p.

TUCKER, M. E., and WRIGHT, V. P., 1990, Carbonate Sedimentology: London, Blackwell, 482 p.

WOOD, R.A., DICKSON, J.A.D., and KIRKLAND, B.L., 1996, New observations on the ecology of the Permian Capitan reef, Texas and New Mexico: Palaeontology, v. 39, p. 733-762.

WOOD, R. A., DICKSON, J. A. D., and KIRKLAND-GEORGE, B., 1994, Turning the Capitan Reef upside down: A new appraisal of the ecology of the Permian Capitan Reef, Guadalupe Mountains, Texas and New Mexico: Palaios, v. 9, p. 422-427.

YUREWICZ, D. A., 1976, Sedimentology, paleoecology, and diagenesis of the massive facies of the lower and middle Capitan Limestone (Permian), Guadalupe Mountains, New Mexico and Texas: Unpublished Ph.D. Dissertation, University of Wisconsin, Madison, 278 p.

YUREWICZ, D. A., 1977, The origin of the Massive Facies of the lower and middle Capitan Limestone (Permian), Guadalupe Mountains, New Mexico and West Texas, *in* Hileman, M. E., and Mazzullo, S. J., eds., Upper Guadalupian Facies, Permian Reef Complex, Guadalupe Mountains, New Mexico and West Texas, Field Conference Guidebook: Midland, Society of Economic Paleontologists and Mineralogists, Permian Basin Section, Publication 77-16, p. 45-92.

DISTRIBUTION, PETROGRAPHY, AND GEOCHEMICAL CHARACTERIZATION OF RADIAXIAL CALCITE AND ASSOCIATED DIAGENETIC EVENTS IN THE CAPITAN FORMATION, WEST TEXAS AND NEW MEXICO

M. A. RAHNIS AND B. L. KIRKLAND

Department of Geological Sciences, The University of Texas at Austin, Austin, Texas 78712

ABSTRACT: The Capitan Formation represents a Permian (Guadalupian) reef complex deposited on the margin of the Delaware Basin. The reef was cemented by aragonite botryoids, radiaxial-fibrous (RFC) and fascicular-optic calcite (FOC), radiaxial-prismatic calcite (RPC), dolomite, and late calcite spar. Variability of fibrous cement textures correlates with the presence or absence of dolomite overgrowths. Zones that contain subequal amounts of RFC and FOC are characterized by radiaxial prismatic terminations lacking dolomite overgrowths, but contain <10 μm diameter dolomite microinclusions. In contrast, zones with <20% FOC and >80% RFC have dolomite cement overgrowths and contain dolomite microinclusions 10–100 μm in diameter.

Trace element concentrations from ICP and stable isotope compositions correlate with stratigraphic depth. Mn concentrations in RFC and FOC increase from an average of 25±2.9 ppm in the upper Capitan Formation to 62±31 ppm in the lower Capitan, whereas average Mg concentration increases from 1.45±0.81 to 2.40±1.37 mole % $MgCO_3$ and average Sr concentration decreases from 210±72 ppm to 63±22 ppm. RFC and FOC also contain covariant carbon and oxygen isotope ratios, which trend towards isotopically lighter values from compositions of $\delta^{18}O=-2.8‰$, $\delta^{13}C=5.2‰$ in the Middle Capitan. Microprobe analyses on individual RFC and FOC crystals include the <10 μm dolomite microinclusions. Point analyses traversing crystals from substrate to termination define apparent trends of increase from 1 to 4 mole % $MgCO_3$, and reflect relative abundance of dolomite microinclusions. Strontium and $MgCO_3$ do not covary, although both generally increase.

Combinations of RFC, FOC, and RPC form complex cement stratigraphies. Anhedral dolomite microinclusions may be either primary (penecontemporaneous with host calcite), or secondary (formed during later diagenesis). Euhedral dolomite inclusions within calcite resemble cement overgrowths. Changes in dolomite and kaolinite microinclusion abundance in these cements may reflect porewater changes during early diagenesis. Early diagenesis was probably driven by reef growth (aggradational and progradational), and sea-level change.

INTRODUCTION

Objectives

The goal of this study is to describe the occurrence of radiaxial-fibrous (RFC) and fascicular-optic calcite (FOC) in the Capitan Formation of the Guadalupe Mountains and to place these cements within the context of the paragenetic sequence of the Capitan Formation. Thorough characterization of RFC and FOC in the Capitan Formation will help: (1) to determine the porewater conditions under which these cements precipitated, (2) to determine whether mineral inclusions in RFC and FOC are primary or secondary, and (3) to attain an understanding of Northwest Shelf paleohydrology.

Setting

The Capitan Formation represents part of a reef complex that is exposed in the Guadalupe Mountains of West Texas and southeastern New Mexico (Fig. 1). The reef complex is preserved virtually as it was deposited. The Capitan Formation includes the Capitan reef (Massive Member) as well as its associated reef-slope deposits (Breccia Member) (Hayes, 1964). Capitan reef growth is divided into lower, middle, and upper portions, which are time-equivalent with the Seven Rivers, Yates, and Tansill formations, respectively (Fig. 2).

The Capitan reef has alternately been interpreted as an organic bank growing below wave base (Achauer, 1969; Baars, 1964; Dunham, 1972; Pray, 1986), or a barrier reef near sea level (King, 1948, Newell et al., 1953; Kirkland and Moore, 1994; B. L. Kirkland, in prep.). Marine cements filled internal cavities formed by organisms, largely bryozoans and sponges. Most shelter cavities were 20–50 cm (8–20 in.) in diameter, but, in some parts of the reef, may have been over 1 m diameter (Wood et al., 1994). The first cement was aragonitic botryoids, followed by up to 4 mm of isopachous early marine calcite, and then, much later, blocky spar cements (Mruk, 1985). The pattern of reef growth may have influenced the formation of these marine cements. Reef growth was dominantly progradational during lower Capitan time, but became increasingly more aggradational during deposition of the middle and upper Capitan Formation (King, 1948; Garber et al., 1989).

Previous Work on Radiaxial-fibrous and Fascicular-optic Calcites

RFC (Bathurst, 1959) refers to calcite crystals with a combination of curved twin planes, convergent optic axes, and diverging subcrystals (Fig. 3). The twin planes are concave in the direction of crystal growth. Another calcite form, FOC (Kendall, 1977), can occur along with RFC, but is generally less abundant. Twin planes in FOC are convex in the direction of growth, and fast-vibration directions diverge from a centerline of crystal bundles. Convergent or divergent optic axes within composite crystals give these cements an unusual extinction pattern. When rotated under crossed polars an extinction brush appears to sweep like the arm of a pendulum through RFC crystals, whereas it sweeps like an inverted pendulum through FOC crystals.

The origin of RFC as a replacement of marine cement was suspected by Bathurst (1959) largely because of its fibrous appearance. Later, Kendall and Tucker (1973) proposed that replacement of an acicular precursor could explain the convergent optic axes of RFC. The hypothesis of Kendall and Tucker (1973) was later rescinded, and in its place a primary origin for

Geologic Framework of the Capitan Reef, SEPM Special Publication No. 65
 ISBN 1-56576-063-8, p. 175-191

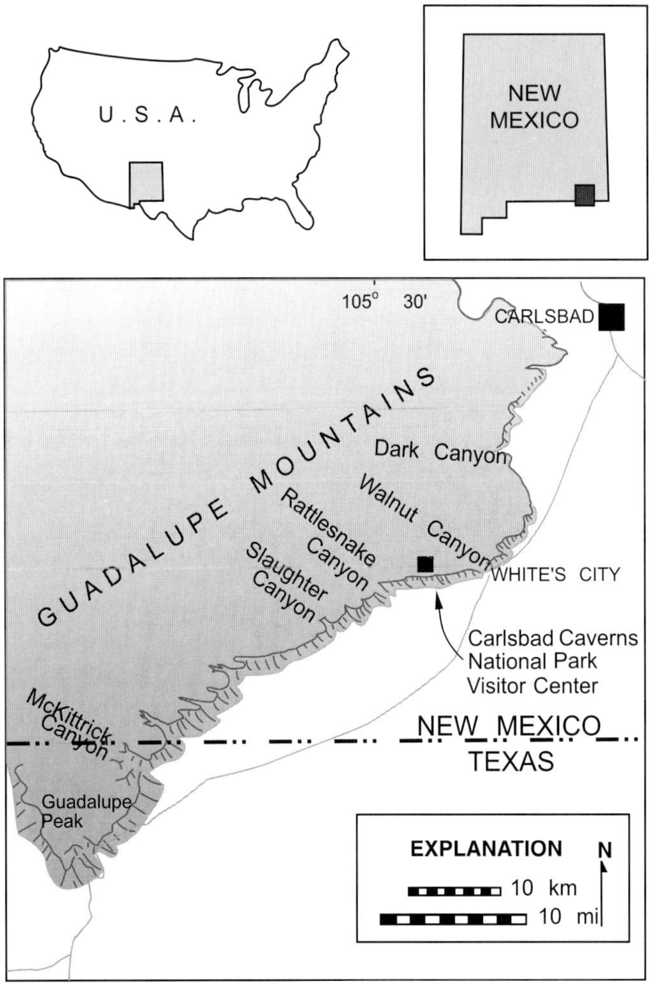

FIG. 1. — Field area map showing the topography that approximates the reef trend and the locations of canyons along that trend from which samples were collected (Modified from King, 1948; Kirkland-George, 1992).

the peculiar cement morphology was proposed (Kendall, 1985). Some authors argue for an initially high-Mg calcite composition of cements based on dolomite microinclusions (Lohmann and Meyers, 1977), or inhomogenous distribution of Mg within the calcite (Videtich, 1985). Others argue that low-Mg RFC at Enewetak Atoll is in isotopic equilibrium with and precipitates directly from seawater below the aragonite lysocline (Saller, 1986). The distinction between high-Mg and low-Mg calcites is now considered arbitrary, however, with the discovery of calcites that have intermediate, 2–11 mole % $MgCO_3$, concentrations (James and Choquette, 1990).

Previous Studies on Diagenesis and Paleohydrology

The general paragenetic sequence for the Capitan Formation is: aragonitic botryoids, RFC, two phases of dolomite (Melim, 1991), three phases of calcite spar (Mruk, 1985), anhydrite, magnesite (Scholle et al., 1992), and kaolinite (Melim, 1991). The following is a discussion of each of these phases in the order of its occurrence.

Neomorphosed aragonite botryoids.—

Marine water precipitated aragonite during and soon after deposition. Relict needle-like morphology with square or feathery terminations, characteristic of aragonite, is still apparent (Loucks and Folk, 1976). Botryoids occur in the reef massive (Rudolph, 1978). Initial layers of botryoids are interlaminated with microbial micrite (B. L. Kirkland, in prep.) and the problematic organism *Archaeolithoporella*, which possibly represents a red alga (Babcock, 1974). Interlamination of botryoids and marine organisms indicates that the aragonite precipitated from marine water. The former aragonite has since transformed to neomorphic cloudy (inclusion-rich) calcite and dolomite, but could provide an estimate of the isotopic composition of Permian marine calcite (Given and Lohmann, 1985).

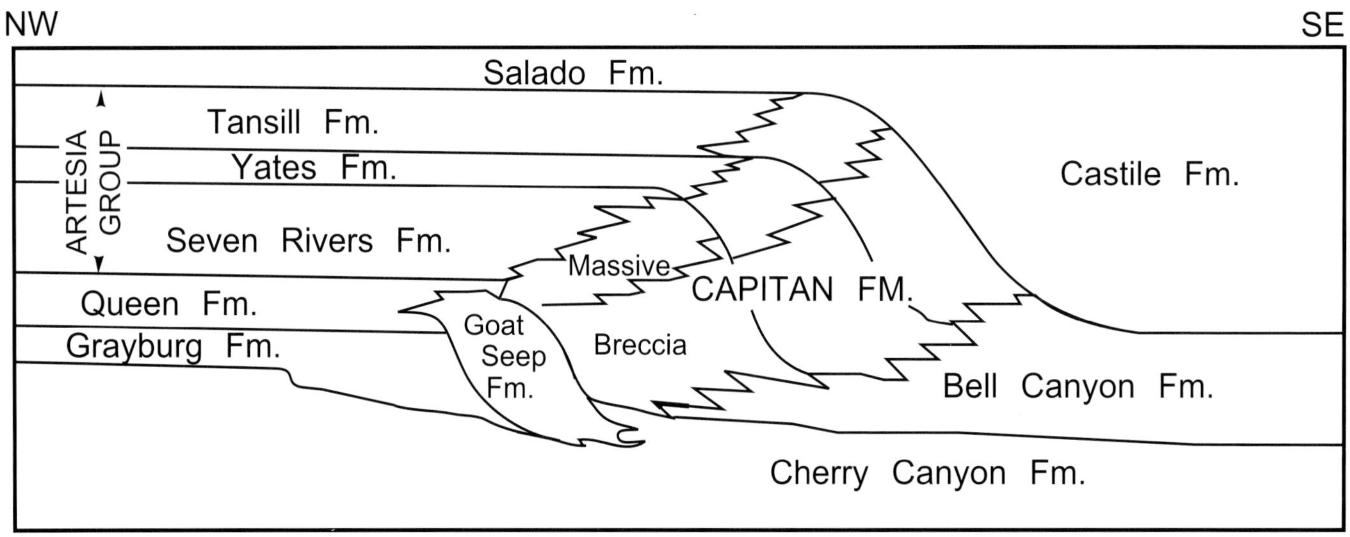

FIG. 2. — Stratigraphic relationships of Permian (Guadalupian and Leonardian) sediments depicted as a shelf to basin cross-section (Modified from King, 1948; Newell et al., 1953; Garber et al., 1989; Kerans et al., 1992).

FIG. 3. — These drawings from Kendall (1985) show the features of various fibrous calcite types. Radial-fibrous calcite has optic axes parallel to the length of fiber crystals. Radiaxial-fibrous calcite has optic axes within bundled subcrystals that converge in the direction of the composite crystal termination. The concave side of curved cleavages faces RFC terminations. Optic axes in fascicular-optic calcite bundles diverge in the direction of the termination. The convex side of curved cleavages faces FOC terminations.

Isopachous calcite.—

Isopachous calcite cement lined remaining pore space after the end of botryoid cementation in the reef. In the forereef, aragonite botryoids are less abundant than in the reef, and isopachous calcite is the most abundant cement. This isopachous zone consists partly of radiaxial-fibrous calcite. Radiaxial-fibrous calcite, as defined by Bathurst, was identified in the Capitan Formation by Dunham (1972), although he preferred the term "drusy cement." Other names have been applied to this type of calcite. Calcite with radiaxial extinction has been called "pigeon toed" in neomorphosed micrite (Folk, 1971), and "isopachous fibrous calcite" (Yurewicz, 1976). Calcite with fascicular-optic extinction has been recognized in a Holocene speleothem from Carlsbad Cavern, where it was given the name "coconut-meat calcite" (Folk and Assereto, 1976).

The isopachous calcite zone has been given many names, but our careful evaluation suggests that it contains both RFC (Bathurst, 1971; Mruk, 1985; Tucker and Wright, 1990), and FOC, along with a more prismatic variety of radiaxial calcite, which we call radiaxial prismatic calcite (RPC). We make these distinctions because, although RFC and FOC grade laterally into each other, they are optically distinct, and the origin of the two optical varieties is unclear. In transmitted light this zone commonly appears to grade in its growth direction into inclusion-rich prismatic calcite (RPC) and then to clear calcite spar (Mruk, 1985). The cement we call RPC occurs at the apparent transition from RFC and FOC to spar, and although RPC behaves optically like RFC, it deserves a distinction. RPC is less fibrous than RFC and FOC, and it contains dolomite and kaolinite microinclusions, whereas RFC and FOC contain only dolomite microinclusions.

Dolomite, first generation.—

Melim (1991) distinguished two types of dolomite in the Capitan Formation. The first, which she called "fabric-preserving Dolomite A," is 5–30 μm in size, commonly with cloudy cores. It is nonluminescent and apparently formed by dolomitization of allochems and earlier marine cements. Dolomite A is most ^{18}O-rich in the far back-reef facies and becomes progressively lighter in the reef and forereef, suggesting that Dolomite A may have precipitated from refluxing fluids originating in the back-reef environment (Melim, 1991).

Calcite spar I.—

The gradational character of isopachous calcite illustrates that progressive overgrowth occurred on RFC and FOC leading up to spar calcite. Spar calcite cements of the Capitan Formation are divided into Spars I, II, and IIb (Mruk, 1985). Spar IIb is referred to here and by others as Spar III. Spar I contains alternating bright and nonluminescent zones. The relationship between spar I and multiple episodes of dolomitization is unclear. Spar I may have formed during sea-level fall that exposed the top of the reef at the end of Capitan time and led to freshwater lens development (Mruk, 1985; Given and Lohmann 1986). Spar I is volumetrically small, with occurrences localized in the back-reef, reef and proximal fore-reef (Mruk, 1985). Mruk (1985) concluded that spar I formed after dolomitization, during burial (Melim, 1991), meaning that dolomitization continued after spar I precipitation. Subaerial exposure caused solution-enlargement of synsedimentary fractures, which later became pathways for dolomitizing fluids (Melim, 1991). The climate was still arid enough, however, to leave intact huge areas of landward evaporites.

Dolomite, second generation.—

Subsequent to spar I precipitation is a second generation of dolomite, called "fabric-destructive Dolomite B," which has grain sizes from 30 μm to 200 μm (Melim, 1991). Unlike Dolomite A, Dolomite B has a distinct red cathodoluminescence. Dolomite B is isotopically depleted in ^{18}O relative to Dolomite A, suggesting that brines attained elevated temperatures or contained a mixture of marine and meteoric water. Fabric-destructive Dolomite B is most abundant deeper within the stratigraphic section and was interpreted to have precipitated during burial (Melim, 1991). The outcrops of Seven Rivers and lower Capitan are largely dolomitized. Moreover, the Tansill, Yates, and lower Capitan formations, penetrated by the PBD-04 core, have been altered to dolomite, magnesite, and anhydrite (Garber et al., 1989). The paleohydrologic history

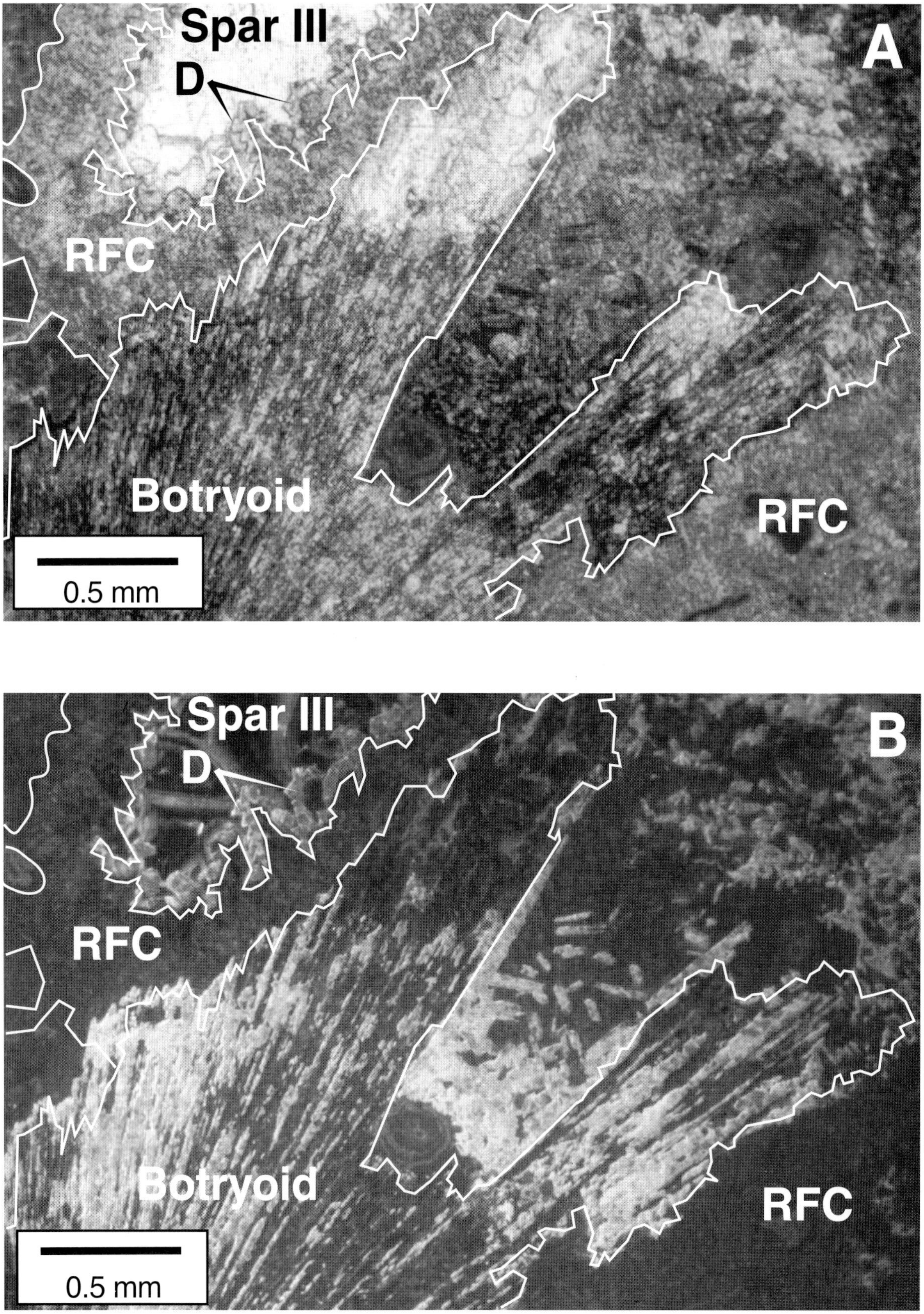

FIG. 4. — (A) Transmitted light photomicrograph shows the generalized sequence of cementation, which occurred as follows: Aragonite botryoids, RFC, dolomite (D), and spar III calcite. (B) CL photomicrograph shows the luminescent character of dolomite, which replaces and preserves the shape of formerly aragonite needles in botryoids and occurs as a cement on the terminations of RFC.

indicates that brine, concentrated in the halite and bittern salt stage, moved downward through the Capitan reef complex during Salado Formation evaporite deposition. These exist today as Ca-Cl brines in subsurface lower Permian and Pennsylvanian sections of the Permian Basin (Bein and Dutton, 1993; Stueber et al., 1998). These brines may play a role in the formation of coarse burial dolomite, magnesite, and pore-occluding anhydrite.

Anhydrite and calcite spars III and III.—

Diagenesis and paleohydrology is poorly constrained during the interval between Permian brine emplacement and Tertiary uplift, which has altered the basin hydrology. Anhydrite and gypsum occlude subsurface porosity in the PDB-04 core (Garber et al., 1989), whereas calcite burial spars II and III occlude pores in outcrop (Mruk, 1985, Scholle et al., 1992). Meteoric influx may have converted much of the anhydrite to calcite spar II and III (Scholle et al., 1992) at burial temperatures as high as 50°C and burial depths of approximately 1 km (Crysdale, 1987). Associated with these spar calcites are kaolinite books (Rudolph, 1978, Scholle et al., 1992), which typically occur near shelf sandstones (Dunham, 1972). Kaolinite is proposed to have accompanied conversion of anhydrite to gypsum and gypsum dissolution just prior to Spar II and III precipitation (Scholle et al., 1992). Meteoric, water-derived Na-Cl brines continue to displace the ancient Ca-Cl brines (Bein and Dutton, 1993; Stueber et al., 1998).

The preceding review of the diagenesis of the Capitan Formation provides the framework for this investigation of RFC, FOC, and RPC in this study. We describe here the petrography and geochemistry of RFC. In so doing we have come to recognize that this thin layer of cement may hold important information about the paleohydrology of the Delaware Basin.

METHODS

Samples were collected in Rattlesnake, Walnut, Slaughter, McKittrick, North McKittrick, and Dark canyons. A total of 47 new samples were collected and 69 thin sections were prepared for this study. By using samples from previous studies, a total of 343 thin sections were available. Petrographic techniques included: etching and staining (Dickson, 1966), cathodoluminescence (Technosyn MK-II CL stage operated at 15–20 kV and 300–600 μA), scanning electron microscopy (Jeol T330A SEM operated at 20 and 30 kV), and back-scattered electron imaging (JEOL Superprobe 733) obtained during electron microprobe analysis. National Institutes of Health (NIH) image software determined the volume percent of dolomite inclusions within calcite from back-scattered electron (BSE) images (White et al., 1998). Photoshop software was used to enhance color and textural variations in photomicrographs.

Samples for stable isotopic and trace element analysis were collected from polished thin-section stubs (not epoxy impregnated) and polished slabs using a miniature drill press. Samples of mixed calcite and dolomite were identified with a Siemens Kristalloflex Diffraktometer and discarded. Twenty-five calcite samples for stable isotopic analysis were reacted with phosphoric acid at 25°C and analyzed using a V.G. Prism gas source mass spectrometer. The same powders were also analyzed for the trace elements Mg, Sr, Mn, and Fe using a Jobin-Yvon 70 inductively coupled plasma atomic emission spectrometer (ICP-AES) with relative percent errors as follows: ±10% Mg, ±2% Sr, ±4% Mn and Fe.

Electron microprobe data were acquired for RFC and FOC from nine carbon-coated thin sections. Back-scatter images and energy dispersive spectra (EDS) spectra identified inclusions prior to analysis of the host calcite. Analyses were run on a JEOL Superprobe 733 at 15 kV with a 14 nA current on Smithsonian calcite standard (12 nA on brass) at 10-μm spot size. Relative percent errors for $MgCO_3$ and Sr are based on counting statistics and displayed as error bars in Fig. 12.

PETROGRAPHY

In the Massive Member of the Capitan Formation, aragonite cement nearly filled many reef framework cavities, locally forming as much as 95% of the rock volume. Aragonite has since neomorphosed to dolomite and mosaic calcite, but still exhibits ghosts of fibrous ray crystals with feathery or square terminations. Thin (1-4 mm) layers of RFC and FOC lined, but did not fill, remaining primary pores (Fig. 4). The first cement phases are zones of RFC and FOC (Mruk, 1985) in forereef slope breccias. Aragonite botryoids are absent from this facies except in float blocks. Radiaxial-prismatic calcite followed RFC and FOC; it is not fibrous in appearance, exhibits only the extinction pattern associated with RFC, and is slightly less cloudy than underlying zones of RFC and FOC, which often appear brown in plane-polarized light. Throughout the Capitan reef massive and forereef, these zones of RFC, FOC, and RPC interlayer with dolomite. In cathodoluminescence, RFC, FOC, and RPC commonly display a dull yellow luminescence. Dolomite microinclusions luminesce red, and are scattered throughout RFC, FOC, and RPC. Kaolinite inclusions luminesce blue and occur near the terminations of RFC and FOC, but are most abundant in RPC. Distribution of these calcite cements in the Capitan Formation fall naturally into three categories: (1) samples containing only radiaxial-fibrous calcite with dolomite overgrowth, (2) samples containing both radiaxial-fibrous calcite and fascicular-optic calcite without dolomite overgrowth, and (3) samples with only radiaxial-prismatic calcite.

Radiaxial-fibrous Calcite with Dolomite Overgrowth

Samples composed of almost exclusively of RFC have luminescent dolomite cement overgrowths on RFC terminations. In addition to dolomite overgrowths, RFC contains dolomite microinclusions (Fig. 5). Dolomite microinclusions in RFC are tens to hundreds of micrometers in diameter (Fig. 5C). Approximately 95% of dolomite microinclusions in RFC are anhedral, whereas the remaining 5% of inclusions appear to represent overgrowths on former calcite terminations. Large

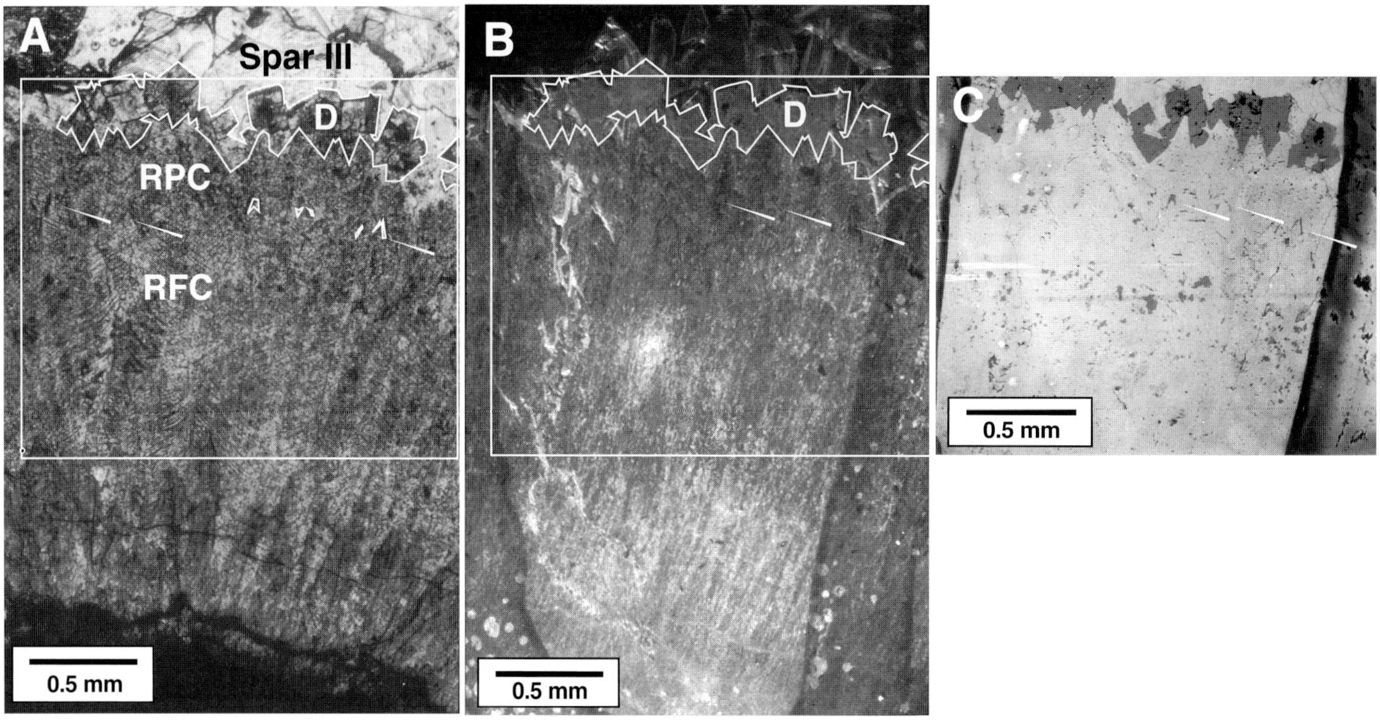

FIG. 5. — (A) Curved cleavages characteristic of radiaxial cements are visible in this sample, which grades from a fibrous (RFC) to a prismatic (RPC) morphology. Coarse, luminescent dolomite (D) overgrows this calcite cement zone. This sample is from the lower Capitan in North McKittrick Canyon. (B) RFC and RPC contain luminescent dolomite microinclusions (pointers) that are visible in CL images. (C) BSE images reveal that some dolomite microinclusions (indicated by pointers) are euhedral, and have shapes that parallel calcite growth faces. Most inclusions are anhedral, and measure tens to hundreds of microns in diameter.

dolomite inclusions comprise approximately 7 volume % within RFC. Dolomite inclusions in RFC are distinctly different from inclusions found in samples where RFC and FOC occur together and lack coarse dolomite overgrowths (Fig. 6).

Radiaxial-fibrous and Fascicular-optic Calcite Without Dolomite Overgrowth

Some pores contain a mixture of the two optically distinct varieties, RFC and FOC (Fig. 6). Samples having mixed FOC and RFC lack dolomite overgrowths on the calcite terminations. Single crystals of RFC and FOC up to 1-mm wide and 4-mm long grade laterally into each other within pore-lining cement zones. Point counting indicates that this textural variety of cement contains a mixture of RFC and FOC optical types, with FOC comprising as much as 45%. In the middle Capitan, RFC is slightly more abundant than FOC. Mixtures of RFC and FOC have also been found in the upper Capitan, but not in the lower Capitan.

These cements have a very fibrous appearance and are characterized by inclusion-poor boundaries separating cloudy, inclusion-rich fibers. BSE images reveal dolomite microinclusions ranging from <1–10 μm (Fig. 6B). In cross-polarized light, dolomite microinclusions are in syntaxial alignment with host calcite. Identifiable dolomite microinclusions are restricted to and abundant in RFC, FOC, RPC, and initially high-Mg echinoderm grains. Dolomite microinclusions increase in abundance toward the terminations of the zone of RFC and FOC and obtain maximum abundances of 8–9.5 volume % (Fig. 6B).

Radiaxial-prismatic Calcite

Radiaxial-prismatic cements are prismatic to equant rather than fibrous, but have the optic axis pattern, sweeping extinction and curved cleavages of RFC. These cements in the Capitan Formation have been referred to as inclusion-rich prismatic cements (Mruk, 1985). Differences in inclusion density and fibrous texture give fibrous calcite cements the appearance of grading into spar calcite (Fig. 7).

Kaolinite inclusions occur within discrete zones at the transition from RFC/FOC to RPC, and are most abundant in RPC. Dolomite and kaolinite inclusions are indistinguishable using BSE imaging; EDS analysis, however, returns Si and Al peaks on some inclusions, suggesting that kaolinite or feldspar is present. RPC in these samples has a distinct blue luminescence (Fig. 7), much like the luminscence of large kaolinite books included in spar II calcite. We were unable to identify inclusions having kaolinite crystal morphology using SEM, however, the EDS and luminescence character provide good evidence for the presence of kaolinite as inclusions in RPC. This difference in microinclusion types indicates that RPC should be distinguished from RFC and FOC.

Dolomite

Dolomite occurs as a replacement of allochems and as a cement phase on the terminations of RFC and RPC. Some dolomite crusts on calcite are discontinuous around pores and vary in thickness (Fig. 8), but other dolomite crusts are of uni-

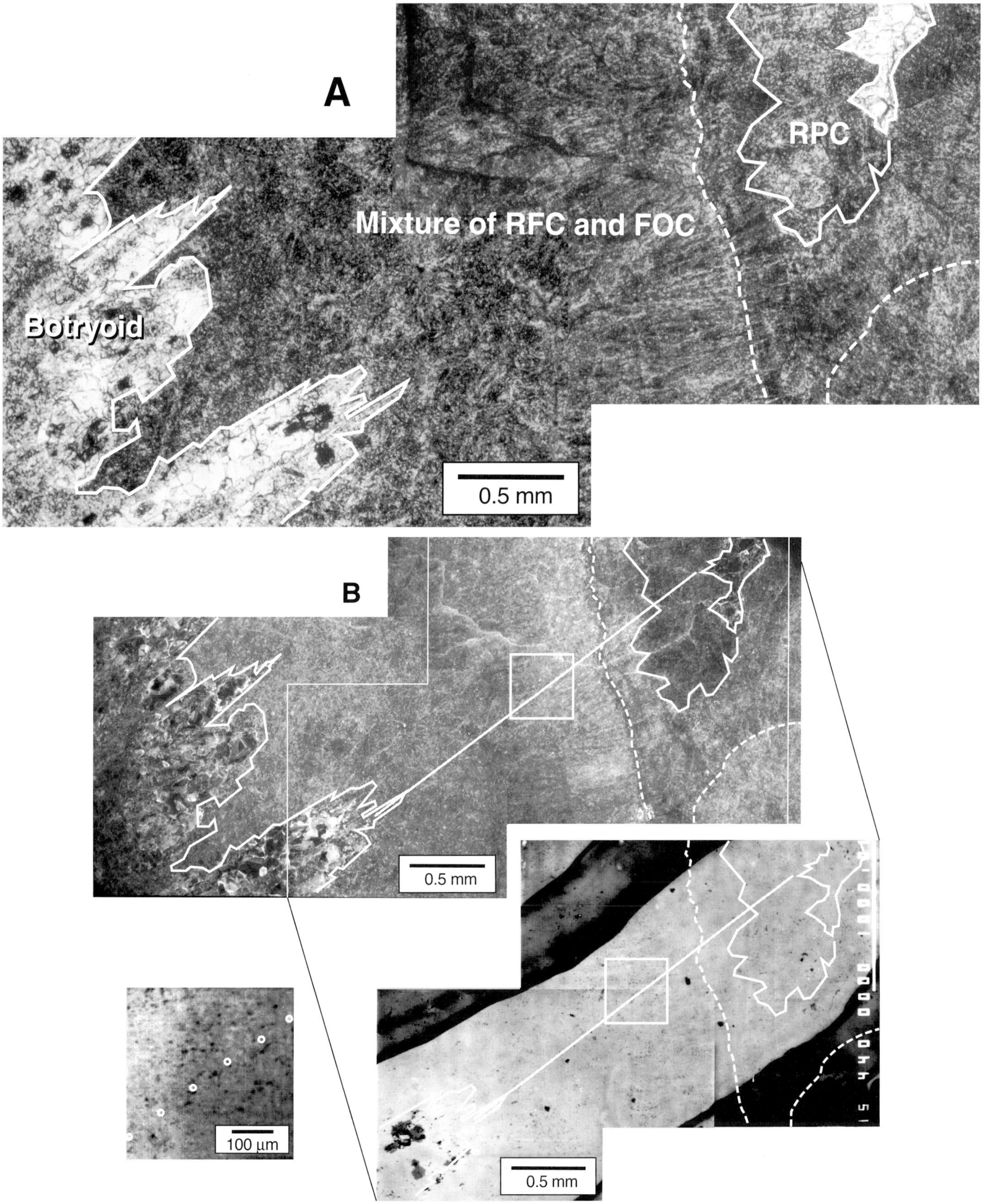

FIG. 6. — (A) Samples that lack coarse dolomite overgrowths grade from RFC/FOC to RPC and finally to spar calcite. This sample is from the middle Capitan in McKittrick Canyon. (B) CL images reveal that these cements contain red-luminescent dolomite microinclusions and blue-luminescent kaolinite microinclusions. Blue-luminescent kaolinite microinclusions occur in discrete bands at the transition (shown as a dashed line) from RFC/FOC to RPC, and are most abundant in RPC. BSE images indicate the <10 µm scale of dolomite microinclusions in mixed RFC/FOC cement.

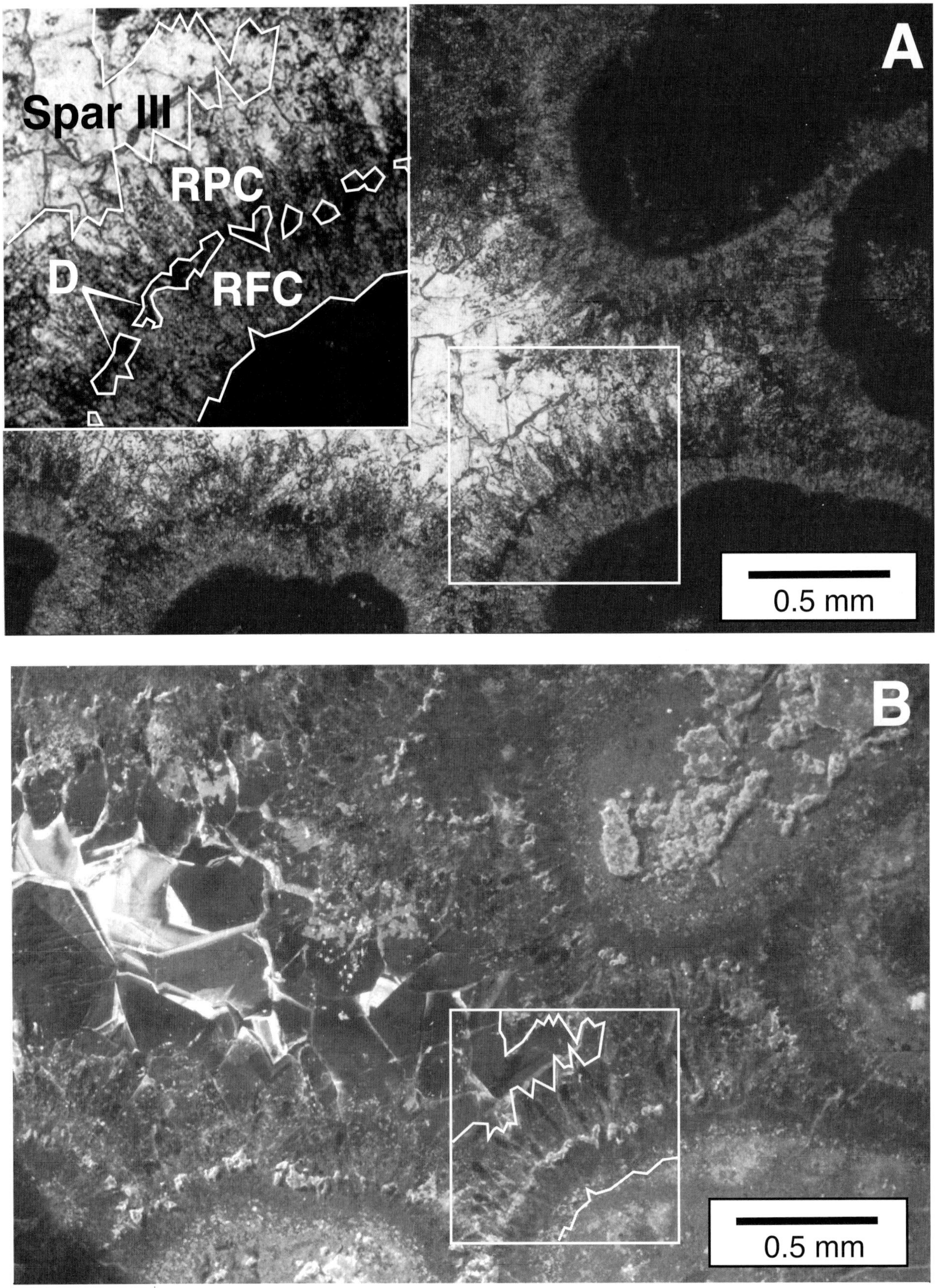

FIG. 7. — (A) Transmitted light image of an etched sample from the upper Capitan in Dark Canyon clearly shows a dolomite cement crust, probably cement, that is included within RPC, which appears to grade into spar calcite. (B) CL image shows the red luminescence of the dolomite crust. Kaolinite-rich RPC has a distinctive blue luminescence in this sample.

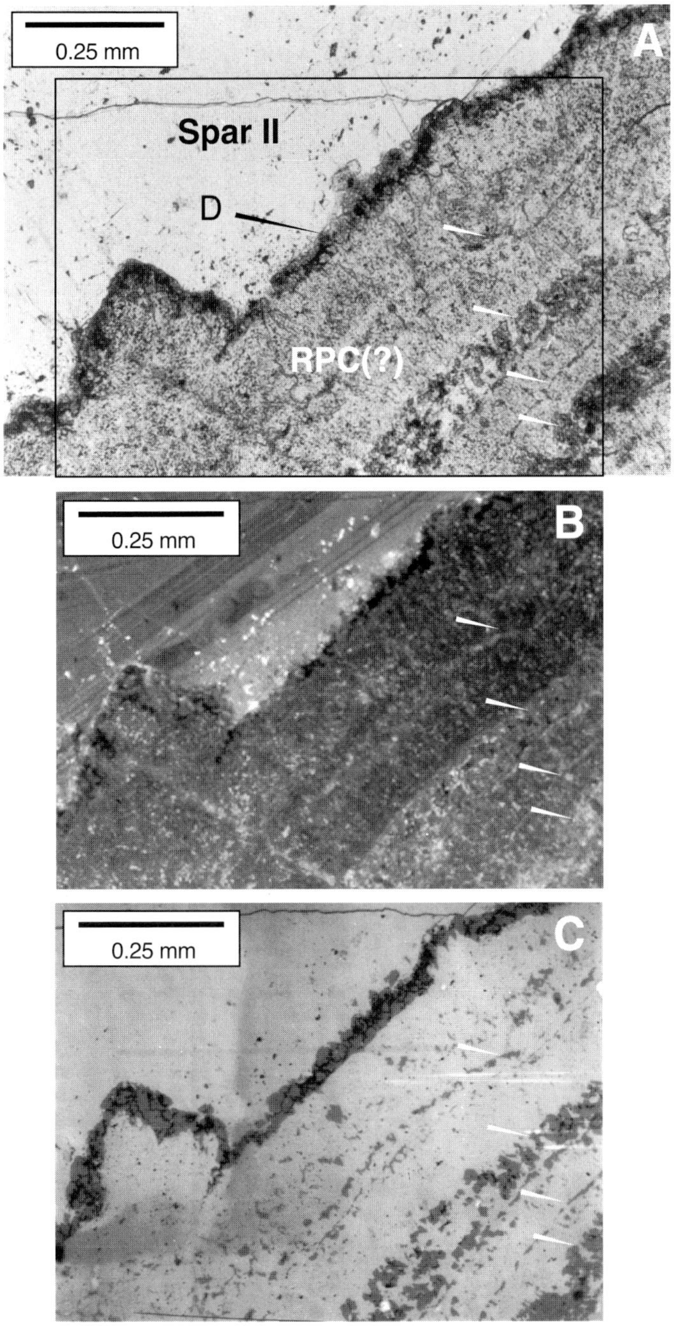

FIG. 8. — (A) Etched thin section in transmitted light shows an abundance of dolomite crusts in and on RPC. The box indicates the areas shown in B and C. Pointers indicate the locations of four different dolomite crusts in each of the images shown here. (B) CL images show the red luminescence of dolomite inclusions and crusts. (C) BSE images reveal the shapes of dolomite microinclusions. Many of these appear anhedral and occur in distinct zones where they are more abundant.

form thickness and completely line pores. This second type of dolomite crust, visible in etched and stained thin sections, either separates RFC from RPC or occurs within the zone of RPC (Fig. 7). It is unclear where these crusts fit into the paragenetic sequence because of the gradational nature of the calcite cements. Dolomite crusts are in optical continuity with the host calcite and preserve the general shape of underlying calcite terminations. Furthermore, some of the inclusions seen in back-

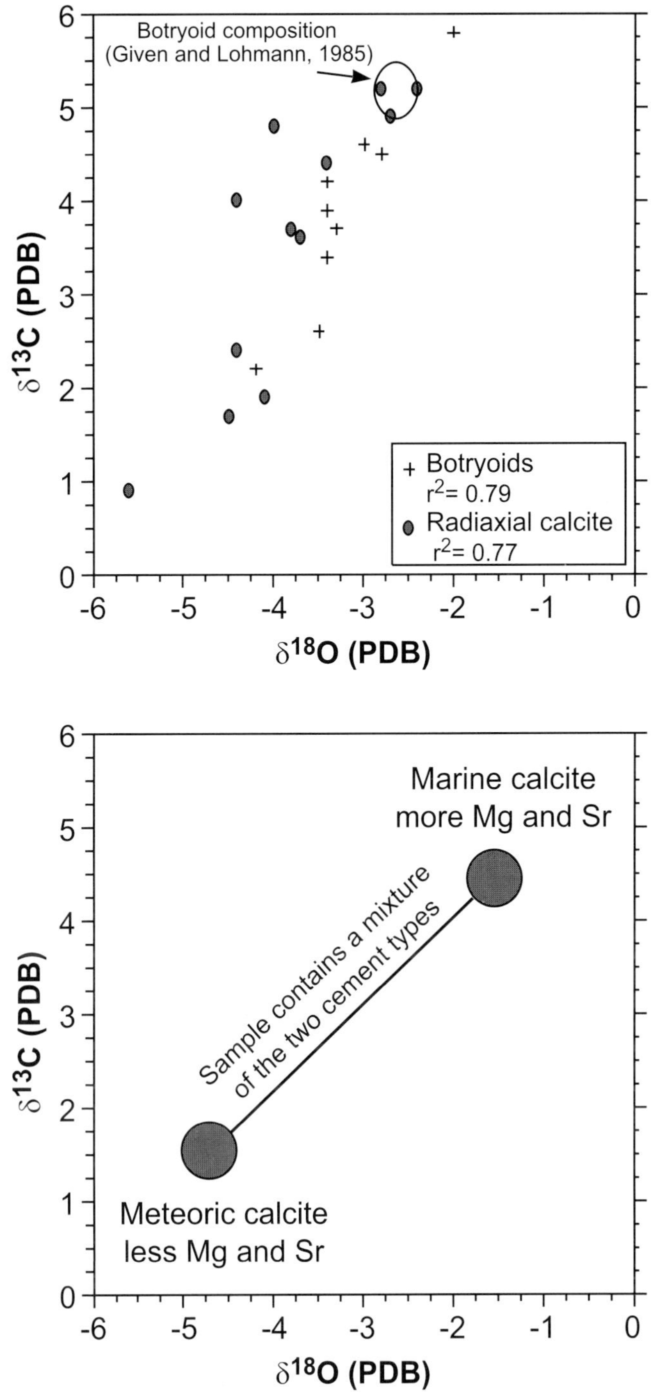

FIG. 9. — (A) Cross-plots of stable isotopic carbon and oxygen reveal positive covariance of the two. Similar trends have already been found for analyses in single hand-samples (Given and Lohmann, 1985). These trends represent the data from many hand-samples taken at various locations. B) Trends are attributed to the mixing of two or more cements that have different compositions. The isotopically heavy endmember is believed to represent marine calcite, whereas the isotopically light endmember represents meteoric calcite (Given and Lohmann, 1985).

scattered electron images are V-shaped and point toward the termination of the RPC crystals (Fig. 5).

GEOCHEMISTRY

Stable Isotope Analysis

Calcitized aragonitic botryoids of the middle Capitan have a wide range of stable isotopic compositions (Mruk, 1985; Given and Lohmann, 1985). Our values fall along trends parallel to those of the middle Capitan fibrous calcites (Fig. 9). The botryoid samples scatter from compositions of $\delta^{18}O$ = -4.4‰ and $\delta^{13}C$ = 2.2‰ to enriched isotopic compositions of $\delta^{18}O$ = -2.0‰ and $\delta^{13}C$ = 5.8‰. Carbon and oxygen isotopic compositions for mixtures of RFC, FOC, and RPC plot as linear covariant trends from an enriched carbon and oxygen composition to relatively lighter isotopic compositions. In the middle Capitan Massive Member the trend ranges from $\delta^{18}O$ = -5.6‰ and $\delta^{13}C$ = 0.9‰ to isotopically heavier compositions of $\delta^{18}O$ = -2.8‰ and $\delta^{13}C$ = 5.2‰. In agreement with the finding of Given and Lohmann (1985), the lighter end of the range falls toward values for the Spar I cement described by Mruk (1985).

ICP Trace Element Data

Trace element concentrations in Capitan RFC, FOC, and RPC vary stratigraphically (Table 1; Fig. 10). Average manganese concentration increases with depth in the stratigraphic section, from 25±3 ppm in the upper Capitan to 62±31 ppm in the lower Capitan. Magnesium shows a similar pattern, increasing from an average of 1.45±0.8 mole % $MgCO_3$ in the upper Capitan to 2.4±1.4 mole % in the lower Capitan. Strontium, however, behaves in an opposite sense with Sr concentration decreasing downward within the reef. Average Sr concentrations are greatest in the upper Capitan at 210±72 ppm, and least in the lower Capitan, where Sr concentrations average 63±22 ppm Sr. One-tailed student's t-tests (Table 1) show the lower and middle Capitan to be significantly different for Mg [P(T>t)≥0.99], Sr [P(T>t)=0.95], and Mn [P(T>t)≥0.99]. Differences between the upper and middle Capitan are significant only for Mn [P(T>t)=0.95]. The real differences may be obscured, however, because these geochemical samples represent a variable mixture of two or more cements.

Trace element compositions vary with respect to oxygen isotopic composition within the stratigraphic section (Fig. 11). Middle Capitan Sr concentrations are highest in samples with enriched $\delta^{18}O$ composition (Fig. 11A). With the exception of one data point at depleted $\delta^{18}O$, lower Capitan samples show a similar trend for strontium. Magnesium does not vary in this sense; instead, it appears to be invariant over the range of oxygen isotopic compositions (Fig. 11B). Stable isotopic compositions were not obtained for upper Capitan samples.

Electron Microprobe Trace Element Analysis

Four of nine samples analyzed with electron microprobe have dolomite overgrowths, and none exhibits trace element trends. When care is taken to avoid dolomite inclusions on traverses from substrate to termination, Mg concentration varies between 0.5–1.5 mole % $MgCO_3$. A few isolated points suggest higher Mg concentrations; but these analyses are probably not representative and contained dolomite microinclusions that were large relative to the beam volume. Ascertaining such an occurrence was not possible because the beam did not produce visible damage spots on dolomite. Sr, Fe, and Mn are typically below detection limits.

TABLE 1.—SUMMARY OF ICP TRACE ELEMENT DATA

	$MgCO_3$ (mole %)	Sr (ppm)	Mn (ppm)
Upper Capitan n=3			
Maximum	2.03	270	29
Minimum	1.05	1.08	22
Average (μ_{upper})	1.64±0.43	210±72	25±2.9
Middle Capitan n=11			
Maximum	4.87	146	62
Minimum	1.12	70	27
Average (μ_{middle})±2 σ	1.64±2.06	110±21	36±12
$H_o:\mu_{upper}=\mu_{middle}$	$H_1:\mu_{upper}<\mu_{middle}$	$H_1:\mu_{upper}>\mu_{middle}$	$H_1:\mu_{upper}<\mu_{middle}$
P(T>t)	0.66	0.90	>0.99
Lower Capitan n=7			
Maximum	3.86	79	94
Minimum	1.78	47	46
Average (μ_{lower}) ±2 σ	2.40±1.37	63±22	62±31
$H_o:\mu_{middle}=\mu_{lower}$	$H_1:\mu_{middle}<\mu_{lower}$	$H_1:\mu_{middle}>\mu_{lower}$	$H_1:\mu_{middle}<\mu_{lower}$
P(T>t)	>0.99	0.95	>0.99

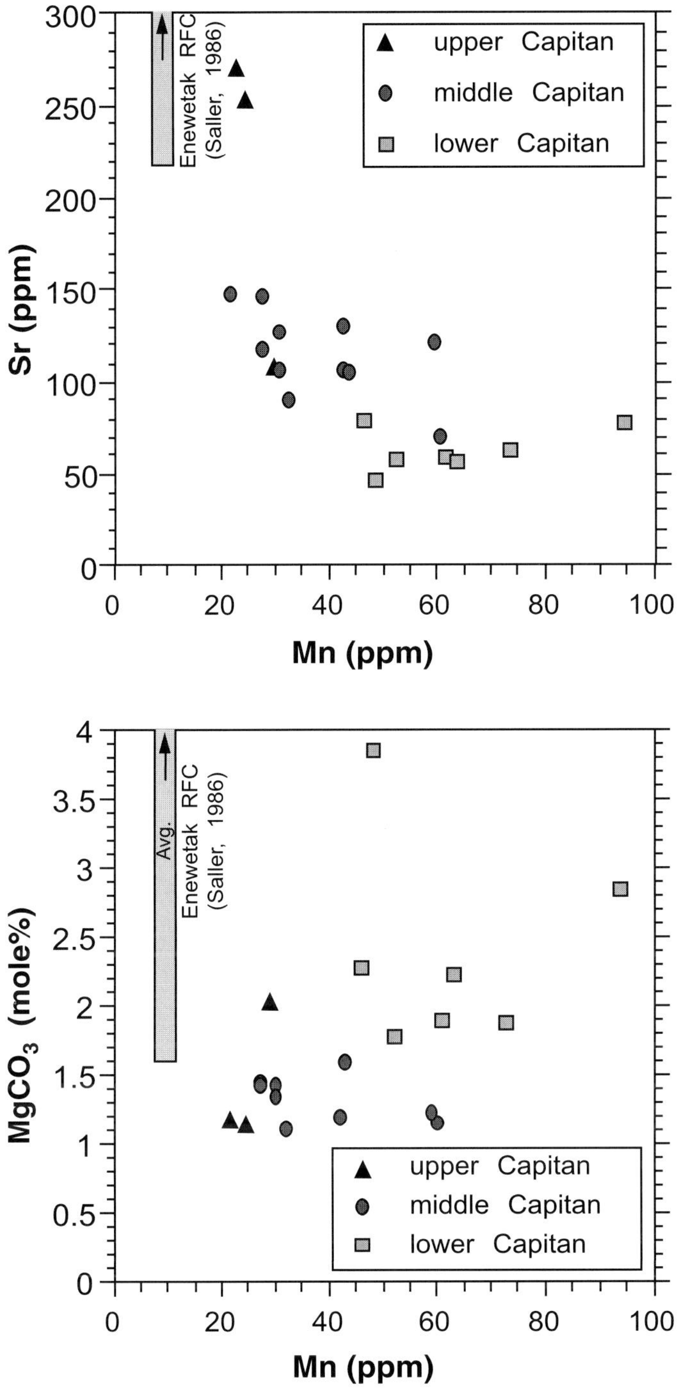

FIG. 10. — (A) Cross-plot of Sr versus Mn shows the stratigraphic variation in these two trace elements. Sr is typically found in marine cements, whereas Mn is typically associated with reducing conditions attained during burial (Veizer, 1983). Here, Sr shows a negative correlation with Mn, and decreases with increasing stratigraphic depth. Sr content is lower than was found in RFC at Enewetak Atoll, where average radiaxial-fibrous calcite contains 330±110 ppm Sr by microprobe analysis (Saller, 1986). (B) This cross-plot shows a positive correlation between Mg and Mn. Both trace elements increase with increasing stratigraphic depth. Down-section increases in Mn and Mg are probably diagenetic and may be associated with recrystallization in dolomitizing fluids. Enewetak Atoll radiaxial-fibrous calcite ranges from 1.6 to 11.1 mole % $MgCO_3$, and averages 3.2 mole % $MgCO_3$ (Saller, 1986).

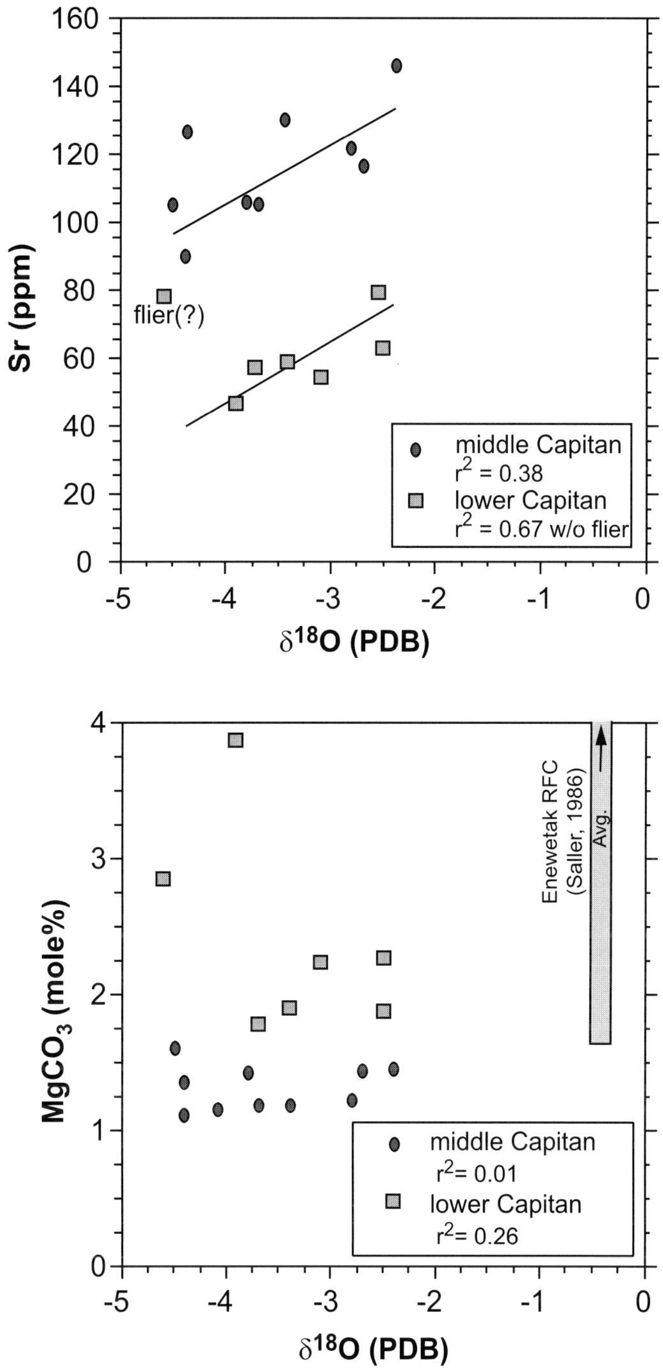

FIG. 11. — (A) Crossplot of Sr versus $\delta^{18}O$ shows a positive correlation in the middle and lower Capitan. The positive correlation supports the conclusion that samples represent mixtures of marine and nonmarine calcite. (Given and Lohmann, 1985). (B) Cross-plot of Mg versus $\delta^{18}O$ shows no correlation in the middle Capitan and a negative correlation in the lower Capitan. This does not fit the model of Given and Lohmann (1985), and has been interpreted two different ways in this study.

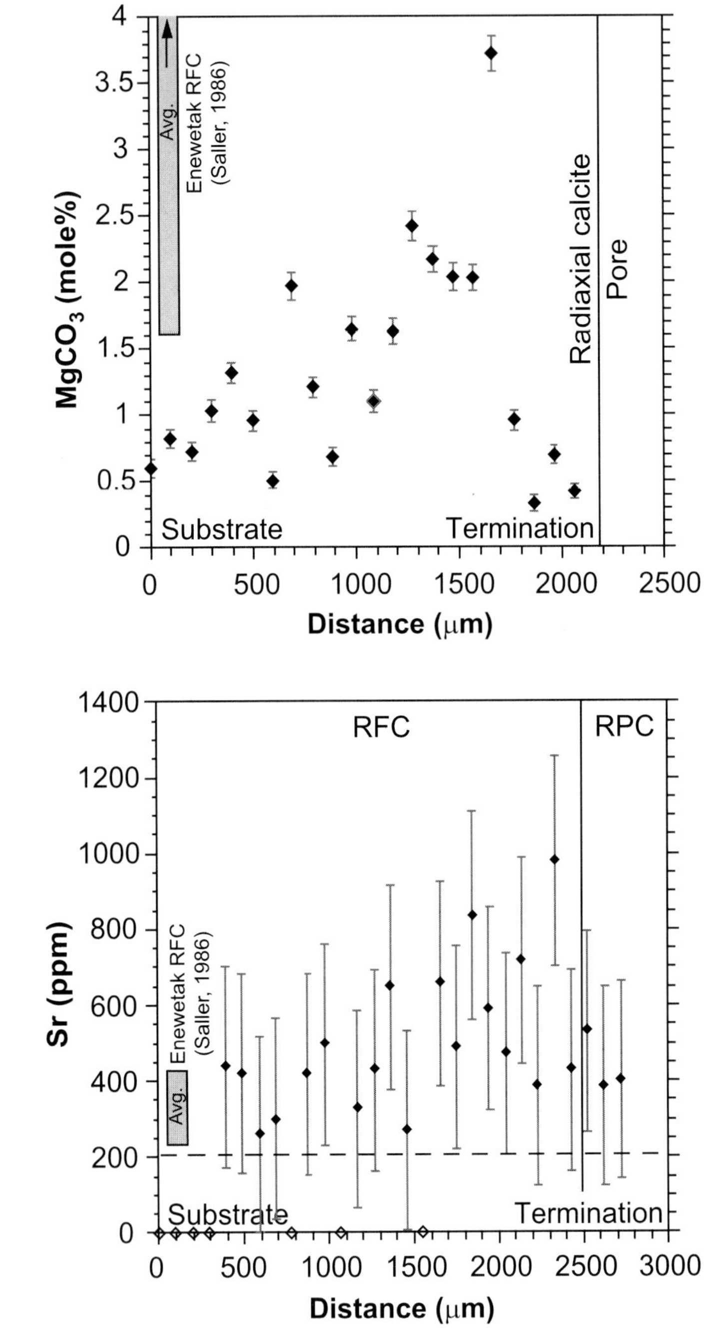

FIG 12. — (A) Microprobe traverse (sample 102393 240) showing an apparent increase in $MgCO_3$ towards the crystal terminations before decreasing near the open pore. We call the concentration apparent because RFC, FOC, and RPC contain dolomite microinclusions. Where dolomite inclusion size is much smaller than the analytical volume of the beam, the resulting data indicate the abundance of dolomite microinclusions. (B) This microprobe traverse (sample R10.2; location shown in Fig. 5B) may indicate an increase in Sr toward RFC and FOC terminations and a sharp decrease in RPC. Uncertainty is high because Sr concentrations are near the detection limit (210 ppm, 95% C.I.) indicated by the horizontal dashed line. We interpret the Sr to reside in calcite rather than microdolomite, because Mg and Sr did not correlate in this sample, even though a parallel increase in dolomite microinclusion abundance is visible in BSE images.

Samples with finely disseminated dolomite microinclusions show evidence for Mg or Sr trace element trends in three of nine samples analyzed with the microprobe. Individual microprobe analyses of inclusion-rich areas average 3.7 mole % $MgCO_3$, whereas inclusion-poor areas average 1.9 mole % $MgCO_3$, which suggests that microprobe analyses reflect the presence of dolomite microinclusions within both 10 µm or 20 µm spot sizes. Traverses reveal an apparent increase in Mg concentration from < 1 mole % near the crystal substrate to nearly 4 mole % Mg at the termination (r^2=0.62; Fig. 12A). The maximum end of this range is roughly in agreement with estimates from BSE images of up to 9.5 volume % of dolomite microinclusions. One sample contains detectable Sr. Strontium contents are below detection near the substrate and generally increase toward the termination of RFC. Strontium values are just above detection in RPC (Fig. 12B). In this sample there is no covariance of Sr and Mg (r^2=0.03). Fe and Mn are below detection for most analyses.

DISCUSSION

Capitan radiaxial calcite comprises a 4-mm thick layer of isopachous calcite, which contains three distinct types of calcite cement (RFC, FOC, and RPC), and represents a complex cement stratigraphy that grades from marine calcite and early dolomite to meteoric calcite, and finally burial dolomite or burial spar calcite (Fig. 13). In particular, the inclusions within these calcite cements appear to be key to understanding pore water evolution, and we present multiple hypotheses for their origin here.

Dolomite Microinclusion Petrography

Dolomite microinclusions are of two types: euhedral and anhedral. We suggest that the euhedral dolomite microinclusions in RPC grew into open pore space and are primary dolomite cements. We therefore recognize two generations of dolomite cement, which are probably equivalent to two generations of replacement dolomite: the first is fine-grained, and the latter is coarse (Melim, 1991). We remain uncertain, however, about the origin of anhedral dolomite microinclusions. Our first hypothesis states that anhedral dolomite microinclusions are primary and formed during precipitation of RFC, FOC, and RPC, whereas our alternate hypothesis states that anhedral dolomite microinclusions are secondary and formed during diagenesis of RFC, FOC, and RPC (Fig. 14).

Primary dolomite microinclusions.—

Our first hypothesis suggests that most dolomite microinclusions are primary. Examples of modern dolomite precipitating in modern reef settings (Saller, 1984, 1986) show that dolomite can precipitate from the same marine waters that precipitate calcite. In particular, the distinct layers of dolomite that accumulated at crystal terminations are strikingly similar to examples of primary marine dolomite cements in modern reefs (Mitchell et al., 1987), and in the Pleistocene Hope Gate Formation (Land, 1991). Based on comparison to modern examples, the cement layers and euhedral inclusions can be interpreted as primary precipitates. If a primary origin is possible for some of the euhderal inclusions, it should also be considered for the small, anhedral inclusions. Their anhedral shape could have been the result of crystal growth in confined spaces or corrosion of small dolomite cement crystals. Variation in dolomite inclusion size may be the result of variation in dolomite precipitation rate in response to changes in saturation state or flow rate.

Secondary dolomite microinclusions.—

In contrast, our alternate hypothesis suggests that anhedral dolomite microinclusions are secondary. Anhedral dolomite microinclusions <10 μm in diameter and in optical continuity with host calcite fit the characteristics of dolomite microinclusions described in Lohmann and Meyers (1977). In our samples anhedral dolomite microinclusions have not been found aligned along calcite growth faces, and so there is no direct evidence that they represent cements like the euhedral inclusions in RPC. Some anhedral crusts of dolomite could also represent preferentially dolomitized zones in a calcite that initially had variable Mg content, or, in the case of crusts of variable thickness, preferentially dolomitized micritic layers (Mruk, 1985). Moreover, anhedral dolomite microinclusion size is large (tens to hundreds of μm) where coarse dolomite cement overgrowths are present, and small (<1–10 μm) where coarse dolomite cement is absent, which suggests that the degree of dolomitization varied. On this petrographic basis one could infer that anhedral dolomite microinclusions are secondary.

This hypothesis remains open to the possibility that there are two origins for dolomite inclusions: formation of small anhedral inclusions during dolomitization of early, Mg-containing calcite, and formation of euhedral inclusions as cements.

Geochemistry

Stable isotopes.—

If dolomite microinclusions are secondary, and had their source of Mg from magnesium-containing calcite, or from the same solutions that removed Sr and added Mn, then we should see evidence to that effect in the calcite. Our samples produced covariant $\delta^{13}C$ and $\delta^{18}O$ trends similar to those described in studies of aragonite and calcite cementation in the Capitan (Given and Lohmann, 1985). XRD of sample powders was used

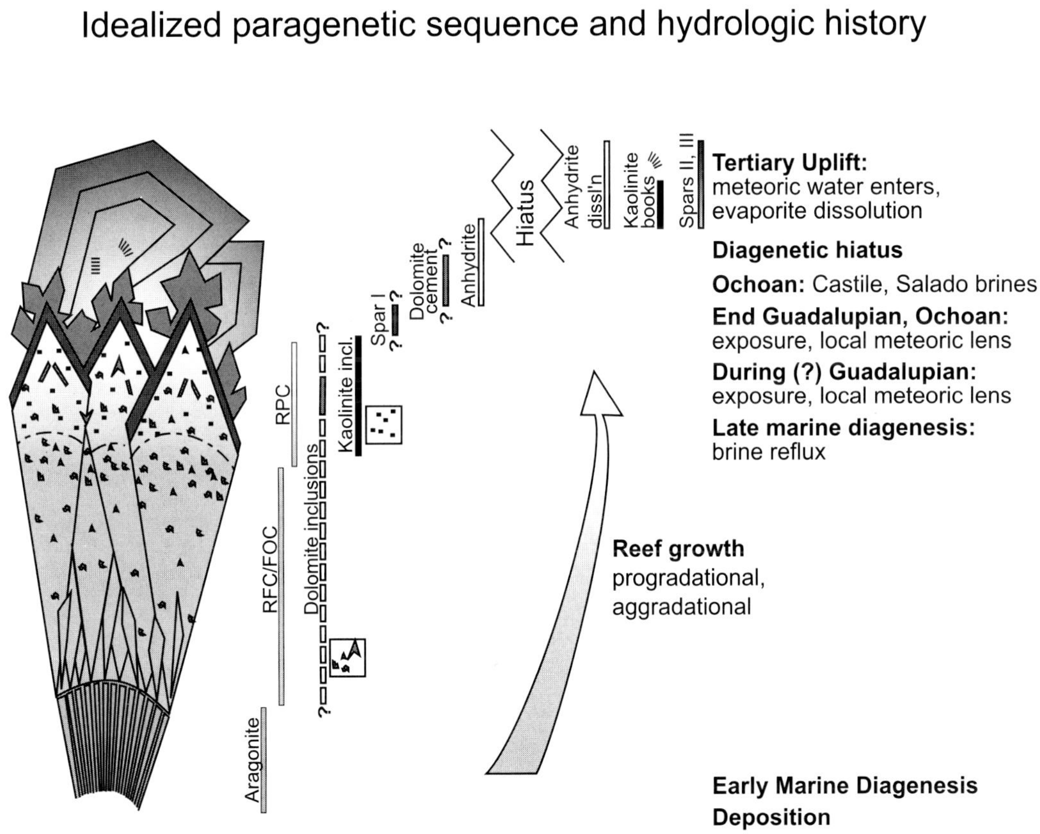

FIG. 13. — This diagram illustrates an idealized complete sequence of cementation up to and including coarse luminescent dolomite cement. It is meant to illustrate the complex package of cement that radiaxial calcite represents, although not all phases are to be found in any one sample. The distribution of inclusions is of primary interest here. Dolomite inclusions (anhedral shapes) increase in abundance toward the terminations of RFC/FOC and decrease after the transition to RPC (shown as a dashed line). Kaolinite inclusions (rectangles) are most abundant in RPC. We show dolomite microinclusions as a dashed line on the paragenetic sequence, because we remain uncertain whether anhedral dolomite microinclusions are primary or secondary. A generalized hydologic history is shown to the right of the paragenetic sequence. RFC and FOC may have passed through hydrologic regimes dominated by reflux brines, marine-meteoric mixing, and meteoric lens to produce this sequence. A combination of reef growth and sea-level fluctuation probably drove these changes during Guadalupian time. During Ochoan time descending Castile and Salado evaporitic fluids precipitated dolomite and porosity-occluding anhydrite. Much later, Tertiary uplift allowed entry of meteoric water.

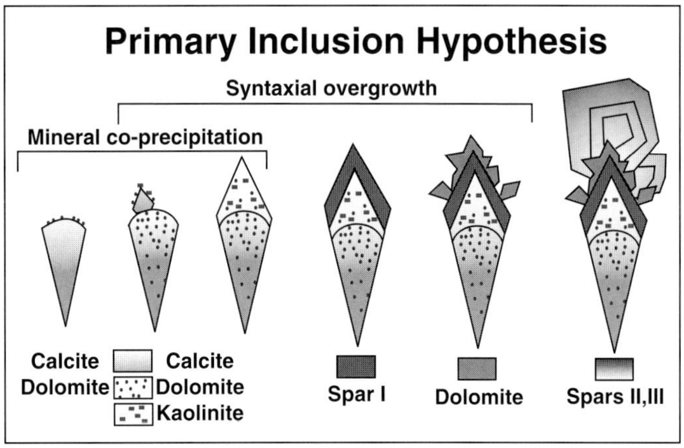

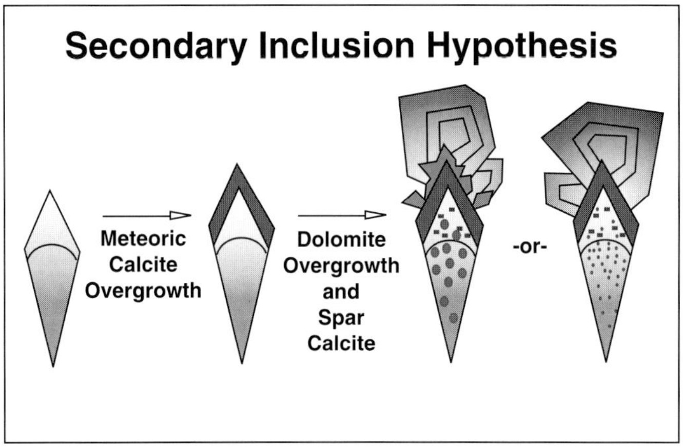

FIG. 14. — (A) The primary inclusion hypothesis explains the presence of dolomite and kaolinite microinclusions as a coprecipitation phenomenon, whereby calcite precipitates with dolomite, and sometimes kaolinite. The result is finely disseminated microinclusions in calcite and the distribution of these inclusions provides information about the precipitating fluids. (B) The secondary inclusion hypothesis proposes that dolomite microinclusions formed during dolomitization, because dolomite microinclusions are large in the presence of coarse dolomite overgrowths but small where they are absent. Geochemical characteristics of these cements potentially agree with formation of dolomite microinclusions during dolomitization.

to identify and discard dolomite-contaminated samples. The sampling scale of the sample drill, and the gradational character of these cements, however, did not allow complete separation of RFC, FOC, RPC, and spar calcite.

We accept the conclusion that covariant $\delta^{13}C$ and $\delta^{18}O$ trends reflect the physical mixing of two (or more) cement types during sampling (Given and Lohmann, 1985). We would also suggest, however, that vertical gradation from RFC/FOC mixtures to spar calcite, and covariant stable isotope trends, reflect the gradual nature of progressive calcite overgrowth during transition from marine to meteoric fluid. Syntaxial overgrowth of meteoric calcite onto RPC makes the petrographic distinction of each of the cements in these layers difficult, and in some cases, impossible.

Trace elements.—

Marine cements are isotopically heavier and generally have higher Mg and Sr concentrations than meteoric calcite (Veizer, 1983). If we sampled two cements—one marine the other meteoric—then they should also produce a positive correlation of $\delta^{18}O$ with Mg and Sr, thus extending the model of Given and Lohmann (1985). For our samples, Sr fits the resulting model, whereas Mg does not. Positive correlation of Sr, but not Mg, with $\delta^{18}O$ could reflect better preservation of Sr than Mg in the marine calcite end-member. Additionally, on trace element cross-plots, Sr decreases and Mg increases with increasing Mn and increasing stratigraphic depth. Similarly, coarse, luminescent dolomite increases in abundance lower in the section (Melim, 1991). This suggests that, during burial, water-rock interaction with dolomite-precipitating fluids mobilized trace elements in preexisting marine and meteoric calcite. Reflux dolomitization, however, could have produced this same pattern. The Capitan reef complex prograded rapidly so that a larger area of evaporitic shelf overlies the lower Capitan than overlies the upper Capitan. Diagenesis in refluxing shelf fluids, therefore, might also account for stratigraphic variations in trace element composition.

Microprobe traverses.—

Microprobe analyses include calcite and dolomite within the beam diameter and provide clues to the source of Mg in microdolomite inclusions. Analyses of inclusion-rich areas produce higher $MgCO_3$ concentrations than do inclusion-poor areas. These suggest that, where inclusions are much smaller than the beam diameter, the microprobe provides a measure of dolomite microinclusion abundance in the form of an apparent $MgCO_3$ concentration. Traverses show increasing abundance of dolomite microinclusions toward the termination of RFC and FOC, followed by a drop in abundance of RPC.

Trace element patterns detected by microprobe could indicate that calcite provided some of the Mg in dolomite microinclusions. Strontium and magnesium should covary in well preserved marine calcite (Carpenter et al., 1991). The traverse that contains Sr did not produce a parallel, covariant $MgCO_3$ trend, even though dolomite microinclusions increase in abundance within this sample, and produced trends in other samples. Most Sr, therefore, must reside in the calcite portion of the cement, whereas most Mg in Capitan RFC, FOC, and RPC resides in dolomite microinclusions.

Origin of Dolomite Microinclusions

The primary inclusion hypothesis proposes that, if dolomite microinclusions formed penecontemporaneously with RFC, FOC, and RPC, then there are several possible origins for such inclusions: (1) a primary marine dolomite origin (Mullins et al., 1988; Land, 1991), in which dolomite and calcite both precipitate in seawater, (2) lysocline shifts relating to sea-level change, which may be responsible for the apparent transition from aragonite to calcite and dolomite (Kirkland and Moore, 1993), (3) oxygen, and organic-stratified (Babcock, 1974) or salinity-stratified (Harms, 1969). Delaware Basin overturn may have promoted massive aragonite precipitation (Grotzinger and Knoll, 1995), and similar processes may have resulted in precipitation of RFC and FOC.

Variations in dolomite microinclusion size, their red luminescence, and the stratigraphic variations of trace elements in calcite hosting the inclusions implicate dolomitization as the origin of these dolomite microinclusions. Even so, the distribution of dolomite microinclusions appears to be characteristic of these cements, unlike a cross-cutting alteration pattern, and may be a relict of trace element distribution in the precursor calcite. According to this hypothesis the unaltered precursor calcite contained Mg and Sr, which increased in the crystal growth direction, and were covariant. This supposed precursor is analogous to Enewetak radiaxial calcites, some of which show trends of increasing Mg on microprobe traverses from substrate to termination (Saller, 1986).

The distribution of a trace element in a growing crystal depends on the amount of trace element in solution, and the crystal growth rate (Lorens, 1981). RFC and FOC, therefore, could record a gradual increase in salinity, or increasing calcite saturation and growth rate. The most likely causes of these changes relate to the flux of evaporation-concentrated fluids from back-reef areas or the vast inner shelf. Dolomitization has since then segregated Mg into anhedral dolomite microinclusions, leaving some Sr remaining within an altered calcite phase. Exsolution, however, seems an unlikely mechanism for redistributing Mg, since maximum burial depth and temperature were only about 1 km and 50°C (Crysdale, 1987). Dissolution and reprecipitation across a replacement front (Wilson and Dickson, 1996) seem more likely. Anhedral dolomite microinclusions, therefore, probably formed during dolomitization, and the presence of dolomite microinclusions in calcite does not necessarily indicate a high-Mg precursor for RFC or FOC.

Origin of Kaolinite Microinclusions

Most kaolinite occurs as booklets included in burial spar II and formed during Tertiary uplift (L. A. Melim, 1991). Kaolinite microinclusions occurring at the transition to RPC may have formed earlier, and might also reflect pore-fluid evolution. Kaolinite microinclusions occur within discrete bands at the transition from RFC/FOC to RPC and are most abundant in RPC. Kaolinite inclusion-banding in RFC/FOC, and the fact that a silicate cannot exsolve or recrystallize from a carbonate, led both authors to agree that kaolinite microinclusions are primary. Placing definite constraints on the timing of this event is difficult. RPC, in addition to kaolinite microinclusions, contains euhedral dolomite inclusions that we equate to fine-grained, fabric-preserving dolomite. Some fine-grained, fabric-preserving dolomite contains elevated $^{87}Sr/^{86}Sr$ (Melim, 1991), which may reflect the presence of a dissolved-feldspar component originating from shelf sandstones (L. A. Melim, pers. comm., 1993). This Sr may have been carried by reflux brines, but another possibility is that some feldspar dissolution occurred after an initial influx of waters with a meteoric component and prior to precipitation of spar I. This accounts for the transition from fibrous calcite morphologies (RFC and FOC) to the more prismatic calcite (RPC), which contains the bulk of kaolinite microinclusions. The timing of this event could coincide with periodic sea-level fall and siliciclastic deposition on the shelf; however, we have not seen evidence that marine and meteoric cementation alternated in the reef. Dunham (1972) proposed a vadose origin for the tepee-pisolite facies and assigned significance to evidence of feldspar dissolution and associated kaolinite books in this facies. Although we now know kaolinite books to have formed much later (Scholle et al., 1992), this earlier occurrence of kaolinite as inclusions in RPC could lend support to Dunham's original hypothesis.

Significance for Porewater Evolution

We interpret microprobe trace element trends to reflect changes in composition of the pore fluids during precipitation. Whether dolomite microinclusions are primary or secondary, the trends themselves appear to be primary and reflect the nature of pore fluid evolution during precipitation of that complex package of gradational cements. The resulting interpretations agree with the hydrologic history of the Permian Basin and help to constrain the timing of events in the paragenetic sequence.

Aragonite precipitation occurred soon after deposition, and was penecontemporaneous with encrusting marine organisms. A change in the precipitating phase—aragonite precipitation ceased and RFC and FOC precipitation began—indicates a change in saturation state. We think the best analog is Enewetak Atoll in the Pacific, where radiaxial-fibrous calcite precipitates below the aragonite lysocline. The Delaware Basin, however, unlike the Pacific, was probably stratified with respect to temperature, oxygen, and/or salinity (Harms, 1969; Babcock, 1977).

As the reef aggraded, older rocks would have come into contact with progressively deeper water. This is consistent with the observation that aragonite is the most abundant cement throughout the massive reef facies, whereas RFC/FOC is most abundant in the forereef (Mruk, 1985). Additionally, more RFC and FOC occur in the lower Capitan than in the upper Capitan (Yurewicz, 1976). As it grew, the reef prograded basinward (Garber et al., 1989), and placed shelf deposits atop older reef rocks. During this time calcite was still growing, but it was losing contact with normal marine water. Normal seawater may have been mixed with or displaced by evaporation-concentrated seawater from the shelf. This produced calcite, which incorporated progressively more Mg as it grew. Reef progradation occurred throughout Capitan time and accounts for the occurrence of trace element increases in RFC and FOC of the middle and upper Capitan reef; poorly preserved examples exist in the lower Capitan.

Continued reflux during RPC precipitation may have culminated in formation of euhedral dolomite microinclusions, and possibly primary kaolinite microinclusions. More likely, the prismatic nature of RPC, euhedral dolomite microinclusions, and kaolinite microinclusions are associated with development of a freshwater lens and mixing zone as proposed by Mruk (1985), and possibly with subaerial exposure in the tepee-pisolite facies (Dunham, 1972). Euhedral dolomite microinclusions in RPC could be analogous to alternating calcite and dolomite precipitation associated with mixing zone diagenesis in

Pleistocene reefal limesones of the Yucatan Peninsula (Ward and Halley, 1985). Volumes of meteoric water may have been small, and mixing probably occurred within a setting that was dominated overall by refluxing fluids. It is not clear whether the meteoric water source was associated with the tepee-pisolite facies, periodic sea-level fluctuation during Guadalupian time, or sea-level fall at the end of Guadaulupian time.

The trace element increases in RFC and FOC are relicts from a precursor calcite and probably formed by stabilization of the precursor calcite during dolomitization by Castile and Salado brines. This second stage of dolomitization is represented by coarse, luminescent dolomite overgrowths on RPC terminations. It is this stage with which we associate the formation of secondary, anhedral dolomite microinclusions in RFC, FOC, and RPC calcite cements. Occurrence of dolomite microinclusions in marine calcite does not indicate that the marine cement had a high-Mg calcite precursor, particularly when it existed prior to a dolomitization event.

CONCLUSIONS

1. RFC, FOC contain dolomite microinclusions. RPC contains dolomite and kaolinite microinclusions. Combinations of RFC, FOC, and RPC form complex cement stratigraphies. One form contains RFC and RPC with anhedral dolomite microinclusions (tens to hundreds µm in diameter) and is overgrown locally by coarse dolomite cement. The second form contains a mixture of RFC, FOC, and RPC with small dolomite microinclusions (<1-10 µm in diameter), and has not been found in the presence of coarse dolomite cement.
2. Anhedral dolomite microinclusions may be either primary (penecontemporaneous with host calcite) or secondary (formed during later diagenesis). Euhedral dolomite inclusions within radiaxial-prismatic calcite or between radiaxial-fibrous calcite and radiaxial-prismatic calcite resemble cement overgrowths.
3. Trace elements and stable isotopes form covariant trends from sample mixing, but do not fit models for unaltered marine cement. Stratigraphic variations in radiaxial calcite trace element compositions parallel down-section increases in burial dolomite abundance.
4. Whether anhedral dolomite microinclusions are primary or exsolved, their distribution may provide information about conditions at the time of RFC and FOC precipitation. Increases in dolomite microinclusion abundance towards RFC and FOC terminations may reflect the influence of fluids refluxing from the shelf.
5. Kaolinite inclusions are distributed at the transition between RFC and FOC to radiaxial-prismatic calcite and are most abundant in radiaxial-prismatic calcite. These might record relatively early feldspar dissolution in shelf sands and could be associated with refluxing fluids or local occurring meteoric lens development. It could also place constraints on timing and location of RPC precipitation.
6. Porewater changes during early diagenesis were probably driven by reef growth (aggradational and progradational) and sea-level change.

ACKNOWLEDGMENTS

Acknowledgment is made to the donors of the Petroleum Research Fund, administered by the American Chemical Society, for support of this research (Grant # 26697-G8). We also acknowledge the support of the Geology Foundation and Amoco for a fellowship that allowed M. A. Rahnis to complete the thesis portion of this research. We thank Guadalupe Mountains National Park and Carlsbad Caverns National Park for granting permission to collect samples used in this study. Thanks go to field assistants P. George and C. Lambert. This paper was reviewed by D. H. Cox, Z. Lasemi, and A. H. Saller. Previous reviewers were also very important, and included M. Coniglio, B. Jones, and A. C. Kendall. Those reading drafts of the manuscript on its way from thesis to publication deserve thanks, including J. L. Beall, R. L. Folk, L. S. Land, F. L. Lynch, R. P. Major, and B. Ward. J. Horowitz assisted with preparation of figures.

REFERENCES

ACHAUER, C. W., 1969, Origin of Capitan Formation, Guadalupe Mountains, New Mexico and Texas: American Association of Petroleum Geologists Bulletin, v. 53, p. 2314-2323.

BAARS, D. L., 1964, Modern carbonate sediments as a guide to old limestones: World Oil, v. 158, p. 95-100.

BABCOCK, J. A., 1974, The role of algae in the formation of the Capitan Limestone (Permian, Guadalupian), Guadalupe Mountains, West Texas—New Mexico: Unpublished Ph.D. Dissertation: University of Wisconsin, Madison, 241 p.

BATHURST, R. G. C., 1959, The cavernous structure of some Mississippian Stromatactis reefs in Lancashire, England: The Journal of Geology, v. 67, p. 506-521.

BATHURST, R. G. C., 1971, Radiaxial fibrous mosaic, *in* Carbonate Cements, Studies in Geology 19: Baltimore, Johns Hopkins University Press, p. 292-293.

BEIN, A. and DUTTON, A. R., 1993, Origin, distribution, and movement of brine in the Permian Basin (U.S.A.): A model for displacement of connate brine: Geological Society of America Bulletin, v. 105, p. 695-707.

CARPENTER, S. J., LOHMANN, K. C., HOLDEN, P., WALTER, L. M., HUSTON, T. J., and HALLIDAY, A. N., 1991, $\delta^{18}O$ values, $^{87}Sr/^{86}Sr$ and Sr/Mg ratios of Late Devonian abiotic marine calcite: Implications for the composition of ancient seawater: Geochimica et Cosmochimica Acta, v. 55, p. 1991-2010.

CRYSDALE, B. L., 1987, Fluid inclusion evidence for the origin, diagenesis and thermal history of sparry calcite cement in the Capitan Limestone, McKittrick Canyon, West Texas: Unpublished M.S. Thesis, University of Colorado, Boulder, 78 p.

DICKSON, J. A. D., 1966, Carbonate identification and genesis as revealed by staining: Journal of Sedimentary Petrology, v. 36, p.491-505.

DUNHAM, R. J., 1972, Capitan Reef, New Mexico and Texas: Facts and Questions to Aid Interpretation and Group Discussion: Midland, Society of Economic Paleontologists and Mineralogists, Permian Basin Section, Publication 72-14., 298 p.

FOLK, R. L., 1971, Unusual neomorphism of micrite, *in* Carbonate Cements, Studies in Geology 19: Baltimore, Johns Hopkins University Press, p. 163-166.

FOLK, R. L., and ASSERTO, R., 1976, Comparative fabrics of length-slow and length-fast calcite and calcitized aragonite in a Holocene speleothem, Carlsbad Caverns, New Mexico: Journal of Sedimentary Petrology, v. 46, p. 486-496.

GARBER, R. A., GROVER, G. A. and HARRIS, P. M., 1989, Geology of the Capitan shelf margin–subsurface data from the northern Delaware Basin, *in* Harris, P. M., and Grover, G. A., eds., Subsurface and Outcrop Examination of the Capitan Shelf Margin, Northern Delaware Basin: Tulsa, Society of Economic Paleontologists and Mineralogists Core Workshop 13, p. 3-269.

GIVEN, R. K., AND LOHMANN, K. C., 1985, Derivation of the original isotopic composition of Permian marine cements: Journal of Sedimentary Petrology, v. 55, p. 430-439.

GIVEN, R. K., AND LOHMANN, K. C., 1986, Isotopic evidence for the early meteoric diagenesis of the reef facies, Permian reef complex of west Texas and New Mexico: Journal of Sedimentary Petrology, v. 56, p. 183-193.

GROTZINGER, J. P., AND KNOLL, A. H., 1995, Anomalous carbonate precipitates: Is the Precambrian the key to the Permian?: Palaios, v. 10, p. 578-596.

HARMS, J. C., 1969, Permian Deep Water Sedimentation by Nonturbid Currents, Guadalupe Mountains, Texas: Boulder, Geological Society of America Special Paper, 127 p.

HAYES, P. T., 1964, Geology of the Guadalupe Mountains, New Mexico: Washington, D.C., U.S. Geological Survey Professional Paper 446, 69 p.

JAMES, N. P. and CHOQUETTE, P. W., 1990, Limestones—The sea-floor diagenetic environment, *in* McIlreath, I. A., and Morrow, D. W., eds., Diagenesis: Geoscience Canada Reprint Series 4, p. 9-34.

KENDALL, A. C., 1977, Fascicular-optic calcite: A replacement of bundled acicular carbonate cements: Journal of Sedimentary Petrology, v. 47, p. 1056-1062.

KENDALL, A. C., 1985, Radiaxial fibrous calcite: A reappraisal, *in* Schneidermann, N., and Harris, P. M., eds., Carbonate Cements: Tulsa, Society of Economic Paleontologists and Mineralogists Special Publication 36, p.59-77.

KENDALL, A. C., and TUCKER, M. E., 1973, Radiaxial fibrous calcite: A replacement after acicular carbonate: Sedimentology, v. 20, p. 365-389.

KERANS, C., FITCHEN, W. M., GARDENER, M. H., SONNENFELD, M. D., TINKER, and S. W., WARDLAW, B. R., 1992, Styles of Leonardian through Guadalupian strata of the Guadalupe Mountains, Texas and New Mexico, *in* Mruk, D. H. and Curran, B. C., eds., Permian Basin Exploration and Production Strategies; Applications of Sequence Stratigraphic and Reservoir Characterization Concepts: Midland, West Texas Geological Society Publication 92-91, p. 1-6.

KING, P. B., 1948, Geology of the Southern Guadalupe Mountains, Texas: Washington, D.C., U.S. Geological Survey Professional Paper 215, 183 p.

KIRKLAND-GEORGE, B., 1992, Distinctions between reefs and bioherms based on studies of fossil algae—*Mizzia*, Permian Capitan reef complex (Guadalupe Mountains, Texas and New Mexico) and *Eugonophyllum*, Pennsylvanian Holder Formation (Sacramento Mountains, New Mexico): Unpublished Ph.D. Dissertation, Louisiana State University, Baton Rouge, 156 p.

KIRKLAND, B. L. and MOORE, C. H., 1993, Radiaxial calcite in the Permian Capitan Formation, Guadalupe Mountains, New Mexico; a possible indicator of sea-level rise (abs.): Tulsa, American Association of Petroleum Geologists Convention Abstracts, p.130.

KIRKLAND, B. L., and MOORE, C. H., 1996, Microfacies analysis of the Tansill outer shelf, Permian Capitan Reef Complex, *in* Martin, R. L., ed., Permian Basin Oil and Gas Fields: Keys to Success that Unlock Future Reserves: Midland, West Texas Geological Society Publication 96-101, p. 99-106.

LAND, L. S., 1991, Dolomitization of the Hope Gate Formation (North Jamaica) by seawater; Reassessment of mixing zone dolomitization, *in* Taylor, H. P., O'Neil, J. R., and Kaplan, I. R., eds., Stable isotope Geochemistry; A Tribute to Samuel Epstein: Geochemical Society, Special Publication, v. 3, p. 121-133.

LOHMANN, K. C., and MEYERS, W. J., 1977, Dolomite microinclusions in cloudy prismatic calcites: A proposed criterion for former high-magnesium calcites: Journal of Sedimentary Petrology, v. 47, p. 1078-1088.

LORENS, R. B., 1981, Sr, Cd, Mn, and Co distribution coefficients in calcite as a function of calcite precipitation rate: Geochimica et Cosmochimica Acta, v. 45, p. 553-561.

MELIM, L. A., 1991, The origin of dolomite in the Permian (Guadalupian) Capitan Formation, Delaware Basin, West Texas and New Mexico: Implications for dolomitization models: Unpublished Ph.D. Dissertation, Southern Methodist University, Dallas, 200 p.

MITCHELL, J. T., LAND, L. S., and MISER, D. E., 1987, Modern marine dolomite cement in a north Jamaican fringing reef: Geology, v. 15, p. 557-560.

MRUK, D. H., 1985, Cementation and Dolomitization of the Capitan Limestone (Permian), McKittrick Canyon, West Texas: Unpublished M.S. Thesis, University of Colorado, Boulder, 125 p.

MULLINS, H. T., GARDULSKI, A. F., HINCHEY, E. J., and HINE, A. C., 1988, The modern carbonate ramp slope of central west Florida: Journal of Sedimentary Petrology, v. 58, p. 273-290.

NEWELL, N. D., RIGBY, J. K., FISHER, A. G., WHITEMAN, A. J., HICKOX, J. E., and BRADLEY, J. S., 1953, The Permian Reef Complex of the Guadalupe Mountains Region, Texas and New Mexico: San Francisco, W. J. Freeman and Company.

PRAY, L. C., 1986, Capitan reef complex (Permian), Guadalupe Mountains, Southwestern United States; reinterpretation of deposition and diagenesis of a classic sedimentologic model (abs.): American Association of Petroleum Geologists Bulletin, v. 70, p. 1762.

RUDOLPH, K. W., 1978, Diagenesis of back-reef carbonates: An example from the Capitan Complex: Unpublished M.A. Thesis, The University of Texas, Austin, 159 p.

SALLER, A. H., 1984, Petrologic and geochemical constraints on the origin of subsurface dolomite, Enewetak Atoll: An example of dolomitization by normal seawater: Geology, v. 12, p. 217-220.

SALLER, A. H., 1986, Radiaxial calcite in Lower Miocene strata, subsurface Enewetak Atoll: Journal of Sedimentary Petrology, v. 56, p. 743-762.

STUEBER, A. M., SALLER, A. H., and ISHIDA, H., 1998, Origin, migration, and mixing of brines in the Permian basin: Geochemical evidence from the eastern Central Basin Platform, Texas: American Association of Petroleum Geologists Bulletin, v. 82, p. 1652-1672.

SCHOLLE, P. A., ULMER, D. S., and MELIM, L. A., 1992, Late-stage calcites in the Permian Capitan Formation and its equivalents, Delaware Basin margin, west Texas and New Mexico: evidence for replacement of precursor evaporites: Sedimentology, v. 39, p.207-234.

TUCKER, M. E. and WRIGHT, V. P., 1990, Diagenetic processes, products and environments (Chapter 7), *in* Carbonate Sedimentology: London, Blackwell Scientific Publications, p. 331.

VEIZER, J., 1983, Trace elements and isotopes in sedimentary carbonates, *in* Reeder, R. J., ed., Carbonates: Mineralogy and Chemistry: Mineralogical Society of America, Reviews in Mineralogy, v. 11, p. 265-299.

VIDETICH, P. E., 1985, Electron microprobe study of Mg distribution in recent Mg calcites and recrystallized equivalents from the Pleistocene and Tertiary: Journal of Sedimentary Petrology, v. 55, p. 421-429.

WARD, W. C. and HALLEY, R. B., 1985, Dolomitization in a mixing zone of near-seawater composition, late Pleistocene, northeastern Yucatan Peninsula: Journal of Sedimentary Petrology, v. 55, p. 407-420.

WHITE, J. V., KIRKLAND, B. L., and GOURNAY, J. P., 1998, Quantitative porosity determination of thin sections using digitized images: Journal of Sedimentary Research, v. 68, p. 220-222.

WILSON, P. A., and DICKSON, J. A. D., 1996, Radiaxial calcite: alteration product and petrographic proxy for magnesian calcite marine cement: Geology, v. 24, p. 945-948.

WOOD, R. A., DICKSON, J. A. D. and KIRKLAND-GEORGE, B., 1994, Turning the Capitan reef upside down: A new appraisal of the ecology of the Permian Capitan reef, Guadalupe Mountains, Texas and New Mexico: Palaios, v. 9, p. 422-427.

YUREWICZ, D. A., 1976, Sedimentology, paleoecology, and diagenesis of the massive facies of the lower and Middle Capitan Limestone (Permian), Guadalupe Mountains, New Mexico and West Texas: Unpublished Ph.D. Dissertation, University of Wisconsin, Madison, 278 p.

DIAGENESIS OF THE CAPITAN FORMATION FOREREEF FACIES (PERMIAN, WEST TEXAS AND NEW MEXICO)

LESLIE A. MELIM

Department of Geology, Western Illinois University, Macomb, Illinois 61455

AND

PETER A. SCHOLLE

Department of Geological Sciences, Southern Methodist University, Dallas, Texas 75275

ABSTRACT: The diagenesis of the Capitan forereef can be divided into four overlapping episodes: 1) early marine diagenesis; 2) early burial diagenesis involving normal marine to hypersaline fluids; 3) late burial diagenesis; and 4) uplift related diagenesis. Early meteoric diagenesis was not identified in the Capitan forereef. Early marine diagenesis is limited to minor isopachous cement in upper forereef grainstones. The greatest amount of diagenesis in the Capitan forereef occurred during early burial diagenesis in marine to hypersaline pore fluids. The lower forereef facies was largely altered by marine pore fluids; fabrics include moldic porosity, fine blocky spar and overgrowth cementation, silicification, and rare aragonite neomorphism. The upper forereef facies, however, was nearly completely dolomitized by mesosaline fluids mainly derived by seepage reflux from the near-backreef carbonate lagoon facies. The middle forereef facies contains both styles of alteration. During and/or subsequent to dolomitization, hypersaline fluids completely cemented the reef and forereef with evaporites (gypsum and/or anhydrite). Deeper burial processes include stylolites and partial recrystallization of the early fabric-preserving dolomite to a fabric-destructive dolomite. During uplift, erosion of overlying Ochoan evaporites allowed the influx of meteoric water into the Capitan Formation. This led first to hydration of anhydrite to gypsum and kaolinitization of feldspars and later to complete dissolution of evaporites and precipitation of coarse blocky spars II and III in the resulting porosity.

Most diagenetic models focus on timing of alteration and the fluid composition. In the case of the Capitan forereef facies, however, the sedimentology was a major factor controlling the distribution of diagenetic fabrics. For example, debris-flow deposits (unsorted packstones to rudstones) are partially to completely dolomitized while interbedded turbidity-current deposits (graded packstones to wackestones) are limestone with fine blocky spar and aragonite alteration. The changing patterns of diagenesis from the upper forereef to the lower forereef are at least partially controlled by the change from a predominance of debris-flow deposits in the upper forereef to turbidity-current deposits in the lower forereef.

INTRODUCTION

The Guadalupe Mountains of west Texas and New Mexico contain one of the most complete exposures of an Upper Permian reef complex in existence. The Capitan Formation includes the reef and forereef facies. Also well exposed in the Guadalupe Mountains are back-reef and basinal sediments equivalent to the Capitan. Although the Capitan Formation has been extensively studied over the last 50 years, the forereef facies largely has been ignored as previous workers focused their attention on the reef facies (e.g., Newell et al., 1953; Babcock, 1974, 1977; Yurewicz, 1976, 1977). However, due to its spectacular exposure, easy access, and relatively simple geologic history, the Capitan Formation provides an unsurpassed opportunity to study depositional and diagenetic processes in large-scale, steep forereef facies.

Although a number of studies have developed depositional models for forereef facies (e.g., Mullins, 1983; Cook and Mullins, 1983; Bosellini, 1984; Mullins and Cook, 1986; Grammer and Ginsburg, 1992; Harris, 1994; Melim and Scholle, 1995), little is published on diagenesis in the forereef setting. Of the few diagenetic studies undertaken, some indicate a strong correlation between diagenesis, particularly dolomitization, and the occurrence of reservoir quality porosity and permeability (Wiggins and Harris, 1985; Mruk, 1985; Garber et al., 1989), while others document a mainly porosity-destructive style of diagenesis (Grammer et al., 1993, Harris, 1994). As the forereef facies remains an active exploration target (e.g., Harris, 1997), a better understanding of the controls on forereef diagenesis should help in evaluating the reservoir potential of specific forereef units. This study reports on the diagenesis of the Capitan forereef facies with an emphasis on limestone fabrics. The reader is referred to Melim (1991) for details on the dolomite in the Capitan Formation.

METHODS

Field work was conducted throughout the eastern Guadalupe Mountains (Fig. 1). Samples were taken every 20–30 m in 14 vertical sampling transects of forereef facies exposure (Fig. 1). Detailed sections were not described due to the limited exposure (generally less than 20%). The wide spacing of sampling transects allowed evaluation of variation in the Capitan Formation forereef facies both parallel and perpendicular to depositional strike (Fig. 1). Precise locations of sampling transects can be found in Melim (1991).

Four hundred samples cover the complete spectrum of the Capitan forereef facies as well as portions of the reef and backreef facies for comparison. Each sample was slabbed, thin-sectioned, and stained for dolomite and iron using the technique of Dickson (1965). Thin sections were examined using standard polarized light, as well as under cathodoluminescence. Since the stain was found to quench luminescence, samples selected for cathodoluminescence were unstained and polished. Individual components for stable isotope analysis were drilled from the slabs using a high-speed micro-drilling machine (Prezbindowski, 1980). Stable isotope compositions of dolomites were determined using the methods of Epstein et al. (1964). No correction was applied for phosphoric acid fractionation (Land, 1980). All results are given relative to the PDB standard. Multiple analyses of a dolomite standard (N=11) indicate reproducibility of ±0.2‰ for $\delta^{18}O$ and 0.05‰ for $\delta^{13}C$.

Geologic Framework of the Capitan Reef, SEPM Special Publication No. 65

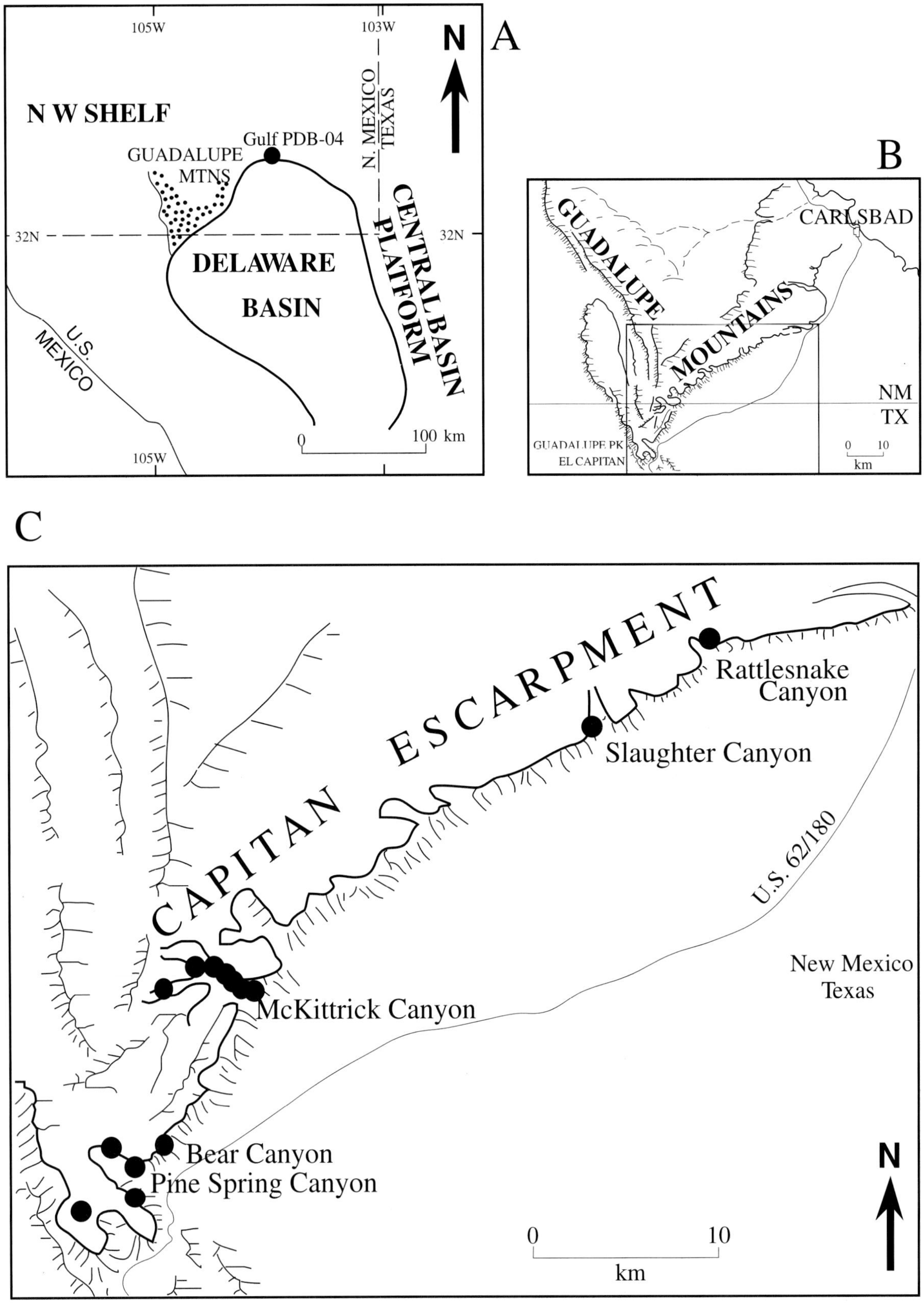

FIG. 1.—Location of the Guadalupe Mountain field area, Texas and New Mexico. (A) Map showing position of Delaware Basin during deposition of the Capitan Formation. Circle marks the location of the Gulf PDB-04 research core (see text). (B) Location map of features in the Guadalupe Mountains. Box shows location of Figure 1C. (C) Sample localities in the Guadalupe Mountains. A vertical transect from the basin to the reef was done at each locality except those in Slaughter Canyon, Rattlesnake Canyon, and the southernmost locality on El Capitan, where the basin facies is not exposed. The Permian Reef Trail is located at the mouth of McKittrick Canyon. (Maps after King 1948).

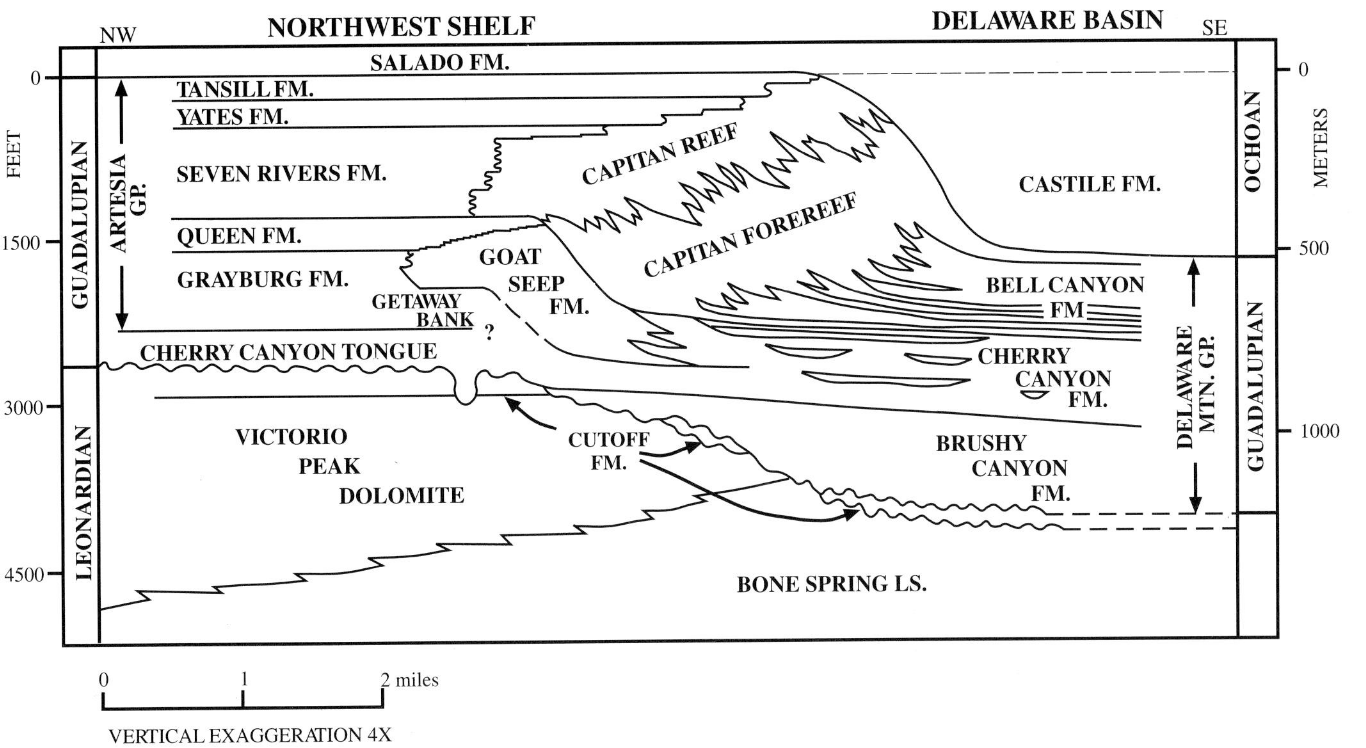

FIG. 2.—Upper Permian stratigraphy of the Delaware Basin (after King 1948; Newell et al. 1953; and Garber et al. 1989)

Stratigraphic Framework

The late Guadalupian (≈253-251 Ma) Capitan reef complex of the Guadalupe Mountains, west Texas and New Mexico, provides the best exposed example of the penultimate stage of evolution of the Permian Delaware Basin (Figs. 1 and 2). The Capitan Formation includes reef and forereef deposits equivalent to the Seven Rivers, Yates, and Tansill Formations in the back-reef and the Bell Canyon Formation in the basin (Newell et al., 1953; Fig. 2). The back-reef units include near-back-reef mesosaline (35–120‰) facies and far-back-reef hypersaline (>120‰) facies (Sarg, 1977, 1989; Hurley, 1978). The Capitan Formation has been informally divided into lower, middle, and upper members based on correlation with the back-reef units (Babcock, 1974; Yurewicz, 1976; Fig. 2). At the end of Guadalupian time, the Delaware Basin became restricted, resulting in draw-down of basin water levels and subaerial exposure of the reef and back-reef (Garber et al., 1989). The Ochoan Castile Formation evaporites infilled most of the basin topography and were followed by deposition of 300–700 m of additional evaporites of the Salado Formation that covered the entire region (King, 1948; Newell et al., 1953).

The Capitan Formation is 500–600 m thick and includes 100–150 m of massive reef facies overlying a bedded forereef facies (King, 1948). Although the origin of the Capitan "reef" remains controversial (e.g., Hurley, 1989; George, 1992), it unquestionably shed large volumes of debris into the deep waters of the Delaware Basin. Sufficient sediment was produced to allow the Capitan margin to prograde 7–19 km (King, 1948; Hayes, 1964; Garber et al., 1989; Melim and Scholle, 1992). This thick forereef package was examined in this study.

SEDIMENTOLOGY OF THE CAPITAN FOREREEF FACIES

Depositional fabrics control the original porosity and permeability of sediments. This, in turn, influences the distribution of diagenetic fabrics that require fluid flow (for example, dolomitization and cementation). The depositional fabric is a function of the processes operating during deposition, which, for carbonate slope facies, are controlled by the available sediment grain-size and the relief (Kenter, 1990; Melim and Scholle, 1995). The Capitan forereef had up to 500 m of relief and the sediment included all grain sizes from boulders to mud, but with an abundance of mud-sized material (Melim and Scholle, 1995). This produced three different facies (upper, middle, and lower forereef) distinguished based on their lithology and depositional dip (Table 1; Melim and Scholle, 1995). The upper forereef interfingers with the reef while the lower forereef facies interfingers with the basin margin facies that is usually considered part of the Bell Canyon Formation (Melim and Scholle, 1995).

Upper Forereef Facies

The majority of the upper forereef facies is composed of steeply dipping (≈30°), poorly bedded and poorly sorted intraclastic packstones to rudstones (Table 1). In most areas, greater

TABLE 1.—SUMMARY OF THE SEDIMENTOLOGY OF THE CAPITAN FORMATION FOREREEF FACIES AND INTERFINGERING BASIN-MARGIN FACIES OF THE BELL CANYON FORMATION*

Facies	Lithology	Approximate abundance	Depositional Dip	Depositional Processes
Upper forereef	Intraclastic packstones to rudstones	>90%	30°–35°	Debris flows, possible minor rock falls
	Fossiliferous to intraclastic grainstones	<10%		Grain flows
Middle forereef	Intraclastic packstones to rudstones	80%–90%	15°–30°	Debris flows and Turbidity currents
	Fossiliferous wackestones to packstones	10%–20%		Turbidity currents
	Siliciclastic sandstones to siltstones	<5%		(not studied)
Lower forereef	Intraclastic packstones and minor rudstones	<20%	5°–15°	Debris flows and Turbidity currents
	Peloidal and fossiliferous to intraclastic packstones to wackestones	70%–80%		Turbidity currents
	Mudstones	10%		Turbidity currents
Basin margin	Intraclastic packstones to rudstones	<10%	10° to <3°	Debris flows and Turbidity currents
	Fossiliferous mudstones, wackestones and packstones	60%–70%		Turbidity currents
	Siliciclastic sandstones to siltstones	20%–40%		(not studied)

*After Melim and Scholle (1995).

than 90% of the upper forereef is composed of this lithofacies. Most of the intraclasts are cobble to boulder-sized and were derived from the reef facies. Lithologies include mudstones, packstones, and boundstones containing typical reef fauna such as calcareous sponges, *Tubiphytes*, foraminifera, bryozoans, brachiopods and crinoids (Mruk and Bebout, 1993; Melim and Scholle, 1995). Rare intraclasts containing the alga *Mizzia* indicate some input from the near back-reef facies (Melim and Scholle, 1995). Reef boundstones containing reworked marine cements (with rotated geopetal fabrics) are common, particularly in the Tansill-equivalent forereef (Melim and Scholle, 1995).

The matrix of the packstones to rudstones is usually a very poorly sorted mixture of carbonate mud and silt to pebble-sized grains. However, in the lower Capitan Formation (Seven Rivers-equivalent), siliciclastic sand is often a major component. Although they are now grain-supported, chemical and mechanical compaction features indicate that they may have started as more matrix-supported (Melim and Scholle, 1995). Bed thickness ranges from a few decimeters up to 10 m thick, but amalgamated beds are common (Mruk and Bebout, 1993; Melim and Scholle, 1995). Melim and Scholle (1995) interpreted these beds as mainly debris-flow deposits based on several factors: most importantly, the poor sorting, the high matrix content, the lack of internal geopetal fabrics, and the lack of internal bedding. This interpretation is slightly different from that of Mruk and Bebout (1993), who interpreted some of the matrix-rich rudstones as rock-fall deposits. We disagree because geopetal fabrics would be expected if the matrix was later infiltrated into a rock-fall deposit (e.g., Harris, 1994).

The second common lithofacies in the upper forereef facies is fossiliferous to intraclastic grainstones. These are typically well-sorted, often parallel- or cross-laminated, and range from a few centimeters to a few meters in thickness. Like the rudstones they are interbedded with, these beds have high primary dips, usually around 30° (Mruk and Bebout, 1993; Melim and Scholle, 1995), but values as high as 66° have been reported (Mruk and Bebout, 1993). Inverse grading is present in some beds (Mruk and Bebout, 1993). Many grainstones are dominated by a single grain type, usually fusulinids but intraclast-rich beds are also seen. Other grains present include coated grains, crinoids, *Mizzia*, and bryozoans (Mruk and Bebout, 1993; Melim and Scholle, 1995). These units are interpreted as grain-flow deposits based on the steep dips, the inverse grading, and the well-sorted nature of the sand grains (Mruk and Bebout, 1993; Melim and Scholle, 1995).

Middle Forereef Facies

The middle forereef facies is composed of intraclastic packstone to rudstone beds with lesser proportions of fossiliferous wackestones and siliciclastic sandstone/siltstone. It is distinguished from the upper forereef based on lower depositional dips (≈15–30°), a smaller average grain size, the absence of grainstone beds, and the presence of fossiliferous wackestones (Table 1). The intraclasts in the middle forereef are reef-derived lithologies similar to those found in the upper forereef. Many of the intraclasts in the middle forereef may have been derived from erosion of the upper forereef rather than directly from the reef, as multiple geopetal orientations are often present. The intraclastic packstone to rudstone beds are very poorly sorted with a matrix of carbonate mud, fine-grained intraclasts, peloids, and fossils. Some of the lower Capitan (Seven Rivers-equivalent) contains significant siliciclastic sand in the matrix. Individual beds are commonly amalgamated into 2–5 m-thick units that give the forereef facies its characteristic large-scale bedding. As in the upper forereef, the packstones to rudstones are interpreted to be mainly debris-flow deposits (Melim and Scholle, 1995).

Interbedded with the coarse packstones to rudstones are fine-grained, often graded and/or laminated, fossiliferous and intraclastic wackestones to packstones interpreted as turbidity current deposits (Table 1; Melim and Scholle, 1995). This lithofacies becomes more common in the lower forereef but also composes 10–20% of the middle forereef. Individual beds are usually 5–15 cm thick and often fill channel features between the coarse rudstone beds (Melim and Scholle, 1995). Intraclasts are similar to those found in the coarser units of the middle forereef but with fewer reef boundstone clasts. Fossils include crinoids, brachiopods, foraminifera, bryozoa, and molluscs. Although some fossils were likely eroded from the reef, many lived on the forereef (for example, crinoids, brachiopods, and bryozoa) based on preservation of delicate features and their greater abundance here than in the reef (Mruk and Bebout, 1993; Melim and Scholle, 1995).

Siliciclastic sandstones to siltstones are found as thin lenses within the middle forereef. Although most of these lenses are < 1 m, units up to 3 m thick have been identified in several localities (Moran, 1955; Garber et al., 1989; Melim and Scholle, 1995). Most sandstone beds are structureless but fine lamination is sometimes present. Based on the small size and limited lateral extent, these sandstone lenses are interpreted as channel fills that probably formed during sea-level lowstands when siliciclastic sediments bypassed the reef to fill the basin (Melim and Scholle, 1995).

Lower Forereef Facies

The lower forereef facies includes intraclastic packstones to rudstones, peloidal and fossiliferous to intraclastic packstones to wackestones, and minor mudstones (Table 1; Melim and Scholle, 1995). The lower forereef is distinguished from the middle forereef based on finer grain size and by a decrease in depositional dip to 5–15°. The intraclastic packstones to rudstones are thinner and finer-grained than those higher up the forereef slope, averaging 0.5–1.5 m thick with clasts usually < 50 cm in diameter (and most < 10 cm). The intraclasts are similar to those found upslope but there are fewer identifiable reef-derived clasts. Matrix-support of the clasts is more common than farther upslope. The packstones to rudstones are interpreted as the distal portions of debris flows (Melim and Scholle, 1995).

Most of the lower forereef is made up of peloidal and fossiliferous to intraclastic packstones to wackestones (Table 1; Melim and Scholle, 1995). These beds are thinner than the rudstones (most are 5-15 cm) and are usually graded and/or laminated. Thin mudstones often cap graded units. A wide range of compositions is found, from intraclast-dominated to fossil-dominated. Many of the fossiliferous beds are mainly brachiopods and/or crinoids derived from communities living on the lower forereef slope (Babcock, 1977; Melim and Scholle, 1995). These finer-grained and more mud-rich units are interpreted as turbidity current deposits based on the grain size, sedimentary structures, and bed thickness (Melim and Scholle, 1995).

Basin Margin Facies

The basin-margin facies is interbedded carbonate and siliciclastic units included in the Bell Canyon Formation (Fig. 2; King, 1948). The carbonate beds are mainly fossiliferous mudstones, wackestones, and packstones with lesser amounts of intraclastic packstones to rudstones (Table 1). The fossiliferous mudstones to packstones are typically well-developed beds, 5–30 cm in thickness, that thin with distance from the forereef slope (Koss, 1977). Near the forereef, the fossils are mainly crinoids, brachiopods, and bryozoans that were derived locally by erosion of the lower forereef and basin margin facies (Babcock, 1977). These units are interpreted as turbidity current deposits (Babcock, 1977; Melim and Scholle, 1995). The intraclastic packstones to rudstones that interfinger with the lower forereef facies are similar to the intraclastic rudstones that are found in the middle to lower forereef and are interpreted as debris-flow deposits that traveled farther than most.

PREVIOUS WORK ON CAPITAN FORMATION FOREREEF FACIES DIAGENESIS

Although early workers noted the presence of significant dolomite in the Capitan Formation forereef facies (King, 1948; Newell et al., 1953), diagenetic studies have largely focused on the reef facies (e.g., Schmidt, 1977; Mazzullo and Cys, 1977). Two more recent studies that looked at the forereef facies are Mruk (1985, 1989) and Reeckmann (1985). In addition, Garber et al. (1989) and Scholle et al. (1992) described forereef facies diagenesis as part of larger-scale studies.

Mruk (1985, 1989) studied a single reef-to-basin transect in the middle to upper Capitan Formation at the mouth of McKittrick Canyon (Fig. 1) and focused on the calcium carbonate cements. She identified seven separate calcium carbonate cementation events within the reef facies. Of these, five (isopachous fibrous, bladed, and inclusion-rich prismatic cements and blocky calcite spars II and III) were found to precipitate in the forereef facies (Mruk, 1985, 1989). The isopachous fibrous and the bladed and inclusion-rich prismatic cements were identified in the proximal forereef, and she interpreted them as in-place marine cements. The blocky spars are found throughout the forereef facies and she believed they formed later in meteoric-derived fluids (Mruk, 1985, 1989). The exact timing of spars II and III is uncertain, but most likely occurred during burial to uplift (Mruk, 1989).

Mruk (1985) also looked at the dolomite in the forereef facies. Dolomite fabrics include isolated rhombs and dolomite mosaics. The dolomite mosaics range from fabric preserving to those where the precursor is unrecognizable (Mruk, 1985). The dolomite formed after the marine cements and before blocky spar II. Based on the most dolomite-rich samples of bulk rock stable isotopes, Mruk (1985) concluded that dolomite formed in a mixing-zone dominated by the marine component.

Reeckmann (1985) described the main diagenetic features of the forereef-to-basin transition at the mouth of McKittrick Canyon (Fig. 1) as part of a study focusing on the sedimentology. Cements found in the forereef include early blocky spar cements in primary and biomoldic pores, and later coarse blocky spar in secondary pores in dolomite and as pseudomorphs after evaporites. Based on the morphology of the early blocky spar and the concurrence of micrite envelopes, she interpreted this early spar as forming in meteoric fluids but noted the difficulty in bringing such fluids into the lower forereef. Reeckmann (1985) also identified reworked blocky spar in megabreccia blocks in the basin and interpreted it as forming prior to transportation during exposure of the platform. It is uncertain how this blocky spar relates to the blocky spars identified by Mruk (1985, 1989). Partial, and rarely complete, dolomitization postdates the early blocky spar but predates the late spar in evaporite molds (Reeckmann, 1985).

Gulf Research and Development Company drilled a continuously cored research well (PDB-04, see Fig. 1A) through the entire Capitan reef complex (back-reef through reef, forereef, and basin facies) that was described by Garber et al. (1989). The reef and forereef facies recovered are lower Capitan (Seven Rivers) in age and both are extensively dolomitized. Dolomite decreases downcore such that the lower forereef facies is dolomitic and the basin margin facies is largely limestone. Based on isotopic and petrographic similarities with the shelf dolomites, Garber et al. (1989) concluded that the Capitan dolomitization also resulted from refluxing of back-reef brines derived from supratidal or sabkha settings. Moldic porosity is minor but ubiquitous throughout the PDB-04 core. Moldic pores are either open or filled with one or more cements, including blocky calcite, dolomite, anhydrite, gypsum, and kaolinite or dickite (Garber et al., 1989). Of these, anhydrite is the most abundant. Marine cements in the Capitan Formation include large botryoids, laminated layers, and isopachous linings (Garber et al., 1989), but it is unclear how common these are in the forereef facies either as primary cements or within transported reef blocks. Garber et al. (1989) noted that the amount of marine cementation in the Capitan Formation of the PDB-04 core is much less than that found in the Guadalupe Mountains. Chert occurs as nodules and beds in the basin margin facies (Garber et al. 1989).

Scholle et al. (1992) looked at the Capitan Formation and equivalent back-reef units and found evidence for the former presence of evaporites. In the forereef facies, this evidence includes calcite pseudomorphs of anhydrite as well as small-scale collapse breccias. Based on petrographic and geochemical data and a comparison with subsurface textures, they interpreted the coarse, blocky calcite spars (II and III of Mruk 1985, 1989) as filling pores formed by the removal of calcium sulfate during uplift-related flushing of the Guadalupe Mountains by meteoric waters.

CAPITAN FOREREEF FACIES DIAGENESIS

Petrography

The diagenesis of the Capitan Formation forereef facies includes a wide variety of fabrics, many of which formed in more than one diagenetic environment. The following description is based largely on petrographic relationships supplemented by stable isotopic data and focuses on the relative timing of the different fabrics.

Calcite Cements.—
Calcite cements that precipitated in the forereef include minor isopachous and overgrowth cements and several generations of blocky spar. Micritic cements may also have formed in the forereef, but the large amount of detrital micrite present makes identification difficult if not impossible. Botryoidal and isopachous cements that are common in the reef (Newell et al., 1953; Kendall, 1969; Babcock, 1974, 1977; Yurewicz, 1976, 1977; Schmidt, 1977; Mazzullo and Cys, 1977; Mruk, 1985; Given and Lohmann, 1985, 1986) are also found in the forereef in reworked boundstones containing rotated geopetal fabrics.

Isopachous cements within the forereef are restricted to well-sorted grainstone beds in the uppermost forereef. Although most are radial fibrous, rare radiaxial samples are present (Mruk, 1985). More than one generation is often present but usually without the interlayered sediment common within the reef facies. Based on morphology and the occurrence of microdolomite inclusions, the Capitan isopachous cements are interpreted as originally high-magnesium calcite (Yurewicz, 1976; Mruk, 1985). In addition, cloudy dolomite cement often occurs in trace amounts as syntaxial overgrowths on the isopachous cements (Fig. 3). Small syntaxial overgrowths on crinoid fragments and brachiopod shells are widespread but volumetrically minor in the Capitan forereef. Syntaxial overgrowths apparently formed at the same time as the isopachous cements, because they are sometimes intergrown.

Blocky or equant calcite spars have long been described from the Capitan Formation and previous workers agree that this spar was the last formed cement in the Capitan (Babcock, 1974; Yurewicz, 1976; Given and Lohmann, 1986; Mruk, 1985, 1989; Crysdale, 1987). Several generations of blocky spar cement (spars I, II, and III) have been identified using a combination of cathodoluminescent petrography and stable isotope analysis (Given and Lohmann, 1986; Mruk, 1985, 1989). Spar I is limited to the reef facies while spars II and III are found throughout the Capitan reef and forereef (Given and Lohmann, 1986; Mruk, 1985, 1989). All three spars have stable isotopic values, which suggests depleted meteoric fluids (Given and Lohmann, 1986; Mruk, 1985, 1989). Spars II and III are com-

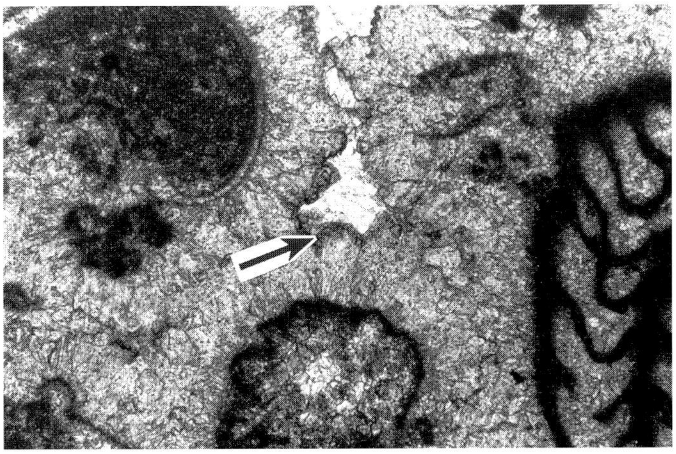

FIG. 3.—Photomicrograph of bioclastic grainstone with bladed marine cements with thin, syntaxial dolomite overgrowths (arrow). Remaining primary porosity is filled with late blocky spar. Sample MT-43-1 (McKittrick Canyon), plane light, long axis of photomicrograph equals 2.5 mm.

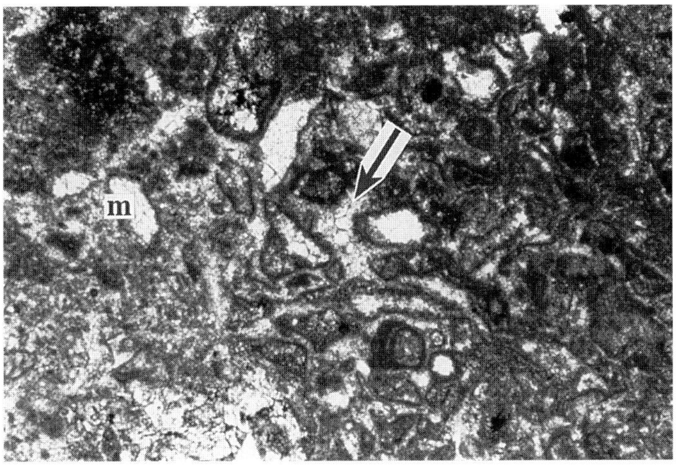

FIG. 4.—Photomicrograph of intraclastic packstone cemented by fine blocky spar (arrow). Micrite envelopes (m) are filled by later coarse calcite spar. Sample MT-F-8 (McKittrick Canyon), plane light, long axis of photomicrograph equals 2.5 mm.

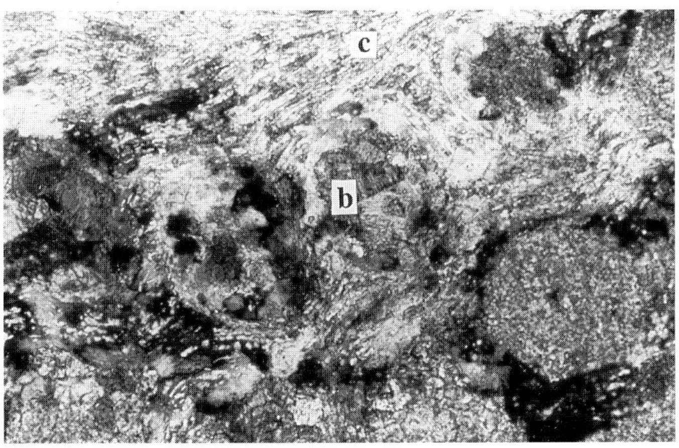

FIG. 5.—Photomicrograph of bryozoan replaced by fibrous chalcedony (c). The silica also replaces earlier formed blocky spar (b) within the primary pores of the bryozoan. Sample MT-F-6-1 (McKittrick Canyon), crossed polarized light, long axis of photomicrograph equals 0.6 mm.

monly found filling molds of anhydrite and gypsum (Scholle et al., 1992).

Within the lower forereef and basin margin facies, a fine blocky calcite spar infills primary and moldic pores in packstone to wackestone facies (Fig. 4). In some samples, early chalcedony replaces this spar (Fig. 5). Garber et al. (1989) analyzed some of these spars for stable isotopes and got values distinctly more enriched than those reported for coarse blocky calcite spars II and III (Given and Lohmann, 1986; Mruk, 1985, 1989). The most enriched of these ($\delta^{18}O$ = −2‰) approaches values reported for Permian Delaware Basin sea water (Given and Lohmann, 1985). We consider this spar to be a separate cementation event from the coarse blocky spars described by Mruk (1985, 1989) and Given and Lohmann (1986).

Aragonite Diagenesis.—

Calcareous sponges, molluscs, and botryoidal cements in the reef were all originally aragonite (Babcock, 1977; Garber et al., 1989) but are found today as low-magnesium calcite or dolomite with various degrees of wall-structure preservation (Fig. 6). Sponges and molluscs are typically dissolved to form micrite envelopes or molds. These moldic pores are often partially to completely filled with blocky spar (Fig. 4), kaolinite, or rarely dolospar. Some calcareous sponges show minor fabric preservation, which suggests replacement by neomorphism rather than dissolution and infilling by blocky spar.

Botryoidal cements in reef blocks in the forereef facies show a greater range of alteration fabrics than skeletal grains. Although dissolution occurs, neomorphism to low-magnesium calcite or replacement by dolomite is more common, often with excellent fabric preservation (Fig. 6). The fibrous botryoidal fabric is preserved as either fine molds of aragonite needles or by elongation of dolomite or calcite crystals. In some instances, the dolomitized area of a sample has better-preserved fabrics than nearby areas of low-magnesium calcite (Fig. 6). This better preservation suggests that some dolomitization occurred prior to neomorphism of the original aragonite to low-magnesium calcite. The lack of fabric retention in other samples suggests continued dolomitization following calcitization. In dolomitized samples, the needle molds are often filled by blocky calcite spar. Well-preserved calcitic botryoids are either nonluminescent under cathodoluminescence or show an intergrowth of nonluminescent and brightly luminescent calcite spar (Mruk, 1985, 1989; Given and Lohmann, 1985). For more altered samples, Mruk (1985, 1989) and Given and Lohmann (1985) found that increasing fabric loss resulted in greater amounts of luminescent spar.

Dolomite.—

The Capitan Formation forereef facies ranges from limestone to dolomite (King, 1948; Newell et al., 1953). The results of a detailed study of field relationships, petrographic fabrics, and stable isotopic data on the forereef dolomite can be found in Melim (1991). Only a brief summary is presented here.

Two types of dolomite are recognized in the Capitan Formation forereef facies. The main dolomite is micritic to coarsely crystalline depending on the texture of the precursor

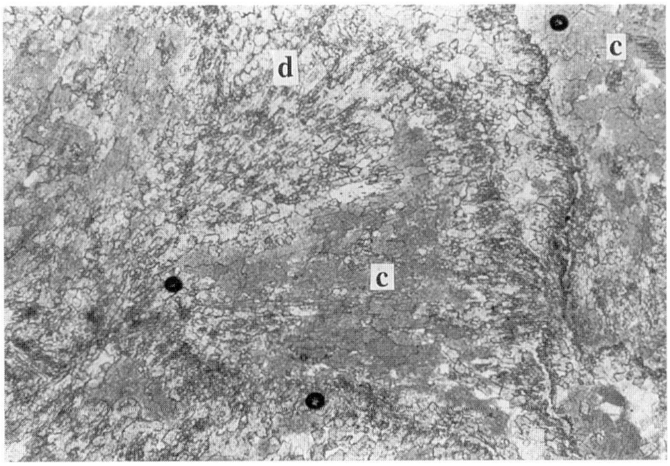

FIG. 6.—Photomicrograph of originally aragonite botryoidal marine cement replaced by fabric-preserving dolomite (d) and calcite (c). The dolomite better preserves the fibrous fabric than the calcite, suggesting the dolomite replaced the original aragonite rather than replacing the botryoid after calcitization. Sample GT-1 (Pine Canyon), stained thin section, plane light, long axis of photomicrograph equals 2.5 mm.

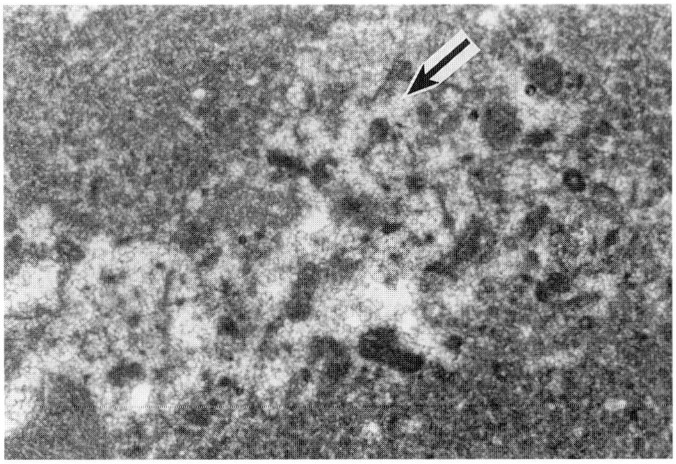

FIG. 7.—Photomicrograph of fabric-preserving dolomite replacing peloidal packstone. Sample is 100% dolomite and the dolomite appears to be replacing fine blocky spar (arrow). Sample GT-19 (Pine Canyon), plane light, long axis of photomicrograph equals 2.5 mm.

FIG. 8.—Photomicrograph of subhedral mosaic of fabric-destroying dolomite. Sample PS-9 (Pine Canyon), plane light, long axis of photomicrograph equals 2.5 mm.

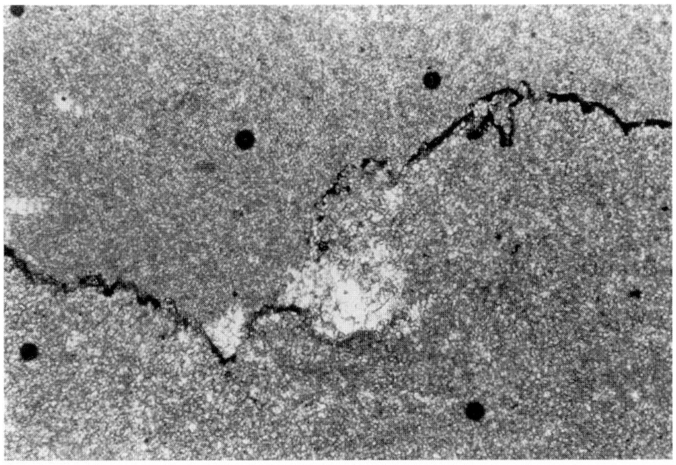

FIG. 9.—Photomicrograph of fine, fabric-preserving dolomite cross-cut by stylolite. Sample MT-6-1 (McKittrick Canyon), plane light, long axis of photomicrograph equals 2.5 mm.

limestone. Micritic material is replaced by micritic dolomite (Fig. 7), while botryoidal cements are replaced by coarse anhedral dolomite (Fig. 6). The fabric-preserving dolomite has enriched oxygen isotopic values ($\delta^{18}O_{PDB} = 0.9\pm1.0‰$, average ± standard deviation). A second phase of more uniform, anhedral, coarsely crystalline dolomite has destroyed all record of the precursor texture (Fig. 8) and has more depleted values ($\delta^{18}O_{PDB} = -3.3\pm1.8‰$). The fabric-destroying dolomite is restricted to the older portions of the Capitan Formation and in some places is observed replacing the fabric-preserving dolomite (Melim, 1991).

Dolomite is most abundant in the upper forereef facies and decreases to zero in the lower forereef to basin-margin facies. The distribution of dolomite is partially controlled by depositional facies with upper forereef grainstone and upper to middle forereef intraclastic rudstones most dolomitized and lower forereef packstones to wackestones least dolomitized. The amount of dolomite also increases with stratigraphic age with nearly all of the lower Capitan Formation forereef dolomitized, but less than half of the upper Capitan forereef facies dolomitized. This trend continues down-section with the underlying Goat Seep Formation being completely dolomitized (Crawford, 1981). Dolomite also occurs in the Capitan reef facies as haloes around some primary reef cavities and early burial fractures (Melim, 1991). Where the reef was not significantly marine-cemented, it is thoroughly dolomitized (Garber et al., 1989; Melim and Scholle, 1992).

Fabric-preserving dolomite formed after marine cements and prior to the blocky spars (Mruk, 1985, 1989; Melim, 1991). Since the fabric-destroying dolomite is found replacing fabric-preserving dolomite, it must have formed later. In addition, fabric-preserving dolomite predates stylolites (Fig. 9) while fabric-destroying dolomite overprints stylolites.

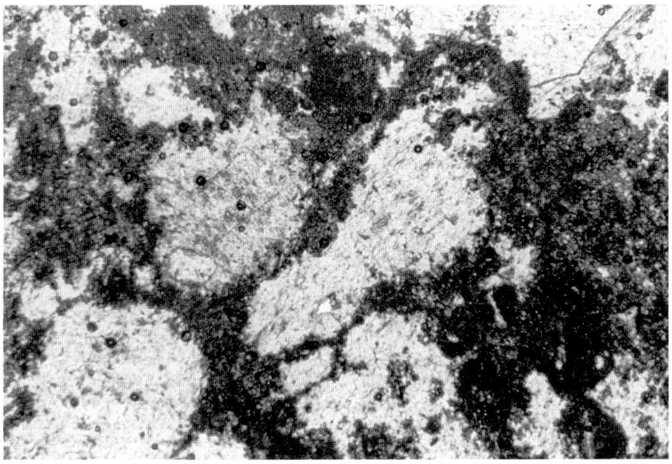

FIG. 10.—Photomicrograph of late calcite spar infilling molds after anhydrite nodules in fine, fabric-preserving dolomite. This irregular, somewhat moth-eaten fabric is still anhydrite in the subsurface (Garber et al. 1989). In this case, the calcite could have formed as a pore-filling phase well after dissolution of the anhydrite. Sample MT-28-2 (McKittrick Canyon), plane light, long axis of photomicrograph equals 5 mm.

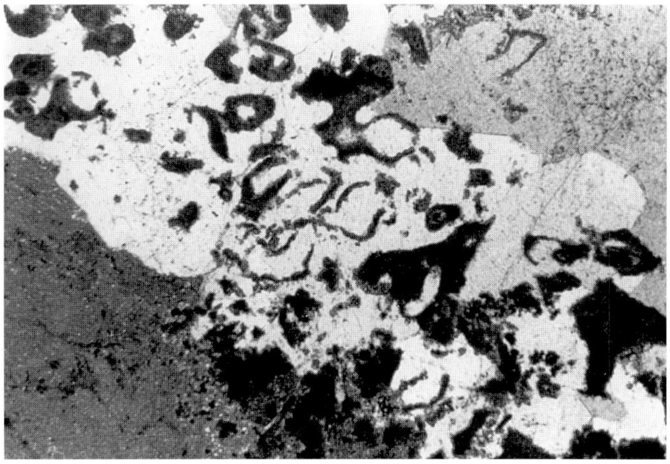

FIG. 11.—Photomicrograph of poikilotopic late calcite spar with floating fragments of calcareous sponge (now fabric-preserving dolomite). In the subsurface, poikilotopic gypsum forms the same floating fabric (Garber et al. 1989). To maintain the floating grains, the calcite likely replaced the gypsum without an intervening void stage. Sample MT-45B (McKittrick Canyon), cross-polarized light, long axis of photomicrograph equals 5 mm.

Silicification.—

Silicification is common in the basin-margin and basin facies where it replaces burrows and fossils (Babcock, 1977; Scholle and Halley, 1980; Garber et al., 1989). Matrix and micritic grains are typically replaced by chert while brachiopods, rugose corals, and other calcitic bioclasts are replaced by a mixture of chert and chalcedony (Fig. 5). The chalcedony may preserve the original fibrous texture of the precursor shell (mainly in brachiopods) or it can destroy the earlier fabric in favor of its own radially oriented fibers. Both chalcedony and euhedral quartz cement locally infill primary porosity. In some samples, earlier-formed fine blocky spar predates the chalcedony (Fig. 5). Silicification is not identified in either the reef or upper forereef facies. The most likely source of silica is the siliceous sponges that inhabited the basin margin to basin facies (Babcock, 1977; Scholle and Halley, 1980; Garber et al., 1989). Timing of chert formation is uncertain but it appears to be fairly early as compaction around chert nodules is common.

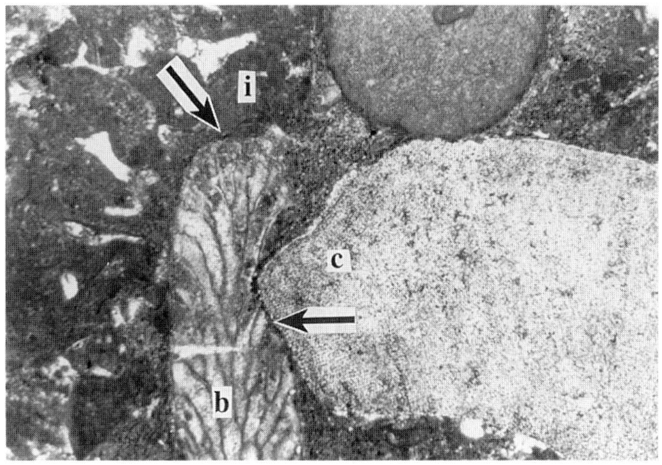

FIG. 12.—Photomicrograph of bioclastic and intraclastic packstone with sutured contacts (arrows). Bryozoan grain (b) shows sutured contact with crinoid (c) and intraclast (i). Sample MT-F-9 (McKittrick Canyon), plane light, long axis of photomicrograph equals 5 mm.

Calcium Sulfate Cementation and Replacement.—

Anhydrite and gypsum are abundant as pore-filling cement and replacive nodules in the Capitan Formation in the subsurface (Garber et al., 1989) but absent in surface outcrops (Scholle et al., 1992). However, calcite-spar-filled evaporite molds are common in surface outcrops, indicating the former presence of extensive evaporite cementation (Reeckmann, 1985; Scholle et al., 1992). Calcitized anhydrite nodules have an irregular ragged edge (Fig. 10) but only rarely show characteristic square terminations. The coarse blocky calcite spars filling the evaporite molds include spars II and III of Mruk (1985, 1989) and Given and Lohmann (1986) (Scholle et al., 1992). Evaporite molds are most common in dolomitized units, but rare examples are found in limestones.

Most of the calcite spar that fills evaporite molds formed as a pore filling cement some time after dissolution of the evaporites, as geopetal accumulations are widespread in the bottom of evaporite molds (Scholle et al., 1992). However, floating grains are often found in poikilotopic calcite crystals (Fig. 11), which suggests a replacement process, as the floating fabric would collapse if not continuously supported (Scholle et al., 1992). In the subsurface, similar cement-supported fabrics are poikilotopic gypsum rather than calcite (Scholle et al., 1992).

Compaction.—

Chemical compaction occurs on several scales in the Capitan Formation forereef facies. Stylolites occur as low amplitude, bedding-parallel stylolites in wackestone to mudstone beds in the lower forereef to basin facies. Sutured contacts, however, are common throughout the forereef (Fig. 12). In the upper forereef, sutured contacts surround most reef intraclasts within poorly sorted intraclast rudstone beds. Larger

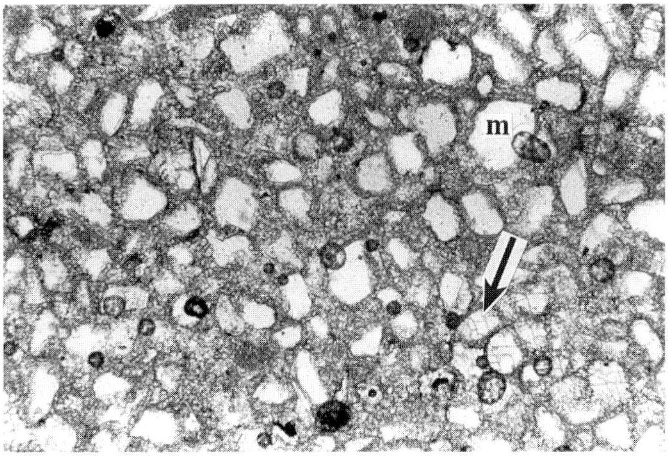

FIG. 13.—Photomicrograph of dolomite-cemented siliciclastic sand with moldic porosity, probably after feldspars. Some molds are filled with blocky calcite spar (note cleavage at arrow) but most are empty (m). Sample MT-45B (McKittrick Canyon), plane light, long axis of photomicrograph equals 1.2 m.

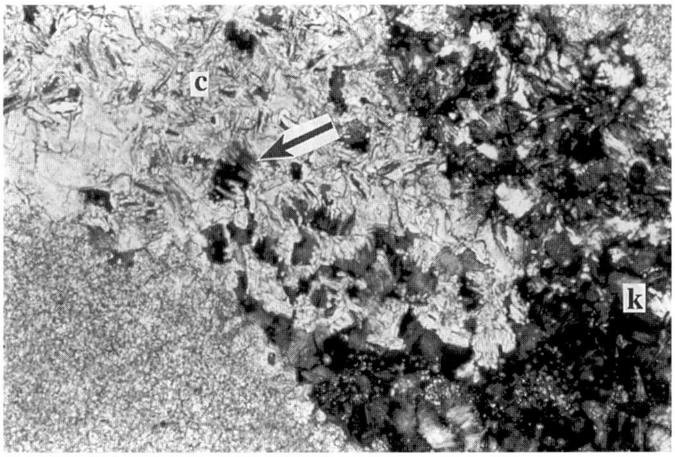

FIG. 14.—Photomicrograph of kaolinite (k) in mold being engulfed by blocky calcite spar (c). Note the calcite extending between books of kaolinite (arrow). Sample MT-45B (McKittrick Canyon), cross-polarized light, long axis of photomicrograph equals 0.6 mm.

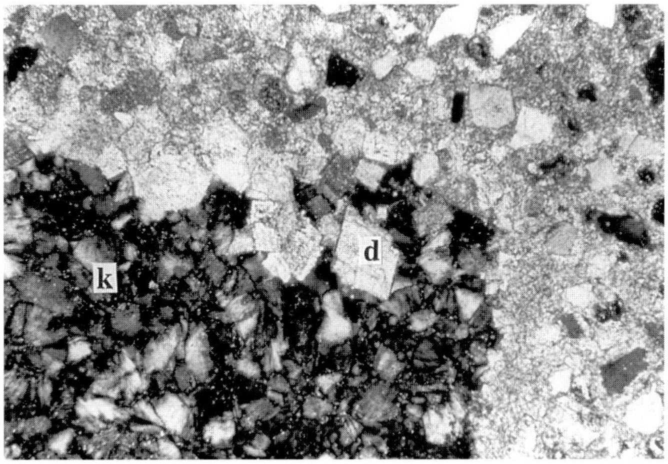

FIG. 15.—Photomicrograph of kaolinite (k) in mold following dolomite cement (d). Note the dolomite does not contain kaolinite (contrast with Fig. 14). Sample GT-25 (Pine Canyon), cross-polarized light, long axis of photomicrograph equals 1.2 mm.

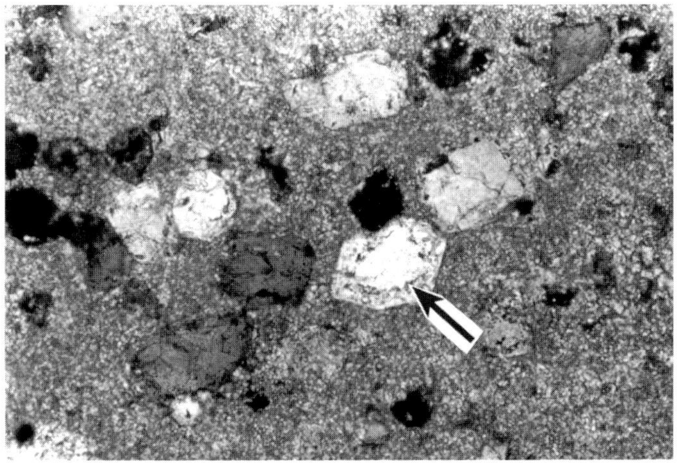

FIG. 16.—Photomicrograph of euhedral quartz overgrowths on very fine quartz sand. Note the rounded ghost of the detrital quartz grain (arrow). Overgrowths are only present on sand disseminated within carbonate samples, they were not observed within sandstone samples. Sample BT-5-1 (Bear Canyon), cross-polarized light, long axis of photomicrograph equals 0.6 mm.

clasts typically have larger amplitude sutures with a maximum amplitude of ≈10 cm. These are easily mistaken for randomly oriented stylolites because of the large size of the reef blocks (>3 m). In many cases it appears that an originally matrix-supported fabric has been transformed into a grain-supported fabric (Melim and Scholle, 1995). Garber et al. (1989) also noted abundant sutured contacts in the forereef facies of the Gulf PBD-04 core.

A consequence of compaction of the poorly cemented forereef facies was extensive fracturing of the overlying well-cemented reef facies (Dunham, 1972; Yurewicz, 1976; Melim and Scholle, 1992; Saller, 1996; Melim, 1991). While some of these fractures extended to the sea floor as neptunian dikes (Dunham, 1972; Yurewicz, 1976; Pray and Stanton, 1992), many—particularly in the middle and lower Capitan Formation—formed in the burial environment and did not reach the sea floor (Melim and Scholle, 1992; Melim, 1991). The neptunian dikes are filled by marine cements, fossils, and sometimes fine sandstones derived from the backreef facies (Dunham, 1972; Yurewicz, 1976). In contrast, the burial fractures are filled by a locally derived breccia (wall collapse) and have dolomite haloes extending 1–4 cm into the surrounding reef limestone. Both of these fracture sets are separate from later tectonic fractures that are often filled with a coarse lithic-rich sandstone that was likely derived from formerly overlying Cretaceous sediments (Dunham, 1972). All of the fracture sets are mainly parallel to the reef trend with a secondary trend perpendicular to the reef trend.

Siliciclastic Sediments.—

Siliciclastic sediments are a widespread, but minor, component in the Capitan forereef (Hull, 1957; Dunham, 1972; Mruk, 1985; Melim and Scholle, 1995). The basin and back-reef sandstones are silt to fine sand feldspathic arenites (20–30% feldspar) with calcite or dolomite cements (Dunham, 1972). The feldspar is absent in the reef and forereef facies, where it has altered to kaolinite, usually leaving silt-sized molds (Fig. 13) (Dunham, 1972; this study). Most of the kaolinite is found as a pore-filling cement within biomoldic pores (Figs. 14 and 15), but it is also found within fractures (Garber et al., 1989; this study) and within the feldspar molds (Dunham, 1972; this study). The kaolinite formed before the late blocky calcite spars, as it is often replaced or engulfed by the calcite (Fig. 14), and is cross-cut by blocky, spar-filled fractures. The kaolinite formed after dolomitization, as dolomite crystals are surrounded by kaolinite (Fig. 15). Kaolinite in the Capitan Formation is mainly found in samples that also contain quartz silt either interbedded or disseminated within the carbonate. Apparently the dissolved feldspar components did not migrate any significant distance before precipitating as kaolinite. Kaolinite is also common in the Gulf PDB-04 research core with similar fabrics (Garber et al., 1989).

Minor doubly terminated euhedral quartz overgrowths are present in the Capitan Formation on detrital quartz silt in the reef and upper forereef facies (Fig. 16). Quartz overgrowths are found in both limestone and dolomite samples, usually where quartz silt occurs as scattered grains rather than as discrete sandstone beds. A typical quartz silt grain is doubled in size by the overgrowth with the overgrowth replacing the surrounding carbonate. Overgrowths were not found growing into open pores or found associated with chert or chalcedony. Kaolinite and quartz overgrowths rarely occurred together.

Paragenesis

The paragenetic sequence for the Capitan Formation is shown in Fig. 17. Many of the events are of uncertain timing or overlap in timing. In addition, no single sample contains all of these fabrics. Although the paragenesis is listed as a "sequence" of events, the reality was multiple processes acting simultaneously on different parts of the large body of limestone that makes up the Capitan Formation forereef facies.

The earliest diagenesis in the Capitan Formation was cementation on the sea floor (e.g., Newell, et al., 1953; Babcock, 1974). This includes micritic, botryoidal, and isopachous fibrous to bladed cements in the reef and in eroded blocks of reef boundstone in the forereef. The isopachous cementation in the upper forereef facies is also interpreted as near-surface marine based on its similarity to that found in eroded reef blocks and other areas of marine cementation (e.g., Newell et al., 1953; Babcock, 1974). At least partly coeval with this cementation was reef-parallel fracturing of the rigid Capitan reef in the Guadalupe Mountains, as botryoidal cements are found lining some of fractures (e.g., Babcock 1977). The earliest fractures were neptunian dikes and sand-filled fractures, which formed early enough to maintain connections to the surface (King, 1948; Yurewicz, 1977). The fracturing, however, continued into the shallow burial environment as compaction of the forereef caused failure of the rigid reef (Melim and Scholle, 1992).

Once in the shallow-burial environment, a wide variety of diagenetic processes acted more-or-less simultaneously (Fig. 17), but probably in different areas of the Capitan Formation. Early burial diagenesis included aragonite neomorphism and dissolution, dolomitization, precipitation of fine blocky calcite spar, and silicification. The relative timing of each of these is difficult to determine, partly due to the overlap of the petrographic fabrics (see above) and partly due to limited concurrence. For example, dolomite is more abundant in the upper portion of the forereef than in the lower forereef while fine blocky spar and chert are generally restricted to the lower forereef to basin margin.

Aragonite dissolution is more common than neomorphism and is present throughout the forereef facies in both limestones and dolomites. A fine, blocky spar is often associated with the moldic porosity in the lower forereef and basin-margin limestones (Fig. 4). The stable isotopic data of Garber et al. (1989) on the fine blocky spars do not support a freshwater lens (Reeckmann 1985), and instead indicate precipitation from a fluid near seawater in composition. Moldic porosity and blocky spar are known to form in marine pore fluids in deeper water environments where seawater is undersaturated with respect to aragonite (Baker et al., 1982; Halley et al., 1984; Saller, 1986; Freeman-Lynde et al., 1986; Dix and Mullins, 1988; McClain et al., 1988). More recently, Melim et al. (1995) and Melim and Massaferro (1997) defined marine-burial diagenesis and documented moldic porosity and aragonite neomorphism forming in marine pore fluids as shallow as 70 m below sea level where the incoming seawater is saturated with respect to aragonite and calcite. Hopkins (1977) also documented fine blocky spar forming in the marine environment. Based on 1) the stratigraphic position (that is, beyond the reach of meteoric fluids), 2) the stable isotopic data, and 3) by analogy with the above examples, we interpret the fine blocky spar and aragonite dissolution in limestones as having formed during marine-burial diagenesis such as described by Melim et al. (1995) at Great Bahama Bank. Aragonite molds in dolomitized samples may have formed in a similar fashion or during the dolomitization process (Melim, 1991).

Dolomitization occurred during early burial. It was coeval with or followed the aragonite dissolution as dolospar often occurs within skeletal molds. Dolomite formed prior to stylolitization as dolomite crystals are regularly cross-cut. The distribution of dolomite indicates a dolomitizing fluid refluxing from above that passed through the cemented reef in fractures and then followed bedding in the forereef (Melim, 1991). Older Capitan forereef facies probably contains more dolomite because it was subject to refluxing brines for a longer period than the younger Capitan forereef facies. Although anhydrite is commonly associated with the dolomite, a comparison of the stable isotopic data from the forereef dolomite, near-back-reef dolomite, and far-back-reef dolomite supports precipitation of the forereef dolomite from near-back-reef mesosaline brines rather than from the hypersaline brines of the far-back-reef (Melim, 1991). However, minor dolomite from reflux of hypersaline brines from either the far-back-reef facies of overlying Ochoan evaporites cannot be ruled out.

Evaporite cementation seems to have followed dolomitization but may have started during dolomitization. Since this cementation apparently occluded all remaining porosity, and it

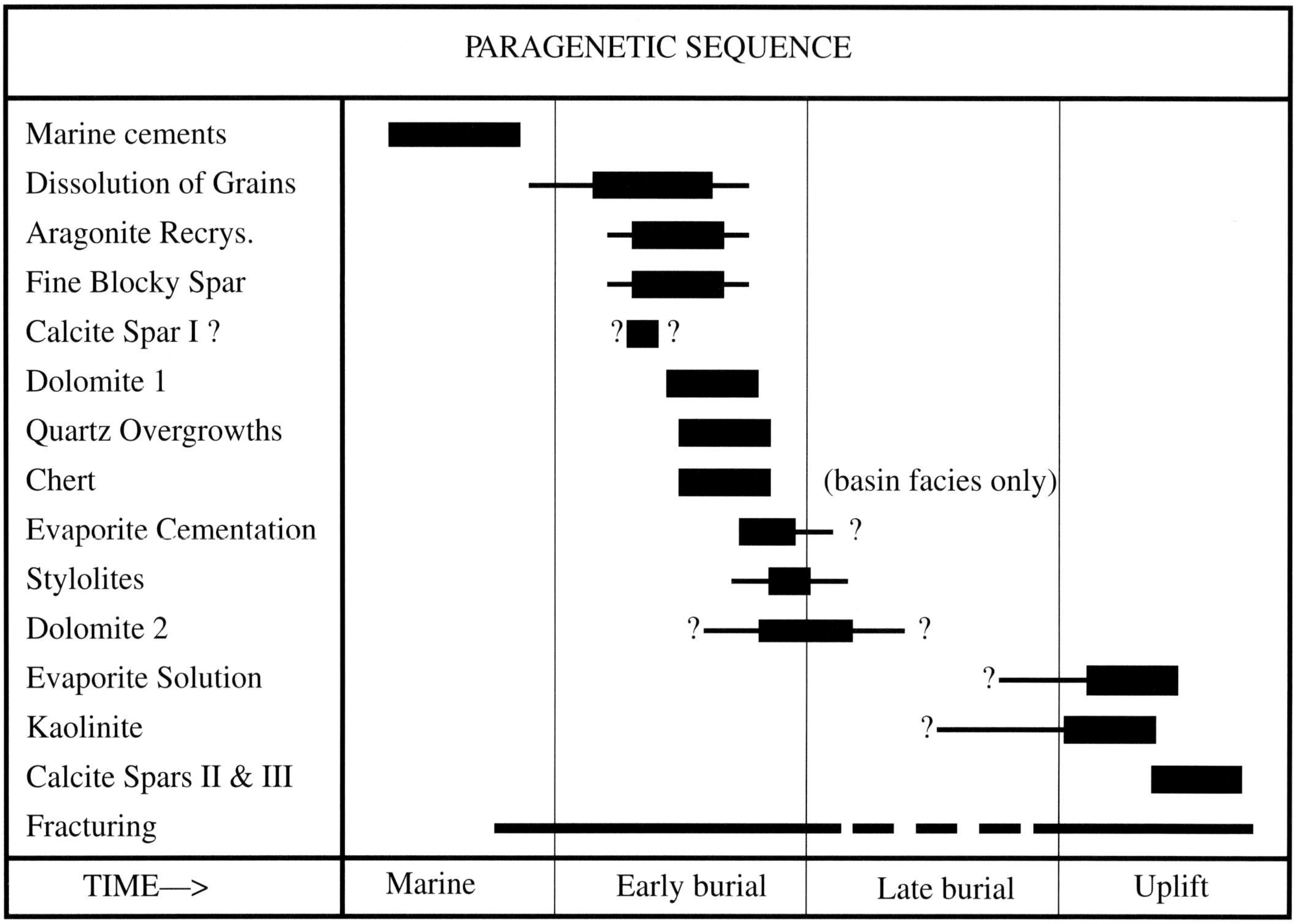

FIG. 17.—Paragenetic sequence for the Capitan Formation. Timing of events is relative. Sea floor is penecontemporaneous; early burial is approximately equal to the Late Permian (Guadalupian-Ochoan); late burial is approximately equal to the Mesozoic; uplift is related to Tertiary tectonic events leading to the formation of the Guadalupe Mountains. Note that older parts of the formation were undergoing early burial diagenesis during deposition of younger units.

likely marked the end of dolomitization where it occurred. Possible sources for the brines precipitating evaporites include the near- to far-back-reef facies as well as the overlying Ochoan evaporites of the Castile and Salado Formations (Garber et al., 1989; Melim, 1991). Since molds are all that remains of evidence for evaporites in the Guadalupe Mountains there is no way to distinguish between these sources, and indeed more than one probably contributed to the evaporite cementation of the Capitan Formation. In the Gulf PDB-04 core, preserved evaporite cements are abundant but the origin of the brines is still equivocal (Garber et al., 1989).

Silicification in the lower forereef and basin facies probably occurred early in the burial history, as most of the chert formed prior to significant compaction. The source of the silica was most likely siliceous sponge spicules (Babcock, 1977). Compaction began during deposition of the Capitan, particularly in the uncemented forereef facies. This early compaction caused the early fracturing of the early-cemented reef facies. Compaction and stylolite formation likely continued during burial in the latest Permian and into the Mesozoic.

Fabric-destructive dolomite followed the fabric-preserving dolomite and could have formed anytime between early burial and the Tertiary uplift of the Guadalupe Mountains. It is also possible that fabric-destructive dolomite was forming in the lower Capitan during formation of fabric-preserving dolomite in the upper Capitan. Constraints on the timing of fabric-destructive dolomite are poor, but a burial origin, perhaps during the Mesozoic, appears most likely. The Capitan had reached near maximum burial by the end of the Permian, when the upper Capitan had 300-700 m of Ochoan evaporites above it (Hills, 1984; Ward et al., 1986; Lowenstein, 1988).

During the Tertiary, faulting and uplift of the Guadalupe Mountains produced new fractures and joints, commonly parallel to the older sets (King, 1948). Some open fractures were filled by coarse, lithic-rich sands eroded from overlying Cretaceous sediments (later removed from the Guadalupe Mountains area) (Dunham, 1972), while others were enlarged into the cave system seen today. Erosion of the overlying Ochoan evaporites allowed input of meteoric waters into the fractured Capitan Formation (the Capitan Aquifer, Hiss, 1980) leading to hydration of anhydrite to gypsum and some dissolution of the evaporites. This is the stage the Gulf PDB-04 core has reached, where both gypsum and anhydrite are abundant (Garber et al., 1989). Kaolinization of the feldspars took place at this time, as kaolinite is occasionally found within evaporite molds and some fractures. In the Guadalupe Mountains, continued flushing by meteoric waters dissolved all of the evaporite cements within the Capitan and precipitated blocky spars II and III in their place (Scholle et al., 1992). These spars engulf and replace the kaolinite.

Sedimentologic Controls on Diagenesis

Most diagenetic models for carbonate sediments focus on the fluids and chemical conditions during diagenesis (for example, marine, meteoric, burial, etc.). However, factors such as depositional environment, paleogeography, relative sea-level fluctuations, and burial history play an important role in determining what fluids are available for diagenesis. An important factor sometimes under-emphasized is the role of the sedimentology in controlling later diagenesis. This is particularly true for large-scale forereef facies, where the wide range of water depths and depositional processes produce sediments with markedly different fluid-flow characteristics (Mruk and Bebout, 1993; Harris, 1994; Melim and Scholle, 1995). The Capitan forereef facies is a heterogeneous mixture deposited by debris flows, grain flows, and turbidity currents (Table 1; Mruk and Bebout, 1993; Melim and Scholle, 1995). Grain-flow deposits are restricted to the upper forereef, where depositional dips were greatest (Table 1). Debris-flow deposits occur throughout the forereef facies and comprise the majority of the sediment in the upper to middle forereef (Table 1). Turbidity current deposits are most abundant in the lower forereef facies to basinal facies but also occur in the middle forereef facies (Table 1; Melim and Scholle, 1995).

Regardless of the exact depth of the Capitan reef facies (e.g., Hurley, 1989; George, 1992, Saller, 1996), it is clear that it was in an environment of high fluid flow, whether by waves or by tidal currents, sufficient to allow formation of the extensive marine cements (e.g., Newell et al., 1953; Babcock, 1974). Further down-slope, in the upper forereef facies, isopachous marine cements formed within some of the grain-flow deposits but left most untouched (Table 2). Debris-flow deposits do not contain these marine cements (except in allochthonous blocks) even when interbedded with marine-cemented grain-flow deposits. Since the poorly sorted, micrite-rich packstones and rudstones of the debris-flow deposits likely had lower permeability than the well-sorted, clean sands of the grain-flow deposits, the depositional fabric may have controlled the distribution of marine cementation.

Alternatively, since many of the grain-flow deposits do not contain marine cements, the upper forereef might have been on the edge of being too deep for the marine cementation. This seems unlikely as the Triassic Latemar forereef facies is extensively cemented by early marine cements to depths approaching 500 m (Harris 1994). The Latemar forereef facies is characterized by grainstones and micrite-free breccias (Goldhammer and Harris, 1989; Harris, 1994), unlike the Capitan forereef facies, where micrite-rich packstones to rudstones dominate. Perhaps the porosity and permeability of the surrounding rocks are as important as those of the individual beds. This makes sense, since potentially cementing fluids need access to the grainstones, and access is difficult if the surrounding rocks are less permeable. Hence, grainstones that had a connected high-permeability pathway for through-flow of water were cemented while the rest of the grainstones, isolated by the surrounding packstones to rudstones, stayed uncemented. The well-cemented reef facies may have had higher average permeability and/or the shallower water depth may have lead to greater flux by wave-driven waters.

Burial diagenetic processes in the Capitan forereef can be divided into two groups: those controlled by grain composition and those apparently controlled by physical properties (that is, porosity and permeability). Alteration of aragonite obviously requires an aragonite-bearing sediment. Similarly, silicification is restricted to the lower forereef to basin (Table 2), where there was a source of silica (probably sponge spicules, Babcock, 1977). Kaolinite occurs throughout the forereef facies (Table 2), but only in sediments that originally contained feldspar. A less obvious compositional control appears to be the distribution of stylolites, as only mud-rich sediments display significant stylolitization (leaving most grainstones unaffected). A common element in the compositionally controlled processes is the closed-system nature—outside components are not necessary for the reactions to occur.

Dolomitization and evaporite cementation, however, require the introduction of new components and therefore imply fluids moving through the forereef and reef facies. The distribution of these components, therefore, was likely controlled by the porosity and permeability of the sediment in the burial environment.

Dolomitization and evaporite cementation both occurred quite early in the shallow burial environment. The porosity and permeability of the sediment in that environment is a result of primary depositional porosity as modified during earlier diagenesis. In the Guadalupe Mountains, the Capitan reef facies was heavily marine cemented and this protected it from significant dolomitization (Melim and Scholle, 1992). In the Gulf PDB-04 research core, the reef largely lacks marine cement and has been completely dolomitized (Garber et al., 1989; Melim and Scholle, 1992). Within the forereef facies, grainstones with significant marine cementation rarely contain more than trace amounts of dolomite (Fig. 3) while those that escaped marine cement are almost always dolomitized. Playford (1980) also noted forereef dolomitization restricted to those units that were not marine cemented.

Debris flow deposits are poorly sorted packstones to rudstones with a micrite-rich matrix that is deposited as the moving mass of material freezes. Turbidity current deposits, on the other hand, are usually graded packstones to wackestones that are deposited by gradual settling out of material from the turbulent cloud of sediment. As both have a micrite-rich matrix, debris flow and turbidity current deposits would appear to have similar physical properties. In support of this, both lack marine cementation (Table 2). However, debris flow deposits are commonly

TABLE 2.—OCCURRENCE OF DIAGENETIC FABRICS FORMING IN THE FOREREEF FACIES

Facies	Lithology	Marine cements	Aragonite alteration	Overgrowths	Fine blocky spar	Calcite spar I?	Dolomite 1	Quartz overgrowths	Chert	Compaction	Evaporite cementation	Stylolites	Dolomite 2	Evaporite solution	Kaolinite	Calcite spars II, III	Fracturing
Upper forereef	Rudstones to packstones		R	C		R	A	R		A	A	A	R	A	R	A	A
	Grainstones	R	R	C			A				C	R		C		A	A
Middle forereef	Packstones to rudstones		R	C			A	R		A	A	A	R	A	R	A	A
	Wackestones		R	R			R			A	R	A		R		R	A
Lower forereef	Packstones to rudstones		R	C	R		C	R		A	C	A	R	C	R	A	A
	Packstones to wackestones		R	C	R		R			A	R	C		R	R	C	A
	Mudstones		R	R			R			A		C				R	A
Basin margin	Packstones to rudstones		R	R	R		R		R	A	R	C		R	R	R	A
	Mudstones to wackestones								R	A		C					A

A = Abundant, <50% beds; C = common, present in 10–50% of beds; R = rare, present in < 10% of beds; Blank = not found.

dolomitized and contain abundant evaporite molds while turbidity current deposits are almost never dolomitized. Perhaps the settling out of a turbidity current leads to lower initial permeability than the freezing of a debris flow. An additional factor could be the relatively distal position of the turbidity current deposits in the lower forereef. However, lower forereef debris flow deposits usually have at least the matrix dolomitized while the interbedded turbidity current deposits only rarely contain dolomite (Table 2). Saller et al. (1989) also recognized preferential dolomitization of debris flow deposits. It is interesting to note that grainstones and rudstones that apparently did not have sufficient fluid flow to allow marine cementation are extensively dolomitized—a process that also requires through-put of large volumes of water. Dolomitization, however, has available long periods of time during early burial for the needed water migration while marine cementation is restricted to the short period of time when the sediment is near the sediment-water interface. Although the volumes of water needed may be similar, the flow rates for dolomitization can be much slower.

PREDICTIONS FOR FOREREEF FACIES DIAGENESIS

The diagenesis of forereef facies is poorly understood, largely due to the difficulty of examining modern examples. In addition, the diagenesis of modern steep forereef facies has been strongly affected by large-scale Pleistocene sea-level fluctuations (Grammer et al., 1993), limiting their applicability to equivalent periods of high-amplitude glacio-eustacy in the ancient rock record. The well-exposed Capitan forereef facies, therefore, can be used to predict diagenetic patterns in large-scale forereef facies during periods of low-amplitude sea-level fluctuations (Table 3). Some of these fabrics can be expected in any forereef deposit, while others will only occur in selected forereef deposits.

Meteoric diagenesis is usually not present since many forereef facies are deposited below the easy reach of meteoric fluids. The uppermost portions of shallower forereef facies, however, may be altered if they are deposited during time periods with large-scale sea-level fluctuations. This interpretation runs counter to that of several workers who have postulated large-scale meteoric flow systems to account for moldic porosity and aragonite neomorphism in deep water facies (Reeckmann, 1985; Enos, 1988). We believe the data support these fabrics forming in sea water during shallow burial (Melim et al., 1995).

Shallow-marine to marine-burial diagenesis, on the other hand, should be present in nearly all forereef deposits, although the relative importance of them may vary (Table 3). The Triassic Latemar forereef facies, for example, has abundant early marine cements (Harris, 1994) while the Capitan forereef facies has very little. This is probably controlled by the depositional fabrics, since the Latemar forereef is dominated by micrite-poor rudstones and grainstones (Harris, 1994) while the Capitan for-

Table 3.—PREDICTED DIAGENETIC PATTERNS FOR LARGE-SCALE FOREREEF FACIES.*

Environment	Importance	Fabrics	Possible examples
Meteoric	Could be present, usually absent	Molds, blocky spar, neomorphism, caves	?
Shallow-Marine	Variable, particularly common in the shallowest water facies	Isopachous, botryoidal, or micritic cements	Capitan Fm.: Mruk 1985, 1989; this paper Triassic Latemar: Harris 1994 Recent Bahamas: Grammer et al., 1993 Devonian Canning Basin: Playford, 1980
Shallow-Burial			
1. Normal marine pore fluids	Abundant, particularly in deepest water facies	Molds, blocky spar, neomorphism, possibly chert, possibly dolomitization	Capitan Fm.: Mruk 1985, 1989; this paper Triassic Latemar(?): Harris 1994 Cretaceous Poza Rica Trend (?): Enos 1988
2. Mesosaline to hypersaline fluids	Expected if restricted platform occurs	Dolomitization, evaporite cementation	Capitan Fm.: this paper Permian Goat Seep Fm.(?): Crawford 1981 Permian Bone Spring Fm.(?): Wiggins and Harris 1985; Saller et al. 1989
Deep-Burial	Common, dependant on burial depth	Mechanical and chemical compaction, late cements	Capitan Fm.: this paper Permian Bone Spring Fm.: Wiggins and Harris 1985

*Possible examples are representative forereef facies drawn from the literature and this study. Question marks after examples indicate the authors interpreted their data differently than indicated or did not interpret the diagenetic environment.

ereef is mainly micrite-rich rudstones to packstones (Melim and Scholle, 1995).

Reflux dolomitization, such as that found in the Capitan Formation, will only occur if the platform configuration permits the development of a large-scale reflux flow system. Factors that should foster reflux dolomitization of forereef facies include a large, restricted platform, an arid climate, a prograding profile (to place the saline facies above the forereef), minimal early marine cementation in the forereef, and relatively stable sea level. In addition, if the refluxing fluids are only mesosaline, evaporite cementation may be absent. Deep burial diagenesis will vary strongly between different forereef deposits depending on the local burial and tectonic history.

CONCLUSIONS

The Capitan Formation forereef facies shows alteration in a number of diagenetic environments including marine diagenesis, marine-burial diagenesis, deeper burial diagenesis, and uplift-related diagenesis. It is important to note that the lower Capitan Formation was already in the shallow-burial environment during deposition of the middle and upper Capitan Formation. The paragenetic sequence is more an average of events in any given location than a simple progression for the entire Capitan Formation forereef facies.

Alteration in specific diagenetic environments includes:

- **Meteoric diagenesis:** Although minor meteoric diagenesis has been suspected for the reef facies (Mruk, 1985; Given and Lohmann, 1986), no evidence of early meteoric diagenesis was found in the forereef facies. We interpret fabrics previously attributed to meteoric diagenesis (Reeckmann, 1985) as having formed during marine-burial diagenesis.
- **Shallow-marine diagenesis:** In contrast to the well-cemented reef facies, marine cements forming in the forereef are limited to minor isopachous radial fibrous and radiaxial in upper forereef grainstones. However, most of the reef fabrics are also found in the forereef as allochthonous blocks of reef boundstone.
- **Marine-burial diagenesis:** Marine-burial diagenesis is ubiquitous in the Capitan forereef facies with normal marine to hypersaline fluids at different times and locations. The lower forereef facies was largely altered by marine pore fluids; fabrics include moldic porosity, fine blocky spar cementation, silicification, and rare aragonite neomorphism. The upper forereef facies, however, was almost completely dolomitized by mesosaline fluids mainly derived by seepage reflux from the near-back-reef carbonate lagoon facies. Aragonite neomorphism apparently predated or was coeval with the dolomitization. Early fracturing of the rigidly cemented reef provided the fluid conduits for dolomitizing fluids to bypass the reef on their way

to the forereef. During and/or subsequent to dolomitization, hypersaline fluids completely cemented the upper forereef with evaporites (gypsum and/or anhydrite).

• **Deeper burial diagenesis:** Includes pressure solution to produce both sutured contacts in intraclastic rudstones and stylolites in finer-grained lithologies. Recrystallization of an early fabric-preserving dolomite to form a fabric-destructive dolomite may also have occurred during burial.

• **Uplift-related diagenesis:** During Tertiary uplift, erosion of overlying Ochoan evaporites allowed influx of meteoric water into the evaporite-cemented Capitan Formation. This led first to hydration of anhydrite to gypsum and kaolinitization of feldspars and later to complete dissolution of evaporites and precipitation of coarse blocky spars II and III in the resulting porosity.

Although traditional diagenetic models focus on the water chemistry, this study also found sedimentology to be an important control (Table 2). The Capitan forereef facies was mainly deposited by three processes, each of which produces a characteristic lithology: grain flows (grainstones), debris flows (unsorted, micrite-rich packstones to rudstones), and turbidity currents (graded packstones to wackestones). Grain-flow deposits are only found in the upper forereef and are either marine cemented or, more commonly, dolomitized. Debris-flow deposits are the main component in the upper to middle forereef but also extend to the lower forereef and basin-margin facies. They are typically partially to completely dolomitized (increasing upslope) and they were cemented by evaporites (now calcitized). Turbidity-current deposits occur in the middle forereef and dominate the lower forereef to basin-margin facies. They are seldom dolomitized and are characterized by fine blocky spar cement, aragonite alteration, and silicification.

ACKNOWLEDGMENTS

The authors wish to thank the National Park Service for permission to sample in Carlsbad Caverns and Guadalupe Mountains National Parks. Special thanks goes to the employees of Carlsbad Caverns and Guadalupe Mountains National Parks for logistical support during field work. Primary support of this research was provided by Grant ACS-PRF No. 20281-AC8 to P. A. S. and acknowledgment is made to the donors of The Petroleum Research Fund, administered by the American Chemical Society, for their support of this research. Additional support to L. A. M. included a Dallas Geological Society Scholarship, a Southwest Section of the American Association of Petroleum Geologists Scholarship, and, from Southern Methodist University, a Champlin Research Award, an R. M. and P. G. Huffington Scholarship, and a C.B. Carpenter Scholarship.

This research benefitted greatly from many discussions with M. J. Shields, R. Koepnick, P. M. (Mitch) Harris, D. Kirkland, N. Hurley, D. (Mruk) Cox, and L. Pray and from reviews by D. (Mruk) Cox, R. Garber, and A. Saller.

REFERENCES

BABCOCK, J. A., 1974, The role of algae in the formation of the Capitan Limestone (Permian, Guadalupian), Guadalupe Mountains, west Texas and New Mexico: Unpublished Ph.D. Dissertation: University of Wisconsin, Madison, 241 p.

BABCOCK, J. A., 1977, Calcareous algae, organic boundstones, and the genesis of the upper Capitan Limestone (Permian, Guadalupian), Guadalupe Mountains, west Texas and New Mexico, *in* Hileman, M. E., and Mazzullo, S. J., eds., Upper Guadalupian Facies, Permian Reef Complex, Guadalupe Mountains, New Mexico and West Texas, 1977 Field Conference Guidebook: Tulsa, Society of Economic Paleontologists and Mineralogists, Permian Basin Section, Publication 77-16, p. 3-44.

BABCOCK, L. C., 1977, Life in the Delaware Basin: The paleoecology of the Lamar Limestone, *in* Hileman, M. E., and Mazzullo, S. J., eds., Upper Guadalupian Facies, Permian Reef Complex, Guadalupe Mountains, New Mexico and West Texas, 1977 Field Conference Guidebook: Tulsa, Society of Economic Paleontologists and Mineralogists, Permian Basin Section, Publication 77-16, p. 357-390.

BAKER, P. A., GIESKES, J. M., and ELDERFIELD, H., 1982, Diagenesis of carbonates in deep-sea sediments-evidence from Sr/Ca ratios and interstitial dissolved Sr^{2+} data: Journal of Sedimentary Petrology, v. 52, p. 71-82.

BOSELLINI, A., 1984, Progradational geometries of carbonate platforms: Example from the Triassic of the Dolomites, northern Italy: Sedimentology, v. 31, p. 1-24.

COOK, H. E., and MULLINS, H. T., 1983, Basin margin, *in* Scholle, P. A., Bebout, D. G., and Moore, C. H., eds., Carbonate Depositional Environments: Tulsa, American Association of Petroleum Geologists Memoir 33, p. 539-617.

CRAWFORD, G. A., 1981, Depositional history and diagenesis of the Goat Seep Dolomite (Permian, Guadalupian), Guadalupe Mountains, west Texas-New Mexico: Unpublished Ph.D. Dissertation, University of Wisconsin, Madison, 300 p.

CRYSDALE, B. L., 1987, Fluid inclusion evidence for the origin, diagenesis and thermal history of sparry calcite cement in the Capitan Limestone, McKittrick Canyon, West Texas: Unpublished Master's Thesis, University of Colorado, Boulder, 78 p.

DICKSON, J. A. D., 1965, A modified staining technique for carbonates in thin section: Nature, v. 205, p. 587.

DIX, G. R., and MULLINS, H. T., 1988, Rapid burial diagenesis of deep-water carbonates: Exuma Sound, Bahamas: Geology, v. 16, p. 680-683.

DUNHAM, R. J., 1972, Capitan Reef, New Mexico and Texas: Facts and Questions to Aid Interpretation and Group Discussion: Tulsa, Society of Economic Paleontologists and Mineralogists, Permian Basin Section, Publication 72-14, 291 p.

ENOS, P., 1988, Evolution of pore space in the Poza Rica Trend (Mid-Cretaceous), Mexico: Sedimentology, v. 35, p. 287-325.

EPSTEIN, S., GRAF, D. L., and DEGENS, E. T., 1964, Oxygen isotope studies on the origin of dolomites, *in* Craig, H., Miller, S. L., and Wasserburg, G. J., eds., Isotopic and Cosmic Chemistry: Amsterdam, North-Holland Publishing Company, p. 169-180.

FREEMAN-LYNDE, R. P., WHITLEY, K. F., and LOHMANN, K. C., 1986, Deep-marine origin of equant spar cements in Bahama escarpment limestones: Journal of Sedimentary Petrology, v. 56, p. 799-811.

GARBER, R. A., GROVER, G. A., and HARRIS, P. M., 1989, Geology of the Capitan shelf margin—subsurface data from the northern Delaware Basin, *in* Harris, P. M., and Grover, G. A., eds., Subsurface and Outcrop Examination of the Capitan Shelf Margin, Northern Delaware Basin: Tulsa, Society of Economic Paleontologists and Mineralogists Core Workshop 13, p. 3-269.

GEORGE, B. K., 1992, Distinctions between reefs and bioherms based on studies of fossil algae: *Mizzia*, Permian Capitan Reef Complex (Guadalupe Mountains, Texas and New Mexico) and *Eugonophyllum*, Pennsylvanian Holder Formation (Sacramento Mountains, New Mexico): Unpublished Ph.D. Dissertation, Louisiana State University, Baton Rouge, 156 p.

GIVEN, R. K., and LOHMANN, K. C., 1985, Derivation of the original isotopic composition of Permian marine cements: Journal of Sedimentary Petrology, v. 55, p. 430-439.

GIVEN, R. K., and LOHMANN, K. C., 1986, Isotopic evidence for the early meteoric diagenesis of the reef facies, Permian Reef Complex of west Texas and

New Mexico: Journal of Sedimentary Petrology, v. 56, p. 183-193.

GOLDHAMMER, R. K., and HARRIS, M. T., 1989, Eustatic controls on the stratigraphy and geometry of the Latemar buildup (Middle Triassic), The Dolomites of Northern Italy, *in* Crevello, P. D., Wilson, J. L., Sarg, J. F. and Read, J. F., eds., Controls on Carbonate Platform and Basin Development: Tulsa, Society of Economic Paleontologists and Mineralogists Special Publication 44, p. 323-338.

GRAMMER, G. M., and GINSBURG, R. N., 1992, Highstand versus lowstand deposition on carbonate platform margins: Insight from Quaternary foreslopes in the Bahamas: Marine Geology, v. 103, p. 125-136.

GRAMMER, G. M., GINSBURG, R. N., and HARRIS, P. M., 1993, Timing of deposition, diagenesis, and failure of steep carbonate slopes in response to a high-amplitude/high-frequency fluctuation in sea-level, Tongue of the Ocean, Bahamas, *in* Loucks, R., and Sarg, J. F., eds., Carbonate Sequence Stratigraphy: Recent Developments and Applications: Tulsa, American Association of Petroleum Geologists Memoir 57, p. 107-132.

HALLEY, R. B., PIERSON, B. J., and SCHLAGER, W., 1984, Alternative diagenetic models for Cretaceous talus deposits, Deep Sea Drilling Project Site 536, Gulf of Mexico, *in* Buffler, R. T., and Schlager, W., et al., eds.: Initial Reports of the Deep Sea Drilling Project, v. 77, p. 397-408.

HARRIS, M. T., 1994, The foreslope and toe-of-slope facies of the Middle Triassic Latemar buildup (Dolomites, northern Italy): Journal of Sedimentary Research, v. B64, p. 132-145.

HARRIS, P. M., 1997, Reservoir potential of the carbonate forereef-slope setting (abs.): Canadian Society of Petroleum Geologists-Society of Economic Paleontologists and Mineralogists Joint Convention, Calgary, Canada. p. 124.

HAYES, P. T., 1964, Geology of the Guadalupe Mountains, New Mexico: Washington, D.C., U.S. Geological Survey Professional Paper 446, 69 p.

HILLS, J. M., 1984, Sedimentation, tectonism, and hydrocarbon generation in Delaware Basin, west Texas and southeastern New Mexico: American Association of Petroleum Geologists Bulletin, v. 68, p. 250-267.

HISS, W. L., 1980, Movement of ground water in Permian Guadalupian aquifer systems, southeastern New Mexico and western Texas, *in* Dickerson, P. W., and Hoffer, J. M., eds., Trans-Pecos Region: New Mexico Geological Society 31st Field Conference: p. 289-294.

HOPKINS, J. C., 1977, Production of foreslope breccia by differential submarine cementation and downslope displacement of carbonate sands, Miette and Ancient Wall Buildups, Devonian, Canada, *in* Cook, H. E., and Enos, P., eds., Deep-Water Carbonate Environments: Tulsa, Society of Economic Paleontologists and Mineralogists Special Publication 25, p. 155-170.

HULL, J. P. D., JR., 1957, Petrogenesis of Permian Delaware Mountain Sandstone, Texas and New Mexico: American Association of Petroleum Geologists Bulletin, v. 41, p. 278-307.

HURLEY, N. F., 1978, Facies mosaic of the lower Seven Rivers Formation (Permian, Guadalupian), Guadalupe Mountains, New Mexico: Unpublished Master's Thesis, University of Wisconsin, Madison, 192 p.

HURLEY, N. F., 1989, Facies mosaic of the lower Seven Rivers Formation, McKittrick Canyon, New Mexico, *in* Harris, P. M., and Grover, G. A., eds., Subsurface and Outcrop Examination of the Capitan Shelf Margin, Northern Delaware Basin: Tulsa, Society of Economic Paleontologists and Mineralogists Core Workshop 13, p. 325-346.

KENDALL, C. G. S. C., 1969, An environmental re-interpretation of the Permian evaporite carbonate shelf sediments of the Guadalupe Mountains: Geological Society of America Bulletin, v. 80, p. 2503-2526.

KENTER, J. A. M., 1990, Carbonate platform flanks: Slope angle and sediment fabric: Sedimentology, v. 37, p. 777-794.

KING, P. B., 1948, Geology of the Southern Guadalupe Mountains, Texas: Washington, D.C., U.S. Geological Survey Professional Paper 215, 183 p.

KOSS, G. M., 1977, Carbonate mass flow sequences of the Permian Delaware Basin, west Texas, *in* Hileman, M. E., and Mazzullo, S. J., eds., Upper Guadalupian facies, Permian Reef Complex, Guadalupe Mountains, New Mexico and West Texas, 1977 Field Conference Guidebook): Midland, Society of Economic Paleontologists and Mineralogists, Permian Basin Section, Publication 77-16, p. 391-408.

LAND, L. S., 1980, The isotopic and trace element geochemistry of dolomite: The state of the art, *in* Zenger, D. H., Dunham, J. B., and Ethington, R. L., eds., Concepts and Models of Dolomitization: Tulsa, Society of Economic Paleontologists and Mineralogists Special Publication 28, p. 87-110.

LOWENSTEIN, T. K., 1988, Origin of depositional cycles in a Permian "saline Giant": The Salado (McNutt zone) evaporites of New Mexico and Texas: Geological Society of America Bulletin, v. 100, p. 592-608.

MAZZULLO, S. J., and CYS, J. M., 1977, Submarine cements in Permian boundstones and reef-associated rocks, Guadalupe Mountains, west Texas and southeastern New Mexico, *in* Hileman, M. E., and Mazzullo, S. J., eds., Upper Guadalupian Facies, Permian reef Complex, Guadalupe Mountains, New Mexico and West Texas, 1977 Field Conference Guidebook: Tulsa, Society of Economic Paleontologists and Mineralogists, Permian Basin Section, Publication 77-16, p. 151-200.

MCCLAIN, W. R., FREEMAN-LYNDE, R. P., and LOHMANN, K. C., 1988, Petrography and stable oxygen and carbon isotope compositions of Campanian grainstones and rudstones, Northeast Providence Channel, Bahamas: ODP Leg 101, Hole 634A, *in* Austin, J. A., and Schlager, W., et al., eds., Proceedings of the Ocean Drilling Program, v. 101, Part B: Scientific Results, p. 245-253.

MELIM, L. A., 1991, The origin of dolomite in the Permian (Guadalupian) Capitan Formation, Delaware Basin, west Texas and New Mexico: Implications for dolomitization models: Unpublished Ph.D. Dissertation, Southern Methodist University, Dallas, 200 p.

MELIM, L. A., and MASSAFERRO, J. L., 1997, Geology of the Bahamas: Subsurface Geology of the Bahamas Banks, *in* Vacher, H. L., and Quinn, T. M., eds., Geology and Hydrogeology of Carbonate Islands: Amsterdam, Elsevier, Developments in Sedimentology, v. 54, p.

MELIM, L. A., and SCHOLLE, P. A., 1992, Controls on dolomitization in the reef facies of the Capitan Formation, *in* Cromwell, D. W., Moussa, M. T., and Mazzullo, L. J. , eds., Southwest Section, American Association of Petroleum Geologists publication 92-90, p. 177-181.

MELIM, L. A., and SCHOLLE, P. A., 1995, The forereef facies of the Permian Capitan Formation: The role of sediment supply versus sea-level changes: Journal of Sedimentary Research, v. B65, p. 107-118.

MELIM, L. A., SWART, P. K., and MALIVA, R. G., 1995, Meteoric-like fabrics forming in marine waters: Implications for the use of petrography to identify diagenetic environments: Geology, v. 23, p. 755-758.

MORAN, W. R., 1955, Sandstone in New Mexico Room of Carlsbad Caverns, New Mexico: American Association of Petroleum Geologists Bulletin, v. 39, p. 256-259.

MRUK, D. H., 1985, Cementation and dolomitization of the Capitan limestone (Permian), McKittrick Canyon, west Texas: Unpublished Master's Thesis, University of Colorado, Boulder, 153 p.

MRUK, D. H., 1989, Diagenesis of the Capitan Limestone, Upper Permian, McKittrick Canyon, west Texas, *in* Harris, P. M., and Grover, G. A., eds., Subsurface and Outcrop Examination of the Capitan Shelf Margin, Northern Delaware Basin: Tulsa, Society of Economic Paleontologists and Mineralogists Core Workshop 13, p. 387-407.

MRUK, D. H., and BEBOUT, D. G., 1993, Slope, *in* Bebout, D. G., and Kerans, C., eds., Guide to the Permian Reef Geology Trail, McKittrick Canyon, Guadalupe Mountains National Park, West Texas, Guidebook 26: Austin, Bureau of Economic Geology, The University of Texas, p. 14-22.

MULLINS, H. T., 1983, Modern carbonate slopes and basins of the Bahamas, *in* Cook, H. E., Hine, A. C., and Mullins, H. T., eds., Platform Margin and Deep Water Carbonates: Tulsa, Society of Economic Paleontologists and Mineralogists Short Course 12, p. 4/1-4/138.

MULLINS, H. T., and COOK, H. E., 1986, Carbonate apron models: Alternatives to the submarine fan model for paleoenvironmental analysis and hydrocarbon exploration: Sedimentary Geology, v. 48, p. 37-79.

NEWELL, N. D., RIGBY, J. K., FISHER, A. G., WHITEMAN, A. J., HICKOX, J. E., and BRADLEY, J. S., 1953, The Permian Reef Complex of the Guadalupe Mountains Region, Texas and New Mexico: San Francisco, W. H. Freeman and Company, 236 p.

PLAYFORD, P. E., 1980, Devonian "Great Barrier Reef" of Canning Basin, Western Australia: American Association of Petroleum Geologists Bulletin, v. 64, p. 814-840.

PRAY, L. C., and STANTON, R. J., JR., 1992, Neptunian fossil-rich carbonate dikes in the Capitan-Massive (Permian) Guadalupe Mountains, New Mexico and Texas, U.S.A.: Tulsa, American Association of Petroleum Geologists Annual Convention, Official Program, v. 1, p. 106.

PREZBINDOWSKI, D., 1980, Microsampling technique for stable isotopic analyses of carbonates: Journal of Sedimentary Petrology, v. 50, p. 643-644.

REECKMANN, S. A., 1985, Geology of the foreslope to basinal transition. Permian Reef Complex, McKittrick Canyon, Guadalupe Mountains: Special study for Guadalupe Mountains National Park, 21 p.

SALLER, A. H., 1986, Radiaxial calcite in Lower Miocene strata, subsurface Enewetak Atoll: Journal of Sedimentary Petrology, v. 56, p. 743-762.

SALLER, A. H., 1996, Differential compaction and basinward tilting of the prograding Capitan reef complex, Permian, west Texas and southeast New Mexico, USA: Sedimentary Geology, v. 101, p. 21-30.

SALLER, A. H., BARTON, J. W., and BARTON, R. E., 1989, Slope sedimentation associated with a vertically building shelf, Bone Spring Formation, Mescalero Escarpe Field, southeastern New Mexico, *in* Crevello, P. D., Wilson, J. L., Sarg, J. F. and Read, J. F., eds., Controls on Carbonate Platform and Basin Development: Tulsa, Society of Economic Paleontologists and Mineralogists Special Publication 44, p. 275-288.

SARG, J. F., 1977, Sedimentology of the carbonate-evaporite facies transition of the Seven Rivers Formation (Guadalupian, Permian) in southeast New Mexico, *in* Hileman, M. E., and Mazzullo, S. J., eds., Upper Guadalupian Facies, Permian Reef Complex, Guadalupe Mountains, New Mexico and West Texas, 1977 Field Conference Guidebook: Tulsa, Society of Economic Paleontologists and Mineralogists, Permian Basin Section, Publication 77-16, p. 451-478.

SARG, J. F., 1989, Stratigraphy and sedimentology of the back-reef upper Queen-lower Seven Rivers strata, Goat Seep-Capitan Reef Complexes (middle-late Guadalupian, Permian) southeast New Mexico, *in* Harris, P. M., and Grover, G. A., eds., Subsurface and Outcrop Examination of the Capitan Shelf Margin, Northern Delaware Basin: Tulsa, Society of Economic Paleontologists and Mineralogists Core Workshop 13, p. 347-364.

SCHMIDT, V., 1977, Inorganic and organic reef growth and subsequent diagenesis in the Permian Capitan Reef Complex, Guadalupe Mountains, Texas, New Mexico, *in* Hileman, M. E., and Mazzullo, S. J., eds., Upper Guadalupian Facies, Permian Reef Complex, Guadalupe Mountains, New Mexico and West Texas, 1977 Field Conference Guidebook: Tulsa, Society of Economic Paleontologists and Mineralogists, Permian Basin Section, Publication 77-16, p. 93-132.

SCHOLLE, P. A., and HALLEY, R. B., 1980, Upper Paleozoic Depositional and Diagenetic Facies in a Mature Petroleum Province (A Field Guide to the Guadalupe and Sacramento Mountains of West Texas and New Mexico): Washington, D.C., U.S. Geological Survey Open-file Report 80-383, 191 p.

SCHOLLE, P. A., ULMER, D. A., and MELIM, L. A., 1992, Late-stage calcites in Permian Capitan Formation and its equivalents, Delaware Basin margin, west Texas and New Mexico: Evidence for replacement of precursor evaporites: Sedimentology, v. 39, p. 207-234.

WARD, R. F., KENDALL, C. G. S. C., and HARRIS, P. M., 1986, Upper Permian (Guadalupian) facies and their association with hydrocarbons—Permian Basin, west Texas and New Mexico: American Association of Petroleum Geologists Bulletin, v. 70, p. 239-262.

WIGGINS, W. D., and HARRIS, P. M., 1985, Burial diagenetic sequence in deep-water allochthonous dolomites, Permian Bone Spring Formation, southeast New Mexico, *in* Crevello, P. D., and Harris, P. M., eds., Deep-Water Carbonates—A core workshop: Tulsa, Society of Economic Paleontologists and Mineralogists Core Workshop 6, p. 140-173.

YUREWICZ, D. A., 1976, Sedimentology, paleoecology, and diagenesis of the massive facies of the lower and middle Capitan Limestone (Permian), Guadalupe Mountains, New Mexico and west Texas: Unpublished Ph.D. Dissertation, University of Wisconsin, Madison, 278 p.

YUREWICZ, D. A., 1977, The origin of the massive facies of the lower and middle Capitan Limestone (Permian), Guadalupe Mountains, New Mexico and west Texas, *in* Hileman, M. E. and Mazzullo, S. J., eds., Upper Guadalupian Facies, Permian Reef Complex, Guadalupe Mountains, New Mexico and West Texas, 1977 Field Conference Guidebook: Tulsa, Society of Economic Paleontologists and Mineralogists, Permian Basin Section, Publication 77-16, p. 45-92.

ORIGIN OF CAVES IN THE CAPITAN

CAROL A. HILL

17 El Arco Drive, Albuquerque, New Mexico 87123

ABSTRACT: Caves exist in the Capitan Formation in the Guadalupe, Apache, and Glass Mountains, southeastern New Mexico and West Texas. Four episodes of karsting have been identified in the Guadalupe Mountains: (1) Stage 1 fissure karst (Late Permian), (2) Stage 2 spongework karst (Mesozoic), (3) Stage 3 thermal karst (Miocene), and (4) Stage 4 sulfuric acid karst (late Miocene to present). The last sulfuric acid episode was responsible for the large cave passages and for the distinctive deposits within these caves (gypsum blocks; native sulfur; the pH-indicator minerals endellite, alunite, and natroalunite; and the uranium minerals tyuyamunite and metatyuyamunite). The hydrogen sulfide responsible for forming sulfuric acid in the Capitan reef originated from hydrocarbons in the Delaware Basin. Stage 4 caves in the Glass Mountains are also of sulfuric acid origin, showing that the basin probably degassed hydrogen sulfide around its entire margin. Caves also exist in the subsurface Capitan on the eastern and northern sides of the Delaware Basin. San Simon Swale/Sink on the east and breccia pipes/domes on the north may stope down to the cavernous Capitan. The same sulfuric acid mechanism that formed caves in the exposed Capitan may have also been partly responsible for caves in the subsurface Capitan.

CONTROLS ON CAVERN DEVELOPMENT

The Capitan Limestone is famous for the caves it hosts — Carlsbad Cavern and Lechuguilla Cave are world-renowned. Most caves in the Capitan exist in the Guadalupe Mountain section of the Delaware Basin, but caves are also present in the Apache and Glass Mountains and in the subsurface on the eastern and northern sides of the basin. Four properties of the Capitan seem to have exerted primary control on cavern development: (1) the arrangement of stratigraphic units, (2) differential dolomitization of those units, (3) joint patterns, and (4) positive structures (anticlines).

Stratigraphy

The Capitan Reef Complex in the Guadalupe Mountains contains three facies: the back-reef, reef and forereef. The back-reef facies is composed of the bedded units of the Queen, Seven Rivers, Yates, and Tansill formations — deposits that formed in a shallow lagoon-evaporite shelf environment and dip towards the reef (Fig. 1). The Capitan reef member is a massive limestone that grades laterally into the back-reef and also into the forereef breccia member. The contacts between these three facies are zones of structural weakness that have been the focal point for water movement and cave development since the Late Permian. While some caves are developed beneath the Yates in the Seven Rivers Formation (Fig. 1), most of the major caves (for example, Carlsbad, Lechuguilla) are developed at the reef-back-reef, reef-forereef contacts (Jagnow, 1977).

Dolomitization

Another major lithologic control on cave development is dolomitization, with caves in the Capitan being preferentially dissolved in reef limestone between the more-dolomitic back-reef and forereef units. A good example of this principle can be seen in the Big Room and along Left Hand Tunnel-Lake of Clouds in Carlsbad Cavern. Where dissolution encountered dolomitic forereef beds in these areas, cave development stopped. Another reason for less dissolution of back-reef and forereef beds is that these units contain a higher content of insoluble residue than the limestone reef.

Dolomitization of the Capitan has also been responsible for less cave development in the Apache and Glass Mountains as compared to the Guadalupe Mountains, and for less cave development in the lower, more-dolomitic Capitan than in the upper, less-dolomitic Capitan. In the Apache Mountains the Capitan reef is, in large part, dolomitized, but trends to limestone in the younger portions (Wood, 1968). In the Glass Mountains the reef facies of the Capitan is also more dolomitized than in the Guadalupe Mountains (King, 1930), with textures varying from somewhat dolomitized and with complete preservation of fossils (for example, in the Old Blue Mountain area), to only matrix replacement, to total replacement (for example, in Jail Canyon) (Haneef, 1993). In the Guadalupe Mountains, the oldest part of the reef massive facies (the lower Capitan) is partially dolomitized whereas the upper Capitan is predominantly limestone. This thin belt of limestone extends northeastward along the mountain front but gradually becomes dolomitized just southwest of Carlsbad, New Mexico (Fig. 2). Thus, a thin "island" of limestone exists only in the upper Capitan in the Guadalupe Mountains, and it is this "island" of rock that hosts the majority of caves. Since dolomitization of the Capitan occurred primarily in the Late Permian (Melim, 1991), it has acted as a control on dissolution ever since.

Joints

Joints are a primary structural control on cavern development in the Capitan, and the major trunk passages are, almost without exception, aligned along joints that are either parallel or perpendicular to the reef front. Jagnow (1977) found that east of Rattlesnake Canyon the major joints along which caves are developed trend in a northeastern direction parallel to the reef front, whereas west of Rattlesnake Canyon, the cave-bearing joints trend in a northwestern direction perpendicular to the reef front.

The age and origin of joints in the Capitan are controversial; however, it appears reasonably certain that the initiation of jointing occurred when the rock was being lithified in the Late

Geologic Framework of the Capitan Reef, SEPM Special Publication No. 65

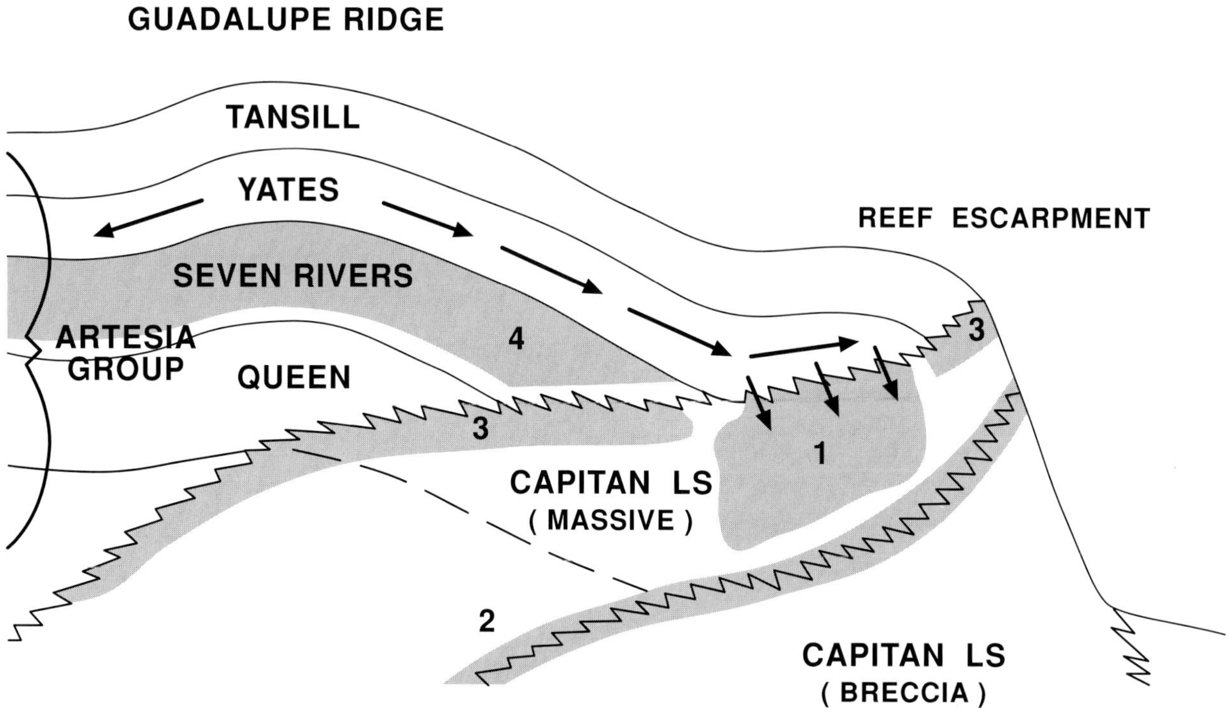

FIG. 1.—Four zones of preferential dissolution (shaded areas) in the Guadalupe Mountains: (1) below the Yates transition into the massive Capitan Limestone, (2) at the contact between the massive and forereef (breccia) members of the Capitan Limestone, (3) at the transition between the back-reef Artesia Group members and Capitan Limestone, and (4) immediately beneath the Yates Formation in the Seven Rivers. Arrows indicate movement of groundwater along impermeable siltstone in the Yates Formation. After Jagnow (1977).

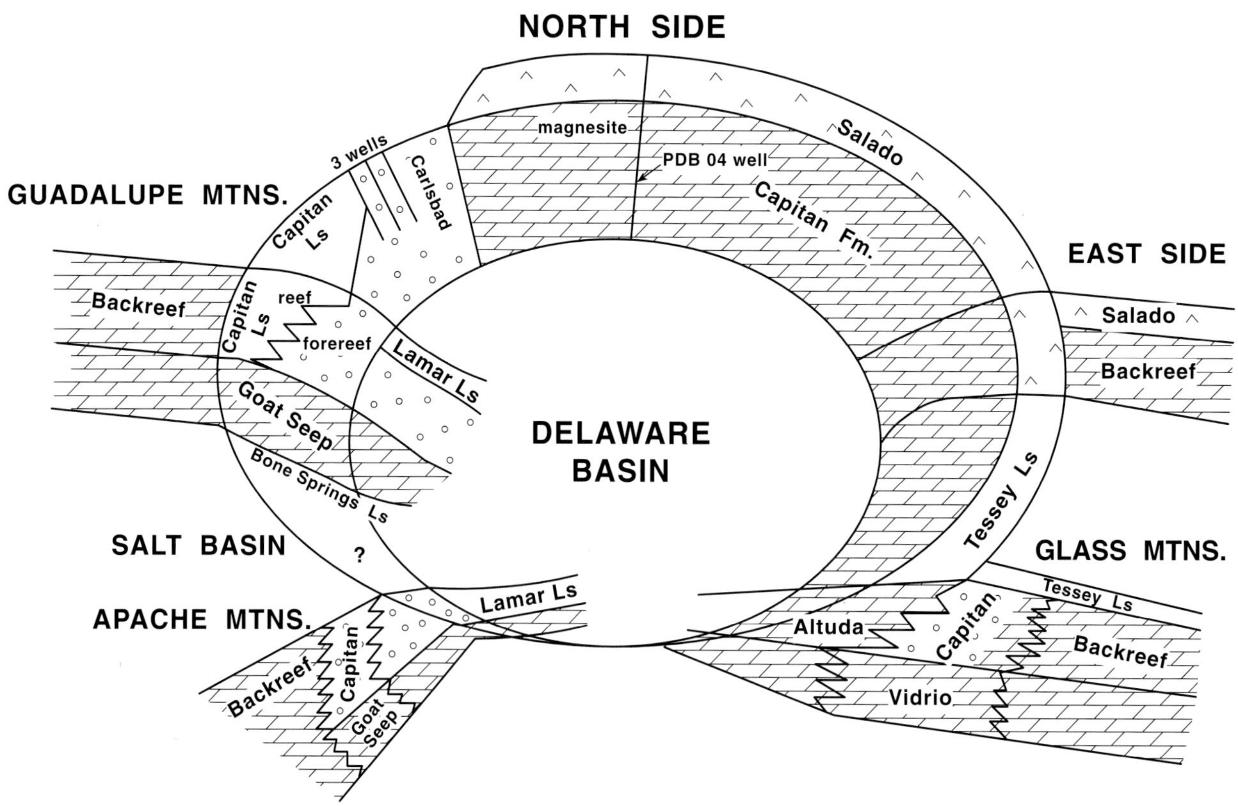

FIG. 2.—Idealized schematic diagram showing the distribution of dolomite in upper Permian units around the Delaware Basin. The upper Capitan in the Guadalupe Mountains is the only location where the Capitan is mostly undolomitized limestone. Diagonal lines show dolomitized rock, circles show rock with patchy dolomitization, white areas show limestone. The location of the Capitan and Goat Seep formations in the Salt Basin is unknown. From Hill (1996).

Permian (Hill, 1996). Initial jointing was probably set up during the compaction phase of diagenesis in the Late Permian, and later jointing took place along these lines of weakness during times of tectonism — possibly during the Laramide and probably during Basin and Range block faulting. Similarly, initial cave development was along these Late Permian joints, with subsequent stages of karstification following along these same lines of weakness.

Positive Structures

Positive structures (for example, anticlines) have been another control on cavern development. Carlsbad Cavern is located along the crest of the Reef anticline, Lechuguilla Cave and Cottonwood Cave are located along the Guadalupe Ridge anticline, the Slaughter Canyon caves are located along the Huapache monocline, and the McKittrick Hill caves are located along the flanks of the McKittrick Hill anticline. The caves are concentrated along positive structures because these structures acted as avenues and traps for hydrogen sulfide ascending from the basin into the reef. This concentrated hydrogen sulfide then oxidized to sulfuric acid, which dissolved out the large cave passages.

CAVES IN THE EXPOSED CAPITAN: GUADALUPE, APACHE, AND GLASS MOUNTAINS

Guadalupe Mountains

The Capitan section in the Guadalupe Mountains is one of the most outstanding karst regions in the world; the caves are renowned for their size, beauty, and their unusual hypogenic (sulfuric acid) mode of speleogenesis. The large cave passages in the Guadalupes, however, represent only the final chapter of cave development, with porosity/karst development having occurred in at least four stages following deposition of the Capitan Reef Complex in the Late Permian.

Stage 1 fissure karst. —
Stage 1 fissure karst represents the earliest episode of cave development in the Guadalupe Mountains. This paleokarst is characterized by small, fissure-like cavities filled with breccia, limy mudstone, siliciclastics, and/or calcite spar. Stage 1 fissure caves are usually located at or near the contact of reef and back-reef sediments, a location that suggests their formation in a zone of structural instability between the Capitan reef core and back-reef shelf members. Hill (1987) was of the opinion that Stage 1 fissure caves (her "Solution Stage I" caves) had formed in the Late Permian.

Stage 1 fissure karst is commonly exposed by Stage 4 sulfuric acid cave passages, but it has also been identified in outcrop. Melim and Scholle (1989) and Scholle et al. (1992) identified an exposure episode in the Guadalupe Mountains, during which an initial stage of meteoric leaching of originally aragonitic material occurred with the development of a large-scale, dissolution-enlarged fracture system. These surface-exposed fractures are filled with clast-supported breccia and blocky calcite, and are oriented mainly parallel to the Capitan reef trend, with a secondary orientation perpendicular to the reef trend (Melim, 1991). Individual fractures average 1 m wide but they can also occur in anastomosing sets averaging 10–20 m wide. Melim (1991) found the breccia fragments in these fractures to be dolomitized and of local origin without evidence of significant transportation, and also identified a dolomite alteration front or halo extending around the fissures for a distance of 5–10 cm into the reef rock. Melim thought that this dolomite (her "Substage B") was Late Permian (Ochoan?) in age.

From this evidence the following origin seems likely for the Stage 1 fissure karst episode. Subsequent to the deposition of the Capitan Limestone and equivalent back-reef facies, most probably in Ochoan time, tectonic activity caused the first uplift and tilting of the Capitan reef. Fractures (joints) that had initially formed in response to differential compaction between the rigid reef and less competent back-reef and forereef became enlarged into fissures due to tectonism and/or dissolution. Locally derived clasts then were displaced downward and partially filled these voids. The anastomosing nature of these fracture sets, with extensive pinching and swelling of individual fractures, is easiest to explain by dissolution-enlargement of already-existing fractures or joint sets (Melim, 1991). However, solutional activity does not appear to have been extensive. Dissolution-enlargement occurred prior to at least one dolomitization stage, because the walls of the Stage 1 fissure caves and breccia clasts within the fissures are dolomitized. Stage 1 fissure caves developed along the zones of weakness between the reef/back-reef or reef/forereef and then became avenues for water movement, porosity enhancement, and dissolutional enlargement during the later three stages of cave development.

Stage 2 spongework karst. —
The second episode of porosity/karst development in the Guadalupe Mountains involved the enlargement of pores and joints in the massive Capitan Limestone into a three-dimensional maze. "Spongework" consists of interconnected dissolution cavities of varied size in a seemingly random, three-dimensional pattern like the pores of a sponge (Palmer, 1991). Essentially, these cavities are freshwater lens voids, where phreatic water under pressure creates a spongework array of passages. Hill (1987) called Stage 2 spongework cavities in the Guadalupe Mountains "Solution Stage II" caves, and described them as being small and randomly oriented, with many being partly filled with montmorillonite clay.

In the early stages of cave-system formation, a complex, three-dimensional array of pores and joints of minimal cross-sectional area can develop in the rock, but the pores and joints are not necessarily integrated so that flow under these conditions is diffuse. As dissolution continues over time, this array expands, and there is a progressive integration of conduits and enlargement of small portions of the array to create a cave system that eventually becomes continuous from input to output (Ford and Ewers, 1978). In such a situation the limestone becomes honeycombed along joints, fractures, and bedding planes, and small caves up to a few meters in extent can form.

Stage 2 spongework caves represent this type of early karst development in the Guadalupe Mountains. These cavities/caves dissolved under phreatic conditions marked by the slow, diffuse circulation of aquifer water during limestone mesogenesis. Stage 2 spongework caves in the Capitan are exposed in the walls of the large cave passages as a three-dimensional array of "Swiss-cheese"-like holes or cavities; sometimes these holes are large enough to crawl through, sometimes they are not. Stage 2 spongework porosity/karst may have formed over a long period of time, or more likely during several episodes from the Mesozoic to early Tertiary. The Triassic-Jurassic was a time of emergence for the Delaware Basin area, and it was also probably a time of extensive dissolution. Montmorillonite clay filling some Stage 2 spongework cavities has a speculative K-Ar date of 188±7 m.y. (early Jurassic; Hill, 1987).

Stage 3 thermal karst. —

The third episode of cave development was named "Stage 3 thermal karst" by Hill (1996), and was defined as those caves lined with "thermal spar" — large calcite spar crystals that have carbon-oxygen isotope compositions of $\delta^{13}C$ = 0 to +3‰ and $\delta^{18}O$ = −14 to −11‰ (PDB), respectively, and fluid-inclusion temperatures of ~30–80°C. Thermal spar is ubiquitous throughout the Capitan reef and parts of the back-reef. It is the large, milky-white, rhombohedral calcite seen everywhere throughout the Guadalupe (and Glass) Mountains (Fig. 3), and it is also the calcite that fills Basin and Range fault zones in the Guadalupe, Delaware, Apache and Glass Mountains (Hill, 1996; Figure 102 and Plate 6). Thermal karst caves are usually small, unconnected chambers sometimes intersected by later, larger, Stage 4 sulfuric acid cave passages, but rarely these spar-lined chambers can be large-size rooms many meters in extent (for example, Spar City in Lechuguilla Cave).

The cooling of thermal water and CO_2 equilibrium processes were probably both involved in the dissolution of thermal caves and the lining of these caves with large dogtooth spar crystals (Hill, 1996). These caves could have begun forming during the thermal episode or, more likely, they represent the enlargement of spongework caves already present from the former Stage 2 episode. Thermal spar, which fills Basin and Range fault zones, has the same isotopic signature as the cave spar (Hill, 1996), and since uplift and spar filling along these fault zones is most likely Miocene in age (Horak, 1985), it is there-

FIG. 3.—Blocky, milky-white thermal spar along Yucca Canyon trail, Guadalupe Mountains. Thermal spar becomes more abundant where the forereef interfingers with the reef. Photo by Carol Hill.

fore probable that Stage 3 thermal karst also dates from the Miocene (~20-15 Ma).

Stage 4 sulfuric acid karst. —

Cutting across all three previous stages of karst development are the large, Stage 4, sulfuric acid caves in the Guadalupe Mountains.

Description. — Over 300 caves are known in the Guadalupe Mountains. Carlsbad Cavern is the most famous of these caves, but it is neither the longest nor the deepest. Lechuguilla Cave is the longest (>150 km) and deepest (486 m) cave in the Guadalupe Mountains, the deepest and third longest cave in the United States, and the fifth longest cave in the world. The main caves/cave areas in the Guadalupe Mountains are: Carlsbad Cavern, Lechuguilla Cave, Cottonwood Cave, the Slaughter Canyon caves, and the McKittrick Hill caves. These are located within 12 km of the Reef Escarpment (most are within 5 km); along the crests or flanks of anticlines or other positive structures (for example, Reef anticline, Guadalupe Ridge anticline, Huapache monocline); at the contact of major facies changes (Fig. 1); and between the reef and more-dolomitic back-reef and forereef beds.

The caves of the Guadalupe Mountains contain large rooms and passages (many are >15 m in height and width) along major elevation levels, with separate levels connected by steeply dipping passages (for example, Main Corridor in Carlsbad Cavern), vertical tubular pits (for example, Bottomless Pit in Carlsbad Cavern), and enlarged vertical fissures (for example, Cable Slot in Carlsbad Cavern). Walls of large rooms and passages sometimes truncate Stage 1 fissure caves, they are honeycombed with Stage 2 spongework caves, or they truncate Stage 3 thermal caves filled with dogtooth spar crystals. Cave passages/rooms end abruptly without breakdown collapse or major passage extensions, and the caves lack a clear genetic relationship to surface topography. Intersections of caves with the land surface (entrances) are random and have no apparent relationship to recharge or resurgence points, either ancient or modern. Well-developed surface karst landforms (for example, sinkholes) are not prominent features.

Stage 4 caves are a result of combined bathyphreatic and water-table development (Hill, 1987). Water-table (shallow-water phreatic) conditions were responsible for the horizontal development of caves along certain levels, and bathyphreatic (deep-water phreatic) conditions were responsible for the strong vertical development of these caves. None of the caves are vadose in origin. Cave sediment, unlike that in vadose caves, is sparse, and where it does occur it is almost always a coarse silt to fine-grained sand derived directly from the dissolution of siliciclastic facies in the bedrock (for example, Sand Passage in Carlsbad Cavern).

Distinctive deposits in Guadalupe Mountain caves are thick floor blocks of massive gypsum and thinner wall rinds of gypsum; native sulfur deposits; and colorful, waxy endellite [$Al_2Si_2O_5(OH)_4.2H_2O$] clay deposits. Alunite [$K_2Al_6(SO_4)_4(OH)_{12}$], natroalunite [$Na_2Al_6(SO_4)_4(OH)_{12}$], tyuyamunite [$Ca(UO_2)_2(VO_4)2{\cdot}5\text{-}8H_2O$], and metatyuyamunite [$Ca(UO_2)_2(VO_4)_2.3\text{-}5H_2O$] are other speleogenesis-related cave minerals (Polyak and Mosch, 1995; Polyak and Güven, 1996).

Origin. — Stage 4 caves are hypogenic; that is, they were formed by acids of deep-seated origin (for example, sulfuric acid) in contrast to epigenic caves, which form in the near surface where carbonic acid is derived from CO_2 in the atmospheric and soil zones (Palmer, 1991). Hypogenic karst displays no relationship to recharge through the overlying surface, and cave passages are typically ramiform, network, or spongework in pattern.

Stage 4 caves in the Guadalupe Mountains were dissolved primarily by sulfuric acid. The source of the sulfuric acid was H_2S derived from hydrocarbons in the Delaware Basin. A number of different lines of evidence point to a sulfuric acid/hydrocarbon origin for Stage 4 caves (Hill, 1987, 1990):

1. Massive gypsum blocks (up to 10 m high) and native sulfur deposits (up to thousands of kilograms) in these caves formed as by-products of a sulfuric acid mode of dissolution. Epigenic, carbonic-acid caves do not contain these types of deposits.
2. The low-pH, sulfuric acid indicator minerals endellite, alunite, and natroalunite occur in these caves.
3. High uranium, radon, and the minerals tyuyamunite /metatyuyamunite in these caves are all indicative of an H_2S system where uranium (and vanadium) have precipitated along a redox boundary interface (Hill, 1995).
4. Other sulfuric acid caves are known worldwide, and these are also associated with hydrocarbons. Some of these caves are actively forming today by a sulfuric acid mechanism; for example, La Cueva de Villa Luz, Tobasco, Mexico, is a sulfuric acid cave related to hydrocarbons in the Gulf of

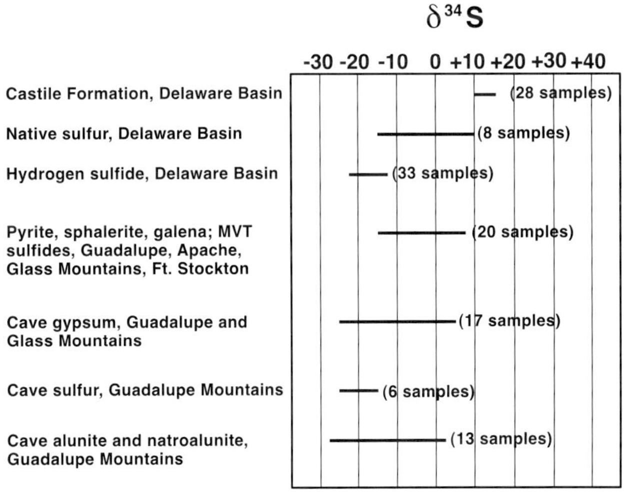

FIG. 4.—Range of sulfur isotope values for various types of deposits, Delaware Basin. Note that the cave gypsum and sulfur are significantly enriched in the light isotope of sulfur compared to the Castile Formation. The isotopically light composition of all the sulfur deposits in the basin (H_2S gas, native sulfur) and reef (sulfides and sulfur/gypsum cave deposits) resulted from bacterial sulfate reduction (BSR) reactions in the basin where Castile Formation sulfate solutions mixed with hydrocarbons to produce isotopically light hydrogen sulfide. Cave alunite and natroalunite values are from Polyak and Güven (1996). After Hill (1995).

Campeche (Pisarowicz, 1994). A milky-white river, with dissolved gypsum and sulfur, flows from the cave, and sulfur crystals are growing in areas where drip water has a measured pH of 1. Sulfur isotope values for sulfur and gypsum in La Cueva de Villa Luz ($\delta^{34}S$ = –26 to –22‰; CDT) are within the same range as the sulfur and gypsum in Guadalupe Mountain caves (Fig. 4).

5. The isotopically light composition of the massive gypsum, sulfur, alunite, and natroalunite deposits in Stage 4 caves is the most convincing evidence for a sulfuric acid origin related to hydrocarbons. Only biologically aided reactions such as occur with hydrocarbons could have produced the large isotopic fractionations found in these deposits. Gypsum and native sulfur deposits in Guadalupe Mountain caves are significantly enriched in ^{32}S, and depletions as great as –25.6‰ for gypsum and –25.8‰ for sulfur have been measured (Fig. 4). The same isotopically light signatures also characterize alunite and natroalunite (Polyak and Güven, 1996; Fig. 4).

The sulfur isotope results are crucial to understanding the hypogenic process of speleogenesis that produced the large Stage 4 cave passages in the Capitan Limestone. The evidence provided by the isotopic data demonstrates that the cave gypsum could not have been derived from Castile anhydrite beds by the local pooling model of Bretz (1949) or by the mixing model of Queen et al. (1977). The average isotopic composition of Castile

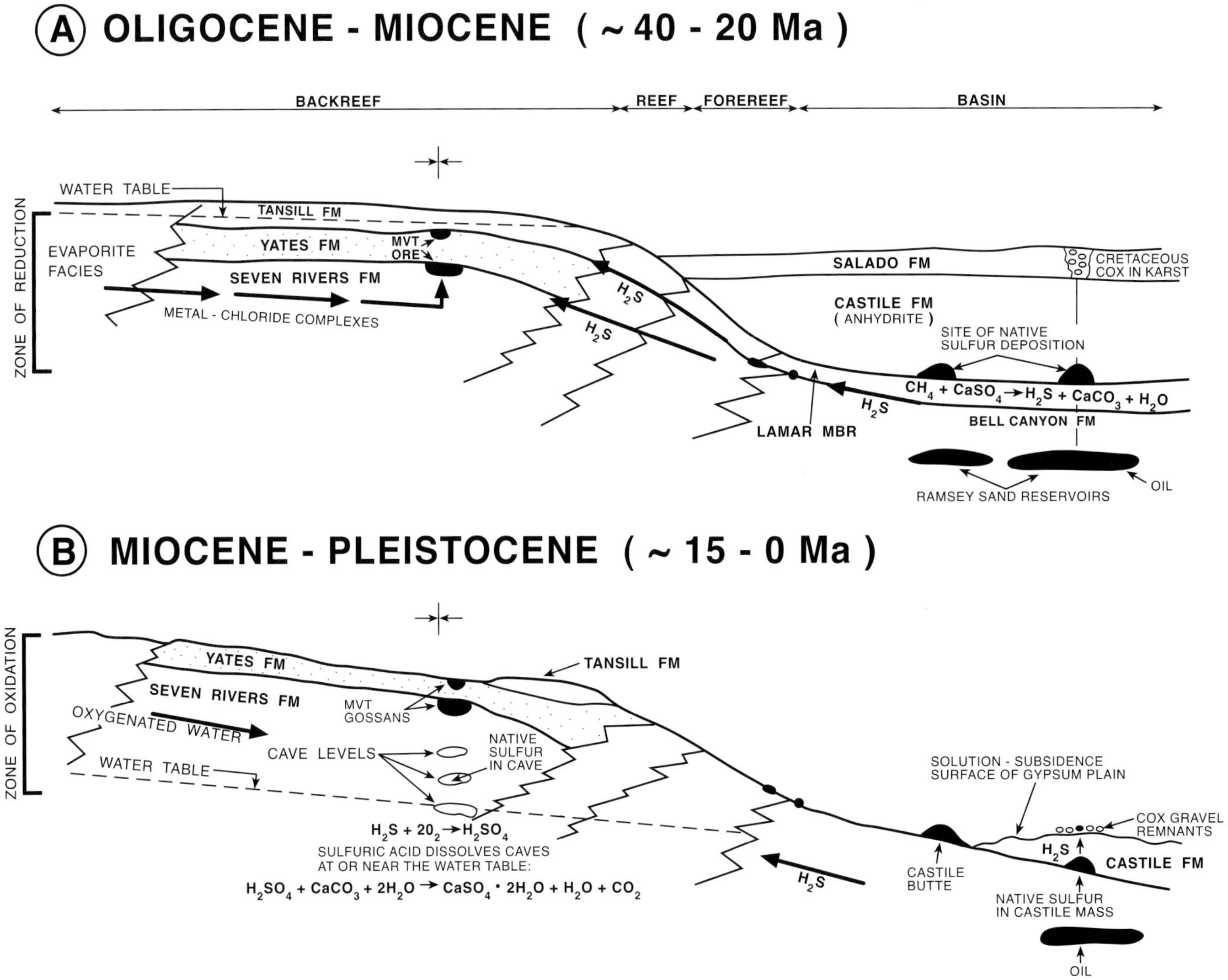

FIG. 5.— Idealized model for the origin of caves, and also for Mississippi Valley-type (MVT) sulfides, in the Guadalupe Mountains. This model proposes a genetic connection between hydrocarbons and native sulfur in the basin, and MVT deposits and sulfuric acid caves in the carbonate-reef margin. (A) In the late Oligocene-early Miocene, during the Tertiary tilting of the Delaware Basin, H_2S was generated in the basin by reactions involving hydrocarbons and anhydrite. The H_2S oxidized to native sulfur in the basin, and also some migrated from basin to reef to accumulate there in structural (anticlinal) and stratigraphic (base of Yates) traps. Metals moved down-dip as chloride complexes from back-reef-evaporite facies, and where they encountered ascending H_2S below the water table in the zone of reduction, they formed MVT deposits. (B) Later in the Miocene and also in the Pliocene to Pleistocene, continued uplift and tilting of the Guadalupe Mountain block and Delaware Basin area caused increased H_2S generation and migration of gas from basin to reef. Cave dissolution occurred in the same structural and stratigraphic position as earlier MVT deposits, where H_2S oxidized to sulfuric acid at or near the water table in the zone of oxidation. Cave levels correspond to a descending base level caused by the regional lowering of the Pecos River. From Hill (1996).

gypsum and anhydrite is +10.3‰ (Fig. 4). If the cave gypsum precipitated amicrobially from Castile brines, as modeled by these authors, then the cave gypsum and Castile Formation gypsum and anhydrite should have virtually identical isotopic compositions. The sulfur isotope data also discount an origin from pyrite as proposed by Jagnow (1977). Sulfide minerals (mostly pyrite) in the Guadalupe Mountains range from $\delta^{34}S = -9.3$ to +12.6‰ (mean = –2.2‰ for 20 samples; Fig. 4). Since there is no significant isotopic fractionation involved in the leaching of sulfides (less than 1‰), it is logical to conclude that the isotopically lighter cave gypsum and sulfur (mean = –16.8‰ for 22 samples) could not have derived from this source. Instead, the isotopically light deposits implicate a hydrocarbon connection for these caves (Hill, 1990).

Cave dissolution by sulfuric acid. — Hydrogen sulfide, generated from hydrocarbon reactions in the basin, migrated into the surrounding Capitan reef and accumulated in structural and stratigraphic traps (Fig. 5). Where it met with oxygenated meteoric groundwater descending to the water table along dipping back-reef beds or joints in the overlying land surface it formed sulfuric acid:

$$H_2S + 2O_2 = HSO_4^- + H^+ \quad (1)$$

$$HSO_4^- + H^+ + CaCO_3 + 2H_2O = Ca^{2+} + SO_4^{2-} + 3H_2O + CO_2 \quad (2)$$

The sulfuric acid produced in (1) dissolved the Capitan reef limestone to produce the cave void, precipitation of gypsum, and generation of CO_2 (2). According to this model, vertical tubes, fissures, and pits in Guadalupe caves are interpreted as having formed along injection points for hydrogen sulfide gas (bathyphreatic dissolution), and horizontal levels are interpreted as forming at the water table, where dissolved oxygen was the most concentrated (water-table dissolution) (Fig. 6). H_2S degassing created the native sulfur deposits and also (ultimately) the precipitation of the secondary, later-stage uranium-vanadium minerals tyuyamunite and metatyuyamunite. A low-pH, sulfuric acid environment also caused clay minerals to reconstitute to endellite, alunite, and natroalunite.

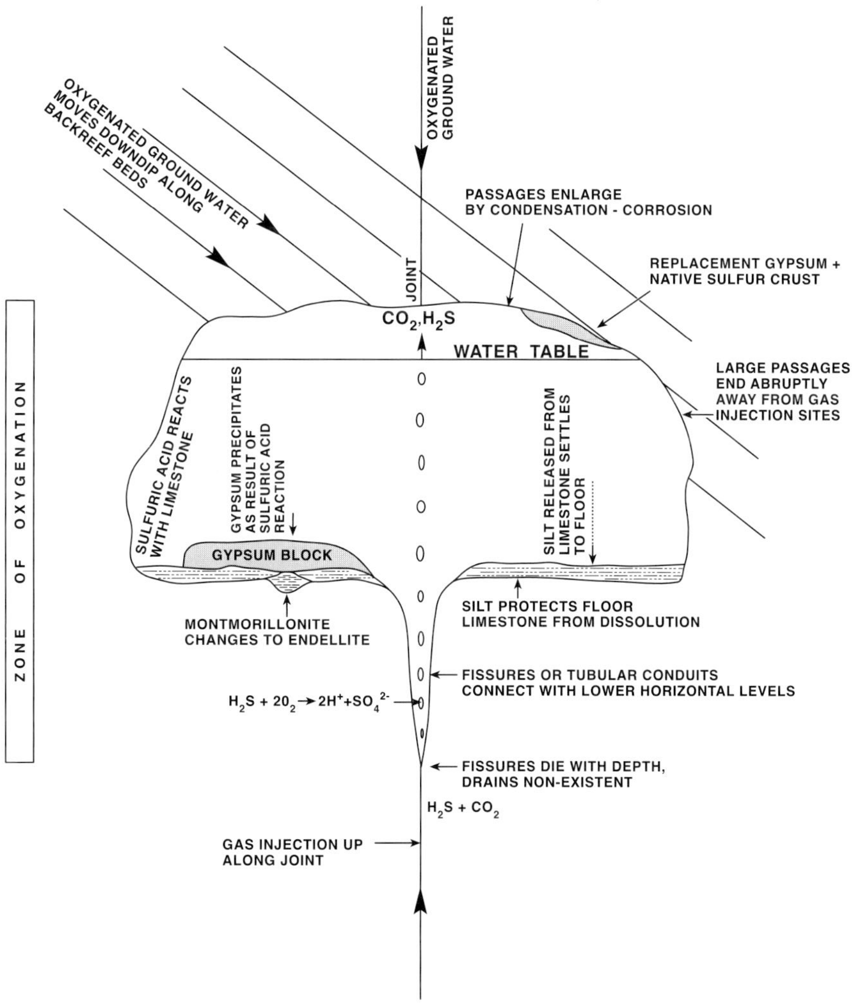

FIG. 6.—Model of hydrogen sulfide reaction with dissolved oxygen near the water table to form the large, Stage 4, sulfuric acid cave passages. Sulfuric acid was neutralized by the limestone away from gas injection points and, therefore, horizontal cave passages in the Guadalupe Mountains end abruptly. The sulfuric acid reaction did not occur below the zone of oxygenation of groundwater and hence vertical passages thin and end with depth below large, horizontal rooms. With successive lowering of base level, new horizontal levels became connected with older horizontal levels by spring shafts and joint chimneys. Silt residue from the limestone settled to the floor; gypsum in solution precipitated over the silt in slack places to form massive gypsum blocks, or directly replaced the limestone bedrock; and CO_2 caused further dissolution beneath the water table or condensation-corrosion of cave passages in the air zone. From Hill (1987, 1996).

Age. — The age of Stage 4 sulfuric acid caves is questionable, but it appears that the beginning of Stage 4 dissolution may be somewhat older than the Pliocene-Pleistocene age ascribed by Hill (1987). Maximum uplift and tilting of the Guadalupe block is now believed to have occurred in the Miocene (15–5 Ma), which means that hydrogen sulfide could have been migrating throughout the middle to late Tertiary with the potential for cave formation (Hill, 1996).

Polyak et al. (1997, 1998) $^{40}Ar/^{39}Ar$-dated the alunite in four Guadalupian caves, and have confirmed that the cave passages formed from about 12 m.y. ago in the southwestern part of the reef (for example, Virgin Cave) to about 4 m.y. ago in the northeastern part of the reef (for example, Carlsbad Cavern and Lechuguilla Cave). These absolute dates are very important because they correlate with the time of major uplift of the Guadalupe Mountains and the proposed time of migration of hydrogen sulfide from the basin into the Capitan reef.

Apache Mountains

The Apache Mountains have not been adequately explored for caves. The owner of the land, Judge Lee Green (pers. comm., 1991) has remarked that there are "holes all over" in these mountains, but the graduate students of Ronald DeFord, who did field work in the 1950s, did not report significant numbers of caves in the area. The inaccessibility and remoteness of this area, and the fact that the land is privately owned, makes cave hunting difficult.

At least one cave is known to exist in the Apache Mountains. Cross Cave is located in the Panther Canyon Quadrangle about 6.5 km southwest of Foster ranch along the margin of the Capitan Reef Escarpment. It is developed along a joint parallel to the escarpment. The cave entrance is 30 m wide by 12 m high, descends at an angle of about 20–30° for about 35 m, and then pinches out as a single small passage 1 m wide, 0.5 m high, and 3 m long (Hill et al., 1988). The steep entrance descent, as is typical for passages created by bathphreatic water, suggests dissolution by ascending water through the Apache Mountain section of the Capitan reef. Cave travertine occurs in the arroyo down the hill from Cross Cave, which shows that the cave was once much more extensive than it is today before canyon retreat by erosion.

In addition to Cross Cave, C. Carrasco (pers. comm., 1992), caretaker for the Apache ranch (the old Munn ranch), related that a well in the Capitan Limestone at "1250 ft" (380 m) "blows air" and "breathes in and out," a good sign that a cave in the subsurface has been intersected. This well is located ~2.4 km east of Apache ranch at the location marked "well" on the Goat Canyon 7.5 min quadrangle.

The relative lack of caves in the Apache Mountains, as compared to the Guadalupe Mountains, may be unsubstantiated (the mountains have not been explored for caves), or it may be due to geologic factors:

1. A major portion of the Capitan reef in the Apache Mountains is dolomitic limestone (Fig. 2), and thus is less likely to be cave-bearing than the Capitan Limestone in the Guadalupe Mountains.
2. The Apache Mountains are not nearly as high or extensive as the Guadalupe Mountains. Therefore there is less Capitan rock exposed (and less cave-hunting ground) in these mountains than in the Guadalupes.

Glass Mountains

The Glass Mountains are even harder to access than the Apaches because the land is strictly controlled by ranchers. King's (1930) map shows a "cave" in his "Capitan massive" unit to the north of Cathedral Mountain, but to the author's knowledge, this cave has never been explored. Hiss (1975) reported that "the blowing and sucking of air, a phenomenon typical of interchange between caves and the atmosphere in response to variations in barometric pressure, has been observed to be associated with wells penetrating the Capitan aquifer in the Glass Mountains."

There are a number of known caves in the Glass Mountains (Fieseler and Kunath, 1975), and most of these are in the Capitan Formation. The longest (>600 m) of these is Four Hundred Foot Cave, which is a three-dimensional spongework-maze cave developed along a series of northwest- and northeast-trending joints and also possibly along a fault. Gypsum blocks have been found in the lower levels of Four Hundred Foot Cave; this gypsum has a sulfur isotope value of $\delta^{34}S = -12.7‰$, equivalent to the massive gypsum blocks in Guadalupe Mountain caves (Hill, 1996; Fig. 4). This find is extremely important because it shows that the H_2S basinal-degassing process should not be thought of as unique to the Guadalupe Mountains, but has also probably occurred around the margins of the entire Delaware Basin. Stage 3 thermal spar caves and Stage 4 sulfuric acid caves are both known to exist in the Glass Mountains (Hill, 1996).

The cave potential in the Glass Mountains is undetermined, but it might also be expected to be less than that in the Guadalupe Mountains for the same reasons as given for the Apaches: the rock is more dolomitized, and the mountains less extensive, than in the Guadalupes.

CAVES IN THE SUBSURFACE CAPITAN: EASTERN AND NORTHERN SIDES OF DELAWARE BASIN

Because the Capitan Formation resides completely in the subsurface on the east and north sides of the Delaware Basin, the presence of caves must be ascertained by subsurface drilling. Sulfur water is known to be present in subsurface sections of the Capitan, and drilling into the Capitan reef in these areas has produced sulfur water where significant cavernous porosity is encountered (Thorton and Gaston, 1967; Conselman, 1987; Cromwell and Mazzullo, 1987). Conselman (1987) reported that the Capitan reef on the north side of the basin had been "...undoubtedly massively invaded by basinal sulfur water...," and Cromwell and Mazzullo (1987) reported that the reef on the west edge of the Central Basin Platform is "...filled with sulphur water which poses severe drilling problems in deep wells. Bits fall 5 to 20 feet in the cavernous porosity, pipe frequently breaks, and lost circulation is common."

Eastern Side of Basin

Cavernous zones encountered in the subsurface on the eastern side of the Delaware Basin are listed in Table 1. The inner margin of the reef, where it is buried around the eastern side, is the locus of a number of deep-seated dissolution features (Fig. 7). The three main dissolution features of this area are: (1) the dissolution troughs of Maley and Huffington (1953), (2) San Simon Swale/Sink, and (3) Wink (Kermit-Wink) Sink. Wink Sink may have an origin related primarily to dissolved salt and oil drilling (Baumgardner et al., 1982), but the other two features appear to be related to cavernous zones in the Capitan.

Belding-San Simon Trough.—

A belt of dissolution troughs on the east side of the Delaware Basin, collectively known as the Belding-San Simon Trough, contains at least three elongate and coalescing lenses of fill (Fig. 7A, B, C). When viewed together, these structures are parallel to the Capitan reef front and were obviously influenced by that feature, dissolution being localized in a belt paralleling the front. The deepest of these depressions is filled with more than 360 m of Cenozoic sediments (Anderson, 1981).

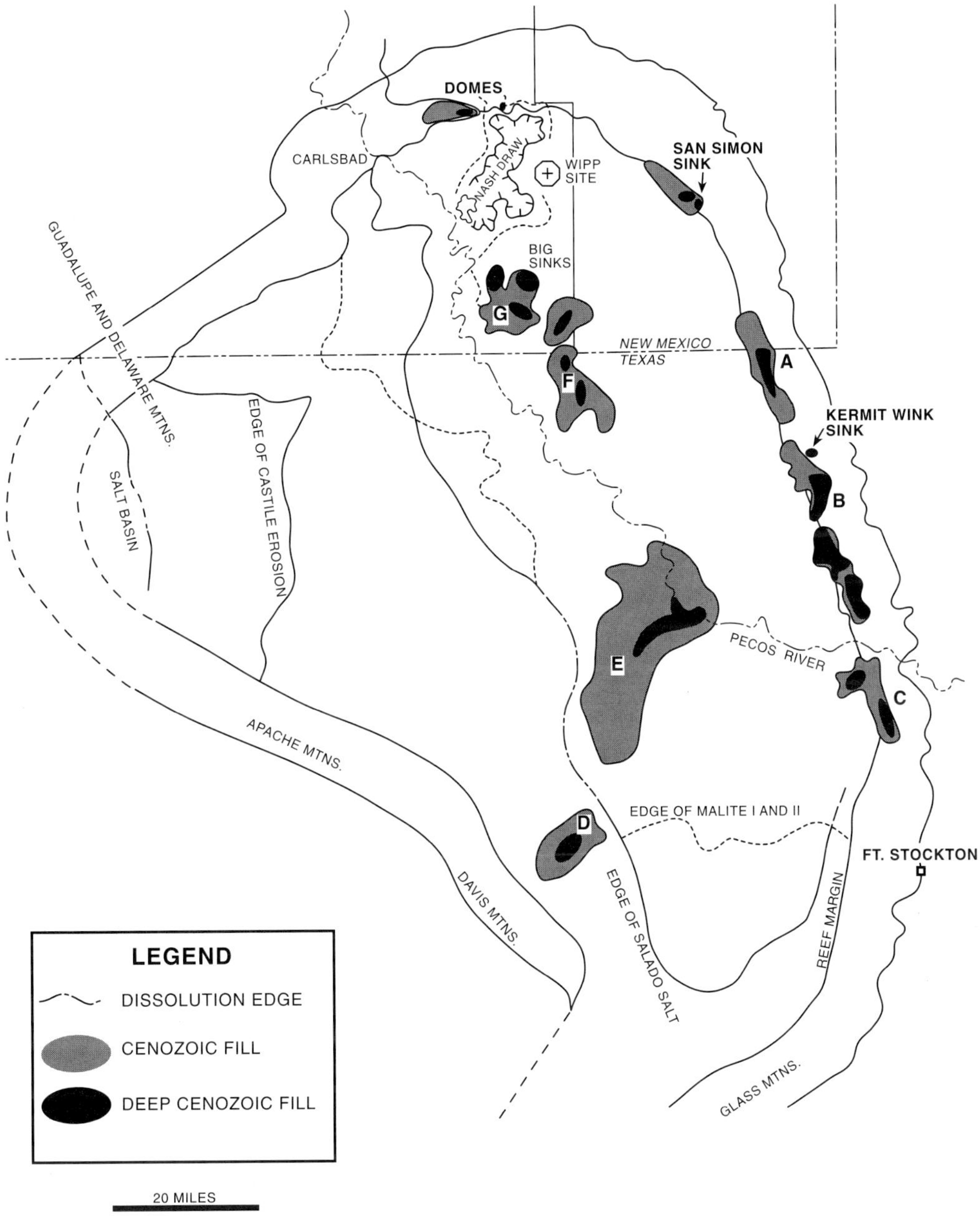

FIG. 7.—Map of the Delaware Basin showing location of the Capitan reef and the major dissolution depressions (troughs). Base map from Anderson (1981). A, B, C, D, E, F, and G correspond to the dissolution troughs of Maley and Huffington (1953) filled with Cenozoic gravel. Note that along the east side of the basin there are a number of dissolution troughs (A, B, C) which lie along the inner margin of the Capitan reef, and that San Simon Swale/Sink and Wink (Wink-Kermit) Sink also lie along the Capitan. On the northern side of the basin Vine's (1960) domes are similarly located along the inner margin of the Capitan reef (area marked "Domes"). From Hill (1996).

Anderson et al. (1978) favored the dissolution of the Belding-San Simon Trough by contact with undersaturated water moving through the Capitan reef. Artesian water from the reef dissolved the salt overlying the Capitan, causing subsidence of Permian, Triassic, and Cretaceous rocks with contemporaneous deposition of Cenozoic sand and gravel in the deepening troughs. It is not known what role sulfuric acid may have had in promoting karst development on the eastern side of the basin, but the possibility exists, since water high in H_2S is found in the area. Wallace (1993) reported that water in the Capitan aquifer on the northeast side of the basin is "...technically fresh, but is impotable due to high concentrations of hydrogen sulfide."

San Simon Swale/Sink. —

San Simon Swale/Sink is located in the northern part of the eastern side of the Delaware Basin (Fig. 7). The feature is roughly 11 km in length, up to 2 km in width, and 260 km^2 in area. During the late Wisconsin glacial (~20,000 yr B.P.) a lake occupied the depression surrounding San Simon Sink, and about 75 m of collapse in the sink has been recorded since then. Climatic dating of sediment in the deeper part of the depression indicates dates for the sink no older than ~50,000 yr B.P. (Anderson, 1980).

Nicholson and Clebsch (1961) were of the opinion that San Simon Swale/Sink had originated as the result of deep-seated collapse. Anderson (1980) thought that San Simon Swale formed as a result of dissolution of halite by waters moving upward from the Capitan along northwest-trending faults and/or fractures near the reef margin. Bachman (1984) presumed that the San Simon Sink was a modern analog of breccia pipe formation in the north part of the basin, related to cavernous zones in the Capitan, with the breccia pipes "rooting" in these caves. Caves are not unknown in the San Simon Swale part of the Capitan aquifer. One open-hole section of Capitan aquifer caved in during completion of Jal Water Supply Well 1, and the Bass Brothers North Custer Mountain Unit No. 1 encountered breaks in drilling varying from 3.5–21 m (Table 1). It is possible that San Simon Sink may be rooted in a large cave — perhaps even as large as the Big Room of Carlsbad Cavern, which would account for the tremendous amount of surface collapse and subsidence.

Northern Side of Basin

Caves also are known to exist in the subsurface on the northern side of the basin (Table 1). Thorton and Gaston (1967) reported that a very vuggy, cavernous, dolomitic, Capitan reef section approximately 210 m thick was intersected by the Lusk Deep Unit, Lea County, on the northern side of the basin. Also, zones of abnormally high porosity exist along the basinal edge of the northern side and represent "cavernous zones" and "honeycomb" solution structures (Gail, 1974). One of the main dissolution features in this area that is possibly related genetically and spatially to caves in the Capitan are the domes/breccia pipes (Fig. 7).

Domes/Breccia Pipes. —

Vine (1960) was the first to describe a series of domes/breccia pipes on the northern side of the Delaware Basin, located about 5.5–6 km east of Carlsbad, adjacent to U.S. Highway 62-180 in the Clayton Basin Quadrangle (Fig. 7; area marked "Domes"). (These are not the same as Motts' (1972) "domes," which are biohermal mounds.) The domes are circular to ellip-

TABLE 1.—CAVERNOUS ZONES IN THE CAPITAN AQUIFER, EASTERN AND NORTHERN SIDES OF THE DELAWARE BASIN*

Well	**Location**	**Depth**	**Reference**
Barton Mobil Fed 1	Sec.24, T21S, R26E, Eddy Co., ~4.8 km north of Carlsbad, Avalon Hills	158–162 m 546–557 m	Hiss (1975)
Pan American Pet. Corp., Big Eddy Unit no. 18	Sec.3, T21S, R29E, Eddy Co., ~1.6 km SW of Breccia Pipe B	792–811 m	Gail (1974), Hiss (1975)
Cities Service Oil Co., Big Eddy Unit no. 17	Sec.2, T21S, R29E, Eddy Co., ~1.6 km south of Breccia Pipe B	640–661 m	Gail (1974)
E.C. Hale Federal 2	Sec.22, T20S, R30E, Eddy Co., ~2.4 km NW of Breccia Pipe A	728–735 m	Gail (1974), Hiss (1975)
Shell Oil Co. & Texas Crude Oil Co., Big Eddy Unit 1-21	Sec.21, T20S, R32E, Lea Co., ~1.6 km north of Laguna Toston	1116–1128 m	Gail (1974)
Bass Brothers, North Custer Mtn., Unit 1	Sec.28, T23S, R35E, Lea Co., ~5.6 km SE of San Simon Sink	1367–1377 m	Hiss (1975)
Skelly Oil Co., Jal Water Supply Well 1	Sec.16, T24S, R36E, Lea Co., ~13 km SE of San Simon Sink	Open-hole, caved-in section. Depth (?)	Hiss (1975)

*From Hill (1996)

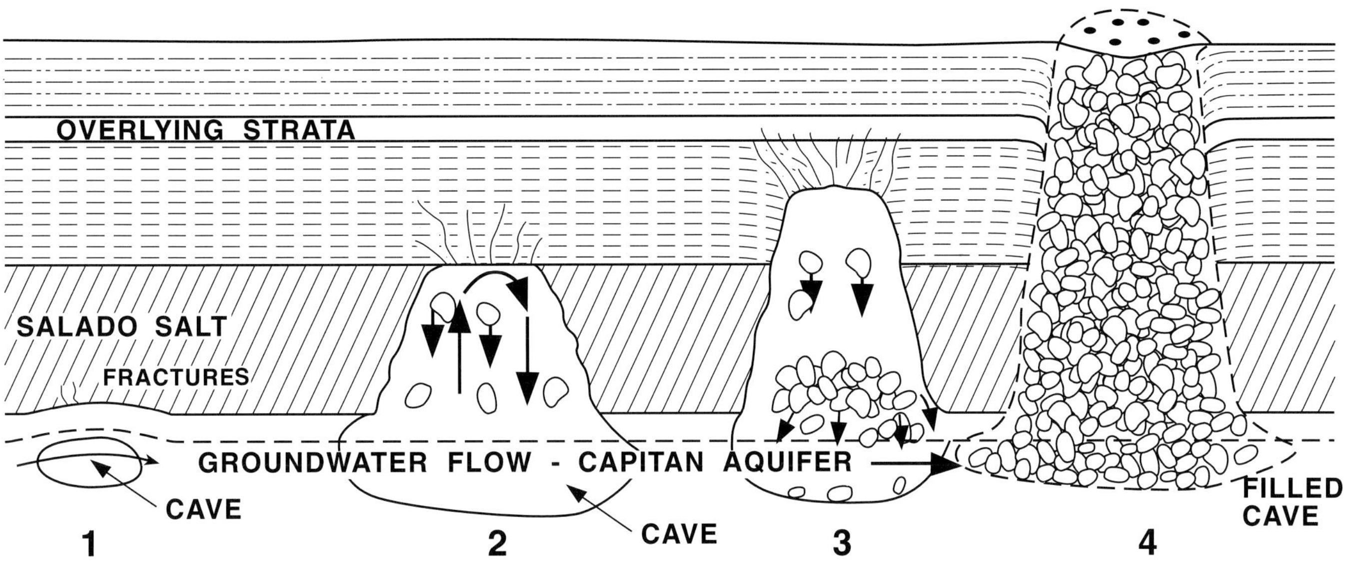

FIG. 8.—Diagrammatic model for the origin of the breccia pipes on the northern side of the basin. (1) Groundwater moving through the Capitan aquifer created cave voids. The water was under artesian head so that once the top of the Capitan was breached (along fractures) the water had access to the soluble Salado salt, which overlies the Capitan. (2) Salt removal was by a brine-density flow mechanism where fresh water moved up and dissolved salt and then descended due to its higher density. (3) As dissolution continued in the salt, overlying non-salt strata collapsed into the forming pipe, occupying the space of the cave void and that created by brine-density flow. Beds dip down toward the forming pipe due to preferred dissolution along the pipe's edges. (4) When the breccia pipe eventually reached the surface it either formed a depression (playa) or a dome if the rock surrounding the pipe was more soluble than the breccia filling the pipe. From Hill (1996).

soidal structures ranging from about 300 m to 450 m in diameter, and the breccia pipes underlying the domes are vertical cylindrical features with sharply faulted boundary walls containing displaced breccia pieces derived by collapse from higher horizons. One of the breccia pipes (Dome C) contains small amounts of oil and H_2S, which were intersected by drilling and appear to have come up along a nearly vertical fault from the Yates Formation (Hill, 1996).

None of the domes/breccia pipes have been drilled in their entirety and so their vertical extent is unknown. However, it is possible that these pipes root in the cavernous Capitan Formation, because they directly overlie the Capitan and because oil in these pipes is known to have come up from the Yates Formation, a unit equivalent in age to the Capitan. Fresh water from the Capitan under artesian pressure could have dissolved the salt in the overlying Salado Formation, with the overlying rock collapsing into the void and with the pipes eventually stoping to the surface (Fig. 8). Alternatively, the features could be explained by partial dissolution and collapse of the Capitan. Whether these are sulfuric acid caves is unknown, but the H_2S encountered in Breccia Pipe C supports this possibility.

Time of breccia pipe/cave formation is speculative, but disconnection of the pipes from the Capitan aquifer appears to have been related to the breaching of the Ancestral Pecos River at Carlsbad about 600,000 yr ago (Bachman, 1980). It may not be a coincidence that a major water table lowering appears to have happened in the Capitan at ~600,000 yr B.P. (Hill, 1996). This was the time when water in the Capitan aquifer stopped exiting at Hobbs and began exiting at Carlsbad (Hiss, 1975). This may have also been the time when the caves in the Guadalupe Mountains experienced a drop in water level, and when the breccia pipes became decoupled from the Capitan aquifer flow system.

CONCLUSIONS

The Capitan Limestone is one of the most important cave-bearing rock units in the world. Properties of the Capitan that have influenced cavern development are stratigraphy, dolomitization, joints, and positive structures. Caves occur in the Capitan in the Guadalupe, Apache, and Glass Mountains, and in the subsurface on the eastern and northern sides of the Delaware Basin, but cave development is most pronounced in the Guadalupe Mountain section of the Capitan reef limestone. The location of the large cave passages is related to injection points for hydrogen sulfide, which are derived from hydrocarbons in the Delaware Basin. As this hydrogen sulfide reacted with oxygenated water slowly diffusing through the Capitan aquifer, it formed sulfuric acid, which dissolved out these caves.

REFERENCES

ANDERSON, R. Y., 1980, Field trip notes for salt dissolution features in the Delaware Basin, *in* Chaturvedi, L., ed., WIPP site and vicinity geological field trip: Albuquerque, Environmental Evaluation Group, EEG-7, p. 112-129.

ANDERSON, R. Y., 1981, Deep-seated salt dissolution in the Delaware Basin, Texas and New Mexico, *in* Wells, S. G., and Lambert, W., eds., Environmental Geology and Hydrology in New Mexico: New Mexico Geological Society Special Publication 10, p. 133-145.

ANDERSON, R. Y., KIETZKE, K. K., and RHODES, D. J., 1978, Development of dissolution breccias, northern Delaware Basin, New Mexico and Texas, *in* Austin, G. S., compiler, Geology and Mineral Deposits of Ochoan Rocks in Delaware Basin and Adjacent Areas: New Mexico Bureau of Mines and Mineral Resources Circular 159, p. 47-52.

BACHMAN, G. O., 1980, Regional Geology and Cenozoic History of Pecos Region, Southeastern New Mexico: Washington, D.C., U.S. Geological Survey Open-file Report 80-1099, 116 p.

BACHMAN, G. O., 1984, Regional Geology of Ochoan Evaporites, Northern Part of Delaware Basin: New Mexico Bureau of Mines and Mineral Resources

Circular 184, 22 p.

BAUMGARDNER, R. W., HOADLEY, A. D., and GOLDSTEIN, A G., 1982, Formation of the Wink Sink, a Salt Dissolution and Collapse Feature, Winkler County, Texas: Austin, Bureau of Economic Geology, University of Texas, Report of Investigation 114, 38 p.

BRETZ, J. H., 1949, Carlsbad Caverns and other caves of the Guadalupe block, New Mexico: Journal of Geology, v. 57, p. 447-463.

CONSELMAN, F. B., 1987, Upper Permian (Guadalupian) facies and their association with hydrocarbons, Permian Basin, West Texas and New Mexico: American Association of Petroleum Geologists Bulletin, v. 71, p. 224-225.

CROMWELL, D. W., and MAZZULLO, L., 1987, Road log from Midland to Fort Stockton and Alpine, *in* Cromwell, D., and Mazzullo, L., eds., The Leonardian Facies in West Texas and Southeastern New Mexico and Guidebook to the Glass Mountains, West Texas: Midland, Society of Economic Paleontologists and Mineralogists, Permian Basin Section, Publication 87-27, p. 1-3.

FIESELER, R. G., and KUNATH, C. E. (eds.), 1975, The caves of Brewster and western Pecos Counties: Texas Speleological Survey, v. 4, 55 p.

FORD, D. D., and EWERS, R. O., 1978, The development of limestone cave systems in the dimensions of length and depth: Canadian Journal of Earth Sciences, v. 15, p. 1783-1798.

GAIL, G. J., 1974, Subsurface geology of Capitan aquifer northeast of Carlsbad, Eddy and Lea Counties, New Mexico: Unpublished M.S. Thesis, University of New Mexico, Albuquerque, 61 p.

HANEEF, M., 1993, Lithofacies and depositional environment of Altuda, Capitan and Tessey formations (Permian), Glass Mountains, West Texas: Unpublished M.S. Thesis, Sul Ross State University, Alpine, Texas, 208 p.

HILL, C. A., 1987, Geology of Carlsbad Cavern and Other Caves in the Guadalupe Mountains, New Mexico and Texas: New Mexico Bureau of Mines and Mineral Resources Bulletin 117, 150 p.

HILL, C. A., 1990, Sulfuric acid speleogenesis of Carlsbad Cavern and its relationship to hydrocarbons, Delaware Basin, New Mexico and Texas: American Association of Petroleum Geologists Bulletin, v. 74, p. 1685-1694.

HILL, C. A., 1995, Sulfur redox reactions, native sulfur, Mississippi Valley-type deposits and sulfuric acid karst, Delaware Basin, New Mexico and Texas: Environmental Geology, v. 25, p. 16-23.

HILL, C. A., 1996, Geology of the Delaware Basin: Guadalupe, Apache and Glass Mountains, New Mexico and West Texas: Midland, Society of Economic Paleontologists and Mineralogists, Permian Basin Section, Publication 96-39, 480 p.

HILL, C. A., MOSCH, C., and REID, M., 1988, Cross Cave, Apache Mountains, Culberson County, Texas: Cave Research Foundation, Annual Report 30, p. 23-24.

HISS, W. L., 1975, Stratigraphy and groundwater hydrology of the Capitan aquifer, southeastern New Mexico and western Texas: Unpublished Ph.D. Dissertation, University of Colorado, Boulder, 396 p.

HORAK, R. L., 1985, Tectonic and hydrocarbon maturation history in the Permian Basin: Oil and Gas Journal, v. 83, p. 124-125.

JAGNOW, D. H., 1977, Geologic factors influencing speleogenesis in the Capitan Reef Complex, New Mexico and Texas: Unpublished M.S. Thesis, University of New Mexico, Albuquerque, 197 p.

KING, P. B., 1930, The geology of the Glass Mountains: Austin, Bureau of Economic Geology, University of Texas, pt. 1, 3038, 167 p.

MALEY, V. C., and HUFFINGTON, R. M., 1953, Cenozoic fill and evaporite solution in the Delaware Basin, Texas and New Mexico: Geological Society of America Bulletin, v. 64, p. 539-546.

MELIM, L. A., 1991, The origin of dolomite in the Permian (Guadalupian) Capitan Formation, Delaware Basin, West Texas and New Mexico — implications for dolomitization models: Unpublished Ph.D. Dissertation, Southern Methodist University, Dallas, Texas, 200 p.

MELIM, L. A., and SCHOLLE, P. A., 1989, Dolomitization model for the forereef facies of the Permian Capitan Formation, Guadalupe Mountains, Texas-New Mexico, *in* Harris, P. M., and Grover, G. A., eds., Subsurface and outcrop examination of the Capitan shelf margin, northern Delaware Basin: Tulsa, Society of Economic Paleontologists and Mineralogists, Workshop 13, p. 407-413.

MOTTS, W. S., 1972, Geology and paleoenvironment of the northern segment, Capitan Shelf, New Mexico and West Texas: Geological Society of America Bulletin, v. 83, p. 701-722.

NICHOLSON, A., and CLEBSCH, H., 1961, Geology and Groundwater Conditions in Southern Lea County, New Mexico: New Mexico Bureau of Mines and Mineral Resources Groundwater Report 6, 123 p.

PALMER, A. N., 1991, Origin and morphology of limestone caves: Geological Society of America Bulletin, v. 103, p. 1-21.

PISAROWICZ, J., 1994, Cueva de Villa Luz — An active case of H_2S speleogenesis (abs.), *in* Sasowsky, I. D., and Palmer, M. V., eds., Breakthroughs in karst geomicrobiology and redox chemistry: Colorado Springs, Karst Waters Institute Special Publication, p. 25-28.

POLYAK, V. J., and GÜVEN, N., 1996, Mineralization of alunite, natroalunite, and hydrated halloysite in Carlsbad Cavern and Lechuguilla Cave, New Mexico: Clays and Clay Minerals, v. 44, p. 843-850.

POLYAK, V. J., and MOSCH, C. J., 1995, Metatyuyamunite from Spider Cave, Carlsbad Caverns National Park: National Speleological Society Bulletin, v. 57, p. 85-90.

POLYAK, V. J., MCINTOSH, W. C., GÜVEN, N., and PROVENCIO, P., 1997, Age of formation of Carlsbad Cavern, Lechuguilla Cave, and other caves of the Guadalupe Mountains based on $^{40}Ar/^{39}Ar$-dating of alunite (abs.): Boulder, Geological Society of America, Abstracts with Programs, v. 29, p. A-347.

POLYAK, V. J., MCINTOSH, W. C., GÜVEN, N., and PROVENCIO, P., 1998, Age and origin of Carlsbad Cavern and related caves from $^{40}Ar/^{39}Ar$ of alunite: Science, v. 279, p. 1919-1922.

QUEEN, J. M., PALMER, A. N., and PALMER, M. V., 1977, Speleogenesis in the Guadalupe Mountains, New Mexico — gypsum replacement of carbonate by brine mixing: Proceedings of the 7th International Congress of Speleology, Sheffield, United Kingdom, p. 332-336.

SCHOLLE, P. A., ULMER, D. S., and MELIM, L. A., 1992, Late-stage calcites in the Permian Capitan Formation and its equivalents, Delaware Basin margin, West Texas and New Mexico — evidence for replacement of precursor evaporites: Sedimentology, v. 39, p. 207-234.

THORTON, D. E., and GASTON, H. H., 1967, Lusk Strawn field, *in* A symposium of oil and gas fields of southeastern New Mexico, 1966 supplement: Roswell Geological Society, p. 15-20.

VINE, J. D., 1960, Recent domal structures in southeastern New Mexico: American Association of Petroleum Geologists Bulletin, v. 44, p. 1903-1911.

WALLACE, M. G., 1993, A total dissolved solids map for the northern portion of the Capitan aquifer, *in* Love, D. W., et al., eds., Carlsbad region, New Mexico and West Texas: New Mexico Geological Society Guidebook, 44th Annual Field Conference, p. 38-39.

WOOD, J. W., 1968, Geology of Apache Mountains, Trans-Pecos Texas: Austin, Bureau of Economic Geology, University of Texas, 32 p., Text accompanying Geologic Quadrangle map no. 35.

INDEX

I

K

L

M

O

P

R

S

T

U